Section Guide

Hollow Structural Sections

Connections Manual

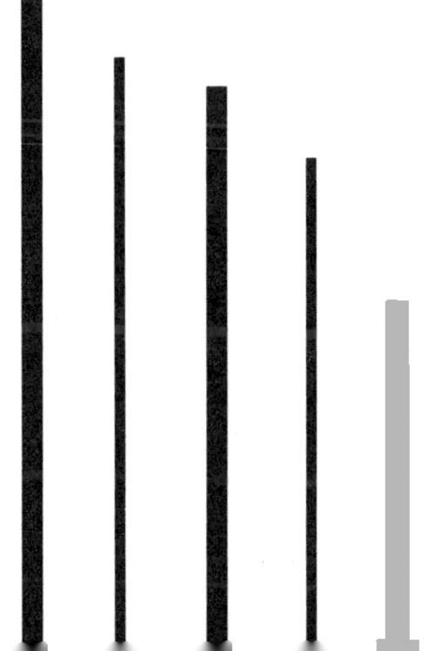

FOREWORD

The American Institute of Steel Construction (AISC) in conjunction with the Steel Tube Institute of North America (STI) is pleased to publish this first edition of the Hollow Structural Sections (HSS) Connections Manual. This is the start of an ongoing effort to provide architects, engineers, fabricators, educators and others interested in the use of HSS in building construction with the most current and practical information available. The development of this new Manual has been generously supported by the American Iron and Steel Institute (AISI) and the member companies of the STI HSS Committee. Additionally, AISC and STI acknowledge the support and assistance provided by the Canadian Institute of Steel Construction (CISC) in the preparation of this Manual.

AISC is the non-profit technical specifying and trade organization for the fabricated structural steel industry in the United States. Founded in 1921, its objectives are to improve and advance the use of fabricated structural steel through research and engineering studies to develop the most efficient and economical design of structures. To accomplish these objectives, the Institute publishes specifications, manuals, design guides and other technical publications, conducts seminars and programs to improve knowledge and product quality, and furthers education.

STI is the not-for-profit trade association for the HSS manufacturing industry in the United States and Canada. Formed in 1930 by a small group of welded steel tubing producers, STI promotes the use of HSS in a cooperative effort to improve manufacturing techniques and by informing consumers about the utility and versatility of HSS. Formally incorporated in Ohio in 1955, STI is dedicated to promoting growth, prosperity and competitiveness of the North American steel tubular industry. The Institute serves as a strong voice for industry, providing accurate and timely information to its membership and external audience.

AISI is a non-profit association whose membership includes a broad range of steel producing companies throughout the western hemisphere. AISI embraces a wide and varied range of activities including education, research and technology, engineering, collection and dissemination of statistics, public dissemination of information about the industry, public affairs and discussions of industrial relations including health and safety.

CISC is a national industry organization representing the structural steel, open-web steel joist and steel plate fabricating industries in Canada. Formed in 1930 and granted a Federal charter in 1942, CISC functions as a non-profit organization promoting the efficient and economical use of fabricated steel in construction.

This inaugural edition of the HSS Connections Manual contains standard HSS dimensions and properties, material grades, and the new AISC *Specification for the Design of Steel Hollow Structural Sections*. Welding and welded fabrication of HSS is covered in Chapter 2. Bolting and bolted fabrication is covered in Chapter 3. Simple shear connections, moment connections and connections for tension and compression are covered in Chapters 4, 5 and 6, respectively. Cap plates, base plates and column splices are covered in Chapter 7. Welded truss connections and design examples are covered in Chapters 8 and 9.

The Committee gratefully acknowledges the contributions of James M. Fisher, LeRoy A. Lutz, Scott Luckiesh, Michael Loescher, and Carol Williams of Computerized Structural Design, Inc. of Milwaukee WI. CSD served as a consultant in the preparation of this Manual. Additionally, the Committee thanks the following individuals, who served as an adjunct task group in the development and review of this

REFERENCED SPECIFICATIONS, CODES, AND STANDARDS

The AISC *Specification for the Design of Steel Hollow Structural Sections*, April 15, 1997, is bound herein in the Appendix.

Part 6 (Volume I) of the LRFD Manual contains the full text of the following:

American Institute of Steel Construction, Inc. (AISC)

Load and Resistance Factor Design Specification for Structural Steel Buildings, December 1, 1993

Specification for Load and Resistance Factor Design of Single-Angle Members, December 1, 1993

Seismic Provisions for Structural Steel Buildings, June 15, 1992

Code of Standard Practice for Steel Buildings and Bridges, June 10, 1992

Research Council on Structural Connections (RCSC)

Load and Resistance Factor Design Specifications for Structural Joints Using ASTM A325 or A490 Bolts, June 3, 1994*

Additionally, the following other documents are referenced in Volumes I and II of the LRFD Manual:

American Association of State Highway and Transportation Officials (AASHTO)

AASHTO/AWS D1.5–88

American Concrete Institute (ACI)

ACI 349–90

American Iron and Steel Institute (AISI)

Load and Resistance Factor Design Specification for Cold-Formed Steel Structural Members, 1996

American National Standards Institute (ANSI)

ANSI/ASME B1.1–82   ANSI/ASME B18.2.2–86
ANSI/ASME B18.1–72   ANSI/ASME B18.5–78
ANSI/ASME B18.2.1–81

American Society of Civil Engineers (ASCE)

ASCE 7-95

American Society for Testing and Materials (ASTM)

ASTM A6–96b	ASTM A490–93	ASTM A617–92
ASTM A27–95	ASTM A500–93	ASTM A618–93
ASTM A36–96	ASTM A501–93	ASTM A668–96
ASTM A53–96	ASTM A502–91	ASTM A687–93
ASTM A148–93b	ASTM A514–94a	ASTM A709–96$^{\varepsilon 1}$
ASTM A153–95	ASTM A529–94	ASTM A770–86
ASTM A193–96b	ASTM A563–94	ASTM A852–94
ASTM A194–96	ASTM A570–96	ASTM B695–91
ASTM A208(A239–89)	ASTM A572–94c$^{\varepsilon 1}$	ASTM C33–90
ASTM A242–93a	ASTM A588–94	ASTM C330–89
ASTM A307–94	ASTM A606–96	ASTM E119–88
ASTM A325–96	ASTM A607–96	ASTM E380–91
ASTM A354–95	ASTM A615–92b	ASTM F436–93
ASTM A449–93	ASTM A616–92	

*Most recent revision is available in latest printing.

American Welding Society (AWS)

AWS A2.4–93	AWS A5.25–91
AWS A5.1–91	AWS A5.28–79
AWS A5.5–81	AWS A5.29–80
AWS A5.17–89	AWS B1.0–77
AWS A5.18–79	AWS D1.1–96
AWS A5.20–79	AWS D1.4–92
AWS A5.23–90	

CHAPTER 1

DIMENSIONS AND PROPERTIES

OVERVIEW

Chapter 1 contains general information for dimensions and properties of HSS. Cross-sectional dimensions, design properties, material specifications, and standard mill practice are covered.

Following is a detailed list of the topics considered.

Chapter 1

DIMENSIONS AND PROPERTIES

HSS SPECIFICATIONS

ASTM specifications for hollow structural sections (HSS) approved for use in building construction are listed in AISC HSS Specification Section 1.2. The availability of HSS used in building construction is shown in Table 1-1. Tensile requirements for the following HSS are also listed in Table 1-1.

ASTM A500 is a cold-formed ERW product specification that covers square, rectangular and round HSS. ASTM A500 grade B and grade C are commonly specified for building construction applications and are available from producers and steel service centers.

ASTM A501 is a hot-formed product specification that covers square, rectangular and round HSS. ASTM A501 is not available in square and rectangular sizes; round sizes are available in mill quantities only from a limited number of producers.

ASTM A847 is a cold-formed ERW high-strength low-alloy atmospheric-corrosion-resistant product specification that covers square, rectangular and round HSS. ASTM A847 is generally available in mill quantities only.

ASTM A618 is a hot-formed high-strength low-alloy product specification that covers square, rectangular and round HSS. ASTM A618 is not available in square and rectangular sizes; round sizes are available in mill quantities only.

PIPE SPECIFICATIONS

ASTM A53 is a pipe product specification that covers a large range of pipe sizes and wall thicknesses. These sizes are also made as pipe-size round HSS under ASTM A500.

COMPARISON OF ASTM A500-93 GRADES B & C, ASTM A847-93 AND CAN/CSA G40.21-92 50W

The following is provided for information purposes only as a general comparison of the specifications listed and in no way should be considered as a complete version of each specification. For more detailed information, the actual specification should be consulted.

ASTM A500-93: Standard Specification for Cold-Formed Welded and Seamless Carbon Steel Structural Tubing in Rounds and Shapes

ASTM A847-93: Standard Specification for Cold-Formed Welded and Seamless High-Strength, Low-Alloy Structural Tubing with Improved Atmospheric Corrosion Resistance

CAN/CSA-G40.20-92: General Requirements for Rolled and Welded Structural Quality Steel (Grade 50W)

CAN/CSA-G40.21-92: Structural Quality Steels (Grade 50W)

Table 1-1.
Availability of Steel Pipe and HSS

Product	ASTM Specification	Grade	F_y Minimum Yield Stress (ksi)	F_u Minimum Tensile Stress (ksi)	Round	Square & Rectangular	Availability
Electric Resistance Welded Pipe	A53 Type E	B	35	60			Note 3
Seamless Pipe	Type S	B	35	60			Note 3
Cold Formed HSS	A500 Round	B	42	58			Note 1
Cold Formed HSS	A500 Round	C	46	62			Note 1
Cold Formed HSS	A500 Shaped	B	46	58			Note 2
Cold Formed HSS	A500 Shaped	C	50	62			Note 4
Hot Formed HSS	A501	—	36	58			Note 1
High-Strength Low-Alloy HSS	A618	I	50	70			Note 1
High-Strength Low-Alloy HSS	A618	II	50	70			Note 1
High-Strength Low-Alloy HSS	A618	III	50	65			Note 1
High-Strength Low-Alloy HSS	A847	—	50	70			Note 1

Notes:
1. Available in mill quantities only; consult with producers.
2. Normally stocked in local steel service centers.
3. Normally stocked by local pipe distributors.
4. Certain sizes stocked in local steel service centers, consult with local steel service centers.

☐ Available

☐ Not Available

Chemical Requirements (Heat Analysis)

Element	ASTM A500B	ASTM A500C	ASTM A847	CSA G40.21 50W
Carbon	0.26 Max	0.23 Max	0.20 Max	0.23 Max
Manganese	—	1.35 Max	1.35 Max	0.50 - 1.50
Phosphorus	0.035 Max	0.035 Max	0.15 Max	0.04 Max
Sulfur	0.035 Max	0.035 Max	0.05 Max	0.05 Max
Silicon	—	—	—	0.40 Max[+]
Copper	0.20 Min*	0.20 Min*	0.20 Min**	0.20 Min*
Grain Refining Elements	—	—	—	0.10 Max[++]

*When specified.

**If chromium and silicon contents are each 0.50 minimum, then copper minimums do not apply. This is an application with improved atmospheric corrosion resistance requiring more than copper additions. For methods of estimating the atmospheric corrosion resistance of low alloy steels refer to ASTM G101 or actual data.

[+]The steel may be made with no minimum silicon content, provided that the steel contains a minimum of 0.015 percent acid-soluble aluminum or 0.020% total aluminum.

[++]The elements columbium, vanadium, may be used singly or in combination up to the total percentage indicated.

—This item is not a requirement of the specification

Tensile Requirements (Mechanical Properties)

Characteristic	Shaped ASTM A500B	Round ASTM A500B	Shaped ASTM A500C	Round ASTM A500C	Shaped & Round ASTM A847	Shaped & Round CSA G40.21 50W
Yield Point Min.	46 ksi	42 ksi	50 ksi	46 ksi	50 ksi	50 ksi
Tensile Strength Min.	58 ksi	58 ksi	62 ksi	62 ksi	70 ksi	65 ksi
Tensile Strength Max.	—	—	—	—	—	90 ksi
Elongation in 2 in. % Min.	23*	23*	21*	21*	19*	22*

*Refer to the complete specification for specific elongation requirements.

—This item is not a requirement of the specification.

Permissible Variations in Dimensions

Characteristics	Shaped ASTM A500* & A847*	Shaped & Round CSA G 40.21 5OW
Outside Dimensions		
2½ inches or under	± 0.020 in.	± 0.020 in.
over 2½ to 3½ inches incl.	± 0.025 in.	± 0.030 in.
over 3½ to 5½ inches incl.	± 0.030 in.	± 0.040 in.
over 5½ inches incl.	± 0.01 × largest flat dimension	± 0.01 × largest flat dimension

*for round ASTM A500 and ASTM A847 the outside diameter shall not vary more than ± 0.5 percent, rounded to the nearest 0.005 in., of the nominal outside diameter size specified for nominal outside diameters 1.900 in. and smaller, ± 0.75 percent, rounded to the nearest 0.005 in., for nominal outside diameters 2 in. and larger.

Characteristics	Shaped & Round ASTM A500 & A847	Shaped & Round CSA G40.21 50W
Wall Thickness (variation from nominal wall)	± 10%	− 0.5% + 10%
Straightness (maximum allowed)	⅛-in. × number of feet divided by 5	⅛-in. × number of feet divided by 5
Outside Corner Radius	3 times specified wall thickness max.	3 times specified wall thickness max.
Squareness of Sides	± 2 degrees	± 2 degrees
Twist (Max per 3 ft of length)		
To 1½ incl	0.050 in.	0.050 in.
Over 1½ to 2½ incl.	0.062 in.	0.062 in.
Over 2½ to 4 incl.	0.075 in.	0.075 in.
Over 4 to 6 incl.	0.087 in.	0.087 in.
Over 6 to 8 incl.	0.100 in.	0.100 in.
Over 8	0.112 in.	0.112 in.
Mass (weight variation)	—	−3.5% or + 10%

—This item is not a requirement of the specification.

COMPARISON OF ASTM A500-93 GRADES B & C, AND ASTM A53-96 TYPES E OR S, GRADE B

The following is provided for information purposes only as a general comparison of the specification listed and in no way should be considered as a complete version of each specification. For more detailed information the actual specification should be consulted.

ASTM A500-93: Standard Specification for Cold-Formed Welded and Seamless Carbon Steel Structural Tubing in Rounds and Shapes

ASTM A53-96: Standard Specification for Pipe, Steel, Black and Hot-Dipped, Zinc-Coated, Welded and Seamless

Chemical Requirements (Heat Analysis)

Element	ASTM A500B	ASTM A500C	ASTM A53 B
Carbon	0.26 Max	0.23 Max	0.30 Max
Manganese	—	1.35 Max	1.20 Max
Phosphorus	0.035 Max	0.035 Max	0.050 Max
Sulfur	0.035 Max	0.035 Max	0.045 Max
Copper**	0.20 Min*	0.20 Min*	0.40 Max.
Nickel**	—	—	0.40 Max
Chromium**	—	—	0.40 Max
Molybdenum**	—	—	0.15 Max
Vanadium**	—	—	0.08 Max

*When specified.

**The combination of these five elements shall not exceed 1.00 percent.

—This item is not a requirement of the specification.

Tensile Requirements (Mechanical Properties)

Characteristic	Round ASTM A500B	Round ASTM A500C	Round ASTM A53B
Yield Point Min.	42 ksi	46 ksi	35 ksi
Tensile Strength Min.	58 ksi	62 ksi	60 ksi
Elongation in 2 in. % Min.	23*	21*	*

*Refer to complete specification for specific elongation requirements.

Permissible Variations in Dimensions

Characteristics	Round ASTM A500	Round ASTM A53B
Outside Diameter		
1.900 in. and under	$\pm 0.5\%$	$\pm \frac{1}{64}$ in.
2 in. and over	$\pm 0.75\%$	$\pm 1\%$
Wall Thickness		
(variation from nominal wall)	$\pm 10\%$	-12.5%
Straightness	$\frac{1}{8}$-in. $\times$ number of	
(maximum allowed)	feet divided by 5	—
Mass (weight variation)	—	$\pm 10\%$

—This item is not a requirement of the specification.

HSS SHAPES DESIGNATIONS, DIMENSIONS AND PROPERTIES

The designations used for HSS listed in this manual are different than those used previously. The development of new designations and the recalculation of the section property data are the result of joint activities of the STI, AISC and the steel industry. The new and old designations correspond as follows:

New Designation	Shape	Old Designation
HSS8×8×$\frac{3}{8}$	Square HSS	TS8×8×$\frac{3}{8}$
HSS5×3×$\frac{3}{8}$	Rectangular HSS	TS5×3×$\frac{3}{8}$
HSS5.563×0.258	Round HSS	P5

Dimensions and section property data are presented for square, rectangular and round HSS. In a change from past practice, section property data are now based upon the HSS design wall thickness t per AISC HSS Specification Section 1.2.2. The design wall thickness for ERW HSS is taken as 0.93 times the nominal (specified) wall thickness. The design wall thickness for SAW HSS is equal to the nominal (specified) wall thickness.

The section property data for square and rectangular ERW HSS are based upon an outside corner radius equal to 2 times the wall thickness. The section property data for square and rectangular SAW HSS are based upon an outside corner radius equal to 3.6 times the wall thickness for SAW HSS with $\frac{5}{8}$-in. nominal wall thickness and 3 times the wall thickness for SAW HSS with nominal wall thicknesses of $\frac{1}{2}$-in. and $\frac{3}{8}$-in.

STANDARD MILL PRACTICE

ERW and SAW Manufacturing Methods

The transformation of a flat strip into an HSS is the result of a series of operations including forming, welding and sizing. Currently three traditional methods are being utilized for the manufacture of HSS in North America. These methods include the formed-from-round process, the weld-square form-square process for ERW HSS and the

SAW process for larger-size HSS. ERW and SAW processes have a product size and gage range that best suits the particular method of production, its own dimensional characteristics and resulting physical properties. Round HSS are always produced by the formed-from-round method, whereas square and rectangular HSS may be produced by any of the three methods. The following is a brief description of the three methods:

1. Formed-From-Round Process: As illustrated in Figure 1-1, the weld mill, flat strip (1) is bent continuously around its longitudinal axis to form an open-seam round by passing the strip through a progressive set of rolls (2-6). The resulting open-seam round (7) is joined by high-frequency induction or contact welding to create a

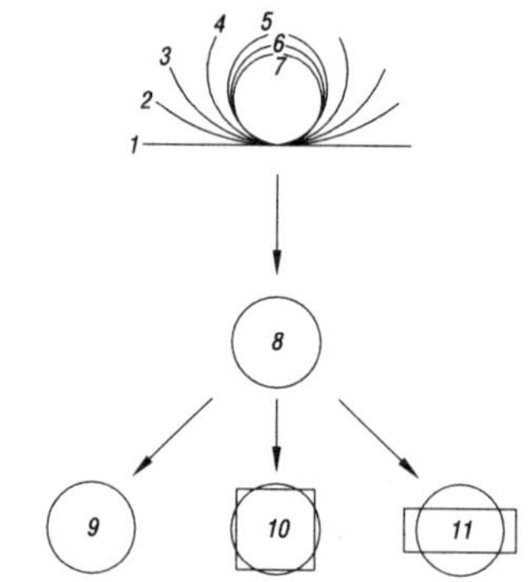

Fig. 1-1. *Formed-from-round process.*

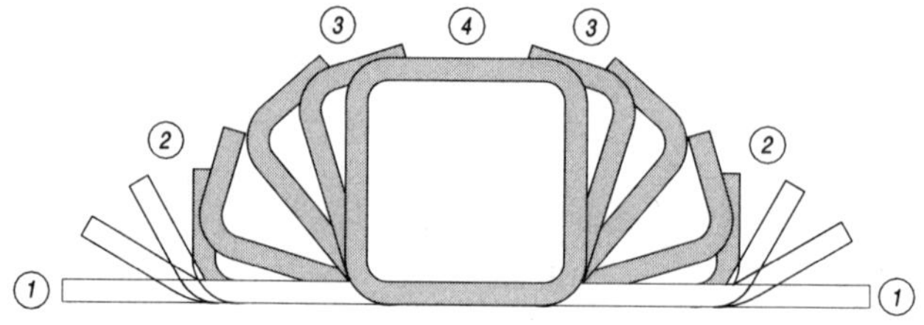

Fig. 1-2. *Form-square weld-square process.*

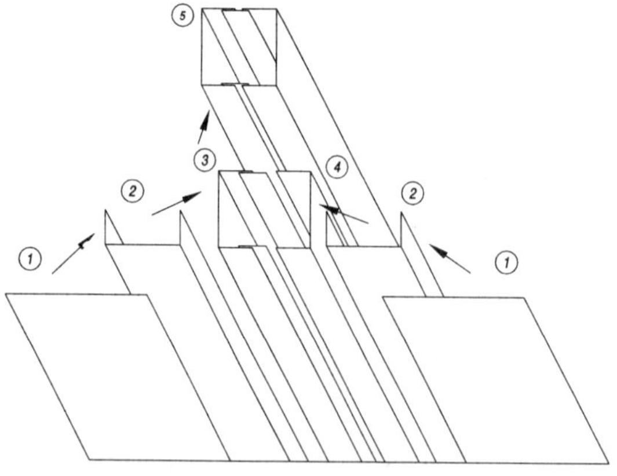

Fig. 1-3. *Brake form process.*

continuous longitudinal weld without the addition of filler metal. Subsequent to the welding process, the round (8) is cooled and then run through an additional set of sizing/shaping rolls to either qualify the final diameter of the round (9) or to form the shape to a square (10) or a rectangle (11).

2. Form-Square Weld-Square Process: As illustrated in Figure 1-2, driven forming dies progressively shape the flat strip (1) forming the top two corners (2) of the square or rectangular HSS in the initial forming stations. Subsequent stations form the bottom two corners (3) of the shape. The seam is welded by high frequency contacts when the HSS is near its final size and shape. The welded HSS (4) is cooled and then driven through a series of sizing stations, which qualify the final dimensions.

3. Submerged Arc Weld (SAW) Process: As illustrated in Figure 1-3, two identical pieces of flat strip (1) are placed in a press brake to form two identical halves (2) of a finished tube size. A backup bar is tack welded to each leg of one of the half sections (3). The two half sections are fitted together toe-to-toe (4) and submerged arc welded together to complete the square or rectangular section (5).

DIMENSIONAL TOLERANCES ON ASTM A500 HSS

Dimensional tolerances are stipulated in ASTM A500 Section 10. The following text is excerpt from "Recommended Methods to Check Dimensional Tolerances on Hollow Structural Sections (HSS) Made to ASTM A500," which is based upon these tolerances. This document, published by the Steel Tube Institute of North America, provides more detailed information on the tools and methods used to check tolerances, including examples for the application of the tolerances herein. A copy can be obtained from STI; 440/974-6990.

Outside Dimensions:

Round HSS—The outside diameter shall not vary more than plus or minus 0.5 percent rounded to the nearest 0.005 in. of the nominal outside diameter specified, for nominal outside diameters 1.900 in. and smaller, and plus or minus 0.75 percent rounded to the nearest 0.005 in. of the nominal outside diameter for nominal outside diameters 2.00 in. and larger. The outside diameter measurements shall be made at positions at least 2 in. from either end of the HSS.

Square and Rectangular HSS—The specified dimensions, measured across the flats at positions at least 2 in. from either end of the HSS and including the allowance for convexity or concavity, shall not exceed the following tolerances:

Outside Large Flat Dimension, in.	Large Flat Dimension Tolerance,[1] plus and minus, in.
2½ or under	0.020
over 2½ to 3½, incl.	0.025
over 3½ to 5½ , incl.	0.030
over 5½	0.01 times large flat dimension

---

[1]Tolerances include allowance for convexity or concavity, For rectangular HSS having a ratio of outside large to small flat dimension less than 1.5, and for square HSS, the tolerance on the small flat dimension shall be identical to the large flat dimension tolerance. For rectangular HSS having a ratio of outside large to small flat dimension in the range of 1.5 to 3.0 inclusive, the tolerance on small flat dimension shall be 1.5 times the large flat dimension tolerance. For rectangular HSS having a ratio of outside large to small flat dimension greater than 3.0, the tolerance on small flat dimension shall be 2.0 times the large flat dimension.

Wall Thickness:

The minimum wall thickness at any point of measurement on the HSS shall be not more than 10 percent less than the nominal wall thickness specified. The maximum wall thickness, excluding the weld seam of welded HSS, shall be not more than 10 percent greater than the nominal wall thickness specified. The wall thickness on square and rectangular HSS is to be measured at the center of the flat.

Length:

HSS are normally produced in random mill lengths 5 ft. and over, in multiple lengths and in specified mill lengths. When specified mill lengths are ordered, the length tolerance shall not exceed the following:

Specified Mill Length, ft	Length Tolerance for Specified Mill Length, in.	
	Over	Under
22 and under	$\frac{1}{2}$	$\frac{1}{4}$
22 to 44, incl.	$\frac{3}{4}$	$\frac{1}{4}$

Straightness:

The permissible variation for straightness of HSS shall be $\frac{1}{8}$-in. times the number of feet of total length divided by 5.

Squareness of Sides:

For square and rectangular HSS, adjacent sides may deviate from 90 degrees by a tolerance of plus or minus 2 degrees maximum.

Radius of Corners:

For square or rectangular HSS, the radius of any outside corner of the HSS shall not exceed 3 times the specified wall thickness.

Twist:

The twist or variation with respect to axial alignment of the HSS for square and rectangular HSS shall not exceed the following tolerances:

Specified Dimension of Longest Side, in.	Maximum Twist in First 3 ft and in Each Additional 3 ft
$1\frac{1}{2}$ or under	0.050
over $1\frac{1}{2}$ to $2\frac{1}{2}$, incl.	0.062
over $2\frac{1}{2}$ to 4, incl.	0.075
over 4 to 6, incl.	0.087
over 6 to 8, incl.	0.100
over 8	0.112

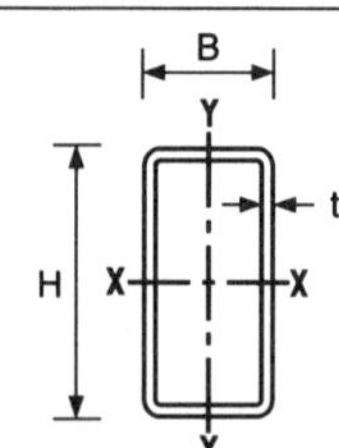

RECTANGULAR HSS
Dimensions and Properties

Dimensions			Properties								
Nominal Size			Wt per foot					X-X Axis			
Depth H	Width B	Wall		t	b/t	h/t	Area	I	S	r	Z
in.	in.	in.	lb.	in.			in.²	in.⁴	in.³	in.	in.³
32	24	⅝	225.80	0.625	35.4	48.2	66.4	9880	617	12.2	733
		½	183.50	0.500	45.0	61.0	53.9	8160	510	12.3	601
		⅜	138.95	0.375	61.0	82.3	40.8	6250	391	12.4	458
30	24	⅝	217.30	0.625	35.4	45.0	63.9	8480	565	11.5	668
		½	176.70	0.500	45.0	57.0	51.9	7010	468	11.6	548
		⅜	133.84	0.375	61.0	77.0	39.3	5380	359	11.7	418
28	24	⅝	208.79	0.625	35.4	41.8	61.4	7210	515	10.8	605
		½	169.89	0.500	45.0	53.0	49.9	5970	426	10.9	497
		⅜	128.74	0.375	61.0	71.7	37.8	4580	327	11.0	379
26	24	⅝	200.28	0.625	35.4	38.6	58.9	6060	466	10.1	545
		½	163.08	0.500	45.0	49.0	47.9	5020	386	10.2	448
		⅜	123.64	0.375	61.0	66.3	36.3	3860	297	10.3	342
24	22	⅝	183.27	0.625	32.2	35.4	53.9	4680	390	9.33	458
		½	149.47	0.500	41.0	45.0	43.9	3900	325	9.42	378
		⅜	113.43	0.375	55.7	61.0	33.3	3000	250	9.49	289
22	20	⅝	166.25	0.625	29.0	32.2	48.9	3530	321	8.51	379
		½	135.86	0.500	37.0	41.0	39.9	2950	269	8.60	313
		⅜	103.22	0.375	50.3	55.7	30.3	2280	207	8.67	240

Shading indicates HSS manufactured by the Submerged Arc Welding (SAW) process.

RECTANGULAR HSS
Dimensions and Properties

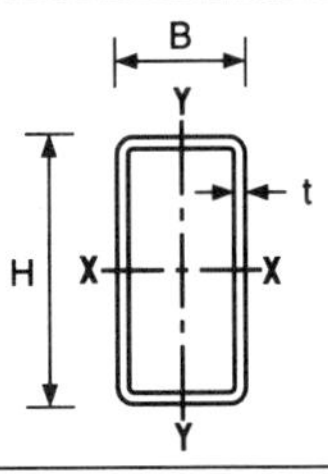

Dimensions			Properties						
Nominal Size			Y-Y Axis				Torsion		Surface Area Per foot
Depth H	Width B	Wall	I	S	r	Z	J	C	
in.	in.	in.	in.^{4}	in.^{3}	in.	in.^{3}	in.^{4}	in.^{3}	ft.^{2}
32	24	5/8	6390	533	9.81	604	12600	913	9.01
		1/2	5280	440	9.89	495	10100	739	9.12
		3/8	4050	337	9.96	378	7670	560	9.17
30	24	5/8	6050	504	9.73	575	11400	854	8.68
		1/2	5000	417	9.82	472	9220	692	8.79
		3/8	3840	320	9.88	360	6990	524	8.84
28	24	5/8	5710	476	9.65	546	10300	796	8.34
		1/2	4730	394	9.73	448	8330	645	8.45
		3/8	3630	302	9.79	342	6320	489	8.51
26	24	5/8	5370	447	9.55	517	9240	737	8.01
		1/2	4450	371	9.64	425	7460	598	8.12
		3/8	3420	285	9.70	324	5660	453	8.17
24	22	5/8	4110	373	8.73	432	7150	621	7.34
		1/2	3420	311	8.82	356	5780	504	7.45
		3/8	2630	239	8.89	273	4390	383	7.51
22	20	5/8	3060	306	7.91	355	5400	514	6.68
		1/2	2560	256	8.00	294	4370	418	6.79
		3/8	1970	197	8.07	225	3330	318	6.84

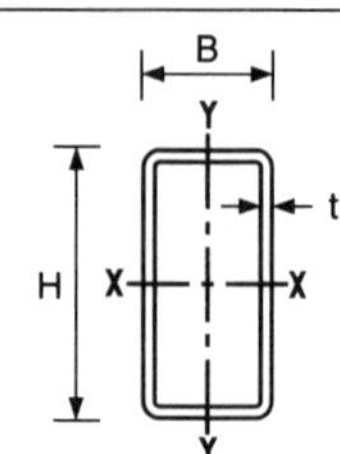

RECTANGULAR HSS
Dimensions and Properties

Dimensions			Properties								
Nominal Size			Wt per foot					X-X Axis			
Depth H	Width B	Wall		t	b/t	h/t	Area	I	S	r	Z
in.	in.	in.	lb.	in.			in.^{2}	in.^{4}	in.^{3}	in.	in.^{3}
20	18	⅝	149.24	0.625	25.8	29.0	43.9	2590	259	7.69	307
		½	122.25	0.500	33.0	37.0	35.9	2180	218	7.78	255
		⅜	93.01	0.375	45.0	50.3	27.3	1690	169	7.85	196
20	16	⅝	140.73	0.625	22.6	29.0	41.4	2360	236	7.55	283
		½	115.45	0.500	29.0	37.0	33.9	1990	199	7.65	236
		⅜	87.91	0.375	39.7	50.3	25.8	1540	154	7.72	181
20	12	⅝	123.72	0.625	16.2	29.0	36.4	1890	189	7.20	234
		½	103.30	0.465	22.8	40.0	28.3	1550	155	7.39	188
		⅜	78.52	0.349	31.4	54.3	21.5	1200	120	7.45	144
		⁵⁄₁₆	65.87	0.291	38.2	65.7	18.1	1010	101	7.48	122
20	8	⅝	110.36	0.581	10.8	31.4	30.3	1440	144	6.89	185
		½	89.68	0.465	14.2	40.0	24.6	1190	119	6.96	152
		⅜	68.31	0.349	19.9	54.3	18.7	926	92.6	7.03	117
		⁵⁄₁₆	57.36	0.291	24.5	65.7	15.7	786	78.6	7.07	98.6
20	4	½	76.07	0.465	5.6	40.0	20.9	838	83.8	6.33	115
		⅜	58.10	0.349	8.5	54.3	16.0	657	65.7	6.42	89.3
		⁵⁄₁₆	48.86	0.291	10.7	65.7	13.4	560	56.0	6.46	75.6
18	12	⅝	115.21	0.625	16.2	25.8	33.9	1450	161	6.55	199
		½	95.03	0.500	21.0	33.0	27.9	1240	138	6.67	168
		⅜	72.59	0.375	29.0	45.0	21.3	971	108	6.75	130
18	6	⅝	93.34	0.581	7.3	28.0	25.7	923	103	6.00	135
		½	76.07	0.465	9.9	35.7	20.9	770	85.6	6.07	112
		⅜	58.10	0.349	14.2	48.6	16.0	602	66.9	6.15	86.4
		⁵⁄₁₆	48.86	0.291	17.6	58.9	13.4	513	57.0	6.18	73.1
		¼	39.43	0.233	22.8	74.3	10.8	419	46.5	6.22	59.4

Shading indicates HSS manufactured by the Submerged Arc Welding (SAW) process.

RECTANGULAR HSS
Dimensions and Properties

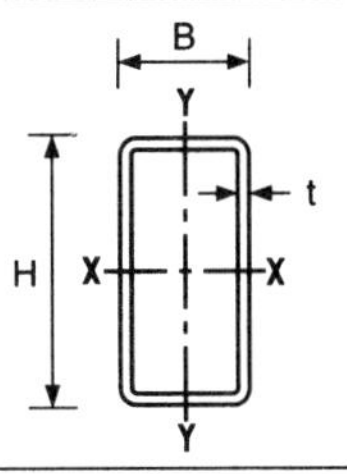

Dimensions			Properties						
Nominal Size			Y-Y Axis				Torsion		Surface Area Per foot
Depth H	Width B	Wall	I	S	r	Z	J	C	
in.	in.	in.	in.^{4}	in.^{3}	in.	in.^{3}	in.^{4}	in.^{3}	ft.^{2}
20	18	$\frac{5}{8}$	2210	245	7.10	286	3960	417	6.01
		$\frac{1}{2}$	1850	206	7.19	238	3220	340	6.12
		$\frac{3}{8}$	1440	160	7.25	183	2450	259	6.17
20	16	$\frac{5}{8}$	1680	210	6.37	243	3280	368	5.68
		$\frac{1}{2}$	1410	177	6.46	203	2670	301	5.79
		$\frac{3}{8}$	1100	137	6.52	156	2040	229	5.84
20	12	$\frac{5}{8}$	864	144	4.87	166	2030	271	5.01
		$\frac{1}{2}$	705	117	4.99	132	1540	209	5.20
		$\frac{3}{8}$	547	91.1	5.04	102	1180	160	5.23
		$\frac{5}{16}$	464	77.3	5.07	85.8	997	134	5.25
20	8	$\frac{5}{8}$	338	84.6	3.34	96.4	916	167	4.50
		$\frac{1}{2}$	283	70.8	3.39	79.5	757	137	4.53
		$\frac{3}{8}$	222	55.6	3.44	61.5	586	105	4.57
		$\frac{5}{16}$	189	47.4	3.47	52.0	496	88.3	4.58
20	4	$\frac{1}{2}$	58.7	29.3	1.68	34.0	195	63.8	3.87
		$\frac{3}{8}$	47.6	23.8	1.73	26.8	156	49.9	3.90
		$\frac{5}{16}$	41.2	20.6	1.75	22.9	134	42.4	3.92
18	12	$\frac{5}{8}$	783	131	4.81	152	1740	243	4.68
		$\frac{1}{2}$	668	111	4.89	127	1430	200	4.79
		$\frac{3}{8}$	524	87.3	4.95	98.6	1100	153	4.84
18	6	$\frac{5}{8}$	158	52.6	2.48	61.0	462	109	3.83
		$\frac{1}{2}$	134	44.6	2.53	50.7	387	89.9	3.87
		$\frac{3}{8}$	106	35.5	2.58	39.5	302	69.5	3.90
		$\frac{5}{16}$	91.3	30.4	2.61	33.5	257	58.7	3.92
		$\frac{1}{4}$	75.1	25.0	2.63	27.3	210	47.7	3.93

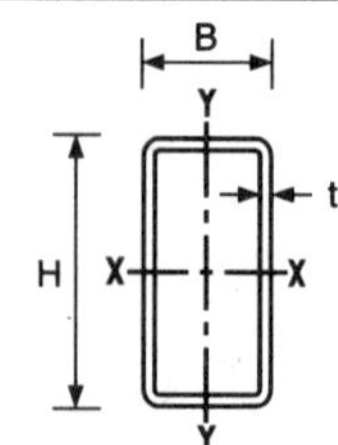

RECTANGULAR HSS
Dimensions and Properties

Dimensions			Properties								
Nominal Size									X-X Axis		
Depth H	Width B	Wall	Wt per foot	t	b/t	h/t	Area	I	S	r	Z
in.	in.	in.	lb.	in.			in.^{2}	in.^{4}	in.^{3}	in.	in.^{3}
16	12	⅝	106.71	0.625	16.2	22.6	31.4	1090	136	5.89	167
		½	89.68	0.465	22.8	31.4	24.6	904	113	6.06	135
		⅜	68.31	0.349	31.4	42.8	18.7	702	87.7	6.12	104
		⁵⁄₁₆	57.36	0.291	38.2	52.0	15.7	595	74.4	6.15	87.7
16	8	⅝	93.34	0.581	10.8	24.5	25.7	815	102	5.63	129
		½	76.07	0.465	14.2	31.4	20.9	679	84.9	5.70	106
		⅜	58.10	0.349	19.9	42.8	16.0	531	66.3	5.77	82.1
		⁵⁄₁₆	48.86	0.291	24.5	52.0	13.4	451	56.4	5.80	69.4
16	4	½	62.46	0.465	5.6	31.4	17.2	455	56.9	5.15	77.3
		⅜	47.90	0.349	8.5	42.8	13.2	360	45.0	5.23	60.2
		⁵⁄₁₆	40.35	0.291	10.7	52.0	11.1	308	38.5	5.27	51.1
14	12	½	81.42	0.500	21.0	25.0	23.9	678	96.9	5.32	116
		⅜	62.39	0.375	29.0	34.3	18.3	534	76.3	5.40	90.0
14	10	⅝	93.34	0.581	14.2	21.1	25.7	687	98.2	5.17	120
		½	76.07	0.465	18.5	27.1	20.9	573	81.8	5.23	98.8
		⅜	58.10	0.349	25.7	37.1	16.0	447	63.9	5.29	76.3
		⁵⁄₁₆	48.86	0.291	31.4	45.1	13.4	380	54.3	5.32	64.6
		¼	39.43	0.233	39.9	57.1	10.8	310	44.3	5.35	52.4
14	6	⅝	76.33	0.581	7.3	21.1	21.0	478	68.2	4.77	88.7
		½	62.46	0.465	9.9	27.1	17.2	402	57.4	4.84	73.6
		⅜	47.90	0.349	14.2	37.1	13.2	317	45.3	4.91	57.3
		⁵⁄₁₆	40.35	0.291	17.6	45.1	11.1	271	38.7	4.94	48.6
		¼	32.63	0.233	22.8	57.1	8.96	222	31.7	4.98	39.6
		³⁄₁₆	24.73	0.174	31.5	77.5	6.76	170	24.3	5.01	30.1
14	4	⅝	67.82	0.581	3.9	21.1	18.7	373	53.3	4.47	73.1
		½	55.66	0.465	5.6	27.1	15.3	317	45.3	4.55	61.0
		⅜	42.79	0.349	8.5	37.1	11.8	252	36.0	4.63	47.8
		⁵⁄₁₆	36.10	0.291	10.7	45.1	9.92	216	30.9	4.67	40.6
		¼	29.23	0.233	14.2	57.1	8.03	178	25.4	4.71	33.2
		³⁄₁₆	22.18	0.174	20.0	77.5	6.06	137	19.5	4.74	25.3

Shading indicates HSS manufactured by the Submerged Arc Welding (SAW) process.

RECTANGULAR HSS
Dimensions and Properties

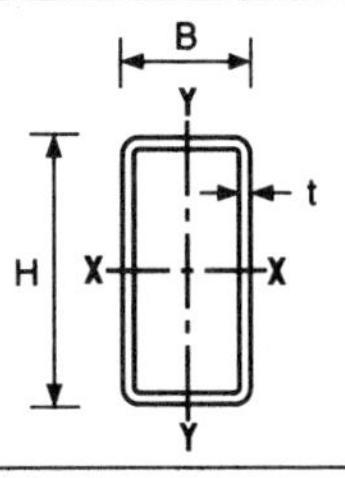

Dimensions			Properties						
Nominal Size			Y-Y Axis				Torsion		Surface Area Per foot
Depth H	Width B	Wall	I	S	r	Z	J	C	
in.	in.	in.	in.^{4}	in.^{3}	in.	in.^{3}	in.^{4}	in.^{3}	ft.^{2}
16	12	5/8	702	117	4.73	137	1470	215	4.34
		1/2	581	96.8	4.86	111	1120	166	4.53
		3/8	452	75.3	4.91	85.5	862	127	4.57
		5/16	384	64.0	4.94	72.2	727	107	4.58
16	8	5/8	274	68.5	3.27	79.2	681	132	3.83
		1/2	230	57.6	3.32	65.5	563	108	3.87
		3/8	181	45.3	3.37	50.8	436	83.4	3.90
		5/16	155	38.7	3.40	43.0	369	70.4	3.92
16	4	1/2	47.0	23.5	1.65	27.4	150	50.7	3.20
		3/8	38.3	19.1	1.71	21.7	120	39.7	3.23
		5/16	33.2	16.6	1.73	18.5	103	33.8	3.25
14	12	1/2	536	89.3	4.73	104	990	154	4.12
		3/8	422	70.4	4.80	81.2	762	118	4.17
14	10	5/8	407	81.5	3.98	95.1	832	146	3.83
		1/2	341	68.1	4.04	78.5	685	120	3.87
		3/8	267	53.4	4.09	60.7	528	91.8	3.90
		5/16	227	45.5	4.12	51.4	446	77.4	3.92
		1/4	186	37.2	4.14	41.8	362	62.6	3.93
14	6	5/8	124	41.2	2.43	48.4	334	83.7	3.17
		1/2	105	35.1	2.48	40.4	279	69.3	3.20
		3/8	84.1	28.0	2.53	31.6	219	53.7	3.23
		5/16	72.3	24.1	2.55	26.9	186	45.5	3.25
		1/4	59.6	19.9	2.58	22.0	152	36.9	3.27
		3/16	45.9	15.3	2.61	16.7	116	28.0	3.28
14	4	5/8	47.1	23.6	1.59	28.5	148	52.6	2.83
		1/2	41.1	20.6	1.64	24.1	127	44.1	2.87
		3/8	33.6	16.8	1.69	19.1	102	34.6	2.90
		5/16	29.2	14.6	1.72	16.4	87.7	29.5	2.92
		1/4	24.4	12.2	1.74	13.5	72.4	24.1	2.93
		3/16	19.0	9.48	1.77	10.3	55.8	18.4	2.95

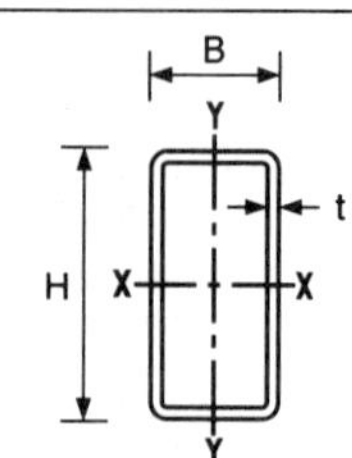

RECTANGULAR HSS
Dimensions and Properties

Dimensions			Properties								
Nominal Size			Wt per foot					X-X Axis			
Depth H	Width B	Wall		t	b/t	h/t	Area	I	S	r	Z
in.	in.	in.	lb.	in.			in.^{2}	in.^{4}	in.^{3}	in.	in.^{3}
12	10	½	69.27	0.465	18.5	22.8	19.0	395	65.9	4.56	78.8
		⅜	53.00	0.349	25.7	31.4	14.6	310	51.6	4.61	61.1
		5/16	44.60	0.291	31.4	38.2	12.2	264	44.0	4.64	51.7
		¼	36.03	0.233	39.9	48.5	9.90	216	36.0	4.67	42.1
12	8	⅝	76.33	0.581	10.8	17.7	21.0	396	66.1	4.34	82.1
		½	62.46	0.465	14.2	22.8	17.2	333	55.5	4.40	68.1
		⅜	47.90	0.349	19.9	31.4	13.2	262	43.7	4.47	53.0
		5/16	40.35	0.291	24.5	38.2	11.1	224	37.4	4.50	44.9
		¼	32.63	0.233	31.3	48.5	8.96	184	30.6	4.53	36.6
		3/16	24.73	0.174	43.0	66.0	6.76	140	23.4	4.56	27.8
12	6	⅝	67.82	0.581	7.3	17.7	18.7	321	53.4	4.14	68.8
		½	55.66	0.465	9.9	22.8	15.3	271	45.2	4.21	57.4
		⅜	42.79	0.349	14.2	31.4	11.8	215	35.8	4.28	44.8
		5/16	36.10	0.291	17.6	38.2	9.92	184	30.7	4.31	38.1
		¼	29.23	0.233	22.8	48.5	8.03	151	25.2	4.34	31.1
		3/16	22.18	0.174	31.5	66.0	6.06	116	19.4	4.38	23.7
12	4	⅝	59.32	0.581	3.9	17.7	16.4	245	40.8	3.87	55.5
		½	48.85	0.465	5.6	22.8	13.5	209	34.9	3.95	46.7
		⅜	37.69	0.349	8.5	31.4	10.4	168	28.0	4.02	36.7
		5/16	31.84	0.291	10.7	38.2	8.76	144	24.0	4.06	31.3
		¼	25.82	0.233	14.2	48.5	7.10	119	19.9	4.10	25.6
		3/16	19.63	0.174	20.0	66.0	5.37	91.8	15.3	4.13	19.6
12	3½	⅜	36.41	0.349	7.0	31.4	10.0	156	26.0	3.94	34.7
		5/16	24.97	0.291	9.0	38.2	8.46	134	22.4	3.98	29.6
12	3	5/16	29.72	0.291	7.3	38.2	8.17	124	20.7	3.90	27.9
		¼	24.12	0.233	9.9	48.5	6.63	103	17.2	3.94	22.9
		3/16	18.35	0.174	14.2	66.0	5.02	79.6	13.3	3.98	17.5
12	2	¼	22.42	0.233	5.6	48.5	6.17	86.9	14.5	3.75	20.1
		3/16	17.08	0.174	8.5	66.0	4.67	67.4	11.2	3.80	15.5

RECTANGULAR HSS
Dimensions and Properties

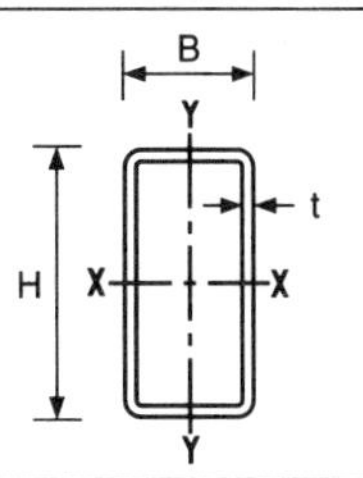

Dimensions			Properties						
Nominal Size			Y-Y Axis				Torsion		Surface Area Per foot
Depth H	Width B	Wall	I	S	r	Z	J	C	
in.	in.	in.	in.^{4}	in.^{3}	in.	in.^{3}	in.^{4}	in.^{3}	ft.^{2}
12	10	$\frac{1}{2}$	298	59.7	3.96	69.6	545	102	3.53
		$\frac{3}{8}$	234	46.9	4.01	54.0	421	78.3	3.57
		$\frac{5}{16}$	200	40.0	4.04	45.7	356	66.1	3.58
		$\frac{1}{4}$	164	32.7	4.07	37.2	289	53.5	3.60
12	8	$\frac{5}{8}$	210	52.5	3.16	61.9	454	97.7	3.17
		$\frac{1}{2}$	177	44.4	3.21	51.5	377	80.4	3.20
		$\frac{3}{8}$	140	35.1	3.27	40.1	293	62.1	3.23
		$\frac{5}{16}$	120	30.1	3.29	34.1	248	52.4	3.25
		$\frac{1}{4}$	98.8	24.7	3.32	27.8	202	42.5	3.27
		$\frac{3}{16}$	75.7	18.9	3.35	21.1	153	32.2	3.28
12	6	$\frac{5}{8}$	106	35.5	2.39	42.1	271	71.1	2.83
		$\frac{1}{2}$	91.1	30.4	2.44	35.2	227	59.0	2.87
		$\frac{3}{8}$	72.9	24.3	2.49	27.7	178	45.8	2.90
		$\frac{5}{16}$	62.8	20.9	2.52	23.6	152	38.8	2.92
		$\frac{1}{4}$	51.9	17.3	2.54	19.3	124	31.6	2.93
		$\frac{3}{16}$	40.0	13.3	2.57	14.7	94.6	24.0	2.95
12	4	$\frac{5}{8}$	40.3	20.1	1.57	24.5	122	44.6	2.50
		$\frac{1}{2}$	35.3	17.6	1.62	20.9	105	37.5	2.53
		$\frac{3}{8}$	28.9	14.5	1.67	16.6	84.1	29.5	2.57
		$\frac{5}{16}$	25.2	12.6	1.70	14.2	72.4	25.2	2.58
		$\frac{1}{4}$	21.0	10.5	1.72	11.7	59.8	20.6	2.60
		$\frac{3}{16}$	16.4	8.20	1.75	9.00	46.1	15.7	2.62
12	$3\frac{1}{2}$	$\frac{3}{8}$	21.3	12.2	1.46	14.0	64.7	25.5	2.48
		$\frac{5}{16}$	18.6	10.6	1.48	12.1	56.0	21.8	2.50
12	3	$\frac{5}{16}$	13.1	8.73	1.27	10.0	41.3	18.4	2.42
		$\frac{1}{4}$	11.1	7.38	1.29	8.28	34.5	15.1	2.43
		$\frac{3}{16}$	8.72	5.81	1.32	6.40	26.8	11.6	2.45
12	2	$\frac{1}{4}$	4.40	4.40	0.845	5.08	15.1	9.64	2.27
		$\frac{3}{16}$	3.55	3.55	0.872	3.97	12.0	7.49	2.28

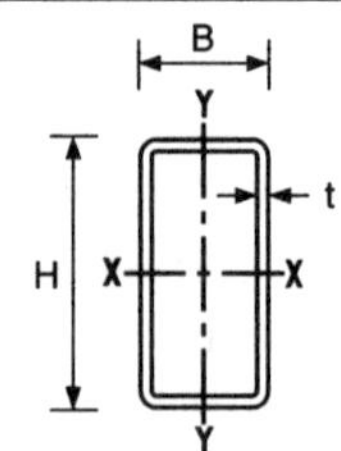

RECTANGULAR HSS
Dimensions and Properties

Dimensions			Properties								
Nominal Size			Wt per foot					X-X Axis			
Depth H	Width B	Wall		t	b/t	h/t	Area	I	S	r	Z
in.	in.	in.	lb.	in.			in.²	in.⁴	in.³	in.	in.³
10	8	½	55.66	0.465	14.2	18.5	15.3	214	42.7	3.73	51.9
		⅜	42.79	0.349	19.9	25.7	11.8	169	33.9	3.79	40.5
		⁵⁄₁₆	36.10	0.291	24.5	31.4	9.92	145	29.0	3.82	34.4
		¼	29.23	0.233	31.3	39.9	8.03	119	23.8	3.85	28.1
		³⁄₁₆	22.18	0.174	43.0	54.5	6.06	91.4	18.3	3.88	21.4
10	6	⅝	59.32	0.581	7.3	14.2	16.4	201	40.2	3.50	51.3
		½	48.85	0.465	9.9	18.5	13.5	171	34.3	3.57	43.0
		⅜	37.69	0.349	14.2	25.7	10.4	137	27.3	3.63	33.8
		⁵⁄₁₆	31.84	0.291	17.6	31.4	8.76	118	23.5	3.66	28.8
		¼	25.82	0.233	22.8	39.9	7.10	96.9	19.4	3.69	23.6
		³⁄₁₆	19.63	0.174	31.5	54.5	5.37	74.6	14.9	3.73	18.0
10	5	⅜	35.13	0.349	11.3	25.7	9.67	120	24.1	3.53	30.4
		⁵⁄₁₆	29.72	0.291	14.2	31.4	8.17	104	20.8	3.56	26.0
		¼	24.12	0.233	18.5	39.9	6.63	85.8	17.2	3.60	21.3
		³⁄₁₆	18.35	0.174	25.7	54.5	5.02	66.2	13.2	3.63	16.3
10	4	⅝	50.81	0.581	3.9	14.2	14.0	149	29.9	3.26	40.3
		½	42.05	0.465	5.6	18.5	11.6	129	25.8	3.34	34.1
		⅜	32.58	0.349	8.5	25.7	8.97	104	20.8	3.41	27.0
		⁵⁄₁₆	27.59	0.291	10.7	31.4	7.59	90.1	18.0	3.44	23.1
		¼	22.42	0.233	14.2	39.9	6.17	74.7	14.9	3.48	19.0
		³⁄₁₆	17.08	0.174	20.0	54.5	4.67	57.8	11.6	3.52	14.6
10	3½	³⁄₁₆	16.44	0.174	17.1	54.5	4.50	53.6	10.7	3.45	13.7
10	3	⅜	30.03	0.349	5.6	25.7	8.27	88.0	17.6	3.26	23.7
		⁵⁄₁₆	25.46	0.291	7.3	31.4	7.01	76.3	15.3	3.30	20.3
		¼	20.72	0.233	9.9	39.9	5.70	63.6	12.7	3.34	16.7
		³⁄₁₆	15.80	0.174	14.2	54.5	4.32	49.4	9.87	3.38	12.8
		⅛	10.71	0.116	22.9	83.2	2.93	34.2	6.83	3.42	8.80
10	2	⅜	27.48	0.349	2.7	25.7	7.58	71.7	14.3	3.08	20.3
		⁵⁄₁₆	23.34	0.291	3.9	31.4	6.43	62.6	12.5	3.12	17.5
		¼	19.02	0.233	5.6	39.9	5.24	52.5	10.5	3.17	14.4
		³⁄₁₆	14.53	0.174	8.5	54.5	3.98	41.0	8.19	3.21	11.1

RECTANGULAR HSS
Dimensions and Properties

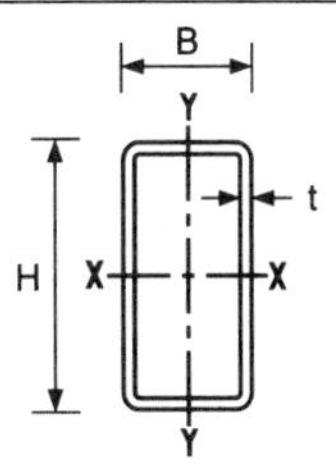

Dimensions			Properties						
Nominal Size			Y-Y Axis				Torsion		Surface Area Per foot
Depth H	Width B	Wall	I	S	r	Z	J	C	
in.	in.	in.	in.^{4}	in.^{3}	in.	in.^{3}	in.^{4}	in.^{3}	ft.^{2}
10	8	$\frac{1}{2}$	151	37.8	3.14	44.5	288	66.4	2.87
		$\frac{3}{8}$	120	30.0	3.19	34.8	224	51.4	2.90
		$\frac{5}{16}$	103	25.7	3.22	29.6	190	43.5	2.92
		$\frac{1}{4}$	84.7	21.2	3.25	24.2	155	35.3	2.93
		$\frac{3}{16}$	65.1	16.3	3.28	18.4	118	26.7	2.95
10	6	$\frac{5}{8}$	89.4	29.8	2.34	35.8	209	58.6	2.50
		$\frac{1}{2}$	76.8	25.6	2.39	30.1	176	48.7	2.53
		$\frac{3}{8}$	61.8	20.6	2.44	23.7	139	37.9	2.57
		$\frac{5}{16}$	53.3	17.8	2.47	20.2	118	32.2	2.58
		$\frac{1}{4}$	44.1	14.7	2.49	16.6	96.7	26.2	2.60
		$\frac{3}{16}$	34.1	11.4	2.52	12.7	73.8	19.9	2.62
10	5	$\frac{3}{8}$	40.6	16.2	2.05	18.7	100	31.2	2.40
		$\frac{5}{16}$	35.2	14.1	2.07	16.0	86.0	26.5	2.42
		$\frac{1}{4}$	29.3	11.7	2.10	13.2	70.7	21.6	2.43
		$\frac{3}{16}$	22.7	9.09	2.13	10.1	54.1	16.5	2.45
10	4	$\frac{5}{8}$	33.4	16.7	1.54	20.6	95.7	36.7	2.17
		$\frac{1}{2}$	29.4	14.7	1.59	17.6	82.6	31.0	2.20
		$\frac{3}{8}$	24.3	12.1	1.64	14.0	66.5	24.4	2.23
		$\frac{5}{16}$	21.2	10.6	1.67	12.1	57.3	20.9	2.25
		$\frac{1}{4}$	17.7	8.87	1.70	9.96	47.4	17.1	2.27
		$\frac{3}{16}$	13.9	6.93	1.72	7.66	36.5	13.1	2.28
10	$3\frac{1}{2}$	$\frac{3}{16}$	10.3	5.89	1.51	6.52	28.6	11.4	2.20
10	3	$\frac{3}{8}$	12.4	8.27	1.22	9.73	37.8	17.7	2.07
		$\frac{5}{16}$	11.0	7.30	1.25	8.42	33.0	15.2	2.08
		$\frac{1}{4}$	9.28	6.18	1.28	6.99	27.6	12.5	2.10
		$\frac{3}{16}$	7.33	4.89	1.30	5.41	21.5	9.64	2.12
		$\frac{1}{8}$	5.16	3.44	1.33	3.74	14.9	6.61	2.13
10	2	$\frac{3}{8}$	4.69	4.69	0.786	5.76	15.9	11.0	1.90
		$\frac{5}{16}$	4.24	4.24	0.812	5.06	14.2	9.56	1.92
		$\frac{1}{4}$	3.67	3.67	0.837	4.26	12.2	7.99	1.93
		$\frac{3}{16}$	2.97	2.97	0.864	3.34	9.74	6.22	1.95

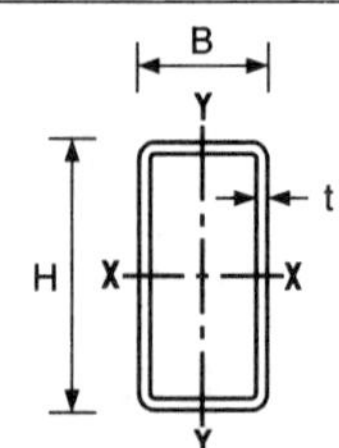

RECTANGULAR HSS
Dimensions and Properties

Dimensions			Properties								
Nominal Size			Wt per foot					X-X Axis			
Depth H	Width B	Wall		t	b/t	h/t	Area	I	S	r	Z
in.	in.	in.	lb.	in.			in.^{2}	in.^{4}	in.^{3}	in.	in.^{3}
9	7	$\frac{5}{8}$	59.32	0.581	9.0	12.5	16.4	174	38.7	3.26	48.3
		$\frac{1}{2}$	48.85	0.465	12.1	16.4	13.5	149	33.0	3.32	40.5
		$\frac{3}{8}$	37.69	0.349	17.1	22.8	10.4	119	26.4	3.38	31.8
		$\frac{5}{16}$	31.84	0.291	21.1	27.9	8.76	102	22.6	3.41	27.1
		$\frac{1}{4}$	25.82	0.233	27.0	35.6	7.10	84.1	18.7	3.44	22.2
		$\frac{3}{16}$	19.63	0.174	37.2	48.7	5.37	64.7	14.4	3.47	16.9
9	5	$\frac{5}{8}$	50.81	0.581	5.6	12.5	14.0	133	29.6	3.08	38.5
		$\frac{1}{2}$	42.05	0.465	7.8	16.4	11.6	115	25.5	3.14	32.5
		$\frac{3}{8}$	32.58	0.349	11.3	22.8	8.97	92.5	20.5	3.21	25.7
		$\frac{5}{16}$	27.59	0.291	14.2	27.9	7.59	79.8	17.7	3.24	22.0
		$\frac{1}{4}$	22.42	0.233	18.5	35.6	6.17	66.1	14.7	3.27	18.1
		$\frac{3}{16}$	17.08	0.174	25.7	48.7	4.67	51.1	11.4	3.31	13.8
9	3	$\frac{1}{2}$	35.24	0.465	3.5	16.4	9.74	80.8	17.9	2.88	24.6
		$\frac{3}{8}$	27.48	0.349	5.6	22.8	7.58	66.3	14.7	2.96	19.7
		$\frac{5}{16}$	23.34	0.291	7.3	27.9	6.43	57.7	12.8	3.00	16.9
		$\frac{1}{4}$	19.02	0.233	9.9	35.6	5.24	48.2	10.7	3.04	14.0
		$\frac{3}{16}$	14.53	0.174	14.2	48.7	3.98	37.6	8.35	3.07	10.8

RECTANGULAR HSS
Dimensions and Properties

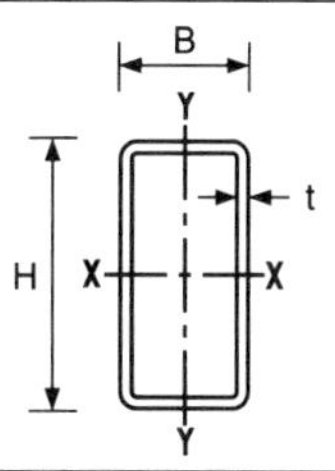

Dimensions			Properties						
Nominal Size			Y-Y Axis				Torsion		Surface Area Per foot
Depth H	Width B	Wall	I	S	r	Z	J	C	
in.	in.	in.	in.^{4}	in.^{3}	in.	in.^{3}	in.^{4}	in.^{3}	ft.^{2}
9	7	$\frac{5}{8}$	117	33.5	2.68	40.5	235	62.0	2.50
		$\frac{1}{2}$	100	28.7	2.73	34.0	197	51.5	2.53
		$\frac{3}{8}$	80.4	23.0	2.78	26.7	154	40.0	2.57
		$\frac{5}{16}$	69.2	19.8	2.81	22.8	131	33.9	2.58
		$\frac{1}{4}$	57.2	16.3	2.84	18.7	107	27.6	2.60
		$\frac{3}{16}$	44.1	12.6	2.87	14.3	81.7	20.9	2.62
9	5	$\frac{5}{8}$	51.9	20.8	1.92	25.3	128	42.5	2.17
		$\frac{1}{2}$	45.2	18.1	1.97	21.5	109	35.6	2.20
		$\frac{3}{8}$	36.8	14.7	2.03	17.1	86.9	27.9	2.23
		$\frac{5}{16}$	32.0	12.8	2.05	14.6	74.4	23.8	2.25
		$\frac{1}{4}$	26.6	10.6	2.08	12.0	61.2	19.4	2.27
		$\frac{3}{16}$	20.7	8.28	2.10	9.25	46.9	14.8	2.28
9	3	$\frac{1}{2}$	13.2	8.79	1.16	10.8	40.0	19.7	1.87
		$\frac{3}{8}$	11.2	7.45	1.21	8.80	33.1	15.8	1.90
		$\frac{5}{16}$	9.88	6.59	1.24	7.63	28.9	13.6	1.92
		$\frac{1}{4}$	8.38	5.59	1.27	6.35	24.2	11.3	1.93
		$\frac{3}{16}$	6.63	4.42	1.29	4.92	18.9	8.66	1.95

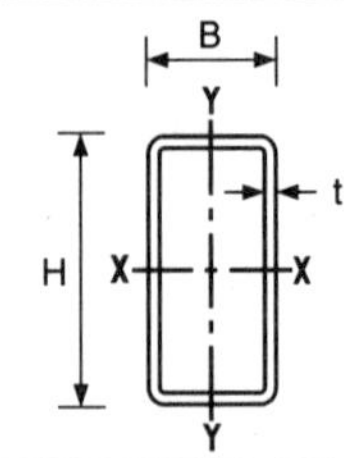

RECTANGULAR HSS
Dimensions and Properties

Dimensions			Properties								
Nominal Size			Wt per foot					X-X Axis			
Depth H	Width B	Wall		t	b/t	h/t	Area	I	S	r	Z
in.	in.	in.	lb.	in.			in.^{2}	in.^{4}	in.^{3}	in.	in.^{3}
8	6	5/8	50.81	0.581	7.3	10.8	14.0	114	28.5	2.85	36.1
		1/2	42.05	0.465	9.9	14.2	11.6	98.2	24.5	2.91	30.5
		3/8	32.58	0.349	14.2	19.9	8.97	79.1	19.8	2.97	24.1
		5/16	27.59	0.291	17.6	24.5	7.59	68.3	17.1	3.00	20.6
		1/4	22.42	0.233	22.8	31.3	6.17	56.6	14.1	3.03	16.9
		3/16	17.08	0.174	31.5	43.0	4.67	43.7	10.9	3.06	13.0
8	4	5/8	42.30	0.581	3.9	10.8	11.7	81.9	20.5	2.64	27.4
		1/2	35.24	0.465	5.6	14.2	9.74	71.7	17.9	2.71	23.5
		3/8	27.48	0.349	8.5	19.9	7.58	58.7	14.7	2.78	18.8
		5/16	23.34	0.291	10.7	24.5	6.43	51.0	12.8	2.82	16.1
		1/4	19.02	0.233	14.2	31.3	5.24	42.5	10.6	2.85	13.3
		3/16	14.53	0.174	20.0	43.0	3.98	33.1	8.27	2.88	10.2
		1/8	9.86	0.116	31.5	66.0	2.70	22.9	5.73	2.92	7.02
8	3	1/2	31.84	0.465	3.5	14.2	8.81	58.5	14.6	2.58	20.0
		3/8	24.93	0.349	5.6	19.9	6.88	48.5	12.1	2.65	16.1
		5/16	21.21	0.291	7.3	24.5	5.85	42.4	10.6	2.69	13.9
		1/4	17.32	0.233	9.9	31.3	4.77	35.5	8.88	2.73	11.5
		3/16	13.25	0.174	14.2	43.0	3.63	27.8	6.94	2.77	8.87
		1/8	9.01	0.116	22.9	66.0	2.46	19.3	4.83	2.80	6.11
8	2	3/8	22.37	0.349	2.7	19.9	6.18	38.2	9.56	2.49	13.4
		5/16	19.08	0.291	3.9	24.5	5.26	33.7	8.43	2.53	11.6
		1/4	15.62	0.233	5.6	31.3	4.30	28.5	7.12	2.57	9.68
		3/16	11.97	0.174	8.5	43.0	3.28	22.4	5.61	2.61	7.51
		1/8	8.16	0.116	14.2	66.0	2.23	15.7	3.93	2.65	5.19

RECTANGULAR HSS
Dimensions and Properties

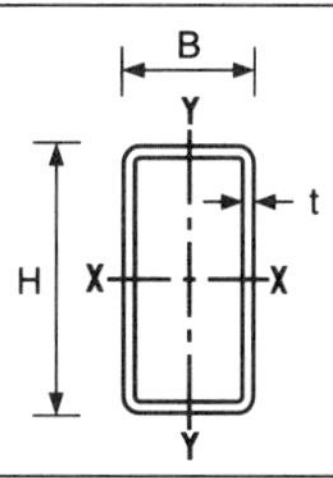

Dimensions			Properties						
Nominal Size			Y-Y Axis				Torsion		Surface Area Per foot
Depth H	Width B	Wall	I	S	r	Z	J	C	
in.	in.	in.	in.^{4}	in.^{3}	in.	in.^{3}	in.^{4}	in.^{3}	ft.^{2}
8	6	5/8	72.2	24.1	2.27	29.5	150	46.0	2.17
		1/2	62.5	20.8	2.32	24.9	127	38.4	2.20
		3/8	50.6	16.9	2.38	19.8	100	30.0	2.23
		5/16	43.8	14.6	2.40	16.9	85.8	25.5	2.25
		1/4	36.4	12.1	2.43	13.9	70.3	20.8	2.27
		3/16	28.2	9.39	2.46	10.7	53.7	15.8	2.28
8	4	5/8	26.6	13.3	1.51	16.6	70.3	28.7	1.83
		1/2	23.6	11.8	1.56	14.3	61.1	24.4	1.87
		3/8	19.6	9.79	1.61	11.5	49.3	19.3	1.90
		5/16	17.2	8.58	1.63	9.91	42.6	16.5	1.92
		1/4	14.4	7.21	1.66	8.20	35.3	13.6	1.93
		3/16	11.3	5.65	1.69	6.33	27.2	10.4	1.95
		1/8	7.90	3.95	1.71	4.36	18.7	7.10	1.97
8	3	1/2	11.7	7.78	1.15	9.64	34.3	17.4	1.70
		3/8	9.94	6.62	1.20	7.88	28.5	14.0	1.73
		5/16	8.81	5.87	1.23	6.84	24.9	12.1	1.75
		1/4	7.49	4.99	1.25	5.70	20.8	9.97	1.77
		3/16	5.94	3.96	1.28	4.43	16.2	7.68	1.78
		1/8	4.20	2.80	1.31	3.07	11.3	5.27	1.80
8	2	3/8	3.72	3.72	0.776	4.61	12.1	8.65	1.57
		5/16	3.38	3.38	0.801	4.06	10.9	7.57	1.58
		1/4	2.94	2.94	0.827	3.43	9.36	6.35	1.60
		3/16	2.39	2.39	0.853	2.70	7.48	4.95	1.62
		1/8	1.72	1.72	0.879	1.90	5.30	3.44	1.63

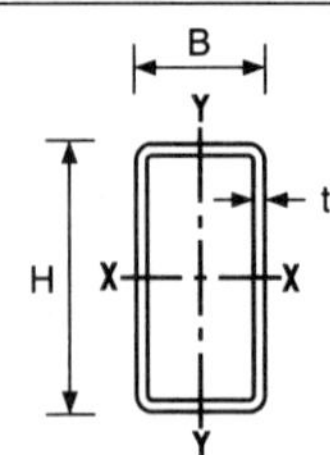

RECTANGULAR HSS
Dimensions and Properties

Dimensions			Properties								
Nominal Size			Wt per foot					X-X Axis			
Depth H	Width B	Wall		t	b/t	h/t	Area	I	S	r	Z
in.	in.	in.	lb.	in.			in.^{2}	in.^{4}	in.^{3}	in.	in.^{3}
7	5	$\frac{5}{8}$	42.30	0.581	5.6	9.0	11.7	69.3	19.8	2.43	25.6
		$\frac{1}{2}$	35.24	0.465	7.8	12.1	9.74	60.6	17.3	2.49	21.9
		$\frac{3}{8}$	27.48	0.349	11.3	17.1	7.58	49.5	14.1	2.56	17.5
		$\frac{5}{16}$	23.34	0.291	14.2	21.1	6.43	43.0	12.3	2.59	15.0
		$\frac{1}{4}$	19.02	0.233	18.5	27.0	5.24	35.8	10.2	2.62	12.4
		$\frac{3}{16}$	14.53	0.174	25.7	37.2	3.98	27.9	7.96	2.65	9.52
		$\frac{1}{8}$	9.86	0.116	40.1	57.3	2.70	19.3	5.52	2.68	6.53
7	4	$\frac{1}{2}$	31.84	0.465	5.6	12.1	8.81	50.6	14.5	2.40	18.8
		$\frac{3}{8}$	24.93	0.349	8.5	17.1	6.88	41.8	11.9	2.46	15.1
		$\frac{5}{16}$	21.21	0.291	10.7	21.1	5.85	36.4	10.4	2.50	13.1
		$\frac{1}{4}$	17.32	0.233	14.2	27.0	4.77	30.5	8.72	2.53	10.8
		$\frac{3}{16}$	13.25	0.174	20.0	37.2	3.63	23.8	6.80	2.56	8.33
		$\frac{1}{8}$	9.01	0.116	31.5	57.3	2.46	16.6	4.73	2.59	5.73
7	3	$\frac{1}{2}$	28.43	0.465	3.5	12.1	7.88	40.7	11.6	2.27	15.8
		$\frac{3}{8}$	22.37	0.349	5.6	17.1	6.18	34.0	9.73	2.35	12.8
		$\frac{5}{16}$	19.08	0.291	7.3	21.1	5.26	29.9	8.54	2.38	11.1
		$\frac{1}{4}$	15.62	0.233	9.9	27.0	4.30	25.2	7.19	2.42	9.22
		$\frac{3}{16}$	11.97	0.174	14.2	37.2	3.28	19.8	5.65	2.45	7.14
		$\frac{1}{8}$	8.16	0.116	22.9	57.3	2.23	13.8	3.95	2.49	4.93

RECTANGULAR HSS
Dimensions and Properties

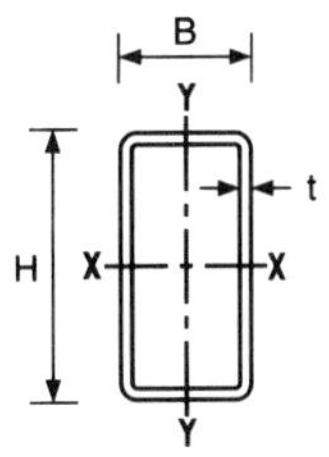

Dimensions			Properties						
Nominal Size			Y-Y Axis				Torsion		Surface Area Per foot
Depth H	Width B	Wall	I	S	r	Z	J	C	
in.	in.	in.	in.^{4}	in.^{3}	in.	in.^{3}	in.^{4}	in.^{3}	ft.^{2}
7	5	$\frac{5}{8}$	40.5	16.2	1.86	20.2	88.5	32.2	1.83
		$\frac{1}{2}$	35.6	14.2	1.91	17.3	75.8	27.2	1.87
		$\frac{3}{8}$	29.2	11.7	1.96	13.8	60.6	21.4	1.90
		$\frac{5}{16}$	25.5	10.2	1.99	11.9	52.1	18.3	1.92
		$\frac{1}{4}$	21.3	8.53	2.02	9.83	42.9	15.0	1.93
		$\frac{3}{16}$	16.6	6.65	2.05	7.57	32.9	11.4	1.95
		$\frac{1}{8}$	11.6	4.63	2.07	5.20	22.5	7.79	1.97
7	4	$\frac{1}{2}$	20.7	10.3	1.53	12.6	50.5	21.1	1.70
		$\frac{3}{8}$	17.3	8.63	1.58	10.2	41.0	16.8	1.73
		$\frac{5}{16}$	15.2	7.58	1.61	8.83	35.4	14.4	1.75
		$\frac{1}{4}$	12.8	6.38	1.64	7.33	29.3	11.8	1.77
		$\frac{3}{16}$	10.0	5.02	1.66	5.67	22.7	9.07	1.78
		$\frac{1}{8}$	7.03	3.51	1.69	3.91	15.6	6.20	1.80
7	3	$\frac{1}{2}$	10.2	6.78	1.14	8.46	28.6	15.0	1.53
		$\frac{3}{8}$	8.70	5.80	1.19	6.95	23.9	12.1	1.57
		$\frac{5}{16}$	7.74	5.16	1.21	6.05	20.9	10.5	1.58
		$\frac{1}{4}$	6.59	4.40	1.24	5.06	17.5	8.68	1.60
		$\frac{3}{16}$	5.24	3.50	1.26	3.94	13.7	6.69	1.62
		$\frac{1}{8}$	3.71	2.48	1.29	2.73	9.48	4.60	1.63

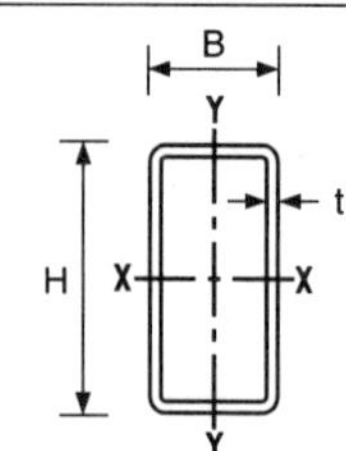

RECTANGULAR HSS
Dimensions and Properties

Dimensions			Properties					X-X Axis			
Nominal Size			Wt per foot								
Depth H	Width B	Wall		t	b/t	h/t	Area	I	S	r	Z
in.	in.	in.	lb.	in.			in.^{2}	in.^{4}	in.^{3}	in.	in.^{3}
6	5	3/8	24.93	0.349	11.3	14.2	6.88	33.9	11.3	2.22	13.8
		5/16	21.21	0.291	14.2	17.6	5.85	29.6	9.85	2.25	11.9
		1/4	17.32	0.233	18.5	22.8	4.77	24.7	8.25	2.28	9.87
		3/16	13.25	0.174	25.7	31.5	3.63	19.3	6.44	2.31	7.62
6	4	1/2	28.43	0.465	5.6	9.9	7.88	33.9	11.3	2.08	14.6
		3/8	22.37	0.349	8.5	14.2	6.18	28.3	9.43	2.14	11.9
		5/16	19.08	0.291	10.7	17.6	5.26	24.8	8.27	2.17	10.3
		1/4	15.62	0.233	14.2	22.8	4.30	20.9	6.96	2.20	8.53
		3/16	11.97	0.174	20.0	31.5	3.28	16.4	5.46	2.23	6.60
		1/8	8.16	0.116	31.5	48.7	2.23	11.4	3.81	2.26	4.56
6	3	1/2	25.03	0.465	3.5	9.9	6.95	26.8	8.94	1.96	12.1
		3/8	19.82	0.349	5.6	14.2	5.48	22.7	7.57	2.04	9.90
		5/16	16.96	0.291	7.3	17.6	4.68	20.1	6.69	2.07	8.61
		1/4	13.91	0.233	9.9	22.8	3.84	17.0	5.66	2.10	7.19
		3/16	10.70	0.174	14.2	31.5	2.93	13.4	4.47	2.14	5.59
		1/8	7.31	0.116	22.9	48.7	2.00	9.43	3.14	2.17	3.87
6	2	3/8	17.27	0.349	2.7	14.2	4.78	17.1	5.71	1.89	7.93
		5/16	14.83	0.291	3.9	17.6	4.10	15.3	5.11	1.93	6.95
		1/4	12.21	0.233	5.6	22.8	3.37	13.1	4.37	1.97	5.84
		3/16	9.42	0.174	8.5	31.5	2.58	10.5	3.49	2.01	4.58
		1/8	6.46	0.116	14.2	48.7	1.77	7.42	2.47	2.05	3.19

RECTANGULAR HSS
Dimensions and Properties

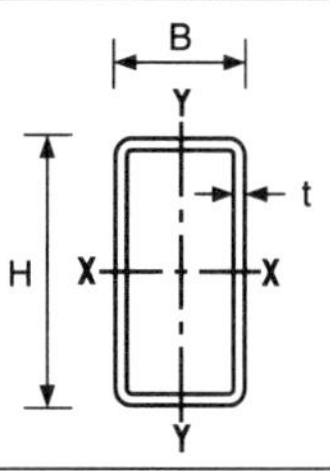

Dimensions			Properties						
Nominal Size			Y-Y Axis				Torsion		Surface
Depth H	Width B	Wall	I	S	r	Z	J	C	Area Per foot
in.	in.	in.	in.^{4}	in.^{3}	in.	in.^{3}	in.^{4}	in.^{3}	ft.^{2}
6	5	$\frac{3}{8}$	25.5	10.2	1.92	12.2	48.1	18.2	1.73
		$\frac{5}{16}$	22.3	8.91	1.95	10.5	41.4	15.6	1.75
		$\frac{1}{4}$	18.7	7.47	1.98	8.72	34.2	12.8	1.77
		$\frac{3}{16}$	14.6	5.84	2.01	6.73	26.3	9.76	1.78
6	4	$\frac{1}{2}$	17.7	8.87	1.50	11.0	40.3	17.8	1.53
		$\frac{3}{8}$	14.9	7.46	1.55	8.94	32.8	14.2	1.57
		$\frac{5}{16}$	13.1	6.57	1.58	7.75	28.4	12.2	1.58
		$\frac{1}{4}$	11.1	5.56	1.61	6.45	23.6	10.1	1.60
		$\frac{3}{16}$	8.76	4.38	1.63	5.00	18.2	7.74	1.62
		$\frac{1}{8}$	6.15	3.08	1.66	3.46	12.6	5.30	1.63
6	3	$\frac{1}{2}$	8.65	5.77	1.12	7.28	23.1	12.7	1.37
		$\frac{3}{8}$	7.47	4.98	1.17	6.03	19.3	10.3	1.40
		$\frac{5}{16}$	6.66	4.44	1.19	5.27	16.9	8.91	1.42
		$\frac{1}{4}$	5.70	3.80	1.22	4.41	14.2	7.39	1.43
		$\frac{3}{16}$	4.55	3.03	1.25	3.45	11.1	5.71	1.45
		$\frac{1}{8}$	3.23	2.15	1.27	2.40	7.73	3.93	1.47
6	2	$\frac{3}{8}$	2.75	2.75	0.759	3.46	8.42	6.35	1.23
		$\frac{5}{16}$	2.52	2.52	0.784	3.07	7.60	5.58	1.25
		$\frac{1}{4}$	2.21	2.21	0.809	2.61	6.55	4.70	1.27
		$\frac{3}{16}$	1.80	1.80	0.835	2.07	5.24	3.68	1.28
		$\frac{1}{8}$	1.31	1.31	0.861	1.46	3.72	2.57	1.30

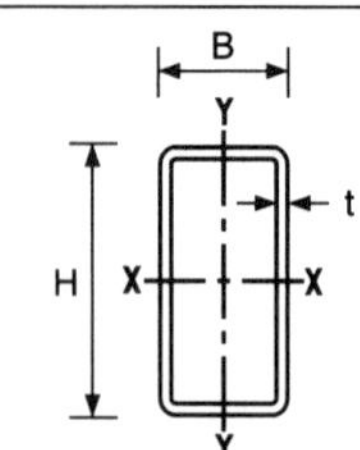

RECTANGULAR HSS
Dimensions and Properties

Dimensions			Properties								
Nominal Size			Wt per foot					X-X Axis			
Depth H	Width B	Wall		t	b/t	h/t	Area	I	S	r	Z
in.	in.	in.	lb.	in.			in.^{2}	in.^{4}	in.^{3}	in.	in.^{3}
5	4	½	25.03	0.465	5.6	7.8	6.95	21.2	8.48	1.75	10.9
		⅜	19.82	0.349	8.5	11.3	5.48	17.9	7.16	1.81	8.96
		5/16	16.96	0.291	10.7	14.2	4.68	15.8	6.32	1.84	7.79
		¼	13.91	0.233	14.2	18.5	3.84	13.4	5.35	1.87	6.49
		3/16	10.70	0.174	20.0	25.7	2.93	10.6	4.22	1.90	5.05
5	3	½	21.63	0.465	3.5	7.8	6.02	16.4	6.56	1.65	8.83
		⅜	17.27	0.349	5.6	11.3	4.78	14.1	5.65	1.72	7.34
		5/16	14.83	0.291	7.3	14.2	4.10	12.6	5.03	1.75	6.42
		¼	12.21	0.233	9.9	18.5	3.37	10.7	4.29	1.78	5.38
		3/16	9.42	0.174	14.2	25.7	2.58	8.53	3.41	1.82	4.21
		⅛	6.46	0.116	22.9	40.1	1.77	6.03	2.41	1.85	2.93
5	2½	¼	11.36	0.233	7.7	18.5	3.14	9.40	3.76	1.73	4.83
		3/16	8.78	0.174	11.4	25.7	2.41	7.51	3.01	1.77	3.79
		⅛	6.03	0.116	18.6	40.1	1.65	5.34	2.14	1.80	2.65
5	2	⅜	14.72	0.349	2.7	11.3	4.09	10.3	4.14	1.59	5.71
		5/16	12.70	0.291	3.9	14.2	3.52	9.34	3.74	1.63	5.05
		¼	10.51	0.233	5.6	18.5	2.91	8.08	3.23	1.67	4.27
		3/16	8.15	0.174	8.5	25.7	2.24	6.50	2.60	1.70	3.37
		⅛	5.61	0.116	14.2	40.1	1.54	4.65	1.86	1.74	2.37
4	3	⅜	14.72	0.349	5.6	8.5	4.09	7.92	3.96	1.39	5.12
		5/16	12.70	0.291	7.3	10.7	3.52	7.13	3.57	1.42	4.51
		¼	10.51	0.233	9.9	14.2	2.91	6.15	3.07	1.45	3.81
		3/16	8.15	0.174	14.2	20.0	2.24	4.93	2.47	1.49	3.00
		⅛	5.61	0.116	22.9	31.5	1.54	3.52	1.76	1.51	2.11
4	2½	5/16	11.64	0.291	5.6	10.7	3.23	6.13	3.06	1.38	3.97
		¼	9.66	0.233	7.7	14.2	2.67	5.32	2.66	1.41	3.38
		3/16	7.51	0.174	11.4	20.0	2.06	4.30	2.15	1.44	2.67
4	2	⅜	12.17	0.349	2.7	8.5	3.39	5.59	2.80	1.28	3.84
		5/16	10.58	0.291	3.9	10.7	2.94	5.12	2.56	1.32	3.43
		¼	8.81	0.233	5.6	14.2	2.44	4.49	2.25	1.36	2.94
		3/16	6.87	0.174	8.5	20.0	1.89	3.66	1.83	1.39	2.34
		⅛	4.75	0.116	14.2	31.5	1.30	2.65	1.32	1.43	1.66

RECTANGULAR HSS
Dimensions and Properties

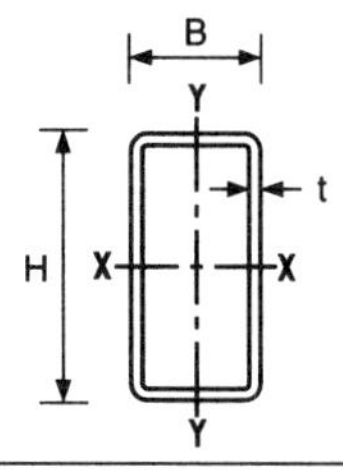

Dimensions			Properties						
Nominal Size			Y-Y Axis				Torsion		Surface Area Per foot
Depth H	Width B	Wall	I	S	r	Z	J	C	
in.	in.	in.	in.^{4}	in.^{3}	in.	in.^{3}	in.^{4}	in.^{3}	ft.^{2}
5	4	1/2	14.8	7.41	1.46	9.35	30.3	14.5	1.37
		3/8	12.6	6.29	1.52	7.67	24.9	11.7	1.40
		5/16	11.1	5.57	1.54	6.67	21.7	10.1	1.42
		1/4	9.46	4.73	1.57	5.57	18.0	8.32	1.43
		3/16	7.48	3.74	1.60	4.34	14.0	6.41	1.45
5	3	1/2	7.14	4.76	1.09	6.10	17.6	10.3	1.20
		3/8	6.23	4.16	1.14	5.10	14.9	8.44	1.23
		5/16	5.59	3.73	1.17	4.48	13.1	7.33	1.25
		1/4	4.81	3.20	1.19	3.77	11.0	6.10	1.27
		3/16	3.85	2.57	1.22	2.96	8.64	4.73	1.28
		1/8	2.75	1.83	1.25	2.07	6.02	3.26	1.30
5	2½	1/4	3.13	2.50	0.998	2.95	7.93	4.99	1.18
		3/16	2.53	2.03	1.02	2.33	6.26	3.89	1.20
		1/8	1.82	1.46	1.05	1.64	4.40	2.70	1.22
5	2	3/8	2.27	2.27	0.746	2.88	6.61	5.20	1.07
		5/16	2.09	2.09	0.771	2.57	5.99	4.59	1.08
		1/4	1.84	1.84	0.796	2.20	5.17	3.88	1.10
		3/16	1.51	1.51	0.822	1.75	4.15	3.05	1.12
		1/8	1.10	1.10	0.848	1.24	2.95	2.13	1.13
4	3	3/8	5.00	3.33	1.11	4.18	10.6	6.59	1.07
		5/16	4.52	3.01	1.13	3.69	9.41	5.75	1.08
		1/4	3.91	2.61	1.16	3.12	7.96	4.81	1.10
		3/16	3.16	2.10	1.19	2.46	6.26	3.74	1.12
		1/8	2.27	1.51	1.21	1.73	4.38	2.59	1.13
4	2½	5/16	2.89	2.31	0.946	2.85	6.77	4.67	1.00
		1/4	2.53	2.02	0.973	2.43	5.78	3.93	1.02
		3/16	2.06	1.65	0.999	1.93	4.59	3.08	1.03
4	2	3/8	1.79	1.79	0.727	2.31	4.83	4.04	0.90
		5/16	1.66	1.66	0.752	2.08	4.40	3.59	0.92
		1/4	1.48	1.48	0.778	1.79	3.82	3.05	0.93
		3/16	1.22	1.22	0.804	1.43	3.08	2.41	0.95
		1/8	0.898	0.898	0.830	1.02	2.20	1.69	0.97

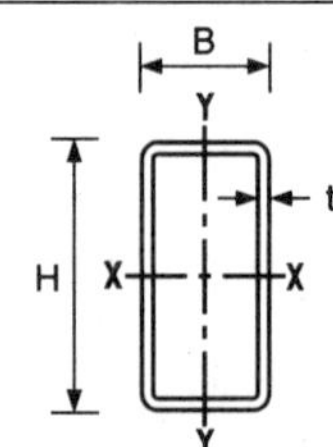

RECTANGULAR HSS
Dimensions and Properties

Dimensions			Properties								
Nominal Size			Wt per foot					X-X Axis			
Depth H	Width B	Wall		t	b/t	h/t	Area	I	S	r	Z
in.	in.	in.	lb.	in.			in.^{2}	in.^{4}	in.^{3}	in.	in.^{3}
$3\frac{1}{2}$	$2\frac{1}{2}$	$\frac{3}{8}$	12.17	0.349	4.2	7.0	3.39	4.74	2.71	1.18	3.59
		$\frac{5}{16}$	10.58	0.291	5.6	9.0	2.94	4.34	2.48	1.22	3.20
		$\frac{1}{4}$	8.81	0.233	7.7	12.0	2.44	3.79	2.17	1.25	2.74
		$\frac{3}{16}$	6.87	0.174	11.4	17.1	1.89	3.09	1.76	1.28	2.18
		$\frac{1}{8}$	4.75	0.116	18.6	27.2	1.30	2.23	1.28	1.31	1.54
3	$2\frac{1}{2}$	$\frac{5}{16}$	9.51	0.291	5.6	7.3	2.64	2.91	1.94	1.05	2.51
		$\frac{1}{4}$	7.96	0.233	7.7	9.9	2.21	2.57	1.71	1.08	2.16
		$\frac{3}{16}$	6.23	0.174	11.4	14.2	1.71	2.11	1.41	1.11	1.73
		$\frac{1}{8}$	4.33	0.116	18.6	22.9	1.19	1.54	1.03	1.14	1.23
3	2	$\frac{5}{16}$	8.45	0.291	3.9	7.3	2.35	2.38	1.58	1.00	2.11
		$\frac{1}{4}$	7.11	0.233	5.6	9.9	1.97	2.12	1.42	1.04	1.83
		$\frac{3}{16}$	5.59	0.174	8.5	14.2	1.54	1.76	1.18	1.07	1.48
		$\frac{1}{8}$	3.90	0.116	14.2	22.9	1.07	1.30	0.866	1.10	1.06
3	$1\frac{1}{2}$	$\frac{1}{4}$	6.26	0.233	3.4	9.9	1.74	1.68	1.12	0.982	1.51
		$\frac{3}{16}$	4.96	0.174	5.6	14.2	1.37	1.42	0.945	1.02	1.24
		$\frac{1}{8}$	3.48	0.116	9.9	22.9	0.96	1.06	0.706	1.05	0.895
3	1	$\frac{1}{8}$	3.05	0.116	5.6	22.9	0.84	0.817	0.545	0.987	0.728
$2\frac{1}{2}$	$1\frac{1}{2}$	$\frac{1}{4}$	5.41	0.233	3.4	7.7	1.51	1.03	0.820	0.825	1.11
		$\frac{3}{16}$	4.32	0.174	5.6	11.4	1.19	0.881	0.705	0.859	0.915
		$\frac{1}{8}$	3.05	0.116	9.9	18.6	0.84	0.668	0.535	0.892	0.671
2	$1\frac{1}{2}$	$\frac{3}{16}$	3.68	0.174	5.6	8.5	1.02	0.494	0.494	0.697	0.639
2	1	$\frac{3}{16}$	3.04	0.174	2.7	8.5	0.84	0.349	0.349	0.643	0.480
		$\frac{1}{8}$	2.20	0.116	5.6	14.2	0.61	0.280	0.280	0.679	0.366

RECTANGULAR HSS
Dimensions and Properties

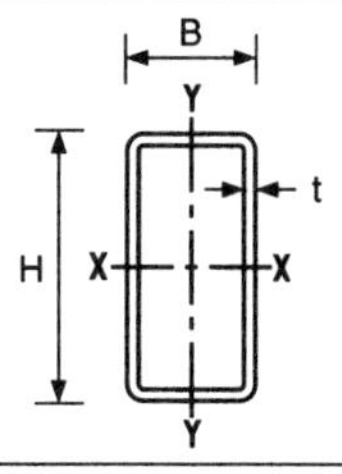

Dimensions			Properties						
Nominal Size			Y-Y Axis				Torsion		Surface Area Per foot
Depth H	Width B	Wall	I	S	r	Z	J	C	
in.	in.	in.	in.^{4}	in.^{3}	in.	in.^{3}	in.^{4}	in.^{3}	ft.^{2}
$3\frac{1}{2}$	$2\frac{1}{2}$	$\frac{3}{8}$	2.75	2.20	0.902	2.82	6.16	4.57	0.90
		$\frac{5}{16}$	2.53	2.03	0.929	2.52	5.53	4.03	0.92
		$\frac{1}{4}$	2.23	1.78	0.956	2.16	4.75	3.40	0.93
		$\frac{3}{16}$	1.82	1.46	0.983	1.72	3.78	2.67	0.95
		$\frac{1}{8}$	1.33	1.06	1.01	1.22	2.67	1.87	0.97
3	$2\frac{1}{2}$	$\frac{5}{16}$	2.17	1.74	0.907	2.20	4.34	3.39	0.83
		$\frac{1}{4}$	1.93	1.54	0.935	1.90	3.74	2.87	0.85
		$\frac{3}{16}$	1.59	1.27	0.962	1.52	3.00	2.27	0.87
		$\frac{1}{8}$	1.16	0.930	0.990	1.09	2.13	1.59	0.88
3	2	$\frac{5}{16}$	1.23	1.23	0.724	1.58	2.87	2.60	0.75
		$\frac{1}{4}$	1.11	1.11	0.750	1.38	2.52	2.23	0.77
		$\frac{3}{16}$	0.931	0.931	0.777	1.12	2.05	1.78	0.78
		$\frac{1}{8}$	0.692	0.692	0.804	0.803	1.47	1.25	0.80
3	$1\frac{1}{2}$	$\frac{1}{4}$	0.541	0.722	0.558	0.911	1.44	1.58	0.68
		$\frac{3}{16}$	0.466	0.621	0.584	0.752	1.21	1.28	0.70
		$\frac{1}{8}$	0.355	0.474	0.610	0.550	0.886	0.920	0.72
3	1	$\frac{1}{8}$	0.138	0.275	0.405	0.325	0.408	0.585	0.63
$2\frac{1}{2}$	$1\frac{1}{2}$	$\frac{1}{4}$	0.447	0.596	0.544	0.764	1.10	1.29	0.60
		$\frac{3}{16}$	0.389	0.519	0.571	0.636	0.929	1.05	0.62
		$\frac{1}{8}$	0.299	0.399	0.597	0.469	0.687	0.759	0.63
2	$1\frac{1}{2}$	$\frac{3}{16}$	0.312	0.416	0.553	0.521	0.664	0.822	0.53
2	1	$\frac{3}{16}$	0.112	0.223	0.364	0.288	0.301	0.505	0.45
		$\frac{1}{8}$	0.092	0.184	0.389	0.223	0.238	0.380	0.47

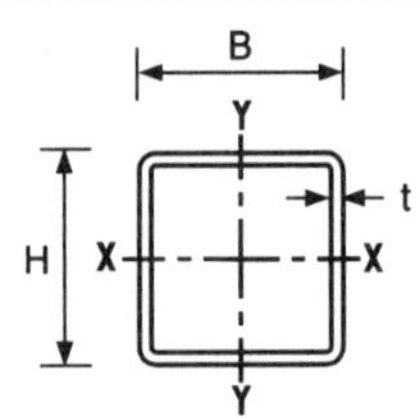

SQUARE HSS
Dimensions and Properties

Dimensions			Properties								
Nominal Size			Wt					X-X Axis			
Depth H	Width B	Wall	per foot	t	b/t	h/t	Area	I	S	r	Z
in.	in.	in.	lb.	in.			in.^{2}	in.^{4}	in.^{3}	in.	in.^{3}
32	32	5/8	259.83	0.625	48.2	48.2	76.4	12300	771	12.7	890
		1/2	210.72	0.500	61.0	61.0	61.9	10100	634	12.8	727
		3/8	159.37	0.375	82.3	82.3	46.8	7750	485	12.9	553
30	30	5/8	242.82	0.625	45.0	45.0	71.4	10100	673	11.9	778
		1/2	197.11	0.500	57.0	57.0	57.9	8320	555	12.0	637
		3/8	149.16	0.375	77.0	77.0	43.8	6370	424	12.1	485
28	28	5/8	225.80	0.625	41.8	41.8	66.4	8140	582	11.1	674
		1/2	183.50	0.500	53.0	53.0	53.9	6730	480	11.2	552
		3/8	138.95	0.375	71.7	71.7	40.8	5150	368	11.2	421
26	26	5/8	208.79	0.625	38.6	38.6	61.4	6460	497	10.3	577
		1/2	169.89	0.500	49.0	49.0	49.9	5350	411	10.4	474
		3/8	128.74	0.375	66.3	66.3	37.8	4110	316	10.4	362
24	24	5/8	191.78	0.625	35.4	35.4	56.4	5030	419	9.44	487
		1/2	156.28	0.500	45.0	45.0	45.9	4170	348	9.53	401
		3/8	118.53	0.375	61.0	61.0	34.8	3210	268	9.60	307
22	22	5/8	174.76	0.625	32.2	32.2	51.4	3820	347	8.62	406
		1/2	142.67	0.500	41.0	41.0	41.9	3190	290	8.72	335
		3/8	108.32	0.375	55.7	55.7	31.8	2460	223	8.78	256
20	20	5/8	157.75	0.625	29.0	29.0	46.4	2830	283	7.81	331
		1/2	129.06	0.500	37.0	37.0	37.9	2370	237	7.90	275
		3/8	98.12	0.375	50.3	50.3	28.8	1830	183	7.97	211
18	18	5/8	140.73	0.625	25.8	25.8	41.4	2020	224	6.99	264
		1/2	115.45	0.500	33.0	33.0	33.9	1700	189	7.08	220
		3/8	87.91	0.375	45.0	45.0	25.8	1320	147	7.15	169

SQUARE HSS
Dimensions and Properties

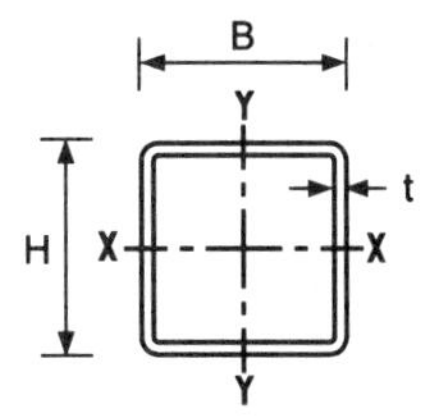

Dimensions			Properties						
Nominal Size			Y-Y Axis				Torsion		Surface Area Per foot
Depth H	Width B	Wall	I	S	r	Z	J	C	
in.	in.	in.	in.^{4}	in.^{3}	in.	in.^{3}	in.^{4}	in.^{3}	ft.^{2}
32	32	5/8	12300	771	12.7	890	19700	1230	10.34
		1/2	10100	634	12.8	727	15900	991	10.45
		3/8	7750	485	12.9	553	12000	750	10.51
30	30	5/8	10100	673	11.9	778	16200	1070	9.68
		1/2	8320	555	12.0	637	13000	869	9.79
		3/8	6370	424	12.1	485	9870	658	9.84
28	28	5/8	8140	582	11.1	674	13100	933	9.01
		1/2	6730	480	11.2	552	10600	755	9.12
		3/8	5150	368	11.2	421	8010	572	9.17
26	26	5/8	6460	497	10.3	577	10500	801	8.34
		1/2	5350	411	10.4	474	8430	649	8.45
		3/8	4110	316	10.4	362	6400	492	8.51
24	24	5/8	5030	419	9.44	487	8180	679	7.68
		1/2	4170	348	9.53	401	6610	551	7.79
		3/8	3210	268	9.60	307	5020	418	7.84
22	22	5/8	3820	347	8.62	406	6260	567	7.01
		1/2	3190	290	8.72	335	5070	461	7.12
		3/8	2460	223	8.78	256	3850	350	7.17
20	20	5/8	2830	283	7.81	331	4670	465	6.34
		1/2	2370	237	7.90	275	3790	379	6.45
		3/8	1830	183	7.97	211	2880	288	6.51
18	18	5/8	2020	224	6.99	264	3370	373	5.68
		1/2	1700	189	7.08	220	2740	305	5.79
		3/8	1320	147	7.15	169	2090	232	5.84

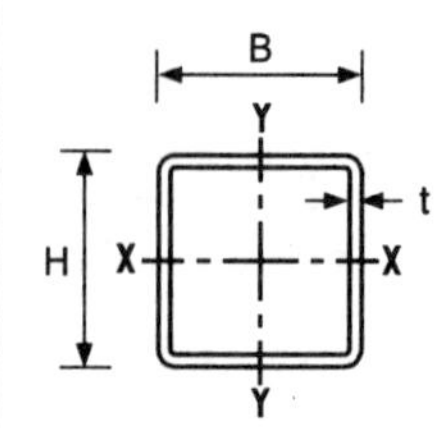

SQUARE HSS
Dimensions and Properties

Dimensions			Properties								
Nominal Size			Wt per foot					X-X Axis			
Depth H	Width B	Wall		t	b/t	h/t	Area	I	S	r	Z
in.	in.	in.	lb.	in.			in.^{2}	in.^{4}	in.^{3}	in.	in.^{3}
16	16	5/8	127.37	0.581	24.5	24.5	35.0	1370	171	6.25	200
		1/2	103.30	0.465	31.4	31.4	28.3	1130	141	6.31	164
		3/8	78.52	0.349	42.8	42.8	21.5	873	109	6.37	126
		5/16	65.87	0.291	52.0	52.0	18.1	739	92.3	6.39	106
14	14	5/8	110.36	0.581	21.1	21.1	30.3	896	128	5.44	151
		1/2	89.68	0.465	27.1	27.1	24.6	743	106	5.49	124
		3/8	68.31	0.349	37.1	37.1	18.7	577	82.5	5.55	95.4
		5/16	57.36	0.291	45.1	45.1	15.7	490	69.9	5.58	80.5
12	12	5/8	93.34	0.581	17.7	17.7	25.7	548	91.3	4.62	109
		1/2	76.07	0.465	22.8	22.8	20.9	457	76.2	4.68	89.6
		3/8	58.10	0.349	31.4	31.4	16.0	357	59.5	4.73	69.2
		5/16	48.86	0.291	38.2	38.2	13.4	304	50.7	4.76	58.6
		1/4	39.43	0.233	48.5	48.5	10.8	248	41.4	4.79	47.6
10	10	5/8	76.33	0.581	14.2	14.2	21.0	304	60.8	3.80	73.2
		1/2	62.46	0.465	18.5	18.5	17.2	256	51.2	3.86	60.7
		3/8	47.90	0.349	25.7	25.7	13.2	202	40.4	3.92	47.2
		5/16	40.35	0.291	31.4	31.4	11.1	172	34.5	3.94	40.1
		1/4	32.63	0.233	39.9	39.9	8.96	141	28.3	3.97	32.7
		3/16	24.73	0.174	54.5	54.5	6.76	108	21.6	4.00	24.8
8	8	5/8	59.32	0.581	10.8	10.8	16.4	146	36.5	2.99	44.7
		1/2	48.85	0.465	14.2	14.2	13.5	125	31.2	3.04	37.5
		3/8	37.69	0.349	19.9	19.9	10.4	99.6	24.9	3.10	29.4
		5/16	31.84	0.291	24.5	24.5	8.76	85.6	21.4	3.13	25.1
		1/4	25.82	0.233	31.3	31.3	7.10	70.7	17.7	3.15	20.5
		3/16	19.63	0.174	43.0	43.0	5.37	54.4	13.6	3.18	15.7

SQUARE HSS
Dimensions and Properties

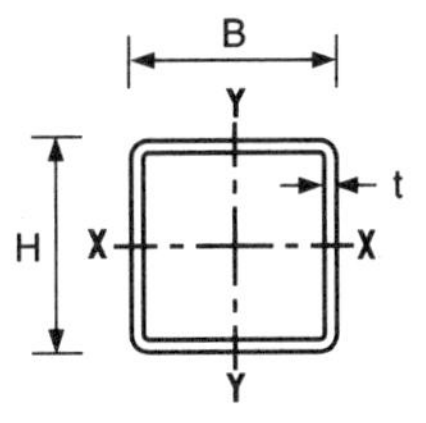

Dimensions			Properties						Surface
Nominal Size			Y-Y Axis				Torsion		Area
Depth H	Width B	Wall	I	S	r	Z	J	C	Per foot
in.	in.	in.	in.^{4}	in.^{3}	in.	in.^{3}	in.^{4}	in.^{3}	ft.^{2}
16	16	5/8	1370	171	6.25	200	2170	276	5.17
		1/2	1130	141	6.31	164	1770	224	5.20
		3/8	873	109	6.37	126	1350	171	5.23
		5/16	739	92.3	6.39	106	1140	144	5.25
14	14	5/8	896	128	5.44	151	1430	208	4.50
		1/2	743	106	5.49	124	1170	170	4.53
		3/8	577	82.5	5.55	95.4	900	130	4.57
		5/16	490	69.9	5.58	80.5	759	109	4.58
12	12	5/8	548	91.3	4.62	109	885	151	3.83
		1/2	457	76.2	4.68	89.6	728	123	3.87
		3/8	357	59.5	4.73	69.2	561	94.6	3.90
		5/16	304	50.7	4.76	58.6	474	79.7	3.92
		1/4	248	41.4	4.79	47.6	384	64.5	3.93
10	10	5/8	304	60.8	3.80	73.2	498	102	3.17
		1/2	256	51.2	3.86	60.7	412	84.2	3.20
		3/8	202	40.4	3.92	47.2	320	64.8	3.23
		5/16	172	34.5	3.94	40.1	271	54.8	3.25
		1/4	141	28.3	3.97	32.7	220	44.4	3.27
		3/16	108	21.6	4.00	24.8	167	33.6	3.28
8	8	5/8	146	36.5	2.99	44.7	244	63.2	2.50
		1/2	125	31.2	3.04	37.5	204	52.4	2.53
		3/8	99.6	24.9	3.10	29.4	160	40.7	2.57
		5/16	85.6	21.4	3.13	25.1	136	34.5	2.58
		1/4	70.7	17.7	3.15	20.5	111	28.1	2.60
		3/16	54.4	13.6	3.18	15.7	84.5	21.3	2.62

SQUARE HSS
Dimensions and Properties

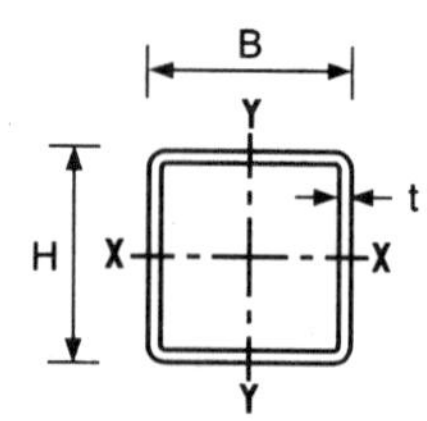

Dimensions			Properties								
Nominal Size								X-X Axis			
Depth H	Width B	Wall	Wt per foot	t	b/t	h/t	Area	I	S	r	Z
in.	in.	in.	lb.	in.			in.^{2}	in.^{4}	in.^{3}	in.	in.^{3}
7	7	5/8	50.81	0.581	9.0	9.0	14.0	93.3	26.7	2.58	33.1
		1/2	42.05	0.465	12.1	12.1	11.6	80.5	23.0	2.63	27.9
		3/8	32.58	0.349	17.1	17.1	8.97	64.9	18.6	2.69	22.1
		5/16	27.59	0.291	21.1	21.1	7.59	56.1	16.0	2.72	18.9
		1/4	22.42	0.233	27.0	27.0	6.17	46.5	13.3	2.75	15.5
		3/16	17.08	0.174	37.2	37.2	4.67	36.0	10.3	2.77	11.9
6	6	5/8	42.30	0.581	7.3	7.3	11.7	55.1	18.4	2.17	23.2
		1/2	35.24	0.465	9.9	9.9	9.74	48.2	16.1	2.23	19.8
		3/8	27.48	0.349	14.2	14.2	7.58	39.4	13.1	2.28	15.8
		5/16	23.34	0.291	17.6	17.6	6.43	34.3	11.4	2.31	13.6
		1/4	19.02	0.233	22.8	22.8	5.24	28.6	9.54	2.34	11.2
		3/16	14.53	0.174	31.5	31.5	3.98	22.3	7.42	2.37	8.63
		1/8	9.86	0.116	48.7	48.7	2.70	15.5	5.15	2.39	5.92
5½	5½	3/8	24.93	0.349	12.8	12.8	6.88	29.7	10.8	2.08	13.1
		5/16	21.21	0.291	15.9	15.9	5.85	25.9	9.43	2.11	11.3
		1/4	17.32	0.233	20.6	20.6	4.77	21.7	7.90	2.13	9.32
		3/16	13.25	0.174	28.6	28.6	3.63	17.0	6.17	2.16	7.19
		1/8	9.01	0.116	44.4	44.4	2.46	11.8	4.30	2.19	4.95
5	5	1/2	28.43	0.465	7.8	7.8	7.88	26.0	10.4	1.82	13.1
		3/8	22.37	0.349	11.3	11.3	6.18	21.7	8.67	1.87	10.6
		5/16	19.08	0.291	14.2	14.2	5.26	19.0	7.61	1.90	9.16
		1/4	15.62	0.233	18.5	18.5	4.30	16.0	6.41	1.93	7.61
		3/16	11.97	0.174	25.7	25.7	3.28	12.6	5.03	1.96	5.89
		1/8	8.16	0.116	40.1	40.1	2.23	8.80	3.52	1.99	4.07

SQUARE HSS
Dimensions and Properties

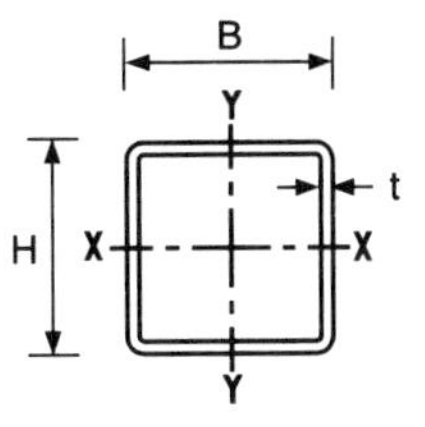

Dimensions			Properties						
Nominal Size			Y-Y Axis				Torsion		Surface
Depth H	Width B	Wall	I	S	r	Z	J	C	Area Per foot
in.	in.	in.	in.^{4}	in.^{3}	in.	in.^{3}	in.^{4}	in.^{3}	ft.^{2}
7	7	$^5/_8$	93.3	26.7	2.58	33.1	158	47.1	2.17
		$^1/_2$	80.5	23.0	2.63	27.9	133	39.3	2.20
		$^3/_8$	64.9	18.6	2.69	22.1	105	30.7	2.23
		$^5/_{16}$	56.1	16.0	2.72	18.9	89.7	26.1	2.25
		$^1/_4$	46.5	13.3	2.75	15.5	73.5	21.3	2.27
		$^3/_{16}$	36.0	10.3	2.77	11.9	56.1	16.2	2.28
6	6	$^5/_8$	55.1	18.4	2.17	23.2	94.9	33.4	1.83
		$^1/_2$	48.2	16.1	2.23	19.8	81.1	28.1	1.87
		$^3/_8$	39.4	13.1	2.28	15.8	64.6	22.1	1.90
		$^5/_{16}$	34.3	11.4	2.31	13.6	55.4	18.9	1.92
		$^1/_4$	28.6	9.54	2.34	11.2	45.6	15.4	1.93
		$^3/_{16}$	22.3	7.42	2.37	8.63	35.0	11.8	1.95
		$^1/_8$	15.5	5.15	2.39	5.92	23.9	8.03	1.97
$5^1/_2$	$5^1/_2$	$^3/_8$	29.7	10.8	2.08	13.1	49.0	18.4	1.73
		$^5/_{16}$	25.9	9.43	2.11	11.3	42.2	15.7	1.75
		$^1/_4$	21.7	7.90	2.13	9.32	34.8	12.9	1.77
		$^3/_{16}$	17.0	6.17	2.16	7.19	26.7	9.85	1.78
		$^1/_8$	11.8	4.30	2.19	4.95	18.3	6.72	1.80
5	5	$^1/_2$	26.0	10.4	1.82	13.1	44.6	18.7	1.53
		$^3/_8$	21.7	8.67	1.87	10.6	36.1	14.9	1.57
		$^5/_{16}$	19.0	7.61	1.90	9.16	31.2	12.8	1.58
		$^1/_4$	16.0	6.41	1.93	7.61	25.8	10.5	1.60
		$^3/_{16}$	12.6	5.03	1.96	5.89	19.9	8.08	1.62
		$^1/_8$	8.80	3.52	1.99	4.07	13.7	5.53	1.63

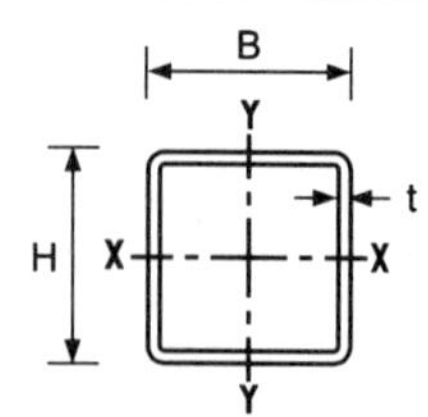

SQUARE HSS
Dimensions and Properties

Dimensions			Properties								
Nominal Size			Wt per foot					X-X Axis			
Depth H	Width B	Wall		t	b/t	h/t	Area	I	S	r	Z
in.	in.	in.	lb.	in.			in.^{2}	in.^{4}	in.^{3}	in.	in.^{3}
$4\frac{1}{2}$	$4\frac{1}{2}$	$\frac{1}{2}$	25.03	0.465	6.7	6.7	6.95	18.0	8.02	1.61	10.2
		$\frac{3}{8}$	19.82	0.349	9.9	9.9	5.48	15.3	6.78	1.67	8.36
		$\frac{5}{16}$	16.96	0.291	12.5	12.5	4.68	13.5	5.99	1.70	7.27
		$\frac{1}{4}$	13.91	0.233	16.3	16.3	3.84	11.4	5.08	1.73	6.06
		$\frac{3}{16}$	10.70	0.174	22.9	22.9	2.93	9.02	4.01	1.75	4.71
		$\frac{1}{8}$	7.31	0.116	35.8	35.8	2.00	6.35	2.82	1.78	3.27
4	4	$\frac{1}{2}$	21.63	0.465	5.6	5.6	6.02	11.9	5.95	1.41	7.70
		$\frac{3}{8}$	17.27	0.349	8.5	8.5	4.78	10.3	5.13	1.46	6.39
		$\frac{5}{16}$	14.83	0.291	10.7	10.7	4.10	9.14	4.57	1.49	5.59
		$\frac{1}{4}$	12.21	0.233	14.2	14.2	3.37	7.80	3.90	1.52	4.69
		$\frac{3}{16}$	9.42	0.174	20.0	20.0	2.58	6.21	3.10	1.55	3.67
		$\frac{1}{8}$	6.46	0.116	31.5	31.5	1.77	4.40	2.20	1.58	2.56
$3\frac{1}{2}$	$3\frac{1}{2}$	$\frac{3}{8}$	14.72	0.349	7.0	7.0	4.09	6.48	3.70	1.26	4.69
		$\frac{5}{16}$	12.70	0.291	9.0	9.0	3.52	5.84	3.34	1.29	4.14
		$\frac{1}{4}$	10.51	0.233	12.0	12.0	2.91	5.04	2.88	1.32	3.50
		$\frac{3}{16}$	8.15	0.174	17.1	17.1	2.24	4.05	2.31	1.35	2.76
		$\frac{1}{8}$	5.61	0.116	27.2	27.2	1.54	2.90	1.66	1.37	1.93
3	3	$\frac{3}{8}$	12.17	0.349	5.6	5.6	3.39	3.77	2.51	1.05	3.25
		$\frac{5}{16}$	10.58	0.291	7.3	7.3	2.94	3.45	2.30	1.08	2.90
		$\frac{1}{4}$	8.81	0.233	9.9	9.9	2.44	3.02	2.01	1.11	2.48
		$\frac{3}{16}$	6.87	0.174	14.2	14.2	1.89	2.46	1.64	1.14	1.97
		$\frac{1}{8}$	4.75	0.116	22.9	22.9	1.30	1.78	1.19	1.17	1.40

SQUARE HSS
Dimensions and Properties

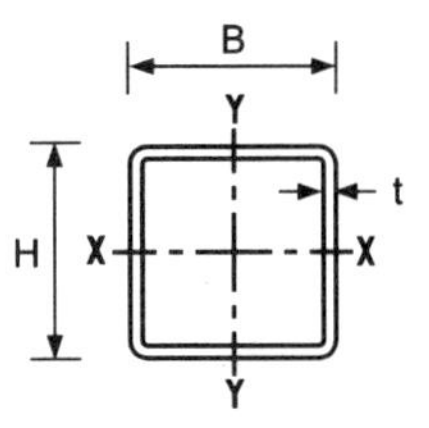

Dimensions			Properties						
Nominal Size			Y-Y Axis				Torsion		Surface Area Per foot
Depth H	Width B	Wall	I	S	r	Z	J	C	
in.	in.	in.	in.^{4}	in.^{3}	in.	in.^{3}	in.^{4}	in.^{3}	ft.^{2}
$4\frac{1}{2}$	$4\frac{1}{2}$	$\frac{1}{2}$	18.0	8.02	1.61	10.2	31.3	14.8	1.37
		$\frac{3}{8}$	15.3	6.78	1.67	8.36	25.7	11.9	1.40
		$\frac{5}{16}$	13.5	5.99	1.70	7.27	22.3	10.2	1.42
		$\frac{1}{4}$	11.4	5.08	1.73	6.06	18.5	8.44	1.43
		$\frac{3}{16}$	9.02	4.01	1.75	4.71	14.4	6.49	1.45
		$\frac{1}{8}$	6.35	2.82	1.78	3.27	9.92	4.45	1.47
4	4	$\frac{1}{2}$	11.9	5.95	1.41	7.70	21.0	11.2	1.20
		$\frac{3}{8}$	10.3	5.13	1.46	6.39	17.5	9.14	1.23
		$\frac{5}{16}$	9.14	4.57	1.49	5.59	15.3	7.91	1.25
		$\frac{1}{4}$	7.80	3.90	1.52	4.69	12.8	6.56	1.27
		$\frac{3}{16}$	6.21	3.10	1.55	3.67	9.96	5.07	1.28
		$\frac{1}{8}$	4.40	2.20	1.58	2.56	6.91	3.49	1.30
$3\frac{1}{2}$	$3\frac{1}{2}$	$\frac{3}{8}$	6.48	3.70	1.26	4.69	11.2	6.77	1.07
		$\frac{5}{16}$	5.84	3.34	1.29	4.14	9.89	5.90	1.08
		$\frac{1}{4}$	5.04	2.88	1.32	3.50	8.35	4.92	1.10
		$\frac{3}{16}$	4.05	2.31	1.35	2.76	6.56	3.83	1.12
		$\frac{1}{8}$	2.90	1.66	1.37	1.93	4.58	2.65	1.13
3	3	$\frac{3}{8}$	3.77	2.51	1.05	3.25	6.64	4.74	0.90
		$\frac{5}{16}$	3.45	2.30	1.08	2.90	5.94	4.18	0.92
		$\frac{1}{4}$	3.02	2.01	1.11	2.48	5.08	3.52	0.93
		$\frac{3}{16}$	2.46	1.64	1.14	1.97	4.03	2.76	0.95
		$\frac{1}{8}$	1.78	1.19	1.17	1.40	2.84	1.92	0.97

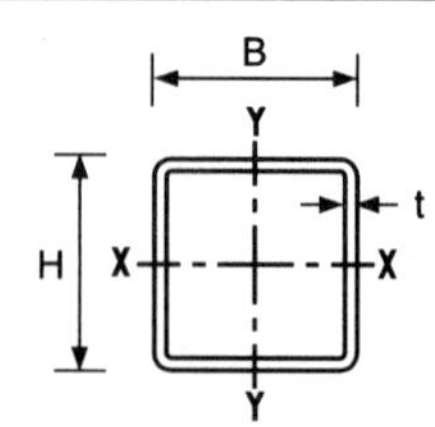

SQUARE HSS
Dimensions and Properties

Dimensions			Properties								
Nominal Size			Wt per foot					X-X Axis			
Depth H	Width B	Wall		t	b/t	h/t	Area	I	S	r	Z
in.	in.	in.	lb.	in.			in.^{2}	in.^{4}	in.^{3}	in.	in.^{3}
2½	2½	5/16	8.45	0.291	5.6	5.6	2.35	1.82	1.45	0.879	1.88
		¼	7.11	0.233	7.7	7.7	1.97	1.63	1.30	0.908	1.63
		3/16	5.59	0.174	11.4	11.4	1.54	1.35	1.08	0.937	1.32
		⅛	3.90	0.116	18.6	18.6	1.07	0.998	0.798	0.965	0.947
2¼	2¼	¼	6.26	0.233	6.7	6.7	1.74	1.13	1.00	0.805	1.28
		3/16	4.96	0.174	9.9	9.9	1.37	0.952	0.847	0.835	1.04
		⅛	3.48	0.116	16.4	16.4	0.96	0.712	0.633	0.863	0.755
2	2	¼	5.41	0.233	5.6	5.6	1.51	0.745	0.745	0.703	0.964
		3/16	4.32	0.174	8.5	8.5	1.19	0.640	0.640	0.732	0.797
		⅛	3.05	0.116	14.2	14.2	0.84	0.486	0.486	0.761	0.584
1¾	1¾	3/16	3.68	0.174	7.1	7.1	1.02	0.405	0.462	0.630	0.585
1⅝	1⅝	3/16	3.36	0.174	6.3	6.3	0.93	0.312	0.384	0.579	0.491
		⅛	2.42	0.116	11.0	11.0	0.67	0.246	0.302	0.608	0.370
1½	1½	3/16	3.04	0.174	5.6	5.6	0.84	0.235	0.314	0.528	0.406
		⅛	2.20	0.116	9.9	9.9	0.61	0.188	0.251	0.556	0.309
1¼	1¼	3/16	2.40	0.174	4.2	4.2	0.67	0.121	0.194	0.425	0.259
		⅛	1.78	0.116	7.8	7.8	0.49	0.101	0.162	0.454	0.204

SQUARE HSS
Dimensions and Properties

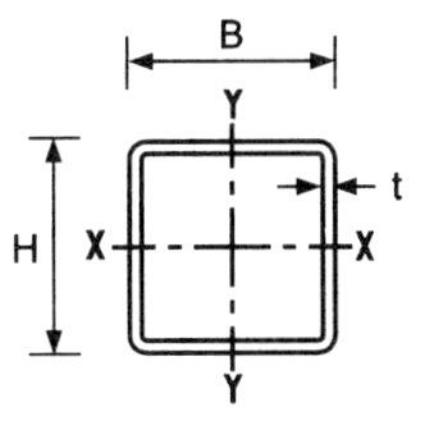

Dimensions			Properties						
Nominal Size			**Y-Y Axis**				**Torsion**		**Surface Area Per foot**
Depth H	**Width** B	**Wall**	I	S	r	Z	J	C	
in.	in.	in.	in.^{4}	in.^{3}	in.	in.^{3}	in.^{4}	in.^{3}	ft.^{2}
$2\frac{1}{2}$	$2\frac{1}{2}$	$\frac{5}{16}$	1.82	1.45	0.879	1.88	3.20	2.74	0.75
		$\frac{1}{4}$	1.63	1.30	0.908	1.63	2.79	2.35	0.77
		$\frac{3}{16}$	1.35	1.08	0.937	1.32	2.25	1.86	0.78
		$\frac{1}{8}$	0.998	0.798	0.965	0.947	1.61	1.31	0.80
$2\frac{1}{4}$	$2\frac{1}{4}$	$\frac{1}{4}$	1.13	1.00	0.805	1.28	1.96	1.85	0.68
		$\frac{3}{16}$	0.952	0.847	0.835	1.04	1.60	1.48	0.70
		$\frac{1}{8}$	0.712	0.633	0.863	0.755	1.15	1.05	0.72
2	2	$\frac{1}{4}$	0.745	0.745	0.703	0.964	1.31	1.41	0.60
		$\frac{3}{16}$	0.640	0.640	0.732	0.797	1.09	1.14	0.62
		$\frac{1}{8}$	0.486	0.486	0.761	0.584	0.796	0.817	0.63
$1\frac{3}{4}$	$1\frac{3}{4}$	$\frac{3}{16}$	0.405	0.462	0.630	0.585	0.699	0.844	0.53
$1\frac{5}{8}$	$1\frac{5}{8}$	$\frac{3}{16}$	0.312	0.384	0.579	0.491	0.544	0.712	0.49
		$\frac{1}{8}$	0.246	0.302	0.608	0.370	0.410	0.522	0.51
$1\frac{1}{2}$	$1\frac{1}{2}$	$\frac{3}{16}$	0.235	0.314	0.528	0.406	0.414	0.592	0.45
		$\frac{1}{8}$	0.188	0.251	0.556	0.309	0.316	0.438	0.47
$1\frac{1}{4}$	$1\frac{1}{4}$	$\frac{3}{16}$	0.121	0.194	0.425	0.259	0.218	0.383	0.37
		$\frac{1}{8}$	0.101	0.162	0.454	0.204	0.174	0.292	0.38

ROUND HSS
Dimensions and Properties

Dimensions					Properties				
Nominal Size									
Outside Diameter	Wall	Wt per foot	t	D/t	Area	I	S	r	Z
in.	in.	lb.	in.		in.^{2}	in.^{4}	in.^{3}	in.	in.^{3}
20.000	0.500	104.13	0.465	43.0	28.5	1360	136	6.91	177
	0.375	78.60	0.349	57.3	21.5	1040	104	6.95	135
18.000	0.500	93.45	0.465	38.7	25.6	985	109	6.20	143
	0.375	70.59	0.349	51.6	19.4	754	83.8	6.24	109
16.000	0.500	82.77	0.465	34.4	22.7	685	85.7	5.49	112
	0.438	72.80	0.407	39.3	19.9	606	75.8	5.51	99.0
	0.375	62.58	0.349	45.8	17.2	526	65.7	5.53	85.5
	0.312	52.28	0.291	55.0	14.4	443	55.4	5.55	71.8
14.000	0.500	72.09	0.465	30.1	19.8	453	64.8	4.79	85.2
	0.375	54.57	0.349	40.1	15.0	349	49.8	4.83	65.1
	0.312	45.61	0.291	48.1	12.5	295	42.1	4.85	54.7
12.750	0.500	65.42	0.465	27.4	17.9	339	53.2	4.35	70.2
	0.375	49.56	0.349	36.5	13.6	262	41.0	4.39	53.7
	0.250	33.38	0.233	54.7	9.16	180	28.2	4.43	36.5
12.500	0.625	79.27	0.581	21.5	21.8	387	62.0	4.22	82.6
	0.500	64.08	0.465	26.9	17.6	319	51.0	4.26	67.4
	0.375	48.56	0.349	35.8	13.3	246	39.4	4.30	51.5
	0.312	40.61	0.291	43.0	11.2	208	33.3	4.32	43.4
	0.250	32.71	0.233	53.6	8.98	169	27.0	4.34	35.1
	0.188	24.72	0.174	71.8	6.74	128	20.5	4.36	26.4

Round HSS
Dimensions and Properties

Dimensions			Properties		
Nominal Size			Torsion		Surface Area Per foot
Outside Diameter	Wall		J	C	
in.	in.		in.^{4}	in.^{3}	ft.^{2}
20.000	0.500		2720	272	5.24
	0.375		2080	208	5.24
18.000	0.500		1970	219	4.71
	0.375		1510	168	4.71
16.000	0.500		1370	171	4.19
	0.438		1210	152	4.19
	0.375		1050	131	4.19
	0.312		886	111	4.19
14.000	0.500		907	130	3.67
	0.375		698	99.7	3.67
	0.312		589	84.2	3.67
12.750	0.500		678	106	3.34
	0.375		523	82.1	3.34
	0.250		359	56.3	3.34
12.500	0.625		774	124	3.27
	0.500		638	102	3.27
	0.375		492	78.7	3.27
	0.312		416	66.6	3.27
	0.250		338	54.1	3.27
	0.188		256	41.0	3.27

ROUND HSS
Dimensions and Properties

Dimensions					Properties				
Nominal Size									
Outside Diameter	Wall	Wt per foot	t	D/t	Area	I	S	r	Z
in.	in.	lb.	in.		in.^{2}	in.^{4}	in.^{3}	in.	in.^{3}
11.250	0.625	70.92	0.581	19.4	19.5	278	49.4	3.78	66.2
	0.500	57.41	0.465	24.2	15.8	229	40.8	3.82	54.1
	0.375	43.56	0.349	32.2	12.0	178	31.6	3.86	41.5
	0.312	36.45	0.291	38.7	10.0	151	26.8	3.88	35.0
	0.250	29.37	0.233	48.3	8.06	122	21.8	3.90	28.3
	0.188	22.21	0.174	64.7	6.05	92.9	16.5	3.92	21.3
10.750	0.500	54.74	0.465	23.1	15.0	199	37.0	3.64	49.2
	0.250	28.04	0.233	46.1	7.70	106	19.8	3.72	25.8
10.000	0.625	62.58	0.581	17.2	17.2	191	38.3	3.34	51.6
	0.500	50.73	0.465	21.5	13.9	159	31.7	3.38	42.3
	0.375	38.55	0.349	28.7	10.6	123	24.7	3.41	32.5
	0.312	32.28	0.291	34.4	8.88	105	20.9	3.43	27.4
	0.250	26.03	0.233	42.9	7.15	85.3	17.1	3.45	22.2
	0.188	19.70	0.174	57.5	5.37	64.8	13.0	3.47	16.8
9.625	0.500	48.73	0.465	20.7	13.4	141	29.2	3.24	39.0
	0.375	37.05	0.349	27.6	10.2	110	22.8	3.28	30.0
	0.312	31.03	0.291	33.1	8.53	93.0	19.3	3.30	25.4
	0.250	25.03	0.233	41.3	6.87	75.9	15.8	3.32	20.6
	0.188	18.95	0.174	55.3	5.17	57.7	12.0	3.34	15.5
8.750	0.500	44.06	0.465	18.8	12.1	104	23.8	2.93	32.0
	0.375	33.54	0.349	25.1	9.21	81.4	18.6	2.97	24.6
	0.312	28.12	0.291	30.1	7.73	69.3	15.8	2.99	20.8
	0.250	22.70	0.233	37.6	6.23	56.6	12.9	3.01	16.9
	0.188	17.19	0.174	50.3	4.69	43.1	9.86	3.03	12.8
8.625	0.500	43.39	0.465	18.5	11.9	99.5	23.1	2.89	31.0
	0.375	33.04	0.349	24.7	9.07	77.8	18.0	2.93	23.9
	0.322	28.55	0.300	28.7	7.85	68.1	15.8	2.95	20.8
	0.250	22.36	0.233	37.0	6.14	54.1	12.5	2.97	16.4
	0.188	16.94	0.174	49.6	4.62	41.3	9.57	2.99	12.4

ROUND HSS
Dimensions and Properties

Dimensions			Properties		
Nominal Size			Torsion		Surface
Outside Diameter	Wall		J	C	Area Per foot
in.	in.		in.^{4}	in.^{3}	ft.^{2}
11.250	0.625		556	98.8	2.95
	0.500		459	81.6	2.95
	0.375		355	63.2	2.95
	0.312		301	53.5	2.95
	0.250		245	43.5	2.95
	0.188		186	33.0	2.95
10.750	0.500		398	74.1	2.81
	0.250		213	39.6	2.81
10.000	0.625		383	76.6	2.62
	0.500		317	63.5	2.62
	0.375		247	49.3	2.62
	0.312		209	41.9	2.62
	0.250		171	34.1	2.62
	0.188		130	25.9	2.62
9.625	0.500		281	58.5	2.52
	0.375		219	45.5	2.52
	0.312		186	38.7	2.52
	0.250		152	31.5	2.52
	0.188		115	24.0	2.52
8.750	0.500		208	47.6	2.29
	0.375		163	37.2	2.29
	0.312		139	31.7	2.29
	0.250		113	25.9	2.29
	0.188		86.2	19.7	2.29
8.625	0.500		199	46.2	2.26
	0.375		156	36.1	2.26
	0.322		136	31.6	2.26
	0.250		108	25.1	2.26
	0.188		89.5	19.1	2.26

ROUND HSS
Dimensions and Properties

Dimensions					Properties				
Nominal Size									
Outside Diameter	Wall	Wt per foot	t	D/t	Area	I	S	r	Z
in.	in.	lb.	in.		in.^{2}	in.^{4}	in.^{3}	in.	in.^{3}
7.625	0.125	10.01	0.116	65.7	2.74	19.3	5.06	2.66	6.54
7.500	0.500	37.38	0.465	16.1	10.3	63.9	17.0	2.49	23.0
	0.375	28.54	0.349	21.5	7.84	50.2	13.4	2.53	17.9
	0.312	23.95	0.291	25.8	6.59	42.9	11.4	2.55	15.1
	0.250	19.36	0.233	32.2	5.32	35.2	9.37	2.57	12.3
	0.188	14.68	0.174	43.1	4.00	26.9	7.17	2.59	9.34
7.000	0.500	34.71	0.465	15.1	9.55	51.2	14.6	2.32	19.9
	0.375	26.53	0.349	20.1	7.29	40.4	11.6	2.35	15.5
	0.312	22.29	0.291	24.1	6.13	34.6	9.88	2.37	13.1
	0.250	18.02	0.233	30.0	4.95	28.4	8.11	2.39	10.7
	0.188	13.68	0.174	40.2	3.73	21.7	6.21	2.41	8.11
	0.125	9.18	0.116	60.3	2.51	14.9	4.25	2.43	5.50

ROUND HSS
Dimensions and Properties

Dimensions				Properties		
Nominal Size				Torsion		Surface
Outside Diameter	Wall			J	C	Area Per foot
in.	in.			in.^{4}	in.^{3}	ft.^{2}
7.625	0.125			38.6	10.1	2.00
7.500	0.500			128	34.1	1.96
	0.375			100	26.8	1.96
	0.312			85.8	22.9	1.96
	0.250			70.3	18.7	1.96
	0.188			53.8	14.3	1.96
7.000	0.500			102	29.3	1.83
	0.375			80.9	23.1	1.83
	0.312			69.1	19.8	1.83
	0.250			56.8	16.2	1.83
	0.188			43.5	12.4	1.83
	0.125			29.7	8.49	1.83

ROUND HSS
Dimensions and Properties

Dimensions					Properties				
Nominal Size									
Outside Diameter	Wall	Wt per foot	t	D/t	Area	I	S	r	Z
in.	in.	lb.	in.		in.^{2}	in.^{4}	in.^{3}	in.	in.^{3}
6.875	0.500	34.04	0.465	14.8	9.36	48.3	14.1	2.27	19.1
	0.375	26.03	0.349	19.7	7.16	38.2	11.1	2.31	14.9
	0.312	21.87	0.291	23.6	6.02	32.7	9.51	2.33	12.6
	0.250	17.69	0.233	29.5	4.86	26.8	7.81	2.35	10.3
	0.188	13.43	0.174	39.5	3.66	20.6	5.99	2.37	7.81
6.625	0.500	32.71	0.465	14.2	9.00	42.9	13.0	2.18	17.7
	0.432	28.57	0.403	16.4	7.88	38.3	11.6	2.20	15.6
	0.375	25.03	0.349	19.0	6.88	34.0	10.3	2.22	13.8
	0.312	21.04	0.291	22.8	5.79	29.1	8.79	2.24	11.7
	0.280	18.97	0.261	25.4	5.22	26.5	7.99	2.25	10.6
	0.250	17.02	0.233	28.4	4.68	23.9	7.22	2.26	9.52
	0.188	12.92	0.174	38.1	3.53	18.4	5.54	2.28	7.24
	0.125	8.68	0.116	57.1	2.37	12.6	3.79	2.30	4.92
6.125	0.500	30.04	0.465	13.2	8.27	33.3	10.9	2.01	14.9
	0.375	23.03	0.349	17.6	6.33	26.5	8.66	2.05	11.7
	0.312	19.37	0.291	21.0	5.33	22.7	7.43	2.07	9.91
	0.250	15.69	0.233	26.3	4.31	18.7	6.12	2.08	8.09
	0.188	11.92	0.174	35.2	3.25	14.4	4.71	2.10	6.16
6.000	0.500	29.37	0.465	12.9	8.09	31.2	10.4	1.96	14.3
	0.375	22.53	0.349	17.2	6.20	24.8	8.28	2.00	11.2
	0.312	18.95	0.291	20.6	5.22	21.3	7.11	2.02	9.49
	0.280	17.11	0.261	23.0	4.71	19.4	6.47	2.03	8.60
	0.250	15.35	0.233	25.8	4.22	17.6	5.86	2.04	7.75
	0.188	11.67	0.174	34.5	3.18	13.5	4.51	2.06	5.91
	0.125	7.84	0.116	51.7	2.14	9.28	3.09	2.08	4.02

ROUND HSS
Dimensions and Properties

Dimensions			Properties		
Nominal Size			**Torsion**		**Surface Area Per foot**
Outside Diameter	**Wall**		J	C	
in.	in.		in.^{4}	in.^{3}	ft.^{2}
6.875	0.500		96.7	28.1	1.80
	0.375		76.4	22.2	1.80
	0.312		65.4	19.0	1.80
	0.250		53.7	15.6	1.80
	0.188		41.1	12.0	1.80
6.625	0.500		85.9	25.9	1.73
	0.432		76.6	23.1	1.73
	0.375		68.0	20.5	1.73
	0.312		58.2	17.6	1.73
	0.280		52.9	16.0	1.73
	0.250		47.9	14.4	1.73
	0.188		36.7	11.1	1.73
	0.125		25.1	7.59	1.73
6.125	0.500		66.7	21.8	1.60
	0.375		53.0	17.3	1.60
	0.312		45.5	14.9	1.60
	0.250		37.5	12.2	1.60
	0.188		28.8	9.41	1.60
6.000	0.500		62.4	20.8	1.57
	0.375		49.7	16.6	1.57
	0.312		42.6	14.2	1.57
	0.280		38.8	12.9	1.57
	0.250		35.2	11.7	1.57
	0.188		27.0	9.02	1.57
	0.125		18.6	6.19	1.57

ROUND HSS
Dimensions and Properties

Dimensions					Properties				
Nominal Size									
Outside Diameter	Wall	Wt per foot	t	D/t	Area	I	S	r	Z
in.	in.	lb.	in.		in.^{2}	in.^{4}	in.^{3}	in.	in.^{3}
5.563	0.375	20.78	0.349	15.9	5.72	19.5	7.02	1.85	9.50
	0.258	14.62	0.241	23.1	4.03	14.3	5.14	1.88	6.83
	0.188	10.79	0.174	32.0	2.95	10.7	3.85	1.91	5.05
	0.134	7.77	0.125	44.5	2.14	7.90	2.84	1.92	3.70
5.500	0.500	26.70	0.465	11.8	7.36	23.5	8.55	1.79	11.8
	0.375	20.53	0.349	15.8	5.65	18.8	6.84	1.83	9.27
	0.258	14.44	0.241	22.8	3.98	13.8	5.02	1.86	6.67
5.000	0.500	24.03	0.465	10.8	6.62	17.2	6.88	1.61	9.60
	0.375	18.52	0.349	14.3	5.10	13.9	5.55	1.65	7.56
	0.312	15.62	0.291	17.2	4.30	12.0	4.79	1.67	6.46
	0.258	13.07	0.241	20.7	3.60	10.2	4.09	1.68	5.46
	0.250	12.68	0.233	21.5	3.49	9.94	3.97	1.69	5.30
	0.188	9.66	0.174	28.7	2.64	7.69	3.08	1.71	4.05
	0.125	6.51	0.116	43.1	1.78	5.31	2.12	1.73	2.77
4.500	0.337	14.98	0.315	14.3	4.14	9.12	4.05	1.48	5.53
	0.237	10.79	0.221	20.4	2.97	6.82	3.03	1.51	4.05
	0.188	8.66	0.174	25.9	2.36	5.54	2.46	1.53	3.26
	0.125	5.84	0.116	38.8	1.60	3.84	1.71	1.55	2.23
4.000	0.337	13.18	0.315	12.7	3.65	6.24	3.12	1.31	4.29
	0.313	12.33	0.291	13.7	3.39	5.87	2.93	1.32	4.01
	0.250	10.01	0.233	17.2	2.76	4.91	2.45	1.33	3.31
	0.237	9.52	0.221	18.1	2.62	4.70	2.35	1.34	3.16
	0.226	9.11	0.211	19.0	2.51	4.52	2.26	1.34	3.03
	0.220	8.88	0.205	19.5	2.44	4.41	2.21	1.34	2.96
	0.188	7.65	0.174	23.0	2.09	3.83	1.92	1.35	2.55
	0.125	5.17	0.116	34.5	1.42	2.67	1.34	1.37	1.75

ROUND HSS
Dimensions and Properties

Dimensions			Properties		
Nominal Size			**Torsion**		**Surface Area Per foot**
Outside Diameter	**Wall**		J	C	
in.	**in.**		**in.**^{4}	**in.**^{3}	**ft.**^{2}
5.563	0.375		39.0	14.0	1.46
	0.258		28.6	10.3	1.46
	0.188		21.4	7.70	1.46
	0.134		15.8	5.68	1.46
5.500	0.500		47.0	17.1	1.44
	0.375		37.6	13.7	1.44
	0.258		27.6	10.0	1.44
5.000	0.500		34.4	13.8	1.31
	0.375		27.7	11.1	1.31
	0.312		24.0	9.58	1.31
	0.258		20.5	8.18	1.31
	0.250		19.9	7.95	1.31
	0.188		15.4	6.15	1.31
	0.125		10.6	4.25	1.31
4.500	0.337		18.2	8.11	1.18
	0.237		13.6	6.06	1.18
	0.188		11.1	4.93	1.18
	0.125		7.68	3.41	1.18
4.000	0.337		12.5	6.24	1.05
	0.313		11.7	5.87	1.05
	0.250		9.82	4.91	1.05
	0.237		9.40	4.70	1.05
	0.226		9.04	4.52	1.05
	0.220		8.83	4.41	1.05
	0.188		7.67	3.83	1.05
	0.125		5.34	2.67	1.05

ROUND HSS
Dimensions and Properties

Dimensions					Properties				
Nominal Size									
Outside Diameter	Wall	Wt per foot	t	D/t	Area	I	S	r	Z
in.	in.	lb.	in.		in.^{2}	in.^{4}	in.^{3}	in.	in.^{3}
3.500	0.313	10.65	0.291	12.0	2.93	3.81	2.18	1.14	3.00
	0.300	10.25	0.280	12.5	2.83	3.70	2.11	1.14	2.91
	0.250	8.68	0.233	15.0	2.39	3.21	1.83	1.16	2.49
	0.216	7.58	0.201	17.4	2.08	2.84	1.63	1.17	2.19
	0.203	7.15	0.189	18.5	1.97	2.70	1.54	1.17	2.07
	0.188	6.65	0.174	20.1	1.82	2.52	1.44	1.18	1.93
	0.125	4.51	0.116	30.2	1.23	1.77	1.01	1.20	1.33
3.000	0.300	8.65	0.280	10.7	2.39	2.24	1.49	0.967	2.08
	0.250	7.34	0.233	12.9	2.03	1.95	1.30	0.982	1.79
	0.216	6.42	0.201	14.9	1.77	1.74	1.16	0.992	1.58
	0.203	6.06	0.189	15.9	1.67	1.66	1.10	0.996	1.50
	0.188	5.65	0.174	17.2	1.54	1.55	1.03	1.00	1.39
	0.152	4.62	0.142	21.1	1.27	1.30	0.870	1.01	1.16
	0.134	4.10	0.125	24.0	1.13	1.17	0.779	1.02	1.03
	0.120	3.69	0.112	26.8	1.02	1.06	0.707	1.02	0.935
2.875	0.250	7.01	0.233	12.3	1.93	1.70	1.18	0.938	1.63
	0.203	5.79	0.189	15.2	1.59	1.45	1.01	0.952	1.37
	0.188	5.40	0.174	16.5	1.48	1.35	0.941	0.957	1.27
	0.125	3.67	0.116	24.8	1.01	0.958	0.667	0.976	0.884
2.500	0.250	6.01	0.233	10.7	1.66	1.08	0.862	0.806	1.20
	0.188	4.64	0.174	14.4	1.27	0.865	0.692	0.825	0.943
	0.125	3.17	0.116	21.6	0.87	0.619	0.495	0.844	0.660
2.375	0.250	5.67	0.233	10.2	1.57	0.910	0.766	0.762	1.07
	0.218	5.02	0.204	11.6	1.39	0.827	0.696	0.771	0.964
	0.188	4.39	0.174	13.6	1.20	0.733	0.617	0.781	0.845
	0.154	3.65	0.143	16.6	1.00	0.627	0.528	0.791	0.713
	0.125	3.00	0.116	20.5	0.82	0.527	0.443	0.800	0.592
1.900	0.145	2.72	0.135	14.1	0.75	0.293	0.309	0.626	0.421
1.660	0.140	2.27	0.130	12.8	0.62	0.184	0.222	0.543	0.305

ROUND HSS
Dimensions and Properties

Dimensions			Properties		
Nominal Size			**Torsion**		**Surface**
Outside Diameter	**Wall**		J	C	**Area Per foot**
in.	in.		in.^{4}	in.^{3}	ft.^{2}
3.500	0.313		7.61	4.35	0.92
	0.300		7.40	4.23	0.92
	0.250		6.41	3.66	0.92
	0.216		5.69	3.25	0.92
	0.203		5.41	3.09	0.92
	0.188		5.04	2.88	0.92
	0.125		3.53	2.02	0.92
3.000	0.300		4.47	2.98	0.79
	0.250		3.90	2.60	0.79
	0.216		3.48	2.32	0.79
	0.203		3.31	2.21	0.79
	0.188		3.10	2.06	0.79
	0.152		2.61	1.74	0.79
	0.134		2.34	1.56	0.79
	0.120		2.12	1.41	0.79
2.875	0.250		3.40	2.37	0.75
	0.203		2.89	2.01	0.75
	0.188		2.70	1.88	0.75
	0.125		1.92	1.33	0.75
2.500	0.250		2.15	1.72	0.65
	0.188		1.73	1.38	0.65
	0.125		1.24	0.990	0.65
2.375	0.250		1.82	1.53	0.62
	0.218		1.65	1.39	0.62
	0.188		1.47	1.23	0.62
	0.154		1.25	1.06	0.62
	0.125		1.05	0.887	0.62
1.900	0.145		0.586	0.617	0.50
1.660	0.140		0.368	0.444	0.43

CHAPTER 2

WELDS

OVERVIEW

Chapter 2 contains general information and design considerations for welding, fabrication and inspection of HSS. The information is based on the 1993 AISC LRFD Specification and ANSI/AWS D1.1-96.

Following is a detailed list of the topics considered.

INTRODUCTION TO WELDING DESIGN, FABRICATION AND INSPECTION

Welding is the primary means of joining HSS at connections. Even bolted connections typically use detail material that is welded to HSS. There are welding design, fabrication, and inspection requirements that are unique to HSS connections. Some of these special considerations along with requirements common to HSS and various types of connections are discussed in this chapter. This brief discussion however, is not intended to replace the full text of the ANSI/AWS D1.1 *Structural Welding Code—Steel*, nor any of the provisions of the AISC *Specification for the Design of Steel Hollow Structural Sections*.

To achieve economical structures, the designer should:

1. Not specify larger welds or more extensive welding than is needed. Oversizing welds is wasteful, costly, and possibly detrimental.
2. Provide actual loads to be transferred rather than elaborate weld symbols. The specifics of the joint depend on the welding process to be used and on the welding position (flat, horizontal, vertical or overhead). It is common practice for the fabricator to provide welded joint cross sections on the shop drawings to compliment the weld symbols if necessary to clarify weld details.
3. Specify fillet welds rather than groove welds whenever possible, unless the size of fillets becomes such that a groove weld is more economical (generally, when fillet legs are around ⅝-in.). When fillets are not feasible, use partial joint penetration (PJP) groove welds where permitted. Avoid complete joint penetration (CJP) groove welds except for butt splices or where required for fatigue or seismic applications.

WELD STRENGTH

Welds are commonly loaded in shear, tension, compression, or a combination thereof, depending on the direction and point of application of external forces and the arrangement of joint components. Design strengths for welds in buildings are given in AISC LRFD Specification Section J2. For bridge work, design weld stresses are provided by AWS, AASHTO and AREA specifications. Because the weld values published by these agencies may differ, it is imperative that applicable specifications be consulted and followed.

As with bolts, the loads in concentrically loaded welded joints are assumed to be distributed uniformly throughout their length. Eccentric loading of joints will stress the welds in varying amounts along their length.

AISC LRFD Specification Section J1.8, (and Commentary) provide that weld groups (as with bolts) should have their center of gravity coincide with the gravity axis of an axially loaded member, or provision must be made for the resulting eccentricity. Certain welded members not subject to fatigue loading are given relief from this provision. The design stress range for welded connections of single angles, loaded either in tension or compression, is very restrictive. Recognition of this provision of not balancing the welds in such statically loaded members as roof or floor trusses, bracing, etc., permits very significant cost savings of material and labor in the fabrication and erection of such members.

Fillet Welds

The critical stress in a fillet weld is always considered to be a shear stress, and may occur: (1) parallel to the weld axis, (2) transverse to the weld axis or (3) at some intermediate angle. Therefore, tension, compression and moments acting upon a fillet-welded joint are always resolved as shear in the weld throat (i.e. the least diagonal dimension as illustrated in Figure 2.2).

Fillet welds loaded transverse to the weld axis are fifty percent stronger than welds loaded parallel to the weld axis. This increased strength is conservatively ignored in AISC LRFD Specification Section J2.4 to avoid a whole range of intermediate values for welds

Table 2-1. (AWS Table II-1) Equivalent Fillet Weld Leg Size Factors for Skewed T-Joints								
Dihedral angle, Ψ	60°	65°	70°	75°	80°	85°	90°	95°
Comparable fillet weld size factor for same strength	0.71	0.76	0.81	0.86	0.91	0.96	1.00	1.03
Dihedral angle, Ψ	100°	105°	110°	115°	120°	125°	130°	135°
Comparable fillet weld size factor for same strength	1.08	1.12	1.16	1.19	1.23	1.25	1.28	1.31

loaded obliquely. AISC LRFD Specification Appendix J2.4 provides an equation that reflects the added strength as dependent on the angle of loading.

Figure 2-1a shows a fillet welded lap joint with welds A in parallel shear and weld B in transverse shear. If the load P is increased enough to exceed the total strength of these welds, rupture will occur on the planes of least resistance. As shown in Figure 2-1b this is taken as the weld throats where the least cross-sectional area is present. Consequently, for design purposes, the effective area of a fillet weld is calculated as the product of its effective throat thickness and its effective length. In certain HSS to HSS welds, the effective length is less than the actual length, see AWS 2.39.4 and 2.39.5. For a typical equal-leg fillet weld, with normal throat size made by the manual shielded metal arc, gas metal arc, or flux cored arc welding processes, the effective area equals $0.707wl$, where w is the weld leg size and l is the effective weld length. Although this expression is not strictly applicable to the throat areas of fillet welds in certain types of skewed work, or to fillet welds with unequal legs, its use in most cases will provide conservative results. However if a critically stressed fillet weld has a dihedral angle substantially greater than 90° the reduced effective throat size should be determined and used in stress computations.

AWS D1.1 provides a table, shown here as Table 2-1, showing equivalent leg size factors for a range of dihedral angles between 60° and 135°, with no root opening. Root

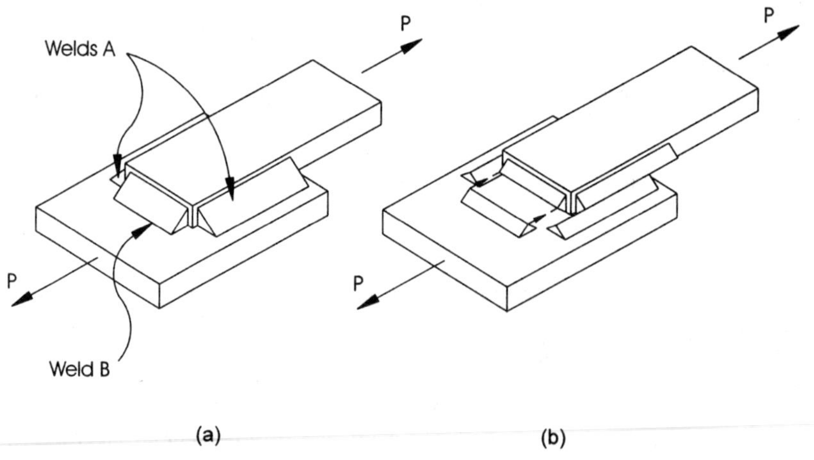

Fig. 2-1. Fillet welds.

opening(s) $\frac{1}{16}$-in. or greater, but not exceeding $\frac{3}{16}$-in. are added to the product of the tabulated values and the leg size to determine the required fillet weld size. The required leg size for fillet welds in skewed joints is calculated using the equivalent leg size factor for the correct dihedral angle. The following example is provided in AWS D1.1.

Example

Given:

Skewed T-joint; 75° angle; $\frac{1}{16}$-in. root opening

Required:

Strength equivalent to 90° fillet weld of $\frac{5}{16}$-in. size

Procedure:

    (1) Factor for 75° from Table 2-1: 0.86
    (2) Equivalent leg size, w, of skewed joint, without root opening:

$$w = 0.86 \times 0.313 = 0.269 \text{ in.}$$

    (3) With root opening of 0.063 in.
    (4) Required leg size $w = 0.269 + 0.063 = 0.332$ in.
    (5) Rounding up to a practical dimension: $w = \frac{3}{8}$-in.

The design strength of a fillet weld is taken as the product of its effective throat area and the design shear stress, ϕF_v. Similarly, the strength of a concentrically loaded fillet weld group is the sum of the strengths of each fillet weld in the group.

Design stresses for fillet welds permitted by the AISC Specification are shown in LRFD Specification Table J2.5.

Fillet Weld Strength

Figure 2-2 shows a diagrammatic view of a 90°fillet weld. The weld leg size w has a throat size of $0.707w$ and the effective throat area is $0.707wl$.

The nominal shear strength of a fillet weld is taken as $0.6 \times$ the nominal strength of the weld metal (specified electrode strength). A resistance factor of 0.75 is applied to the nominal shear strength, resulting in a design shear stress of $0.45 \times$ specified electrode strength.

AWS D1.1, Table 2.5, shown here as Table 2-2, lists matching electrodes for the various ASTM specified steels. The AISC LRFD Specification refers to this table but permits the use of an electrode one strength level stronger than the AWS tabulated "matching" electrode (LRFD Specification Table J2.5). Tests have demonstrated that, in this case, a fillet weld will fail through its effective throat before the base material will fail along the fillet weld leg.

The typical HSS steel grades with F_y up to 50 ksi are normally welded with an electrode having a nominal tensile strength of 70 ksi, indicated as E70XX for SMAW or its equivalent.

Complete-Joint-Penetration (CJP) Groove Welds

AISC LRFD Specification Section J2.4 establishes the limiting tensile stress in CJP groove welds as being equal to the limiting stress of the base metal, providing "matching" base and weld metal are used. These values are contained in Table 2-2 for HSS connection welds. Exceptions are made as follows:

1. When the stress is always compressive and normal to the effective area, weld metal with a strength level equal to or less than "matching" weld may be used.
2. For any condition of stress, weld metal one strength level stronger than "matching" weld metal is permitted.
3. In joints involving base metal of two different yield stress values, the weld metal applicable to the lower strength material may be used, except that low-hydrogen electrodes must be used if required by the higher strength material.

Given compliance with these requirements, it can be said that CJP groove welds develop the strength of the base metal and that no allowance for the presence of such welds need be made in proportioning the connections of structural members for any type of static loading. However, where members of unequal cross section or differing yield strength are joined, the splice is limited to the strength of the weaker member. Additionally, some HSS-to-HSS and similar connections with differing relative flexibility along the weld are such that an effective length concept limits the joint strength. See page 2-23.

CJP groove welds onto backing bars can be used to splice members. Accurate fitting of the backing bars is particularly important and proper inspection at that stage prior to welding is advised. CJP-groove-welded butt splices without backing bars should be avoided because they are seldom justified economically or structurally. Additionally, they are expensive for several reasons: welder qualifications are appreciably more demanding, the resulting joints are not prequalified, and preparation and fit-up are more exacting. Inspection of such welds can be difficult and potentially controversial when attempts are made to confirm acceptable complete penetration. Weld repairs are intricate and may damage other portions of the joint unless care is used.

Round HSS joints are very complex. Fig. 2-3 depicts joint geometry variations progressing around the perimeter of a member framing at an angle θ. Initially, the intersection is expressed as the line where the inner surface of the joining member meets

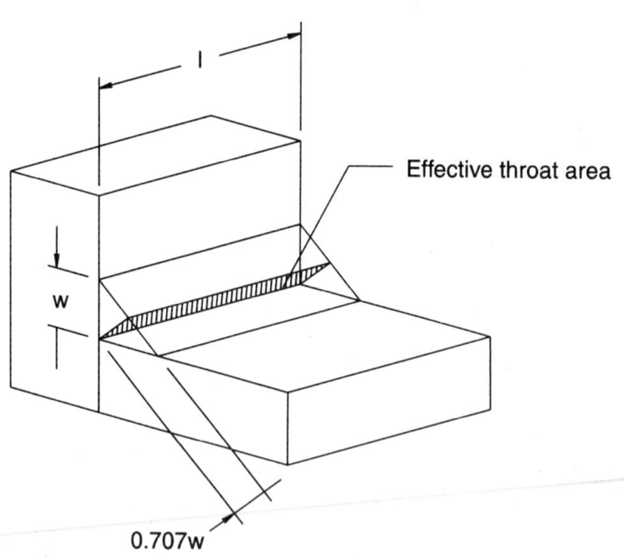

Fig. 2-2. Fillet weld.

Table 2-2.
Design Strength in HSS Connection Welds (See AWS 2.36.3)

Type of Weld: Complete-Joint-Penetration (CJP) Groove Weld

Kind of Stress	Resistance Factor ϕ	Nominal Strength	Required Filler Metal Strength Level[1]
Application: Longitudinal butt joints (longitudinal seams)			
Tension or compression parallel to axis of the weld[2]	0.9	$0.6F_y$	Filler metal with strength equal to or less than matching filler metal may be used
Beam or torsional shear	Base metal 0.9 Weld metal 0.8	$0.6F_y$ $0.6F_{EXX}$	
Application: Circumferential butt joints (girth seams)			
Compression normal to the effective area[2]	0.9	F_y	Matching filler metal shall be used
Shear on effective area	Base metal 0.9 Weld metal 0.8	$0.6F_y$ $0.6F_{EXX}$	
Tension normal to the effective area	0.9	F_y	
Application: Weld joints in T-, Y-, or K-connections in structures designed for critical loading such as fatigue, which would normally call for CJP groove welds			
Tension, compression or shear on base metal adjoining weld conforming to detail of AWS Figures 3.6, 3.8–3.10 (HSS weld made from outside only without backing)	Same as for base metal or as limited by connection geometry (see AWS 2.40)		Matching filler metal shall be used
Tension, compression, or shear on effective area of groove welds, made from both sides or with backing			

Type of Weld: Fillet Weld

Kind of Stress	Resistance Factor ϕ	Nominal Strength	Required Filler Metal Strength Level[1]
Application: Longitudinal joints of built-up HSS			
Tension or compression parallel to axis of the weld	0.9	F_y	Filler metal with a strength equal to or less than matching filler metal may be used.
Shear on effective area	0.75	$0.6F_{EXX}$	

For Notes, see page 2-8.

Table 2-2. (Continued) Design Strength in HSS Connection Welds (See AWS 2.36.3)			
Type of Weld: Fillet Weld			
Kind of Stress	**Resistance Factor ϕ**	**Nominal Strength**	**Required Filler Metal Strength Level[1]**
Application: Joints in T-, Y-, or K-connections, in circular lap joints and joints or attachments to HSS			
Shear on effective throat regardless of direction of loading (see AWS 2.39 and 2.40.1.3)	0.75 or as limited by connection geometry (see AWS 2.40)	$0.6F_{EXX}$	Filler metal with a strength level equal to or less than matching filler metal may be used.[4]
Type of Weld: Plug and Slot Welds			
Kind of Stress	**Resistance Factor ϕ**	**Nominal Strength**	**Required Filler Metal Strength Level[1]**
Shear Parallel to faying surfaces (on effective area)	Not Applicable to HSS		
Type of Weld: Partial-Joint-Penetration (PJP) Groove Weld			
Kind of Stress	**Resistance Factor ϕ**	**Nominal Strength**	**Required Filler Metal Strength Level[1]**
Application: Circumferential and longitudinal joints that transfer loads			
Compression normal to the effective area, joint designed to bear or not to bear	0.9	F_y	Filler metal with a strength level equal to or less than matching metal may be used
Shear on effective area	0.75	$0.6F_{EXX}$	Filler metal with a strength level equal to or less than matching filler metal may be used
Tension on effective area	base metal 0.9 filler metal 0.8	F_y $0.6F_{EXX}$	
Application: T-, Y-, or K-connection in ordinary structures not requiring CJP groove welds			
Load transfer across the weld as stress on the effective throat (see AWS 2.39 and 2.40.1.3)	base metal 0.9 filler metal 0.8 or as limited by connection geometry (see AWS 2.40)	F_y $0.6F_{EXX}$	Matching filler metal shall be used

Notes:
1. For matching filler metal see AWS Table 3.1.
2. Beam or torsional shear up to 0.30 minimum specified tensile strength of filler metal is permitted, except that shear on adjoining base metal shall not exceed $0.40F_y$.
3. Groove and fillet welds parallel to the longitudinal axis of tension or compression members, except in connection areas, are not considered as transferring stress and hence may take the same stress as that in the base metal, regardless of electrode (filler metal) classification. Where the provisions of AWS 2.40.1 are applied, seams in the main member within the connection area shall be complete joint penetration groove welds with matching filler metal, as defined in AWS Table 3.1
4. See AWS 2.40.1.3.
5. Alternatively, see AWS 2.14.4 and 2.14.5.

the outer surface of the other member; then bevels on the joining member are defined by the local dihedral angle Ψ to permit clearances for the welds.

There are usually four zones as illustrated in Fig. 2-3. Common practice was to produce a profile template to wrap around and mark the joining member for a cut made perpendicular to the HSS axis. A second cut was then made with varying bevels for the weld preparation. Computer-controlled cutting machinery is now available for producing the final cut in a single operation, but geometry ranges are limited and few fabrication shops are so equipped. Obviously the complexities rapidly multiply with overlapping connections involving multiple members of varying size.

Structurally, as seen in Chapter 8, HSS connections quite often cannot be configured to develop the full strength of the HSS. Thus, an arbitrary design requirement for CJP groove welds is questionable in many cases.

Joint profiles of CJP groove welds are shown in AWS D1.1, and also are reproduced with minor modification in this Manual courtesy of AWS.

Partial-Joint-Penetration (PJP) Groove Welds

The design shear strength parallel to the axis of the weld is the same as that specified for tensile strength normal to the effective area:

$$0.75 \times 0.6 \times \text{nominal tensile strength of the weld metal}$$

The effective area of a PJP groove weld is the product of the effective length of the weld times the effective throat thickness of the weld. These quantities are determined as follows.

For butt splices the effective length is the width of the part joined. For other splices see AWS D1.1 Section 2.39.5. The effective throat thickness E is as determined from AISC LRFD Specification Table J2.1, but not less than that specified in AISC LRFD Specification Table J2.3.

Nomenclature for PJP groove welds is shown in Figure 2-4. Note that the effective throat thickness shown is less than the dimensioned groove-weld size. The effective throat dimension E shown in these tables differs from that in AWS D1.1. AWS prequalified PJP groove welds establish for each joint an effective throat E as a function of the material thickness, weld-preparation size, or depth S along with weld process and position. E equals S minus the distance at the root of the weld that is considered ineffective due to the likely presence of impurities or lack of fusion. Thus, the design drawings must either

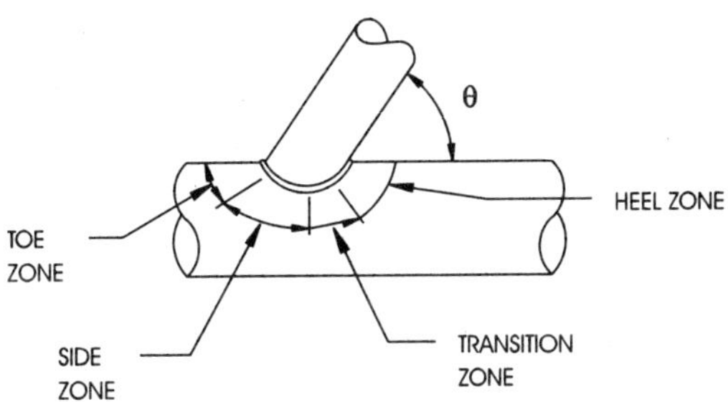

Fig. 2-3. Round HSS joint.

specify the effective weld length and the required effective throat or show the required strength. The shop drawings must specify the effective weld length and the required effective throat.

The three exceptions listed in the paragraph of the previous section headed "Complete-Joint-Penetration-Groove (CJP) Welds" also apply to PJP groove welds. The comments regarding the difficulties of making connections of round HSS to HSS apply as well.

Flare Welds

A flare weld is a special case of the PJP groove weld wherein the convex surface of the part creates the joint preparation. This convexity may be the result of an edge preparation, but more often one or both joint components consists of a round rod or shape with a rounded bend or corner radius created by bending or rolling. The deposition of sound weld metal to the bottom of the flare is very difficult, as with many of the PJP groove

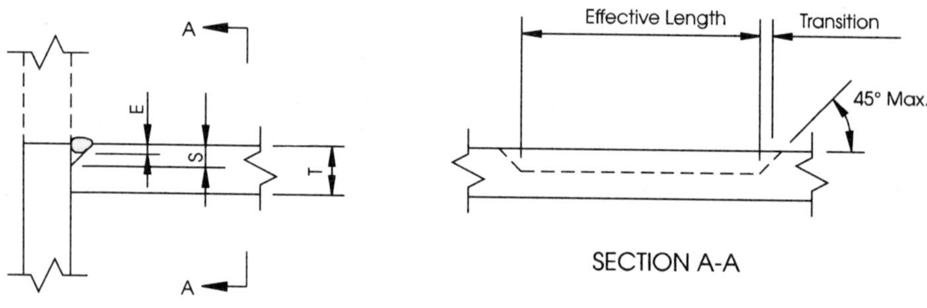

Fig. 2-4. Partial-Joint-Penetration groove weld nomenclature.

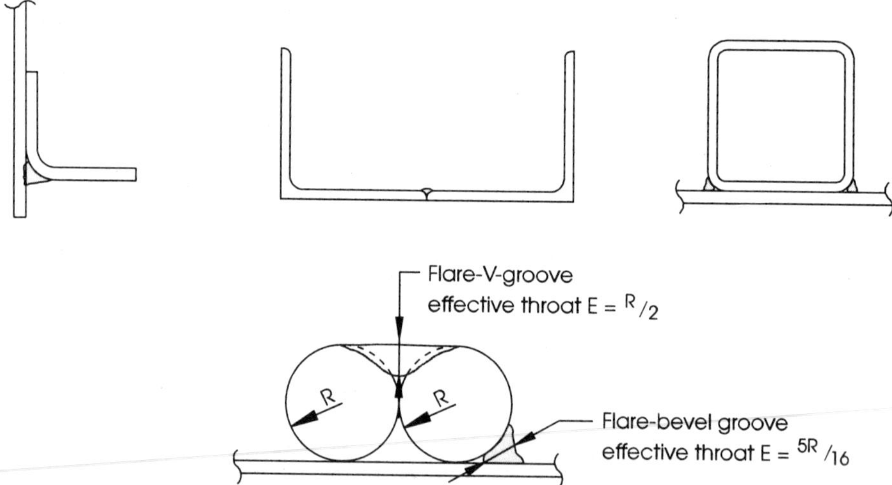

Fig. 2-5. Flare weld nomenclature.

welds, because the welding arc short-circuits across the surfaces due to the sharp angular slopes.

AISC LRFD Specification Section J2.1a establishes the effective throat thickness of a flare groove weld, when the weld is flush (tangent) to the surface of the solid section, as shown in Figure 2-5. AISC LRFD Specification Table J2.2 defines the ratio of radius to effective throat. The quality of this weld is difficult to control and the Specification permits examination and adjustment of the weld strength based on testing and special qualification.

The effective area of a flare weld A_w is the product of the effective length of the weld times the effective throat thickness of the flare weld; the effective length is the width of the part joined and the effective throat thickness E is as determined from AISC LRFD Specification Table J2.2.

A common flare bevel configuration, which occurs when equal-width sections are joined, is illustrated in Fig. 2-6a. The easiest arrangement for welding occurs with sections of equal wall thickness. However, when the corner radius increases due to differing wall thickness or manufacturing tolerances, the root gap may need to be adjusted by profile shaping, building out with weld metal or by the use of backing. See Fig. 2-6b.

APPLICATION OF THE AWS D1.1 TO HSS

ANSI/AWS D1.1 *Structural Welding Code—Steel* is the national standard for welding in building structures and will normally be used unless specified otherwise in the contract documents. AWS uses the terminology "tubular" for all hollow members including pipe, hollow structural sections, and fabricated box sections. AWS reformatted AWS D1.1 in 1996 into the sections listed below:

Section 1

General Requirements contains basic information on the scope and limitations of the code. It states that most provisions of the code are mandatory when the code is specified. Optional provisions apply only when specified in the contract documents. The optional provisions generally are inspection requirements that the engineer may specify.

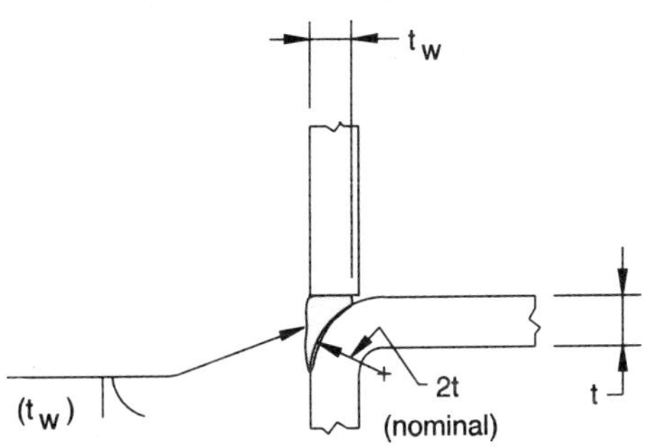

Flare-bevel weld

Fig. 2-6a. Equal width HSS weld joint.

Section 2

Design of Welded Connections contains requirements for the design of welded connections applied to tubular or non-tubular members.

Part A

Common Requirements of Non-tubular and Tubular Connections covers basic weld sizing information for all types of members.

Part D (Parts B & C do not apply to HSS)

Specific Requirements for Tubular Connections covers requirements for tubular cross-sections regardless of the type of loading. The commentary to this section states that "the tubular provisions of this code originally evolved from a background of practices and experience with fixed off-shore platforms of welded tubular construction." The range of section sizes covered is well beyond the sections covered herein; thus only the provisions applicable to HSS have been used.

The commentary explains that "In commonly used types of tubular connections, the weld itself may not be the factor limiting the [strength] of the joint. Such limitations as local failure (punching shear), general collapse of the main member, and lamellar tearing are discussed because they are not adequately covered in other codes." Because of these various limit states, the design of HSS-to-HSS connections must be part of the member sizing process. The members selected must be capable of transmitting the required design forces or adequate reinforcement must be shown on the design documents.

Differences in the relative stiffness across the walls of an HSS loaded normal to its surface can make the load transfer highly non-uniform. To prevent progressive failure and to ensure ductile behavior of the joint, minimum welds must be provided in T-, Y-,

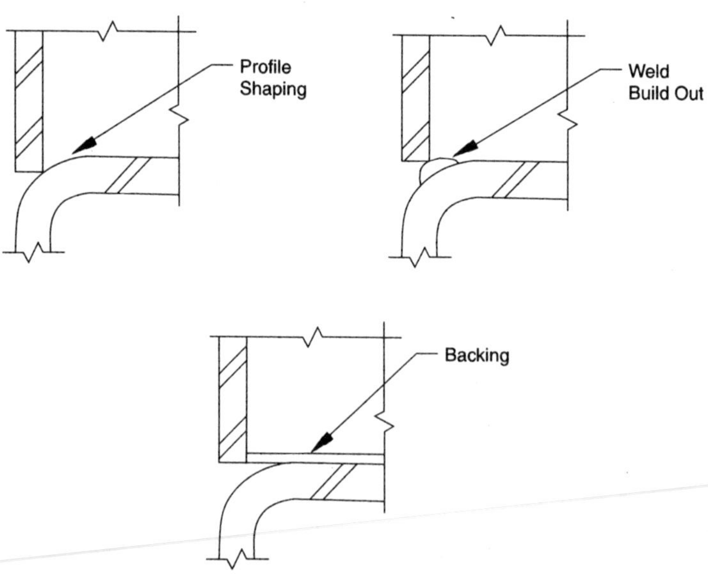

Fig. 2-6b. Weld joint for unequal HSS wall thickness.

and K-connections to transmit the factored load in the branch or web member. For normal building applications, fillet welds and PJP groove welds can be used.

While Part D deals primarily with HSS-to-HSS connections, some of these provisions are applicable to welded attachments that deliver a load normal to the wall of a tubular member.

Section 3
Prequalification of WPS contains the requirements for exempting a Welding Procedure

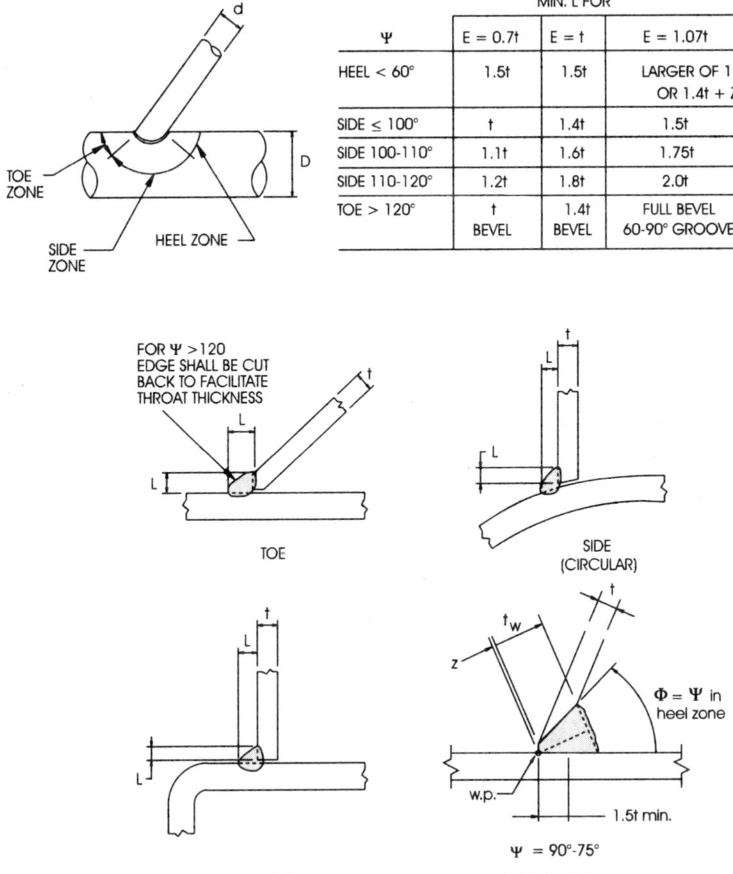

Ψ	MIN. L FOR		
	E = 0.7t	E = t	E = 1.07t
HEEL < 60°	1.5t	1.5t	LARGER OF 1.5t OR 1.4t + Z
SIDE ≤ 100°	t	1.4t	1.5t
SIDE 100-110°	1.1t	1.6t	1.75t
SIDE 110-120°	1.2t	1.8t	2.0t
TOE > 120°	t BEVEL	1.4t BEVEL	FULL BEVEL 60-90° GROOVE

Notes from AWS D1.1-96:

1. t = thickness of thinner part.
2. L = minimum size (see 2.40.1.3 which may require increased weld size for combinations other than 36 ksi (250 MPa) base metal and 70 ksi (480 MPa) electrodes).
3. Root opening 0 to 3/16 in. (5mm) - see 5.22.
4. φ = 15° min. Not prequalified for under 30°. For φ < 60°, loss dimension (Table 2.8) and special welder qualifications (Table 4.8) apply.
5. See 2.39.1.2 for limitations on β = d/D.

Fig. 2-7. Fillet welded prequalified tubular (HSS) joints made by shielded metal arc, gas metal arc, and flux cored arc welding.

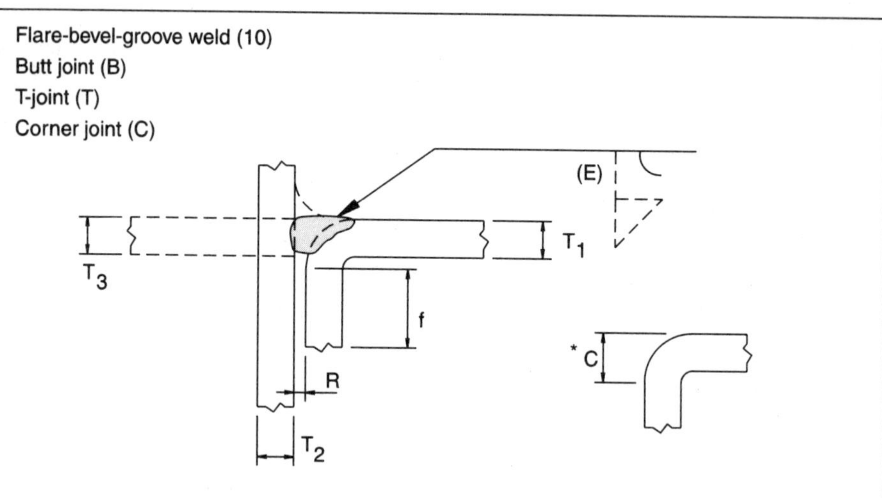

| Welding Process | Joint Designation | Base Metal Thickness (U=unlimited) | | | Groove Preparation | | | Permitted Welding Positions | Weld Size (E) | Notes |
| | | | | | Root Opening Root Face Bend Radius | Tolerances | | | | |
		T_1	T_2	T_3		As Detailed (see AWS)	As Fit-Up (see AWS)			
SMAW	BTC-P10	3/16 min	U	T_1 min	R=0 f=3/16 min * $C=\dfrac{3T_1}{2}$ min	+1/16,-0 +U,-1/16 -0,+Not-Limited	+1/8,-1/16 +U,-1/16 -0,+Not-Limited	All	$5/8T_1$	D,J,N,Z
GMAW FCAW	BTC-P10-GF	3/16 min	U	T_1 min	R=0 f=3/16 min * $C=\dfrac{3T_1}{2}$ min	+1/16,-0 +U,-0 -0,+Not-Limited	+1/8,-1/16 +U,-1/16 -0,+Not-Limited	All	$5/8T_1$	A,J,N,Z
SAW	T-P10-S	1/2 min	1/2 min	N/A	R=0 f=1/2 min * $C=\dfrac{3T_1}{2}$ min	0 +U,-0 -0,+Not-Limited	+1/16,-0 +U,-1/16 -0,+Not-Limited	F	$5/8T_1$	J,N,Z

* For cold formed (A500) rectangualar tubes, C dimension is not limited. See the following:

Effective Weld Size of Flare-Bevel-Groove Welded Joints. Tests have been performed on cold formed ASTM A500 material exhibiting a "C" dimension as small as T_1 with a nominal radius of 2t. As the radius increases, the "C" dimension also increases. The corner curvature may not be a quadrant of a circle tangent to the sides. The corner dimension, "C", may be less than the radius of the corner.

Notes

A: Not prequalified for gas metal arc welding using short circuiting transfer nor GTAW. Refer to AWS Annex A.

D: SMAW detailed joints may be used for prequalified GMAW (except GMAW-S) and FCAW.

J: If fillet welds are used in statically loaded structures to reinforce groove welds in corner and T-joints, these shall be equal to $1/4T_1$, but need not exceed 3/8-in. Groove welds in corner and T-joints of cyclically loaded structures shall be reinforced with fillet welds equal to $1/4T_1$, but not more than 3/8-in.

N: The orientation of the two members in the joints may vary from 135° to 180° for butt joints, or 45° to 135° for corner joints, or 45° to 90° for T-joints.

Z: Weld size (E) is based on joints welded flush.

Fig. 2-8. Prequalified partial joint penetration (PJP) groove welded joint details.

Specification (WPS) from the qualification requirements of AWS D1.1. A WPS is a written description of the detailed methods and practices involved in the production of a weldment. Certain welding processes such as shielded metal arc (SMAW), submerged arc (SAW), gas metal arc (GMAW), and flux cored arc (FCAW) welding have been thoroughly tested in similar joints and have a record of satisfactory performance. These welds are exempt from further testing when they comply with the requirements of AWS D1.1 Section 3. These requirements can be divided into welding variables and joint

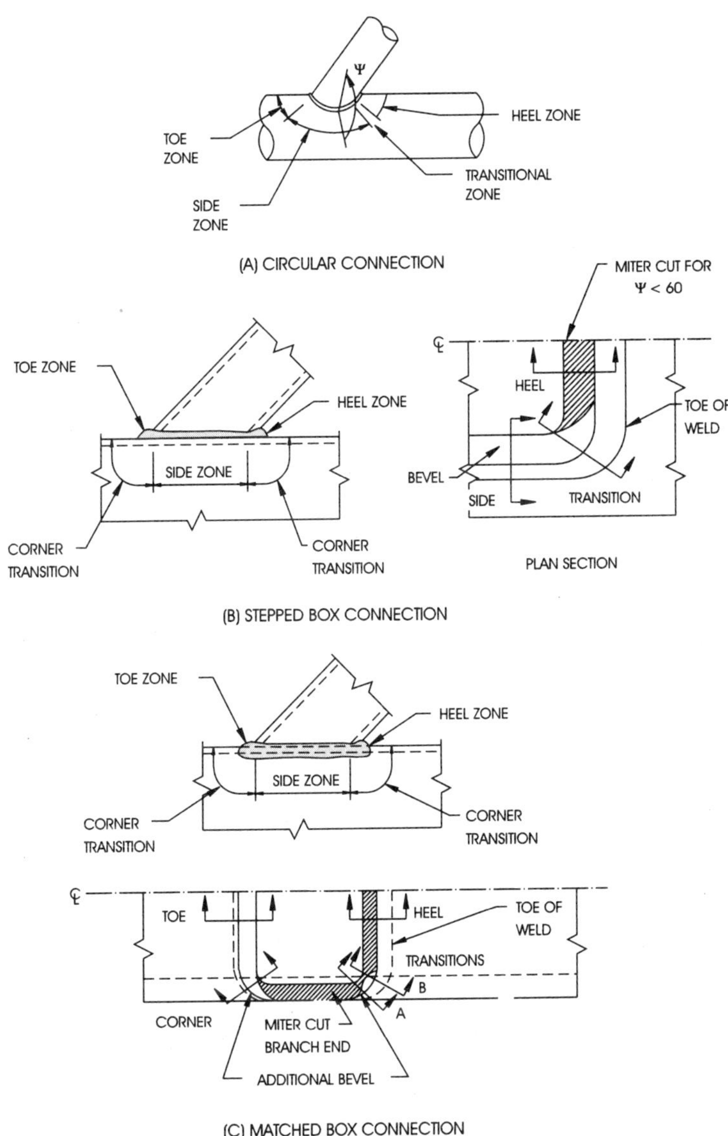

Fig. 2-9. Prequalified joint details for PJP T-, Y-, and K-Tubular (HSS) connections

details. The welding variables consist of items such as process, base metal and electrode combinations, preheat and interpass temperatures, amperage, voltage, travel speed, flux or shielding gas, pass size and other special requirements. The joint details that are used in conjunction with these welding variables show the preparation and fitup required for each process and welding position. While some of the joints shown can be used for tubular or non-tubular connections, there are a number of joints that are unique to tubular members. Figure 2-7 (AWS Fig. 3.2) shows prequalified fillet weld details for tubular

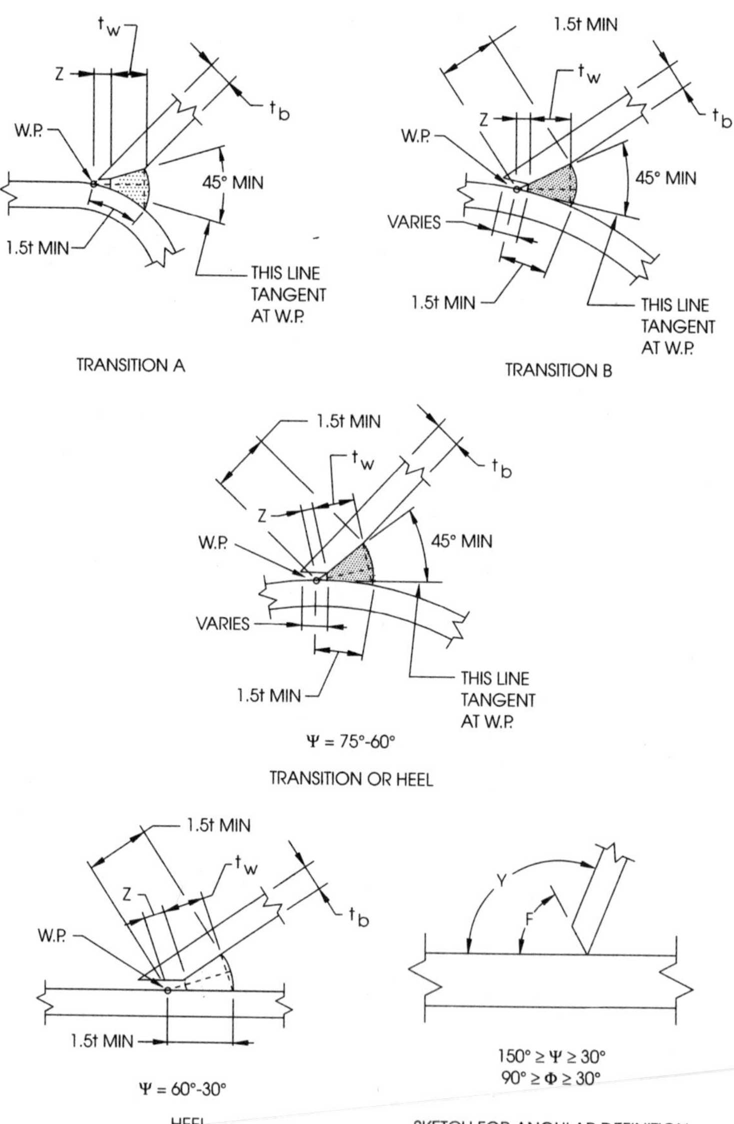

Fig. 2-10. Prequalified joint details for PJP T-, Y-, and K-Tubular (HSS) connections.

joints that differ from details for non-tubular skewed T-joints. These details will provide the minimum weld strength needed to ensure ductile joint behavior.

Figure 2-8 (AWS Fig. 3.3) shows the joint detail and the effective throat for a flare bevel PJP groove weld that is commonly used for welding connection material to the face of an HSS. Groove weld joint details for HSS are designed to accommodate both the geometry of the section and the lack of access to the back side of the joint.

Figures 2-9, 2-10 and 2-11 (AWS Fig. 3.5) show various PJP groove welded joint

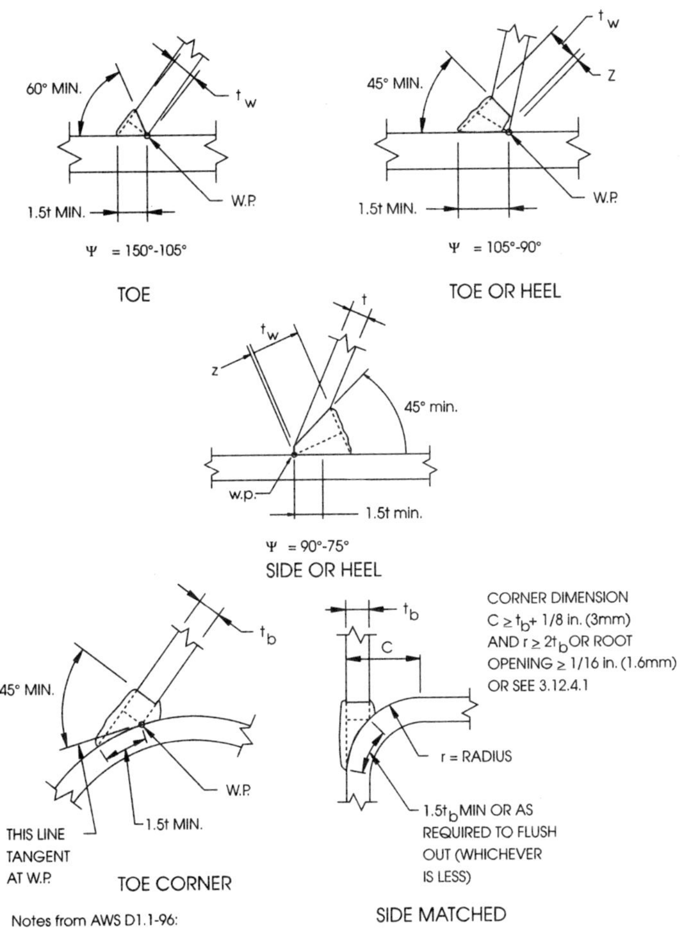

Notes from AWS D1.1-96:

1. t = thickness of thinner section.
2. Bevel to feather edge except in transition and heel zones.
3. Root opening: 0 to 3/16 in. (5 mm)
4. Not prequalified for under 30°.
5. Weld size (effective throat) $t_w \geq t$; Z Loss Dimensions shown in Table 2.8.
6. Calculations per 2.40.1.3 shall be done for leg length less than 1.5t, as shown.
7. For Box Section, joint preparation for corner transitions shall provide a smooth transition from one detail to another. Welding shall be carried continuously around corners, with corners fully built up and all weld starts and stops within flat faces.
8. See Annex B for definition of local dihedral angle, Ψ.
9. W.P. = work point.

Fig. 2-11. Prequalified joint details for PJP T-, Y-, and K-Tubular (HSS) connections

details and Figures 2-12 thru 2-15 (AWS Figs. 3.6, 3.8, 3.9 and 3.10) show CJP groove welded joint details. The joint preparation and weld sizing are complex and critical to obtain a sound weld. These details also provide the weld strength needed to ensure ductile joint behavior.

AWS Table 3.6 summarizes the dimensional requirements for prequalified CJP groove welds for the various processes. In addition, all prequalified WPS must comply with the

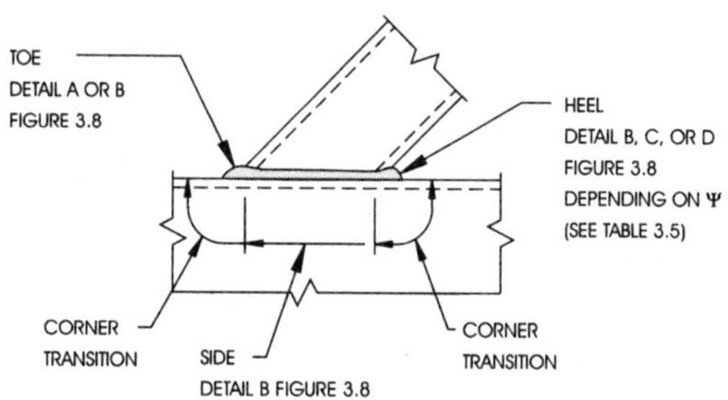

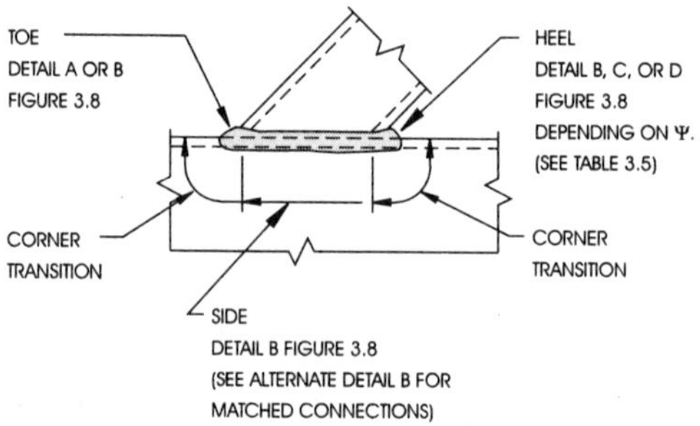

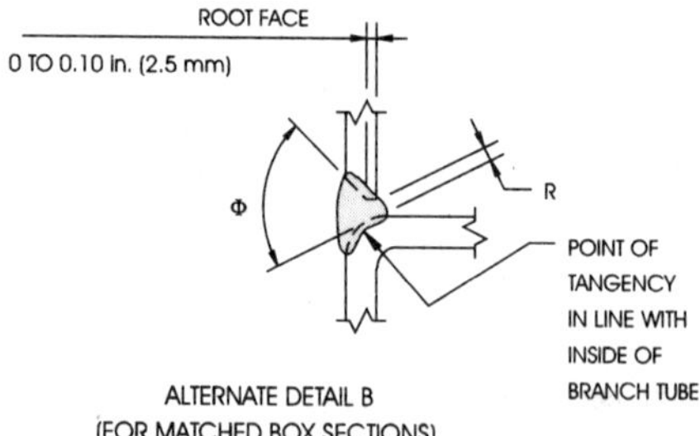

Notes from AWS D1.1-96:

1. Details A, B, C, D as shown in Figure 3.8 and all notes from Table 3.6 apply.
2. Joint preparation for corner welds shall provide a smooth transition from one detail to another. Welding shall be carried continuously around corners, with corners fully built up and all arc starts and stops within flat faces.
3. References to Figure 3.8 include Figures 3.9 and 3.10 as appropriate to thickness (see 2.36.6.7)

Fig. 2-12. Prequalified joint details for CJP T-, Y-, and K-Tubular (HSS) connections.

AWS Table 3.7 which gives maximum electrode diameter, maximum current, and maximum size for various passes.

Section 4

Qualification covers the requirements for qualification testing of Welding Procedure Specifications (WPS) and performance testing of the welder's ability to produce sound welds. HSS connections may not always meet the requirements for a prequalified WPS because of unique geometry, connection access or for other reasons. This section also

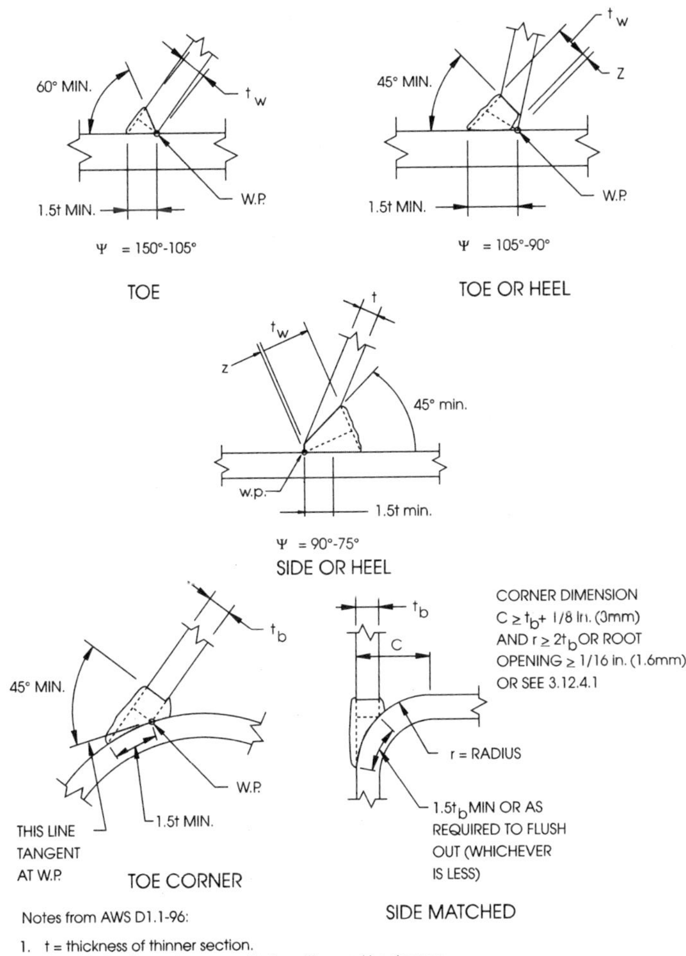

Notes from AWS D1.1-96:

1. t = thickness of thinner section.
2. Bevel to feather edge except in transition and heel zones.
3. Root opening: 0 to 3/16 in. (5 mm)
4. Not prequalified for under 30°.
5. Weld size (effective throat) $t_w \geq t$; Z Loss Dimensions shown in Table 2.8.
6. Calculations per 2.40.1.3 shall be done for leg length less than 1.5t, as shown.
7. For Box Section, joint preparation for corner transitions shall provide a smooth transition from one detail to another. Welding shall be carried continuously around corners, with corners fully built up and all weld starts and stops within flat faces.
8. See Annex B for definition of local dihedral angle, Ψ.
9. W.P. = work point.

Fig. 2-13. Prequalified joint details for complete joint penetration groove welds in tubular (HSS) T-, Y-, and K-Connections—standard flat profiles for limited thickness.

gives the requirements for a Procedure Qualification Record (PQR) which is the basis for qualifying a WPS.

The performance testing of welders and welding operators considers process, material thickness, position and non-tubular or tubular joint access. AWS Tables 4.8 and 4.9 list the required qualifications needed for each type of joint. Most welders are qualified for a particular process and position in plate (non-tubular) joints. These qualifications will allow the welder to make similar fillet, PJP groove, and backed CJP groove welds on

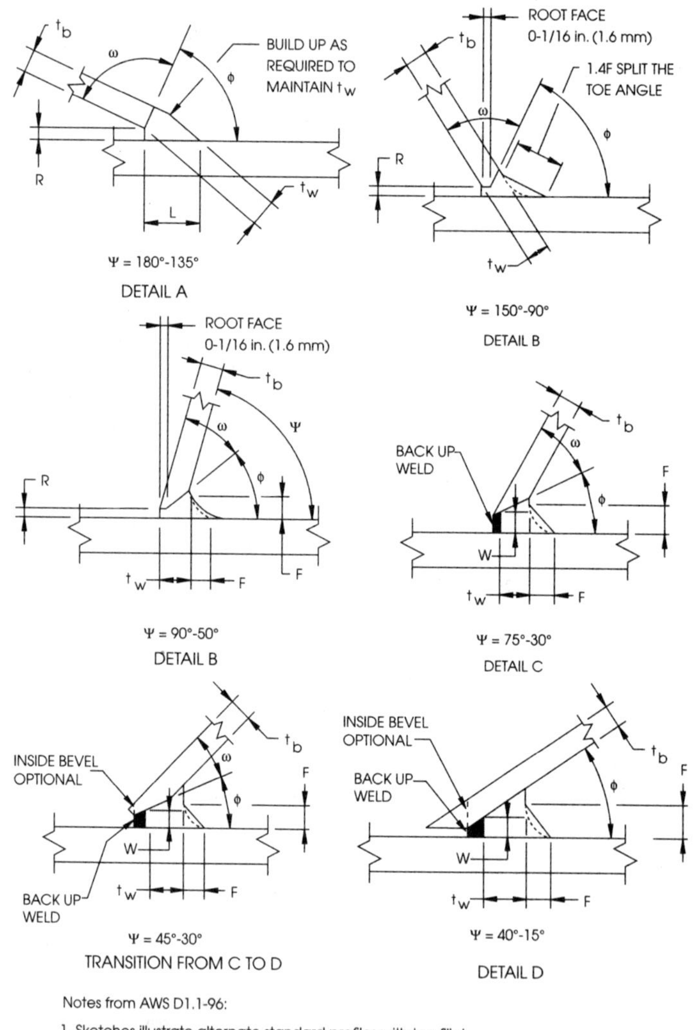

Notes from AWS D1.1-96:

1. Sketches illustrate alternate standard profiles with toe fillet.
2. See 2.36.6.7 for applicable range of thickness t_b.
3. Minimum fillet weld size, $F=t_b/2$, also subject to limits of Table 5.8.
4. See Table 3.6 for dimensions t_w, L, R, W, ω, φ.
5. Convexity and overlap are subject to the limitations of 5.24.
6. Concave profiles, as shown by dashed lines are also acceptable.

Fig. 2-14. Prequalified joint details for complete joint penetration groove welds in tubular (HSS) T-, Y-, and K-Connections—profile with toe fillet for intermediate thickness.

tubular members. However, certain types of tubular connections, such as unbacked T-, Y-, and K-connections, require special welder certifications because the lack of access to the back of the joint, the position of the connection, and the access to the connection require special skill to produce a sound connection.

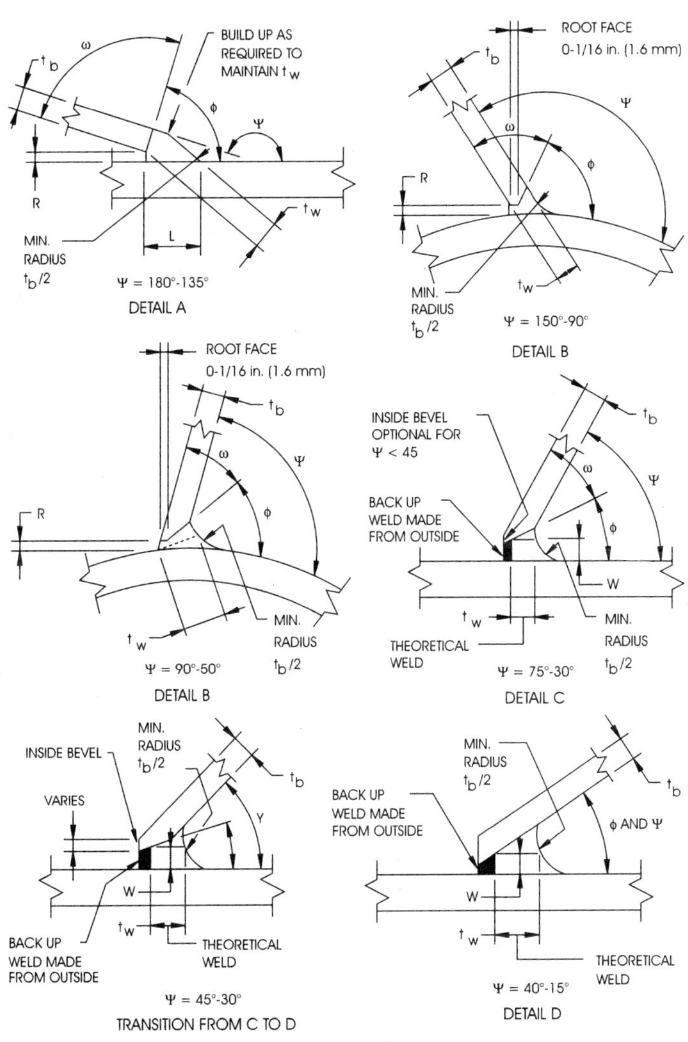

Notes from AWS D1.1-96:

1. Illustrating improved weld profiles for 2.36.6.6(1) as welded and 2.36.6.6(2) fully ground.
2. For heavy sections or fatigue critical applications as indicated in 2.36.6.6.7.
3. See Table 3.6 for dimensions t_b, L, R, W, ω, φ.

Fig. 2-15. Prequalified joint details for complete joint penetration groove welds in tubular (HSS) T-, Y-, and K-Connections—concave improved profile for heavy sections or fatigue.

Section 5

Fabrication covers the requirements for the preparation, assembly, and workmanship of welded steel structures.

AWS Table 5.5 Tubular Root Opening Tolerances give the acceptable fitup for unbacked groove welds. AWS Table 5.8 Minimum Fillet Weld Size gives the minimum weld pass size based on material thickness and process.

Section 6

Inspection contains the requirements for the inspector's qualifications and responsibilities, acceptance criteria for discontinuities, and procedures for non-destructive examination (NDE). AWS D1.1 considers fabrication/erection inspection and testing a separate function from verification inspection and testing. Fabrication/erection inspection and testing is usually the responsibility of the contractor and is performed as appropriate prior to assembly, during assembly, during welding and after welding to ensure the requirements of the contract documents are met. Verification inspection and testing are the prerogatives of the owner. The extent of NDE and verification inspection must be specified in the contract documents.

If non-destructive testing other than visual is not specified in the contract documents, but is subsequently requested, the owner is responsible per AWS D1.1 for all costs associated with this testing including handling, surface preparation, and repairs (if required).

The inspection covers WPS qualification, equipment, welder qualification, joint preparation, joint fitup, welding techniques, and weld size length and location. It is especially important when inspecting HSS-to-HSS joints that joint preparation and fitup be checked prior to welding.

In addition to inspecting the above items, AWS requires all welds to be visually inspected for conformance to the standards in AWS Table 6.1 Visual Acceptance Criteria.

Four types of non-destructive testing can be used to supplement visual inspection. They are penetrant testing (PT), magnetic particle testing (MT), radiographic testing (RT) and ultrasonic testing (UT).

The AWS UT acceptance criteria for non-HSS type groove welds starts at $^5/_{16}$-in. thick material. The procedures for HSS T-, Y-, and K-connections have a minimum applicable thickness of $^1/_2$-in., and diameter of $12^3/_4$-in. AWS does however make provision for qualifying UT procedures for smaller size applications. It is possible to UT portions of butt-type splices with backing bars using the non-HSS criteria, however, the corners of rectangular HSS cannot be inspected.

The AWS Code does make provision for using alternate acceptance criteria based upon an evaluation of suitability for service using past experience, experimental evidence, or engineering analysis. This can be especially important when deciding if and how to make any repairs.

Section 7

Stud Welding covers the requirements for welding steel studs to steel. AWS Table 7.1 Mechanical Property Requirements for studs gives the tensile strength requirement for Type A studs which includes the threaded type stud that can be used for HSS erection connections.

Section 8

Strengthening and Repairing existing structures contains basic information pertinent to this type of work.

WELD SIZING AND DETAILING OF HSS-TO-HSS

Weld Sizing for Uneven Distribution of Loads

The connection strength for a member welded normal to a wall of an HSS is a function of their geometric parameters and is often less than the strength of the member. This strength cannot be increased by increasing the weld strength. Due to the differing relative flexibility of an HSS wall loaded normal to its surface and the axial stiffness of the connected member, the transfer of load along the weld line is highly non-uniform. To prevent progressive failure of the weld, "unzipping", it is important to provide welds that are adequate to ensure ductile behavior of the joint.

Welds that satisfy this ductility requirement can be proportioned for the required strength using an effective width criteria similar to that used for checking the axial strength of the branch member or plate.

For effective weld length in K- and N-connections (See Fig. 2-16) of rectangular HSS members:

for $\theta \leq 50°$

$$L_e = 2H_b + 2B_b$$

for $\theta \geq 60°$

$$L_e = 2H_b + B_b$$

For $50° < \theta < 60°$ interpolate between these values.

For effective weld length in T-, Y- and Cross-connections (see Fig. 2-16) of rectangular HSS members:

for $\theta \leq 50°$

$$L_e = 2H_b + B_b$$

for $\theta \geq 60°$

$$L_e = 2H_b$$

For $50° < \theta < 60°$ interpolate between these values

For the effective weld length on a transverse plate welded to face of an HSS member:

$$L_e = 2\,\frac{10}{B\,/\,t}\,\frac{F_y\,t}{F_{y1}\,t_1}\,b_1 \leq 2b_1$$

where the above variables are as defined in AISC HSS Specification Section 9.2.

Similar effective length formulas for round HSS can be found in AWS D1.1. An alternative to the above effective length procedure is the use of the prequalified fillet and PJP groove weld details in AWS D1.1 that are sized to ensure ductile behavior. In addition, fillet welds with an effective throat of 1.1 times the thickness of the branch member can be used. Either of these two alternatives will in most cases be overly conservative.

AWS D1.1 and the AISC HSS Specification Section 9.2 provide the basis for the above requirements.

Detailing Considerations

1. Butt joints will require a groove weld detail. Where possible the joint should be a prequalified PJP groove weld sized for actual load or a CJP groove weld with steel backing.
2. T-, Y-, and K-connections should, where possible, use either fillet welds or PJP

groove welds sized for the design forces and checked for the minimum size needed to insure ductile joint behavior. Where CJP groove welds are required, joint details using steel backing should be used whenever possible. For a detailed discussion of various types of backing and the advantages of using backing see Post (1990).

WELD INSPECTION METHODS

Table 2-3 summarizes the characteristics and capabilities of the five most commonly used methods for welding inspection. These methods are discussed below and in AWS B1.0 *Guide for the Non-Destructive Inspection of Welds*. The designer must specify in the

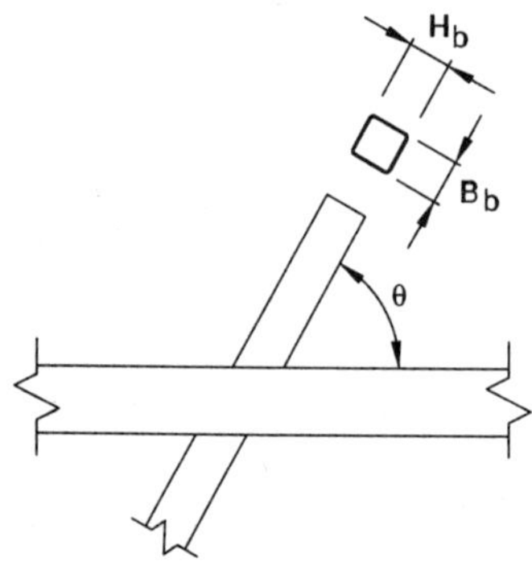

TYPE T-, Y- AND CROSS-CONNECTION

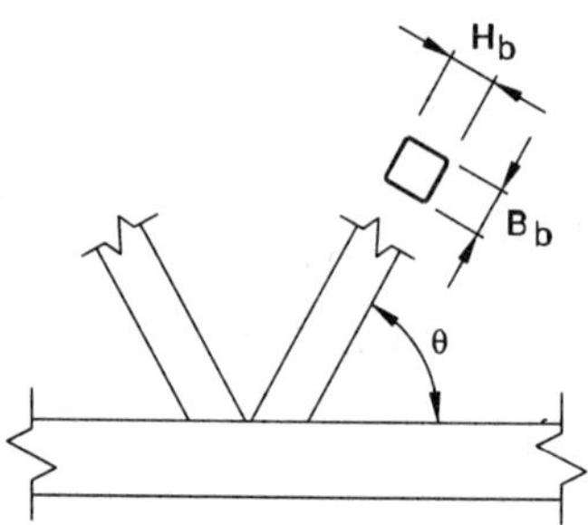

TYPE K- AND N-CONNECTION

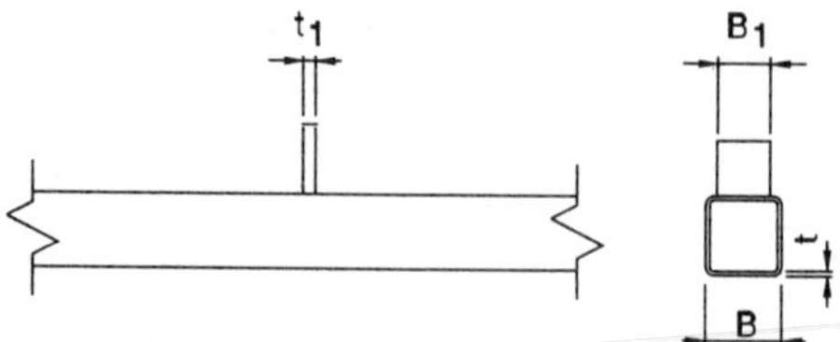

TYPE K- AND N-CONNECTION

Figure 2-16.

contract documents the type of weld inspection required as well as the extent and application of each type of inspection. In the absence of instructions for weld inspection, the fabricator or erector is only responsible for those weld discontinuities found by visual inspection (see AWS D1.1). If additional inspection more stringent than visual is later required, the owner is normally responsible for the costs of inspection and of weld repairs other than those identified by the visual inspection requirements. Weld repairs that may be difficult to perform and that may potentially damage other aspects of the connection are best referred to the engineer of record to determine the necessity of the correction with due consideration of fitness for purpose.

Visual inspection is the most commonly required inspection process. The designer must realize that more stringent requirements for inspection can needlessly add significant cost to the project and should specify them only in those instances where they are essential to the integrity of the structure.

Visual (VT)

Visual inspection provides the most economical way to check weld quality. If the welder were to check the fit-up for accuracy and cleanliness before welding, and comply with the welding procedure requirements, most weld defects would be prevented.

If there is a fit-up gap for a fillet weld, the leg size of the fillet should be increased by the amount of the gap with a permissible maximum increase (and gap) of $\frac{3}{16}$-in.

Dye (Liquid) Penetrant Test (PT)

This test uses a red dye penetrant applied to the work from a pressure spray can. The dye penetrates any crack or crevice open to the surface. Excess dye is removed and white developer is sprayed on. Dye seeps out of the crack, producing a red image on the white developer (See Figure 2-17).

Magnetic Particle (MT)

A magnetizing current is introduced with a yoke, or contact prods, into the weldment to be inspected, as sketched in Fig. 2-18 (prods shown). This induces a magnetic field in the work, which will be distorted by any cracks, seams, inclusions, etc. located on or near (within approx. 0.1 in. of) the surface. A dry magnetic powder, blown lightly on the surface by a rubber squirt bulb, will be picked up at such discontinuities, making a distinct mark. The magnetically held particles show the location, size, and shape of the discontinuity.

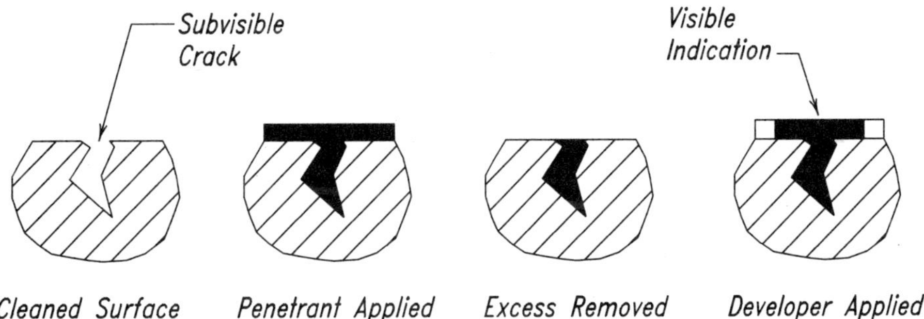

Fig. 2-17. Schematic diagram of liquid dye penetrant test for cracks, courtesy American Welding Society.

The method will indicate surface cracks that might be difficult for liquid penetrant to enter. Records may be kept by picking up the powder pattern with clear plastic tape. Cleanup is easy, but demagnetizing, if necessary, may not be. If the magnetizing prod is lifted from the work while the current is still on, an arc strike which could lead to cracking could result. If arc strikes occur, they should be ground out.

Ultrasonic (UT)

The ultrasonic (UT) inspection process is analogous to radar. A short pulse of high frequency sound is broadcast from a crystal into a metal, after which the crystal waits to receive reflections from the far end of the metal member and from any voids encountered on the way through. The technique is called pulse echo. The sound beam is produced by a piezoelectric transducer energized by an electric current which causes the crystal to vibrate and transmit through a liquid couplant into the metal. Any reflections are displayed as pips on a cathode ray tube (CRT) grid whose horizontal scale represents distance through the metal. The vertical scale represents the strength (or area) of the reflecting surface. The system is shown schematically in Figure 2-19.

The accuracy of ultrasonic inspection is highly dependent upon the skill and training of the operator and frequent calibration of the instrument. There is a "dead" area beneath most transducers that makes it difficult to inspect members less than $\frac{5}{16}$-in. in thickness. Austenitic stainless steels, and extremely coarse-grained steels, e.g., electroslag welds, are difficult to inspect; but on structural carbon and low alloy steels the process can detect flat discontinuities (favorably oriented for reflection) smaller than $\frac{1}{64}$-in. The crystal, which is $\frac{3}{8}$-in. to 1 inch in size, can be readily moved about to check many orientations and can project the beam into the metal at angles of 90°, 70°, 60°, and 45°. With the latter three angles, the beam can be bounced around inside the metal, producing echoes from any discontinuity on the way. For more information see Krautkramer (1977) and Institute of Welding (1972).

Ultrasonic testing is a more versatile, rapid, and economical inspection method than radiography, but it does not provide a permanent record like the X-ray negative. The operator, instead, makes a written record of discontinuity indications appearing on his CRT. Certain joint geometries limit the use of the ultrasonic method.

Radiographic (RT)

Radiographic testing is basically an X-ray film process. To be detected by radiography, a crack must be oriented roughly parallel to the impinging radiation beam, and occupy

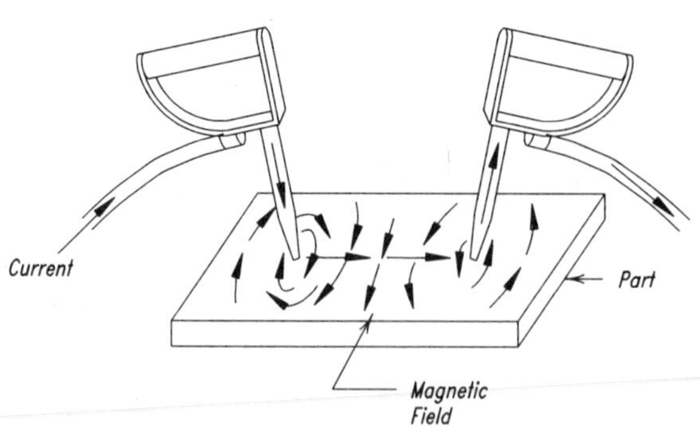

Fig. 2-18. Schematic diagram of magnetic particle test, courtesy American Welding Society.

Table 2-3.
Characteristics of Common Weld Inspection Methods

Inspection Method	Characteristics and Applications	Limitations
Visual (VT)	Most common, most economical. Particularly good for single pass welds.	Detects surface imperfections only, except when VT is done before welding and determination of acceptable fit-up, root gap, bevel angles, surface condition, etc. can be made to improve overall quality.
Dye Penetrant (PT)	Will detect tight cracks, open to surface.	Detects surface imperfections only. Deep weld ripples and scratches may give false indications.
Magnetic Particle (MT)	Will detect surface cracks and subsurface cracks to about 0.1-in. depth with proper magnetization. Indications can be "lifted" and preserved on clear plastic tape.	Requires relatively smooth surface. Careless use of magnetizing prods may leave arc-strike defects.
Ultrasonic (UT)	Detects cracks in any orientation, slag, lack of fusion, inclusions, lamellar tears, voids. Can detect a favorably oriented planar reflector smaller than $\frac{1}{32}$-in. Regularly calibrate on $\frac{1}{16}$-in. dia. drilled hole. Can scan almost any commercial thickness.	Surface must be smooth. Equipment must be frequently calibrated. Operator must be qualified. Certain geometric configurations give false indications of flaws.
Radiographic (RT)	Detects porosity, slag, voids, cracks, irregularities, lack of function. Film negative is permanent record.	Defects must occupy more than about $1\frac{1}{2}$ percent of thickness to register. Only cracks parallel to impinging beam register. Radiation hazard. Exposure time increases with thickness.

about $1\frac{1}{2}$ percent of the metal thickness along that beam. There are problems with radiographs of fillets, tee, and corner joints, however, because the radiation beam must penetrate varying thicknesses.

Precautions for avoiding radiation hazards interfere with shop work, and equipment and film costs make it the most expensive inspection method. Ultrasonic systems have gradually supplemented and even supplanted radiography.

APPLICATION OF WELDING INSPECTION METHODS TO HSS CONSTRUCTION
The following is a summary of the application of welding inspection method to HSS.

1. Visual inspection is the most commonly used method for achieving specified structural quality in HSS fabrication. Critical joints are examined prior to the

commencement of welding to check fit-up, preparation bevels, gaps, alignment, and other variables. After the joint is welded, it is then visually inspected in accordance with AWS D1.1. If a discontinuity is suspected, either the weld is repaired or other inspection methods are used to validate the integrity of the weld. In most cases, timely visual inspection by an experienced inspector is sufficient and offers the most practical and effective inspection alternative to other, more costly methods.

2. Dye penetrant examination tends to be messy and slow, but can be helpful when determining the extent of a defect found by visual inspection. This is especially true when a defect is being removed by gouging or grinding for the repair of a weld to assure that the defect is completely removed.

3. Magnetic particle examination can be useful when a defect is suspected from visual inspection or when the absence of cracking in areas of high restraint must be confirmed.

4. Ultrasonic examination has limited applicability in HSS fabrication. Relatively thin sections and variations in joint geometry can lead to difficulties in interpreting the signals, although technicians with specific experience on weldments similar to those to be examined may be able, in some instances, to decipher UT readings. Similarly,

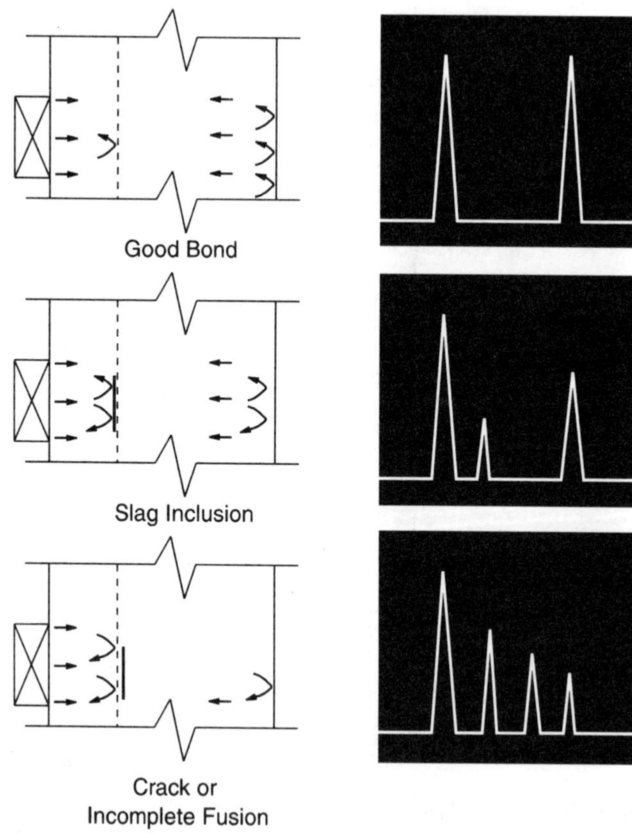

Fig. 2-19. Variations in UT reflections due to differences in acoustic properties caused by defects at the boundary.

UT is usually not suitable for use with fillet welds and smaller PJP groove welds. CJP groove welds with and without backing bars also give readings that are subject to differing interpretations. Ultrasonic examination may be specified to validate the integrity of CJP groove welds that are subject to tension. Ultrasonic examination has largely replaced radiographic examination for the inspection of critical CJP groove welds.

5. Radiographic examination has very limited applicability in HSS fabrication because of the irregular shape of common joints and the resulting variations in thickness of material as projected onto film. RT can be used successfully for butt splices, but can only provide limited information about the condition of fusion at backing bars near the root corners. The general inability to place either the radiation source or the film inside the HSS means that exposures must usually be taken through both the front and back faces of the section with the film attached to the outside of the back face. Several such shots progressing around the member are needed to examine the complete joint.

REFERENCES

American Welding Society, 1986, AWS B1.10, *Guide for the Non-Destructive Inspection of Welds*, AWS, Miami, FL.

ANSI/AWS D1.1-96 1996. *Structural Welding Code—Steel*, American Welding Society, Miami, FL.

Institute of Welding, 1972, *Procedures and Recommendations for the Ultrasonic Testing of Butt Welds*, London, England.

Krautkramer, J., 1977. *Ultrasonic Testing of Materials*, 2nd. ed., Springer-Verlag, Berlin, Germany.

Packer, J. A., and Henderson, J. E., 1993, *Design Guide for Hollow Structural Section Connections*, Canadian Institute of Steel Construction.

Post, J. W., 1990, *Box Tube Connections; Choices of Joint Details and their Influence on Cost*, Proceedings of the 1990 AISC National Steel Conference, AISC, Chicago, IL.

International Institute of Welding, 1984, *Welding of Tubular Structures—Proceedings of the Second International Conference*, Pergamon Press, New York, NY

Kloiber, L. A., 1993, "Fabricating Architecturally Exposed Tube Structures", *Proceedings of the 1993 AISC National Steel Construction Conference*, AISC, Chicago, IL

CHAPTER 3

BOLTS

OVERVIEW

Chapter 3 contains general information and design considerations for bolting and other mechanical fasteners relative to their application to HSS.

Following is a detailed list of the topics considered.

BOLTING TO HSS

Although welding is the most frequently used method for making attachments to HSS, there are a variety of methods for using mechanical fasteners. These have had varying degrees of research and field use. The methods that could be considered for a particular application depend on the magnitude of the load to be transferred and the thickness of the HSS. For the methods that are described, there are three limit states that may apply to the HSS in addition to the limit states for fasteners.

1. Shear bearing

$$\phi R_n = \phi 2.4 n F_u dt$$

where

ϕ = 0.75
n = number of fasteners
d = fastener diameter
F_u = tensile strength of HSS
t = thickness of HSS

2. Pull-out from tension in one fastener

$$\phi R_n = \phi 0.85 d_w t F_u$$

where

ϕ = 0.5
d_w = diameter of part in contact with the HSS
F_u = tensile strength of the HSS

3. HSS wall distortion

This strength limit state is determined from yield-line theory applied to the HSS wall. Since it depends on the pattern of fasteners, connected elements and type of loading, a general equation cannot be given. However, a distortion limit is likely to control before the strength limit is developed. Testing or rational analysis are suitable to determine the strength.

The limit state for shear bearing is from Section J3 of the LRFD Specification (AISC, 1993). The pull-out criteria is similar to that used for screws in Section E4.4.1 of the AISI Cold-Formed Specification (AISI, 1996). Although AISI limits d_w to ½-in., tests have shown (Korol et al, 1993) that a similar limit state and equation for the strength is also applicable for larger contact diameters for fasteners in HSS.

FASTENERS

The AISC LRFD Specification, in Section A3.3, Bolts, Washers, and Nuts, approves the use of fasteners conforming to the following ASTM specifications:

ASTM A325 *High Strength Bolts*:

Intended for general use in structural joints and available in two types with diameters from ½-in. to 1½-in.:

Type 1: Medium-carbon steel for general purpose and elevated temperatures.
Type 3: Bolts with improved atmospheric corrosion resistance and weathering characteristics compatible with those of A242 or A588 steels (Type 2 has been eliminated).

Matching ASTM A563 nuts and ASTM F436 washers should be specified including weathering qualities.

ASTM A490 *Quenched and Tempered Alloy Steel Bolts:*
Intended for general use in structural joints and available in two types with diameters ½-in. to 1½-in.:

Type 1: Bolts made of alloy steel.
Type 3: Bolts of alloy steel and having atmospheric corrosion resistance and weathering characteristics comparable to those of A242 or A588 steels (Type 2 has been eliminated).

Matching ASTM A563 nuts and ASTM F436 washers should be specified including weathering qualities. A490 bolts should not be hot-dip galvanized and caution should be exercised in using them in highly corrosive environments.

A325 and A490 bolts, A563 nuts and F436 washers are given identifying marks as indicated in Figure 3-1. A detailed description of identifying marks is given in the RCSC LRFD Specification, which is bound in the AISC LRFD Manual.

A325 and A490 bolts are produced as heavy hex bolts with dimensions as given in Table 3-1.

The ordered length of a bolt should be calculated as the grip (See Figure 3-1) plus an allowance for washers plus the allowance from Table 3-1, and rounded to the next ¼-in. Bolts longer than five inches generally are available only in ½-in. increments, except by special arrangement with the bolt manufacturer or vendor. Bolts generally are cold-headed up to about nine inches in length and hot-headed when over that length. While longer lengths may be ordered, a length of eight inches generally is the maximum stock length available.

Table 3-1 lists the standard hardened washer dimensions for use with either A325 or A490 bolts. A thickness allowance of $^5/_{32}$-in. for circular and $^5/_{16}$-in. for beveled washers should be provided for each washer used. Clipped washers are available for use in areas of tight clearance. RCSC requires a $^5/_{16}$-in. minimum thickness washer under A490 bolts over 1-in. diameter when used in slotted or oversize holes in external plies. Additionally, the RCSC Specification has specific requirements on the use of washers, e.g., on A490 bolts, on slotted holes, and on beveled surfaces. The user should make reference to the appropriate section.

High strength bolts in tension devices, such as hangers, and bolts in certain shear connections must be pretensioned. RCSC specifies the pretension as a function of bolt type and diameter. These are tabulated in Table 3-2. RCSC approves four different installation methods and the reader is referred to that specification for details. It is important to note, however, that RCSC disapproves of any published torque-tension relationship.

ASTM A307 *Carbon Steel Externally and Internally Threaded Standard Fasteners*:
Intended for general use and available in two grades with diameters of ¼-in. to four inches:

Grade A: For general applications.
Grade B: For flanged piping joints where one or both flanges are cast iron.

Use of this fastener is discussed in the AISC LRFD Specification in the following sections:

Table J3.2: Design Strength of Fasteners
Sect. J3.7: When used in combined shear and tension.
Sect. J1.9: Used in combination with welds

Three radial lines @ 60°
Three radial lines @ 120°, optional
A325
A325
A325
Type 3 **
Type 2
Type 1
BOLT HEAD MARKING
See council specification for rules giverning use of washers
Grip
Washer face
Hex head
Length
Thread
Dia.
Hex nut *
ASTM A325 BOLT
Standard marking
Type 1 & 2
Type 3 **
NUT MARKING
Alt. marking "D" or "DH" for Type 1 & 2 "DH3" for Type 3
Six radial lines @ 30°
Standard bolt marking
A490
A490
A490
Type 3 **
Type 2
Type 1
Grip
Grip
See Council Specification for rules governing use of hardened washers
Thread
Length
Dia.
Hex Nut *
ASTM A490 BOLT
Type 1, 2 & 3 **
Nut marking "DH" for Type 1 & 2 "DH3" for Type 3
HARDENED WASHER (Beveled)
12
2
HARDENED WASHER (Plain) **
Marking for washer used with Type 3 bolt
* Nuts may be washer faced as shown, or chamfered on both sides.
** All bolt heads, Nuts & Washers shall include mftr's. identification symbol. The mftr. may also add other marks indicating that the nut or bolt is corrosion resistant and of a weathering type. Nuts are mftrd. to ASTM A563. Washers are mftrd. to ASTM F436.

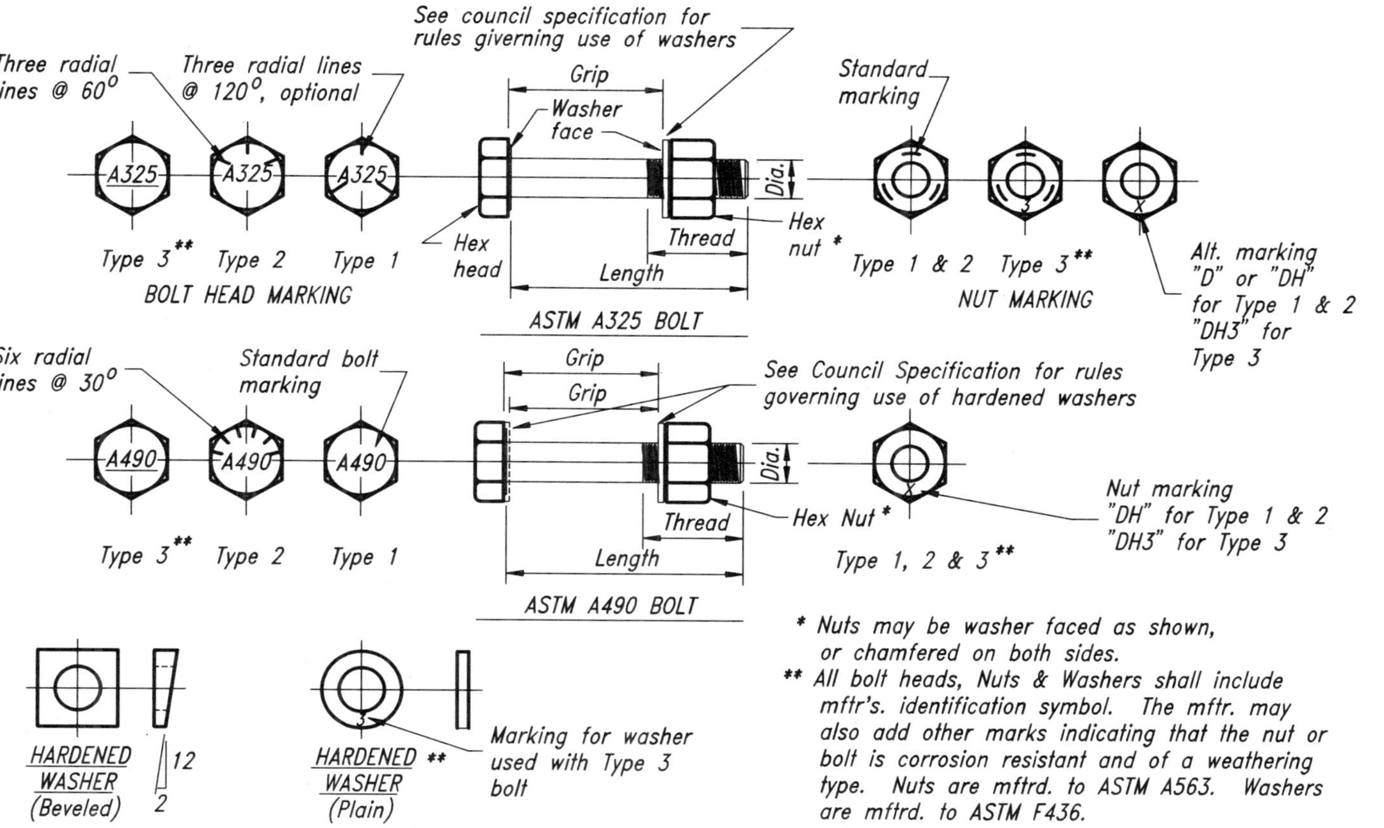

Figure 3-1.

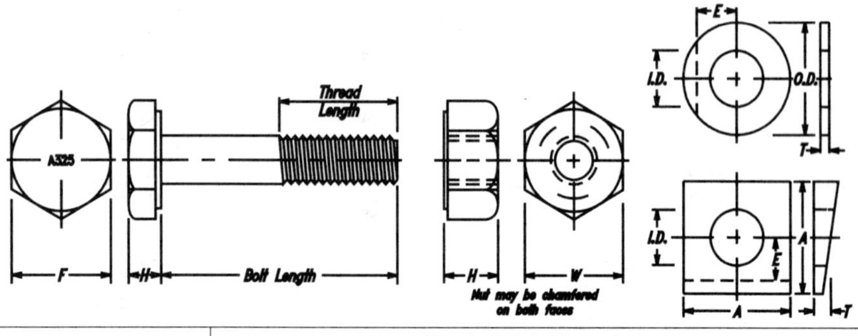

Table 3-1.
Dimensions of High-Strength Fasteners, in.

	Measurement	½	⅝	¾	⅞	1	1⅛	1¼	1⅜	1½
		\multicolumn{9}{c}{Nominal Bolt Diameter, in.}								
A325 and A490 Bolts [a]	Width Across Flats F	⅞	1¹⁄₁₆	1¼	1⁷⁄₁₆	1⅝	1¹³⁄₁₆	2	2³⁄₁₆	2⅜
	Height H	⁵⁄₁₆	²⁵⁄₆₄	¹⁵⁄₃₂	³⁵⁄₆₄	³⁹⁄₆₄	¹¹⁄₁₆	²⁵⁄₃₂	²⁷⁄₃₂	¹⁵⁄₁₆
	Thread Length	1	1¼	1⅜	1½	1¾	2	2	2¼	2¼
	Bolt Length [f] =Grip + →	¹¹⁄₁₆	⅞	1	1⅛	1¼	1½	1⅝	1¾	1⅞
A563 Bolts [b]	Width Across Flats W	⅞	1¹⁄₁₆	1¼	1⁷⁄₁₆	1⅝	1¹³⁄₁₆	2	2³⁄₁₆	2⅜
	Height H	³¹⁄₆₄	³⁹⁄₆₄	⁴⁷⁄₆₄	⁵⁵⁄₆₄	⁶³⁄₆₄	1⁷⁄₆₄	1⁷⁄₃₂	1¹¹⁄₃₂	1¹⁵⁄₃₂
F436 Circular Washers [c]	Nom. Outside Diameter OD	1¹⁄₁₆	1⁵⁄₁₆	1¹⁵⁄₃₂	1¾	2	2¼	2½	2¾	3
	Nom. Inside Diameter ID	¹⁷⁄₃₂	¹¹⁄₁₆	¹³⁄₁₆	¹⁵⁄₁₆	1⅛	1¼	1⅜	1½	1⅝
	Thckns. T — Max.	0.097	0.122	0.122	0.136	0.136	0.136	0.136	0.136	0.136
	Thckns. T — Min.	0.177	0.177	0.177	0.177	0.177	0.177	0.177	0.177	0.177
	Min. Edge Distance E [d]	⁷⁄₁₆	⁹⁄₁₆	²¹⁄₃₂	²⁵⁄₃₂	⅞	1	1³⁄₃₂	1⁷⁄₃₂	1⁵⁄₁₆
F436 Square or Rect. Washers [c,e]	Min. Side Dimension A	1¾	1¾	1¾	1¾	1¾	2¼	2¼	2¼	2¼
	Mean Thckns. T	⁵⁄₁₆	⁵⁄₁₆	⁵⁄₁₆	⁵⁄₁₆	⁵⁄₁₆	⁵⁄₁₆	⁵⁄₁₆	⁵⁄₁₆	⁵⁄₁₆
	Taper in Thickness	2:12	2:12	2:12	2:12	2:12	2:12	2:12	2:12	2:12
	Min. Edge Distance E [d]	⁷⁄₁₆	⁹⁄₁₆	²¹⁄₃₂	²⁵⁄₃₂	⅞	1	1³⁄₃₂	1⁷⁄₃₂	1⁵⁄₁₆

a   Tolerances as specified in ASTM A325 and A490.
b   Tolerances as specified in ASTM A563.
c   ASTM F436 Washer Tolerances, in.:
    Nominal Outside Diameter   −1/32; +1/32
    Nominal Diameter of Hole   −0; +1/32
    Flatness: max. deviation from straight-edge placed on cut side shall not exceed   0.010
    Concentricity: center of hole to outside diameter (full indicator runout)   0.030
    Burr shall not project above immediately adjacent washer surface more than   0.010
d   For clipped washers only.
e   For use with American standard beams (S) and channels (C).
f   Tabular value does not include thickness of washer(s).

<table>
<tr><td colspan="3" align="center">Table 3-2.
Fastener Tension</td></tr>
<tr><td rowspan="2" align="center">Nominal Bolt Size,
Inches</td><td colspan="2" align="center">Minimum Fastener Tension* in Thousands
of Pounds (kips)</td></tr>
<tr><td align="center">A325 Bolts</td><td align="center">A490 Bolts</td></tr>
<tr><td align="center">½</td><td align="center">12</td><td align="center">15</td></tr>
<tr><td align="center">⅝</td><td align="center">19</td><td align="center">24</td></tr>
<tr><td align="center">¾</td><td align="center">28</td><td align="center">35</td></tr>
<tr><td align="center">⅞</td><td align="center">39</td><td align="center">49</td></tr>
<tr><td align="center">1</td><td align="center">51</td><td align="center">64</td></tr>
<tr><td align="center">1⅛</td><td align="center">56</td><td align="center">80</td></tr>
<tr><td align="center">1¼</td><td align="center">71</td><td align="center">102</td></tr>
<tr><td align="center">1⅜</td><td align="center">85</td><td align="center">121</td></tr>
<tr><td align="center">1½</td><td align="center">103</td><td align="center">148</td></tr>
<tr><td colspan="3">* Equal to 70 percent of specified minimum tensile strengths of bolts, rounded off to the nearest kip.</td></tr>
</table>

Sect. J1.11: Limitations on Bolted and Welded Connections
Sect. J3.11: Long grips

Typical bolt head and nut dimensions are given in AISC LRFD Manual Part 8. Thread lengths are also listed and are calculated as $2D + \frac{1}{4}$-in. for bolts up to six inches in length and $2D + \frac{1}{2}$-in. for bolts over six inches long, where D is the bolt diameter. These thread lengths are longer than for high-strength bolts.

ASTM A449 *Quenched and Tempered Steel Bolts and Studs*:
This is a fastener that may be hot-dip galvanized and is available headed or unheaded in diameters to three inches.

The AISC Specification permits the use of this material in bearing-type connections for bolt diameters greater than $1\frac{1}{2}$-in., i.e., for diameters not available as A325 or A490 bolts. This material is acceptable for high-strength anchor rods or threaded rods in any available diameter for tension applications (AISC LRFD Specification Section A3.4).

If this material is used as a bolt in tension or in a bearing-type shear connection and is to be tensioned to more than 50 percent of its minimum specified tensile strength (dependent on diameter), the nuts (ASTM A563) must meet the requirements for those for A325 bolts and a hardened washer (ASTM F436) must be installed under the head (AISC LRFD Specification Sect. J3.1).

While this material appears to be the equal of A325, there are two important differences in the fasteners: (a) A449 bolts are not produced to the same inspection and quality assurance requirements as A325, and (b) they are not produced to the same heavy hex head and nut dimensions.

Two other ASTM specifications which are frequently used for their strength or size are:

ASTM A354 *Quenched and Tempered Alloy Steel Bolts, Studs, and Other Externally Threaded Fasteners*:
Available to 4-in. diameter.

ASTM A687 *High-Strength Nonheaded Steel Bolts and Studs*:

Available in diameters of 1¼-in. to 3 inches and with enhanced Charpy V-notch impact properties.

ASTM A354, A449, and A687 fasteners are not generally recognized under the AISC LRFD Specification for structural applications requiring full pretensioning during installation. Thus, for fully tensioned applications, the specifier should carefully review these three ASTM specifications to determine suitability for the intended application. Nuts and washers also should be selected from the above referenced ASTM Specifications A563 and F436 to match the strength requirements. It also may be appropriate to specify the dimensions of the heavy hex head and the heavy hex nut for certain applications.

Section 2(d) of the RCSC specifications permits the use of other fasteners when they meet the requirements as outlined therein. One type is a tension-control bolt which is installed by a special tool which twists off the splined end when the proper tension is achieved.

BOLT SPECIFICATIONS

The reader is encouraged to review and understand the RCSC *LRFD Specification for Structural Joints Using ASTM A325 or A490 Bolts.*

Two points in the RCSC specifications that will simplify inspection requirements are worthy of note:

1. Section 2(b): ". . . The length of bolts shall be such that the end of the bolt will be flush with or project beyond the face of the nut when properly installed." Inspection controversy will be reduced by recognizing that bolts intentionally have a limited thread length, a manufacturing tolerance and limited length increments. The requirement for "stick-through" as is sometimes written in project specifications increases the risk of the nut jamming on thread run-out, thus preventing tightening. It will not enhance the performance of the bolt.
2. Commentary Sect. C9, first paragraph: "It is apparent from the commentary on installation procedures that the inspection procedures giving the best assurance that bolts are properly installed and tensioned is provided by inspector observation of the calibration testing of the fasteners using the selected installation procedure followed by monitoring of the work in progress to assure that the procedure that was demonstrated to provide the specified tension is routinely adhered to. When such a program is followed, no further evidence of proper bolt tension is required."

Attention is directed to the LRFD Specification for Structural Joints Using ASTM A325 or A490 Bolts (June 3, 1994). There are many revisions since the Specification of August 14, 1980. Two of the most important are:

a. The calibrated wrench method of installation was reinstated as an accepted procedure. Although the method is the least accurate, it remains very popular with erectors.
b. "Snug-tight" bearing-type connections were introduced.

Economical Considerations

Since the ratio of material cost to strength of ASTM A490 bolts is nearly equal to that of ASTM A325 bolts, it might seem more cost effective to reduce the number of bolts in a given connection by specifying ASTM A490 bolts. However, ASTM A490 bolts are more difficult to tighten and raise inventory and quality control issues associated with the use of multiple fastener grades; mixing of ASTM A325 and A490 bolts of the same diameter should be avoided to assure that the ASTM A490 bolts are installed in the proper locations. Thus, the net benefit of specifying ASTM A490 bolts may be less than expected.

Similarly, cost ratios between grades of alternative design bolts will vary from those of conventional high-strength bolts. Thus, the decisions regarding fastener selection will be influenced accordingly.

Regardless of the bolt type selected, the normal sizes of $\frac{3}{4}$-in., $\frac{7}{8}$-in., and 1-in. diameter are usually preferred. Diameters above one inch are not commonly available, nor are they practical since special tools may be required to achieve fully tensioned installation.

Bearing-type connections should be specified whenever possible. Slip-critical connections with coatings other than clean mill scale incur appreciable extra costs associated with blasting, painting, drying, assembling, reblasting, and abrasion touch-up. If slip-critical connections are required for the proper serviceability of the structure, care should be taken to avoid requiring the faying surfaces to be masked as this also contributes great expense; coatings which provide a Class A or Class B slip coefficient may be an economical alternative to masking.

Entering and Tightening Clearances

The assembly of high-strength bolted connections requires clearance for entering and tightening the bolts with an impact wrench. The clearance requirements for conventional high-strength bolts are as given in Table 3-3. When high-strength tension-control bolts are specified, the entering and tightening clearances are as specified in Table 3-4.

DESIGN STRENGTH OF BOLTS

The design strength of bolts is determined in accordance with the provisions of LRFD Specification Section J3, which are based upon the provisions of the RCSC Specification.

For bolts in bearing-type connections subjected to shear only, the limit states of bolt shear strength and bearing strength at bolt holes must be checked. For bolts in bearing-type connections subjected to tension only, the limit state of bolt tensile strength, including the effect of prying action, must be checked. For bolts in bearing-type connections subjected to combined shear and tension, the limit states of bolt tensile strength, including the effects of both the bolt shear stress present and prying action, and bearing strength at bolt holes must be checked.

For bolts in slip-critical connections subjected to shear only, the limit states of slip resistance, bolt shear strength, and bearing strength at bolt holes must be checked. For bolts in slip-critical connections subjected to combined shear and tension, the limit states of slip resistance, including the effect of the tensile force present, bolt shear strength, and bearing strength at bolt holes must be checked.

Bolt Shear Strength

As illustrated in Figure 3-2a, this limit state considers a shear failure of the bolt shank on plane **cdef**. Since there is one shear plane, the bolt is in single shear (S). Additional plies of material may increase the number of shear planes and, therefore, the shear strength of the bolt. This condition, as illustrated in Figure 3-2b, is called double shear (D).

High-strength bolts may be specified with the threads included (N) or excluded (X) from the shear plane of the connection. Note that the shear strength of bolts with the threads included is about 25 percent less than that of bolts with the threads excluded. In spite of this, many designers prefer to specify N bolts when possible due to the difficulty in assuring that threads are excluded from the shear plane in the as-built condition. If however, the threads are to be excluded from the shear plane, care must be taken to provide a bolt of sufficient overall length given the thread length and required bolt length from Table 3-1. Note that washers may be required to accomplish this.

From LRFD Specification Section J3.6, the design bolt shear strength is ϕR_n, where $\phi = 0.75$ and:

Table 3-3.
Entering and Tightening Clearances, in.
Conventional ASTM A325 and A490 Bolts

Aligned Bolts

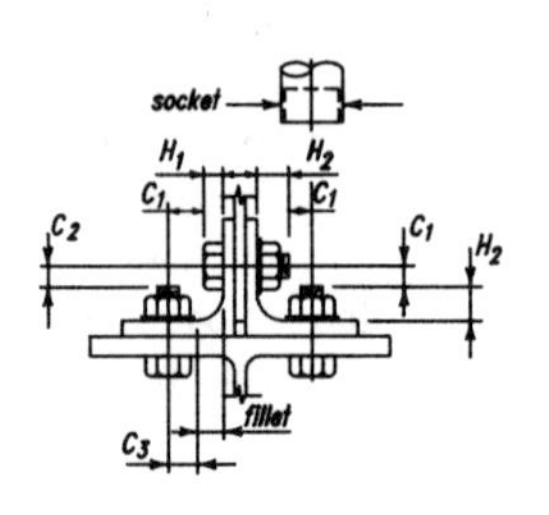

Nominal Bolt Dia., in.	Socket Dia., in.	H_1	H_2	C_1	C_2	C_3 Circular	C_3 Clipped
$5/8$	$1\tfrac{3}{4}$	$25/64$	$1\tfrac{1}{4}$	1	$11/16$	$11/16$	$9/16$
$3/4$	$2\tfrac{1}{4}$	$15/32$	$1\tfrac{3}{8}$	$1\tfrac{1}{4}$	$3/4$	$3/4$	$11/16$
$7/8$	$2\tfrac{1}{2}$	$35/64$	$1\tfrac{1}{2}$	$1\tfrac{3}{8}$	$7/8$	$7/8$	$13/16$
1	$2\tfrac{5}{8}$	$39/64$	$1\tfrac{5}{8}$	$1\tfrac{7}{16}$	$15/16$	1	$7/8$
$1\tfrac{1}{8}$	$2\tfrac{7}{8}$	$11/16$	$1\tfrac{7}{8}$	$1\tfrac{9}{16}$	$1\tfrac{1}{16}$	$1\tfrac{1}{8}$	1
$1\tfrac{1}{4}$	$3\tfrac{1}{8}$	$25/32$	2	$1\tfrac{11}{16}$	$1\tfrac{1}{8}$	$1\tfrac{1}{4}$	$1\tfrac{1}{8}$
$1\tfrac{3}{8}$	$3\tfrac{1}{4}$	$27/32$	$2\tfrac{1}{8}$	$1\tfrac{3}{4}$	$1\tfrac{1}{4}$	$1\tfrac{3}{8}$	$1\tfrac{1}{4}$
$1\tfrac{1}{2}$	$3\tfrac{1}{2}$	$15/16$	$2\tfrac{1}{4}$	$1\tfrac{5}{16}$	$1\tfrac{5}{16}$	$1\tfrac{1}{2}$	$1\tfrac{5}{16}$

Staggered Bolts

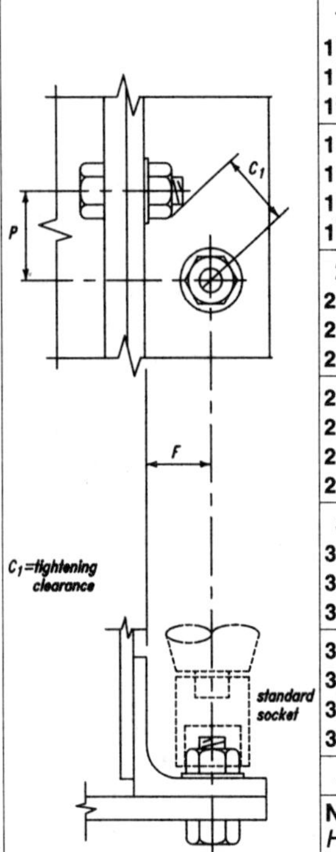

	Stagger P, in.							
	Nominal Bolt Diameter, in.							
F	$5/8$	$3/4$	$7/8$	1	$1\tfrac{1}{8}$	$1\tfrac{1}{4}$	$1\tfrac{3}{8}$	$1\tfrac{1}{2}$
1	$1\tfrac{5}{8}$							
$1\tfrac{1}{8}$	$1\tfrac{1}{2}$							
$1\tfrac{1}{4}$	$1\tfrac{1}{2}$	$1\tfrac{15}{16}$						
$1\tfrac{3}{8}$	$1\tfrac{7}{16}$	$1\tfrac{7}{8}$	$2\tfrac{3}{16}$					
$1\tfrac{1}{2}$	$1\tfrac{1}{4}$	$1\tfrac{13}{16}$	$2\tfrac{1}{8}$	$2\tfrac{5}{16}$				
$1\tfrac{5}{8}$	$1\tfrac{1}{4}$	$1\tfrac{3}{4}$	$2\tfrac{1}{16}$	$2\tfrac{5}{16}$	$2\tfrac{9}{16}$			
$1\tfrac{3}{4}$	$1\tfrac{3}{16}$	$1\tfrac{11}{16}$	2	$2\tfrac{1}{4}$	$2\tfrac{9}{16}$	$2\tfrac{13}{16}$	3	
$1\tfrac{7}{8}$	$1\tfrac{1}{8}$	$1\tfrac{9}{16}$	$1\tfrac{15}{16}$	$2\tfrac{3}{16}$	$2\tfrac{1}{2}$	$2\tfrac{3}{4}$	3	$3\tfrac{3}{4}$
2	1	$1\tfrac{1}{2}$	$1\tfrac{13}{16}$	$2\tfrac{1}{8}$	$2\tfrac{7}{16}$	$2\tfrac{3}{4}$	$2\tfrac{15}{16}$	$3\tfrac{1}{4}$
$2\tfrac{1}{8}$	$13/16$	$1\tfrac{3}{8}$	$1\tfrac{11}{16}$	2	$2\tfrac{3}{8}$	$2\tfrac{11}{16}$	$2\tfrac{15}{16}$	$3\tfrac{3}{16}$
$2\tfrac{1}{4}$		$1\tfrac{1}{4}$	$1\tfrac{9}{16}$	$1\tfrac{7}{8}$	$2\tfrac{1}{4}$	$2\tfrac{5}{8}$	$2\tfrac{7}{8}$	$3\tfrac{3}{16}$
$2\tfrac{3}{8}$		$1\tfrac{1}{8}$	$1\tfrac{1}{2}$	$1\tfrac{3}{4}$	$2\tfrac{1}{8}$	$2\tfrac{1}{2}$	$2\tfrac{13}{16}$	$3\tfrac{1}{8}$
$2\tfrac{1}{2}$		$7/8$	$1\tfrac{3}{8}$	$1\tfrac{5}{8}$	2	$2\tfrac{7}{16}$	$2\tfrac{3}{4}$	$3\tfrac{1}{16}$
$2\tfrac{5}{8}$			$1\tfrac{3}{16}$	$1\tfrac{1}{2}$	$1\tfrac{15}{16}$	$2\tfrac{5}{16}$	$2\tfrac{7}{8}$	3
$2\tfrac{3}{4}$			$15/16$	$1\tfrac{3}{8}$	$1\tfrac{7}{8}$	$2\tfrac{1}{8}$	$2\tfrac{1}{2}$	$2\tfrac{7}{8}$
$2\tfrac{7}{8}$				$1\tfrac{3}{16}$	$1\tfrac{3}{4}$	$2\tfrac{1}{16}$	$2\tfrac{3}{8}$	$2\tfrac{13}{16}$
3				$7/8$	$1\tfrac{5}{8}$	2	$2\tfrac{1}{4}$	$2\tfrac{11}{16}$
$3\tfrac{1}{8}$					$1\tfrac{1}{2}$	$1\tfrac{7}{8}$	$2\tfrac{1}{8}$	$2\tfrac{1}{2}$
$3\tfrac{1}{4}$					$1\tfrac{1}{4}$	$1\tfrac{3}{4}$	2	$2\tfrac{3}{8}$
$3\tfrac{3}{8}$					$15/16$	$1\tfrac{5}{8}$	$1\tfrac{15}{16}$	$2\tfrac{1}{4}$
$3\tfrac{1}{2}$						$1\tfrac{3}{8}$	$1\tfrac{3}{4}$	$2\tfrac{1}{8}$
$3\tfrac{5}{8}$						$1\tfrac{1}{16}$	$1\tfrac{9}{16}$	2
$3\tfrac{3}{4}$							$1\tfrac{5}{16}$	$1\tfrac{7}{8}$
$3\tfrac{7}{8}$								$1\tfrac{11}{16}$
4								$1\tfrac{3}{8}$

Notes:
H_1 = height of head, in.
H_2 = maximum shank extension,* in.
C_1 = clearance for tightening, in.
C_2 = clearance for entering, in.
C_3 = clearance for fillet,* in.
P = bolt stagger, in.
F = clearance for tightening staggered bolts, in.
*Based on one standard hardened washer.

Table 3-4.
Entering and Tightening Clearances, in.
Tension-Control ASTM A325 and A490 Bolts

Aligned Bolts

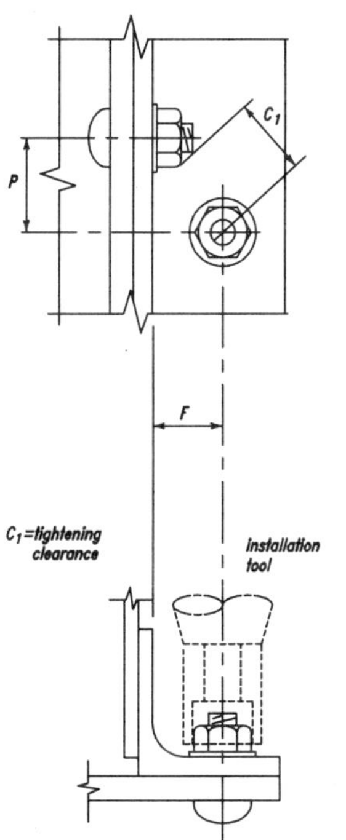

Tools	Nominal Bolt Dia, in.	H_1	H_2	C_1	C_2	C_3 Circular	C_3 Clipped
Large Tools	$3^3/_8$-in. Diameter Critical						
	$3/_4$	$1/_2$	$1^3/_8$	$1^7/_8$	$7/_8$	$3/_4$	—
	$7/_8$	$9/_{16}$	$1^1/_2$	$1^7/_8$	1	$7/_8$	—
	1	$5/_8$	$1^3/_4$	$1^7/_8$	$1^1/_8$	1	—
	$2^1/_2$-in. Diameter Critical						
	$3/_4$	$1/_2$	$1^3/_8$	$1^3/_8$	$7/_8$	$3/_4$	—
	$7/_8$	$9/_{16}$	$1^1/_2$	$1^3/_8$	1	$7/_8$	—
	1	$5/_8$	$1^3/_4$	$1^3/_8$	$1^1/_8$	1	—
Small Tools	3-in. Diameter Critical						
	$5/_8$	$7/_{16}$	$1^1/_4$	$1^5/_8$	$13/_{16}$	$11/_{16}$	—
	$3/_4$	$1/_2$	$1^3/_8$	$1^5/_8$	$7/_8$	$3/_4$	—
	$7/_8$	$9/_{16}$	$1^1/_2$	$1^5/_8$	1	$7/_8$	—
	$2^3/_{16}$-in. Diameter Critical						
	$5/_8$	$7/_{16}$	$1^1/_4$	$1^1/_8$	$13/_{16}$	$11/_{16}$	—
	$3/_4$	$1/_2$	$1^3/_8$	$1^1/_8$	$7/_8$	$3/_4$	—
	$7/_8$	$9/_{16}$	$1^1/_2$	$1^1/_8$	1	$7/_8$	—

Staggered Bolts

F	Stagger P, in. — Nominal Bolt Diameter, in. $5/_8$	$3/_4$	$7/_8$	1
$1^1/_4$	$1^{13}/_{16}$			
$1^3/_8$	$1^3/_4$	$2^1/_{16}$	$2^1/_4$	$2^7/_{16}$
$1^1/_2$	$1^{11}/_{16}$	2	$2^3/_{16}$	$2^3/_8$
$1^5/_8$	$1^9/_{16}$	$1^7/_8$	$2^1/_{16}$	$2^1/_4$
$1^3/_4$	$1^1/_2$	$1^{13}/_{16}$	2	$2^3/_{16}$
$1^7/_8$	$1^7/_{16}$	$1^3/_4$	$1^7/_8$	$2^1/_8$
2	$1^5/_{16}$	$1^5/_8$	$1^3/_4$	2
$2^1/_8$	$1^1/_4$	$1^9/_{16}$	$1^{11}/_{16}$	$1^{15}/_{16}$
$2^1/_4$	$1^3/_{16}$	$1^1/_2$	$1^9/_{16}$	$1^7/_8$
$2^3/_8$	$1^1/_8$	$1^3/_8$	$1^1/_2$	$1^3/_4$
$2^1/_2$	1	$1^5/_{16}$	$1^3/_8$	$1^{11}/_{16}$
$2^5/_8$		$1^3/_{16}$	$1^5/_{16}$	$1^9/_{16}$
$2^3/_4$		$1^1/_8$	$1^3/_{16}$	$1^1/_2$
$2^7/_8$			$1^1/_8$	$1^3/_8$
3				$1^5/_{16}$
$3^3/_8$				$1^5/_{16}$

Notes:

H_1 = height of head, in.
H_2 = maximum shank extension,* in.
C_1 = clearance for tightening, in.
C_2 = clearance of entering, in.
C_3 = clearance for fillet,* in.
P = bolt stagger, in.
F = clearance for tightening staggered bolts, in.

*Based on one standard hardened washer.

$$R_n = (F_v A_b)n$$

In the above equation, n is the number of bolts in the connection, F_v is the nominal shear strength, and A_b is the nominal bolt area. For convenience, the design bolt shear strengths of various bolts are summarized in Table 3-5; design bolt shear strengths of vertical rows of n bolts are summarized in Table 3-6.

Bearing Strength at Bolt Holes

As illustrated in Figure 3-3, this limit state considers both a tear fracture of the connected material and deformation around the bolt holes. Bearing strength is a function of the material being connected, the type of bolt hole, and the spacing and edge distance; it is independent of both the type of bolt and the presence or absence of threads on the bearing area.

From LRFD Specification Section J3.10, when deformation around the bolt holes is a design consideration for standard holes, oversized holes, short-slotted holes, and long-slotted holes parallel to the line of force, the design bearing strength at bolt holes is ϕR_n, where $\phi = 0.75$ and, for two or more bolts in the line of force, when $L_e \geq 1.5d$ and $s \geq 3d$:

$$R_n = (2.4dtF_u)n$$

For a single bolt in the line of force or when $L_e < 1.5d$ or $s < 3d$:

$$R_n = L_e tF_u \text{ for end bolt}$$

$$= \left(s - \frac{d}{2}\right)tF_u \text{ for each remaining bolt}$$

but no individual bolt bearing strength can exceed $R_n = (2.4dtF_u)$.

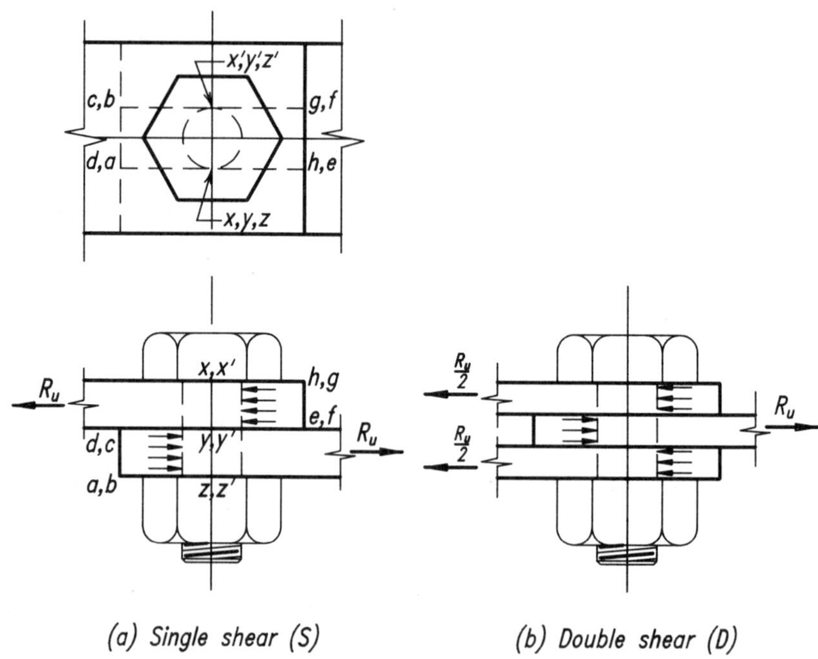

(a) Single shear (S)   (b) Double shear (D)

Fig. 3-2. Bolt Shear.

Table 3-5.
Design Shear Strength of One Bolt, kips

ASTM Desig.	Thread Cond.	ϕF_v (ksi)	Loading	Nominal Bolt Diameter d, in.							
				5/8	3/4	7/8	1	1 1/8	1 1/4	1 3/8	1 1/2
				Nominal Bolt Area, in.^{2}							
				0.3068	0.4418	0.6013	0.7854	0.9940	1.227	1.485	1.767
A325	N	36.0	S	11.0	15.9	21.6	28.3	35.8	44.2	53.5	63.6
			D	22.1	31.8	43.3	56.5	71.6	88.4	107	127
	X	45.0	S	13.8	19.9	27.1	35.3	44.7	55.2	66.8	79.5
			D	27.6	39.8	54.1	70.7	89.5	110	134	159
A490	N	45.0	S	13.8	19.9	27.1	35.3	44.7	55.2	66.8	79.5
			D	27.6	39.8	54.1	70.7	89.5	110	134	159
	X	56.3	S	17.3	24.9	33.9	44.2	56.0	69.1	83.6	99.5
			D	34.5	49.7	67.7	88.4	112	138	167	199
A307	—	18.0	S	5.52	7.95	10.8	14.1	17.9	22.1	26.7	31.8
			D	11.0	15.9	21.6	28.3	35.8	44.2	53.5	63.6

N = Threads included in shear plane
X = Threads excluded from shear plane
S = Single shear
D = Double shear

Table 3-6.
Design Shear Strength of n Bolts in Double Shear*

	ASTM A325						ASTM A490					
	N			X			N			X		
n	3/4	7/8	1	3/4	7/8	1	3/4	7/8	1	3/4	7/8	1
12	382	520	679	477	649	848	477	649	848	596	812	1060
11	350	476	622	437	595	778	437	595	778	547	744	972
10	318	433	565	398	541	707	398	541	707	497	676	884
9	286	390	509	358	487	636	358	487	636	447	609	795
8	254	346	452	318	433	565	318	433	565	398	541	707
7	223	303	396	278	379	495	278	379	495	348	474	619
6	191	260	339	239	325	424	239	325	424	298	406	530
5	159	216	283	199	271	353	199	271	353	249	338	442
4	127	173	226	159	216	283	159	216	283	199	271	353
3	95.4	130	170	119	162	212	119	162	212	149	203	265
2	63.6	86.6	113	79.5	108	141	79.5	108	141	99.4	135	177
1	31.8	43.3	56.5	39.8	54.1	70.7	39.8	54.1	70.7	49.7	67.7	88.4

N = Threads included in shear plane
X = Threads excluded in shear plane
*For design strength of bolts in single shear, divide tabular value by 2.

Table 3-7.
Design Bearing Strength at Bolt Holes, kips/in. thickness
Two or more holes in line of force with $L_e \geq 1.5d$
and $s \geq 3d$; hole deformation considered*

Hole Type	F_u, ksi	Nominal Bolt Diameter d, in.							
		$5/8$	$3/4$	$7/8$	1	$1\tfrac18$	$1\tfrac14$	$1\tfrac38$	$1\tfrac12$
		1.5d							
		$15/16$	$1\tfrac18$	$1\tfrac{5}{16}$	$1\tfrac12$	$1\tfrac{11}{16}$	$1\tfrac78$	$2\tfrac{1}{16}$	$2\tfrac14$
		3d							
		$1\tfrac78$	$2\tfrac14$	$2\tfrac58$	3	$3\tfrac38$	$3\tfrac34$	$4\tfrac18$	$4\tfrac12$
STD, OVS SSL, LSLP	58	65.3	78.3	91.4	104	117	131	144	157
	65	73.1	87.8	102	117	132	146	161	176
	70	78.8	94.5	110	126	142	158	173	189
LSLT	58	54.4	65.3	76.1	87.0	97.9	109	120	131
	65	60.9	73.1	85.3	97.5	110	122	134	146
	70	65.6	78.8	91.9	105	118	131	144	158

STD = Standard Hole
OVS = Oversized Hole
SSL = Short-Slotted Hole
LSLP = Long-Slotted Hole parallel to line of force
LSLT = Long-Slotted Hole transverse to line of force
*When $s < 3d$, or when hole deformation is not a design consideration, refer to LRFD Specification Section J3.10. When $L_e < 1.5d$ or for one hole in the line of force, refer to Table 3-8.

Table 3-8.
Design Bearing Strength at Bolt Holes, kips/in. thickness
One hole in line of force or top bolt with $L_e < 1.5d$*

F_u, ksi	Edge Distance L_e, in.							
	1	$1\tfrac18$	$1\tfrac14$	$1\tfrac38$	$1\tfrac12$	$1\tfrac58$	$1\tfrac34$	$1\tfrac78$
58	43.5	48.9	54.4	59.8	65.3	70.7	76.1	81.6
65	48.8	54.8	60.9	67.0	73.1	79.2	85.3	91.4
70	52.5	59.1	65.6	72.2	78.8	85.3	91.9	98.4

*Design strength from Table 3-8 shall not exceed tabular value from Table 3-7. For remaining bolts, when $s - d/2 > 2.4d$, refer to Table 3-7; otherwise refer to LRFD Specification Section J3.10.

In the above equations

n = number of bolts in the connection
d = nominal bolt diameter
t = thickness in bearing
L_e = edge distance

If deformation around the bolt hole is not a design consideration, or for long-slotted holes perpendicular to the line of force, refer to LRFD Specification Section J3.10.

For convenience, the design bearing strength at bolt holes is tabulated for the foregoing conditions in Tables 3-7 and 3-8, respectively. Note that these tables may be applied to bolts with countersunk heads, by subtracting one-half the depth of the countersink from the material thickness t. This is equivalent to subtracting one-quarter the diameter of the bolt from the material thickness t.

Table 3-9.
Design Tensile Strength of Bolts, kips

ASTM Desig.	ϕF_b ksi	Nominal Bolt Diameter d, in.							
		$\frac{5}{8}$	$\frac{3}{4}$	$\frac{7}{8}$	1	$1\frac{1}{8}$	$1\frac{1}{4}$	$1\frac{3}{8}$	$1\frac{1}{2}$
		Nominal Bolt Area, in.^{2}							
		0.3068	0.4418	0.6013	0.7854	0.9940	1.227	1.485	1.767
A325	67.5	20.7	29.8	40.6	53.0	67.1	82.8	100	119
A490	84.8	26.0	37.4	51.0	66.6	84.2	104	126	150
A307	33.8	10.4	14.9	20.3	26.5	33.5	41.4	50.1	59.6

Bolt Tensile Strength

From LRFD Specification Section J3.6, when subjected to tension only, the design bolt tensile strength is ϕR_n,

where

$$\phi = 0.75$$
$$R_n = (F_t A_b)n$$

In the above equation, n is the number of bolts in the connection. For convenience, the design bolt tensile strengths of various bolts are summarized in Table 3-9. When subjected to combined shear and tension, the design bolt tensile strength is reduced by a function of the shear stress present in the bolt as specified in LRFD Specification Section J3.7.

LRFD Specification Section J3.6 states that any tension resulting from prying action must be considered in determining the required strength of the bolts. Prying action is a phenomenon (in bolted construction only) whereby the deformation of a fitting under a tensile force increases the tensile force in the bolt. The required strength per bolt is the sum of r_{ut}, the factored force per bolt due to the tensile force, and q_u, the additional tension per bolt resulting from prying action produced by deformation of the connected parts.

While the effect of prying action is considered in the design of the bolts, it is primarily a function of the connected elements; thus, the connected elements must possess adequate flexural strength and it is their stiffness which is the key to satisfactory performance. Refer to "Hanger Connections" in Part 11 of the AISC Manual for a treatment of prying action.

Slip Resistance

In slip-critical connections, the fully tensioned bolt creates resistance to slip through friction on the faying surfaces between two connected parts. This slip resistance is a function of the slip coefficient μ of the faying surface.

Clean mill scale with no coating is defined as a Class A surface with $\mu = 0.33$. Blast-cleaned surfaces with no coatings are defined as Class B surface with $\mu = 0.50$. Hot-dip galvanized and roughened surfaces are defined as Class C surfaces with $\mu = 0.35$.

Slip coefficients for all other coated blast-cleaned surfaces must be determined by the *Testing Method to Determine the Slip Coefficient Used in Bolted Joints*. Refer to Appendix A of the RCSC Specification. When the tests results in $0.33 \leq \mu < 0.50$, the coating is a Class A coating and the design slip coefficient is $\mu = 0.33$. If the test results in $\mu \geq 0.50$, the coating is a Class B coating and the design slip coefficient is $\mu = 0.50$. The surface requirements for slip-critical connections apply only to the faying surfaces and do not include the surfaces under the bolt, washer, or nut.

Bolts in slip-critical connections may be designed at either service loads or factored loads with the provisions of LRFD Specification Section J3.8. From LRFD Specification

Section J3.8a, when subjected to shear only, the resistance to slip for comparison with service loads is ϕR_n,

where

$$R_n = (F_v A_b)n$$

and $\phi = 1.0$ for standard holes, oversized holes, short-slotted holes, and long-slotted holes perpendicular to the direction of the load; $\phi = 0.85$ for long-slotted holes parallel to the direction of the load. In the above equation, n is the number of bolts in the connection.

Note that the values of F_v tabulated in LRFD Specification Table J3.6 for bolts in slip-critical connections assume Class A surfaces with $\mu = 0.33$. As stated in LRFD Specification Section J3.8a, it is permissible to increase F_v to the applicable value in the RCSC Specification for other surfaces. When subjected to combined shear and tension, the slip resistance for comparison with service loads must be reduced by the factor:

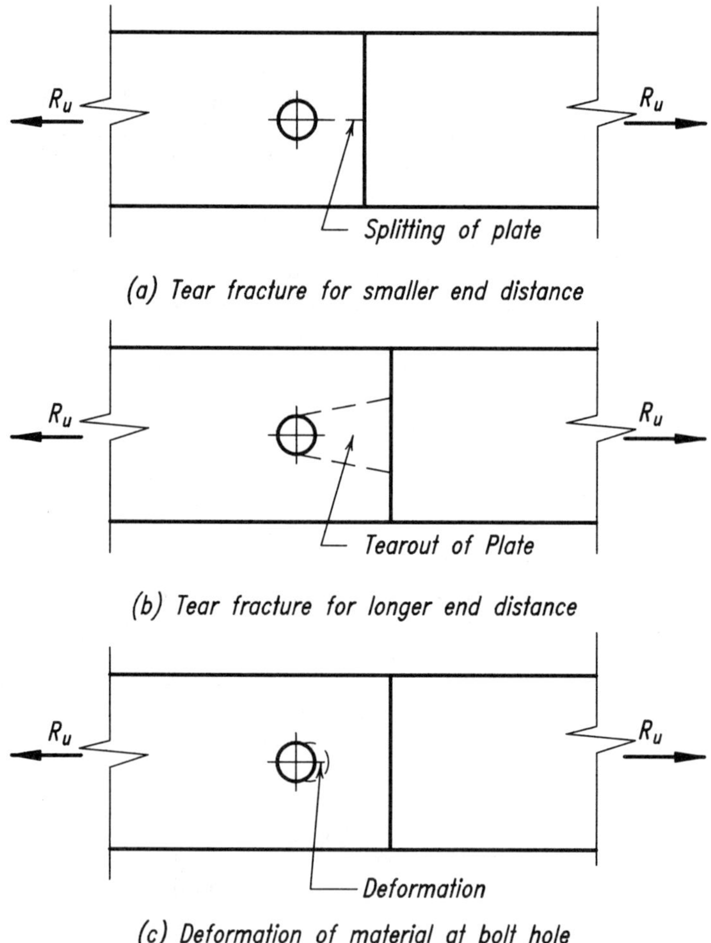

(a) Tear fracture for smaller end distance

(b) Tear fracture for longer end distance

(c) Deformation of material at bolt hole

Fig. 3-3. Bearing strength at bolt holes.

$$\left(1 - \frac{T}{T_b}\right)$$

as specified in LRFD Specification Section J3.9a, where

T = unfactored force on the connection
T_b = minimum bolt tension from LRFD Specification Table J3.1.

From LRFD Specification Appendix J3.8b, the design slip resistance for comparison with factored loads is ϕR_{str},

where

$$R_{str} = 1.13\mu T_m N_b N_s$$

and ϕ is equal to 1.0 for standard holes, 0.85 for oversized and short-slotted holes, 0.70 for long-slotted holes perpendicular to the direction of the load, and 0.60 for long-slotted holes parallel to the direction of the load. When subjected to combined tension and shear, the design slip resistance for comparison with factored loads must be reduced by the factor:

$$\left(1 - \frac{T_u}{1.13 T_m N_b}\right)$$

as specified in LRFD Specification Appendix J3.9b. In the above equations, T_u is the factored force on the connection, T_m is the minimum bolt tension from LRFD Specification Table J3.1, and N_b is the number of bolts in the connection.

ACCESS HOLES

A325, A490 or A307 bolts can be used in an HSS wall if access holes (or handholes) are cut in one of the other walls or in the bolted wall away from the connection to permit installation of the washer and nut inside the HSS. Since all plies of material are in contact within the grip of the bolt, the connection can be designed with criteria and procedures for bearing, slip-critical or tension connections. The only limit states that concern the HSS are bolt bearing in shear, tear-out due to tension and wall distortion. Although the strength of the member will be reduced due to loss of cross section at the hole, this may not be a critical concern at locations that do not require the full strength of the HSS. The access hole can be covered and sealed in a manner to reinforce the section and develop the original strength.

INSPECTION OF FULLY TENSIONED HIGH-STRENGTH BOLTS

When a joint with fully tensioned high-strength bolts is assembled, the RCSC Specification requires that all joint surfaces, including surfaces adjacent to the bolt head and nut be free of scale, except tight mill scale, and of dirt or other foreign material. Burrs need not be removed unless they prevent solid seating of the connected parts in the snug-tightened condition.

ASTM A6 lists tolerances for straightness and flatness. These tolerances can prevent the faying surfaces from sufficiently contacting in medium- to large-size connections. Section C8 of the Commentary on the RCSC Specification states: ". . . Even after being fully tightened, some thick parts with uneven surfaces may not be in contact over the entire faying surface. In itself, this is not detrimental to the performance of the joint. As long as the specified bolt tension is present in all bolts of the completed connection, the clamping force equal to the total of the tensions in all bolts will be transferred at the locations that are in contact and be fully effective in resisting slip through friction."

OTHER MECHANICAL FASTENERS

Through Bolting

Long bolts that extend through the entire HSS are satisfactory for shear connections that do not require a fully tensioned installation. The flexibility of the walls of the HSS precludes installation of fully tensioned bolts. Standard structural bolts may be used, although A449 may be required for longer lengths. The bolts are designed for shear and the only limit state involving the HSS is bearing strength for pin-type applications per LRFD Specification Section J8.

Shear bearing

$$\phi R_n = \phi 1.8 n F_y dt$$

where

ϕ = 0.75
n = number of fasteners
d = fastener diameter
F_y = yield strength of HSS
t = thickness of HSS

Pull-out from tension in one fastener

$$\phi R_n = \phi 0.85 d_w t F_u$$

where

ϕ = 0.5
d_w = diameter of part in contact with the inner surface of HSS
F_u = ultimate strength of the HSS

Blind Bolts

Special fasteners are available that eliminate the need for access to install a nut (Korol et al, 1993; Henderson, 1996.) The shank of the fastener is inserted through holes in the parts to be connected until the head bears on the outer ply. See Figure 3-4. A special wrench is used on the open side to keep the outer part of the shank from rotating and simultaneously turn the threaded part of the shank. A wedge or other mechanism on the blind side causes the fixed part of the shank to expand and form a contact with the inside of the HSS. Some fasteners contain a break-off mechanism when the fastener is fully tightened. Recent versions of these fasteners meet the requirements for a fully tensioned A325 bolt (Henderson, 1996) and could be used in slip-critical or tension conditions. HSS limit states are bolt bearing in shear, tear-out of the bolt in tension and wall distortion. Manufacturer's literature must be consulted to determine design strength values.

Flow-Drilling

Flow-drilling is a process that can be used to produce a threaded hole in an HSS to permit blind bolting when the inside of the HSS is inaccessible (Sherman, 1995; Henderson, 1996.) The process is to force a hole through the HSS with a carbide conical tool rotating at sufficient speed to produce high rapid heating which softens the material in a local area. The material that is displaced as the tool is forced through the plate forms a truncated hollow cone (bushing) on the inner surface and a small upset on the outer surface. Tools can be obtained with a milling collar so that the material on the outer surface is removed, producing a flat surface allowing parts to be brought in close contact. A cold-formed tap is then used to roll a thread into the hole without any chips or removal of material. The

| Table 3-10. HSS Thickness and Bolt Diameter Combinations | | | | | |
| HSS Thickness (in.) | Bolt Diameter (in.) | | | | |
	$\frac{1}{2}$	$\frac{5}{8}$	$\frac{3}{4}$	$\frac{7}{8}$	1
$\frac{3}{16}$	X	X			
$\frac{1}{4}$	X	X	X		
$\frac{5}{16}$		X	X	X	
$\frac{3}{8}$			X	X	X
$\frac{1}{2}$					X
X indicates that shear and tensile strengths of A325 bolts can be developed.					

resulting threaded hole has the approximate dimensions and hardness of a heavy hex nut. Shear and tension strengths of A325 bolts can be developed for certain combinations of bolt size and HSS thickness. See Table 3-10.

Drilling equipment with suitable rotational speed, torque and thrust is required, but with small sizes and thicknesses field installation with conventional tools is possible. The bolts are designed with the normal criteria and the HSS limit states are bolt bearing in shear and distortion of the HSS wall in tension. HSS strength is not affected by the process except for the reduction in area due to the holes.

Threaded Studs

Threaded studs are available in $\frac{3}{8}$-in. to $\frac{7}{8}$-in. diameters and can be shop or field welded to an HSS with a stud welding gun. The connection would be similar to a bolted connection with an external nut. The strength of the stud in tension or shear is based on manufacturers recommendations and tests. The HSS limit state is distortion of the wall.

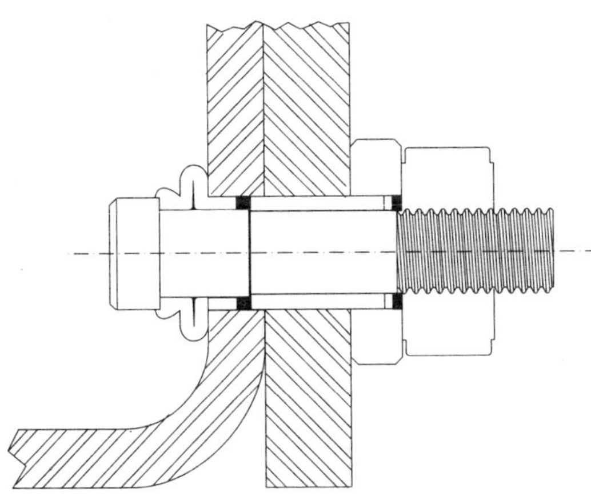

Figure 3-4.

When using threaded studs, countersunk holes must be used in the attached element to clear the weld fillet at the base of the stud.

Nailing

Power-driven nails that are installed with a powder-actuated gun are satisfactory for pure shear connections where the combined thickness of the attachment and the HSS does not exceed ½-in. This system was tested for splices between telescoping round HSS loaded with an axial force (Packer, 1996). The shear resistance of the fasteners is taken as the number of nails times the shear strength of a single nail and ignores any secondary contribution from a dimpling effect between the materials. The limit state for the HSS is shear bearing. See Packer (1996).

Screws

Self-tapping screws with or without self-drilling points are available for connecting materials with combined thicknesses up to ½-in. The screws have diameters from 0.08 to 0.25 inches. The limit states for connections in the AISI *Specification for the Design of Cold-Formed Steel Members* (AISI, 1996) are associated with bearing failure of the material or pull-out of the screw either in direct tension or after tilting occurs in a shear load. Failure of the screws themselves is prevented by requiring that the product is 25 percent stronger than the shear or tension resistance based on the material limit states. Edge distances and spacing of screws should not be less than 3 times the screw diameter, d. For attaching material with thickness t_1 and ultimate strength F_{u1} to an HSS with thickness t and strength F_u, the design strengths are ϕP_n with $\phi = 0.5$ and P_n determined as follows.

Connection shear per screw

For $t/t_1 \leq 1$, P_n is the smallest of

$$4.2(t^3 d)^{1/2} F_u$$

$$2.7 t_1 d F_{u1}$$

$$2.7 t d F_u$$

For $t/t_1 \geq 2.5$, P_n is the smaller of

$$2.7 t d F_u$$

$$2.7 t_1 d F_{u1}$$

For $1 < t/t_1 < 2.5$, P_n shall be determined by linear interpolation between the above two cases.

Connection tension per screw, P_n is the smaller of

$$0.85 t_c d F_u$$

$$1.5 t_1 d_w F_{u1}$$

where

t_c = lesser of the depth of penetration and the HSS thickness
d_w = larger of the screw head or washer diameter, and shall not be taken larger than ½-in.

REFERENCES

American Institute of Steel Construction, 1993, *Load and Resistance Design Specification for Structural Steel Buildings*, AISC, Chicago IL.

American Iron and Steel Institute, 1996, *Specification for the Design of Cold-Formed Steel Members*, AISI, Washington, D. C.

Henderson, J. E., 1996, "Bending, Bolting and Nailing of Hollow Structural Sections," *Proc. International Conference on Tubular Structures*, pp. 150–161, American Welding Society.

Korol, R. M., Ghobarah, A. and Mourad, S., 1993, "Blind Bolting W-Shape Beams to HSS Columns," *J. of Structural Engineering*, ASCE, Vol. 119, No. 12, pp. 3463–3481.

Packer, J. A., 1996, "Nailed Tubular Connections under Axial Loading," *J. of Structural Engineering*, ASCE, Vol. 122, No. 8, pp. 867–872.

Sherman, D. R., 1995, "Simple Framing Connections to HSS Columns," *Proc. National Steel Construction Conference*, American Institute of Steel Construction, 30-1 to 30-16.

CHAPTER 4

SIMPLE SHEAR CONNECTIONS

OVERVIEW

Chapter 4 contains general information, design examples, and design tables for simple shear connections between wide-flange beams and HSS columns. All of the connection types can be used with rectangular HSS, while round HSS can only accommodate the single-plate and through-plate connections. The information is based on the limit states used in Part 9 of the AISC LRFD Manual, related sections of the HSS Specification and references for applications to HSS columns.

Following is a detailed list of the general information and the connection types considered.

INTRODUCTION

Many of the familiar simple shear connections that are used to connect wide-flange beams to wide-flange columns can be used with HSS columns. These include double and single angles, unstiffened and stiffened seats, single plates (shear tabs) and tee connections. One additional connection that is unique for HSS columns is the through-plate; note that this alternative is seldom required structurally and presents a significant economic penalty when a single plate connection would otherwise suffice. Variations in attachments are more limited with HSS columns since the connecting element will typically be shop welded to the HSS and bolted to the supported beam. Except for seated connections, the bolting will be to the web of a wide-flange or other open profile section. Coping is not required except for bottom-flange copes that facilitate knifed erection with double angle connections.

To assist with stability during erection, it is recommended that the minimum length of elements connected to a beam web be one-half the T-dimension of the beam to be supported. The maximum length must be compatible with the T-dimension. The connecting element may encroach on the fillets by $\frac{1}{8}$-in. to $\frac{5}{16}$-in., depending upon the radius of the fillets; refer to Table 4-1.

The advantages of the various types of connections, erectability considerations, shop and field practices that are contained in Part 9 of the AISC LRFD Manual also apply when HSS columns are used. However, the erection difficulties when beams frame opposite each other and share the same open holes are not a factor with HSS columns.

In selecting the type of connection, consideration must be given to the fact that small HSS columns may not have the width of typical W-shape column flanges or webs, thereby restricting the width of the connecting element to something less than standard. This applies to the length of an angle leg connected to the column or the width of the flange of a tee. Since these situations will involve relatively small shear forces, a single plate connection is a good choice. Tests have shown (Sherman, 1995) that, if the flat width to thickness ratio of the HSS does not exceed $1.4(E/F_y)^{0.5}$ so that the section is not classified as slender, it is not necessary to provide a through-plate to maintain the column strength.

Tables presented herein for the design strength of a particular type of connection are based on the least applicable limit state. Since the tables do not indicate the controlling limit state it is not always possible to make a simple proportional correction for variations of the standard dimensions used to generate a table; e.g. the dimension from the face of the HSS to the bolt line. In these situations or when design strength tables do not exist, the connection design must be based on an evaluation of the appropriate limit states.

LIMIT STATES FOR SIMPLE SHEAR CONNECTIONS

There are a number of limit states associated with the bolts, connecting elements, welds and beam webs that are considered in the design of the simple shear connections in Part 9 of the AISC LRFD Manual. One limit state for the HSS that applies to all the connections is the shear strength of the wall adjacent to a weld. Punching shear through the HSS wall is a limit state that applies to single plate connections. A yield line mechanism, with one exception, is not a limit state since the end rotations of beams supported at both ends are limited and are insufficient to develop the mechanism in the HSS. However, tests of stiffened seated connections to the webs of wide-flange columns (Sputo and Ellifritt, 1991) have shown that a yield line mechanism may be an applicable limit state and forms the basis of limitations of column sizes in Part 9 of the AISC LRFD Manual. This situation (connecting to a web that is supported for a long length on two edges) is similar to that for a rectangular HSS column. Thus the yield line mechanism is considered as a limit state for the stiffened seated connection.

Methods of designing the various types of connections have been developed separately based on experimental programs, judgment and experience. A major factor in establishing

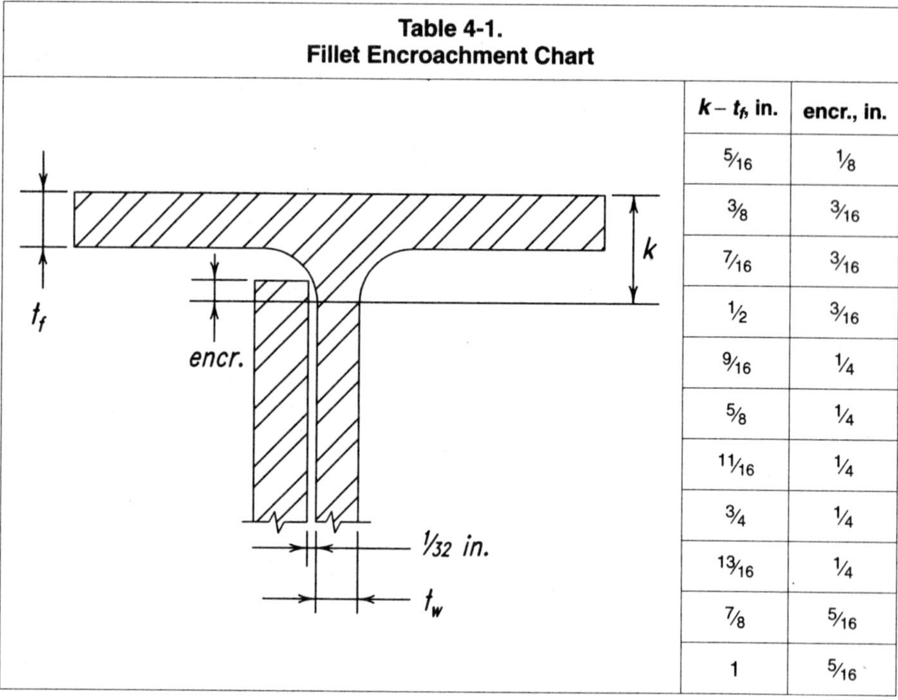

Table 4-1.
Fillet Encroachment Chart

$k - t_f$, in.	encr., in.
$\frac{5}{16}$	$\frac{1}{8}$
$\frac{3}{8}$	$\frac{3}{16}$
$\frac{7}{16}$	$\frac{3}{16}$
$\frac{1}{2}$	$\frac{3}{16}$
$\frac{9}{16}$	$\frac{1}{4}$
$\frac{5}{8}$	$\frac{1}{4}$
$\frac{11}{16}$	$\frac{1}{4}$
$\frac{3}{4}$	$\frac{1}{4}$
$\frac{13}{16}$	$\frac{1}{4}$
$\frac{7}{8}$	$\frac{5}{16}$
1	$\frac{5}{16}$

the design procedures is how the eccentricity is treated; whether the shear is considered to be at the weld, the bolt line or some other location. When the welds carry the eccentric moment in addition to the shear, a linear vector analysis is used in some cases while an ultimate strength analysis is used in others. Shear yielding of the connecting material is a limit state for all of the connections, as is net section shear rupture for connections bolted to the web of the beam. Limit states associated with flexure in the connecting material are considered in only a few of the connections. Table 4-2 summarizes the limit states that are considered in the design of the various connections. The numbers in Table 4-2 refer to the following list of equations that are used for the limit states. The notations used in the equations are defined in Figure 4-1 for connections that are bolted to the beam web and in Figure 4-2 for the unstiffened and stiffened seated connections.

The following equations are used to evaluate the limit states indicated in Table 4-2. Table 4-3 indicates the reference sources for the various equations.

Bolts

For n bolts in single line:

Shear with no eccentricity

$$\phi R_n = \phi k n A_b F_v \tag{4-1}$$

where

$\phi = 0.75$
$A_b = \pi d_b^2 / 4$
F_v = shear strength for threads -N or -X
$k = 2$ for double angles
$k = 1$ for single angle

Table 4-2.
Limit States for the Various Connections

Connection Type	Tee	Double Angle	Unstiffened Seat	Stiffened Seat	Single Plate	Through Plate	Single Angle[b]
Bolts							
Shear with no eccentricity		1[a]					1
Shear by ultimate analysis	2				2	2	
Connection Material							
Bolt Bearing, $L_{ev} \geq 1.5d,\ s \geq 3d$	4	3			4	4	3
Gross shear at yield	5	5	5	5	5	5	5
Net section shear rupture	6	6			6	6	6
Flexural yield	7		8				
Flexural rupture	9						
Block shear rupture	10	10			10	10	10
Rotation ductility	11				11	11	
Buckling					12	12	
Welds							
Direct shear	13					13	
Shear by vector analysis		14	15	16			
Shear by ultimate analysis					17[c]		17
Ductility	18						
Tube Wall							
Shear at weld	19	19	19	19	19	19	19
Punching shear					20		
Yield line				21			
Beam Web							
Bolt bearing	4	4			4	4	3
Web local yield			8	22			
Web crippling			23	23			
Block shear rupture			never controls with no cope				
Strength at cope			see LRFD Manual pages 8-225 to 8-236				

[a] numbers refer to equation numbers in this chapter (e.g. 4-1).
[b] single line of bolts.
[c] See discussion on page 4-97.

<table>
<tr><td colspan="2" align="center">Table 4-3.
Basis for Equations</td></tr>
<tr><td>1.</td><td>LRFD (AISC, 1993) J3.6 for n bolts.</td></tr>
<tr><td>2.</td><td>Manual (AISC, 1994) page 8-28 and associated tables</td></tr>
<tr><td>3.</td><td>LRFD (AISC, 1993) J3.10 for n bolts. Also used for beam web.</td></tr>
<tr><td>4.</td><td>Manual (AISC, 1994) page 8-28 and tables using LRFD (AISC 1993) J3.10 for bolt resistance. Also used for beam web.</td></tr>
<tr><td>5.</td><td>LRFD (AISC, 1993) J5.3</td></tr>
<tr><td>6.</td><td>LRFD (AISC, 1993) J4.1</td></tr>
<tr><td>7.</td><td>Manual (AISC, 1994) Example 9-16</td></tr>
<tr><td>8.</td><td>Carter et al. (1997)</td></tr>
<tr><td>9.</td><td>Manual (AISC, 1994) Example 9-16</td></tr>
<tr><td>10.</td><td>LRFD (AISC, 1993) J4.3 (b) since shear rupture will control in practical designs.</td></tr>
<tr><td>11.</td><td>Manual (AISC, 1994) page 9-170</td></tr>
<tr><td>12.</td><td>Manual (AISC, 1994) page 9-148</td></tr>
<tr><td>13.</td><td>LRFD (AISC, 1993) J2.4</td></tr>
<tr><td>14.</td><td>(Salmon and Johnson, 1995, page 806, Equation 13.2.8) model proposed by Blodgett and used in Manual (AISC, 1994)</td></tr>
<tr><td>15.</td><td>(Salmon and Johnson, 1995, page 807, Equation 13.2.16)</td></tr>
<tr><td>16.</td><td>(Salmon and Johnson, 1995, page 820, Equation 13.4.13)</td></tr>
<tr><td>17.</td><td>Manual (AISC, 1994) page 8-154 and associated tables</td></tr>
<tr><td>18.</td><td>Manual (AISC, 1994) page 9-170</td></tr>
<tr><td>19.</td><td>Manual (AISC, 1994) page 9-15 develops a minimum thickness below which the shear strength of the base metal controls. The solution for $(D_{eff})_{max}$ can be used in the weld limit state equations for vector or ultimate analysis to determine limit state for the HSS wall.</td></tr>
<tr><td>20.</td><td>(Sherman and Ales, 1991) develops a criteria for the maximum thickness of a single plate to prevent punching shear in the HSS wall. This has been modified to include resistance factors.</td></tr>
<tr><td>21.</td><td>In the design procedure for stiffened seat connections to W-column webs, (Sputo and Ellifritt, 1991) includes a yield line limit state based on an analysis by (Abolitz and Warner, 1965). This has been applied to the HSS wall which is also supported on two edges. However, since the HSS side supports are the same thickness rather than much heavier as in the case of W-section flanges, the equation (Abolitz and Warner, 1965) for rotationally free edge supports has been used instead of fixed.</td></tr>
<tr><td>22.</td><td>LRFD (AISC, 1993) K1.3</td></tr>
<tr><td>23.</td><td>LRFD (AISC, 1993) K1.4</td></tr>
</table>

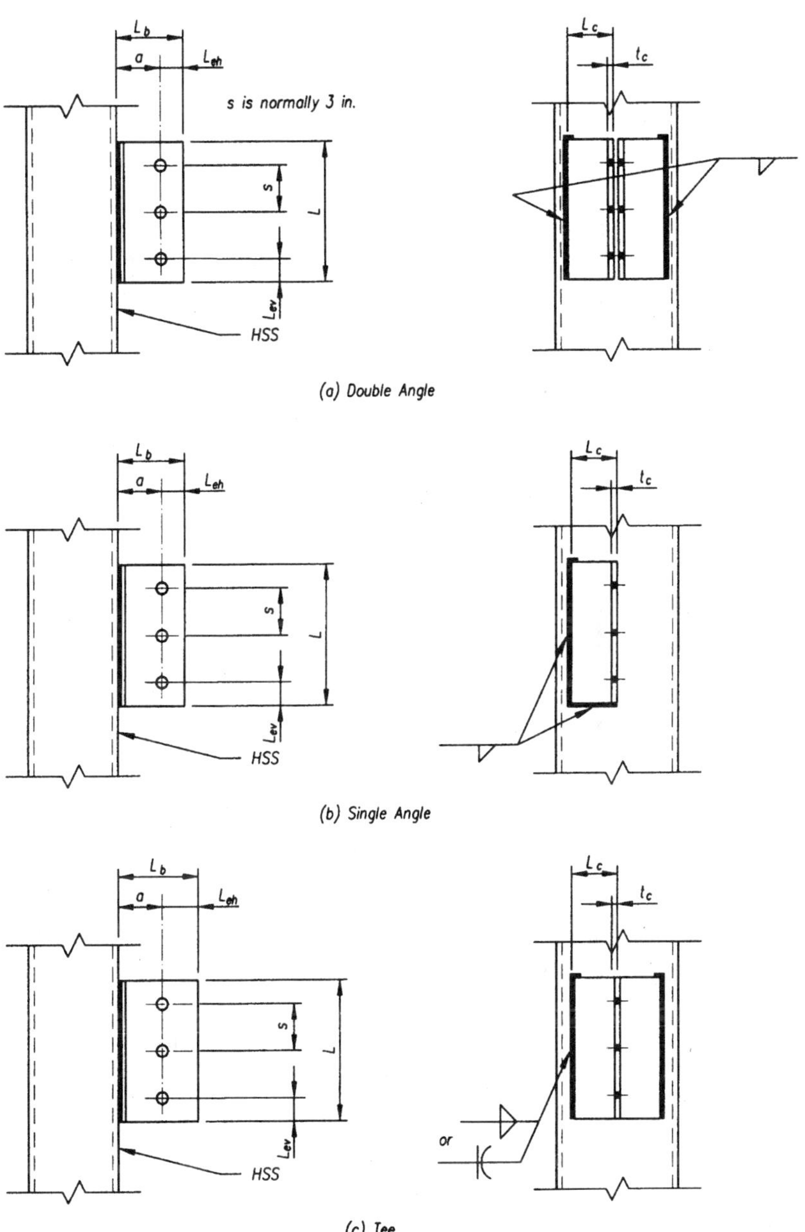

*Fig. 4-1. Notation for connections bolted to the beam web
(continued on next page).*

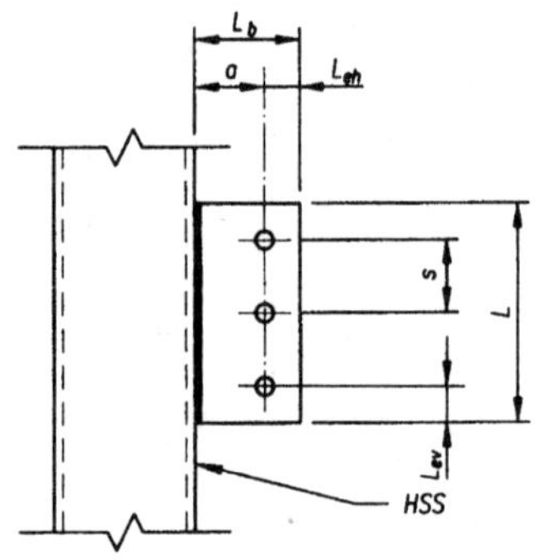

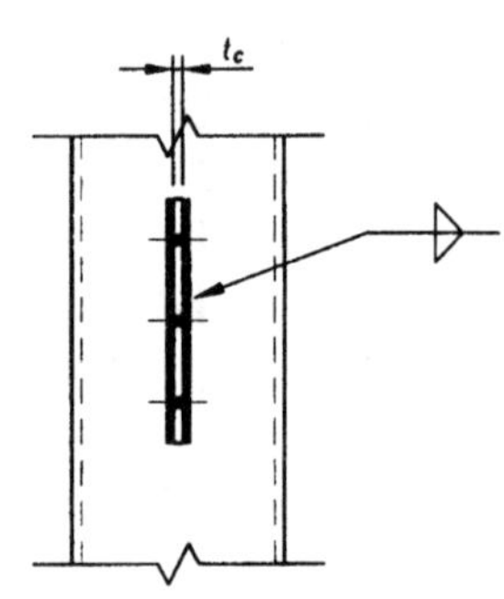

(d) Single Plate

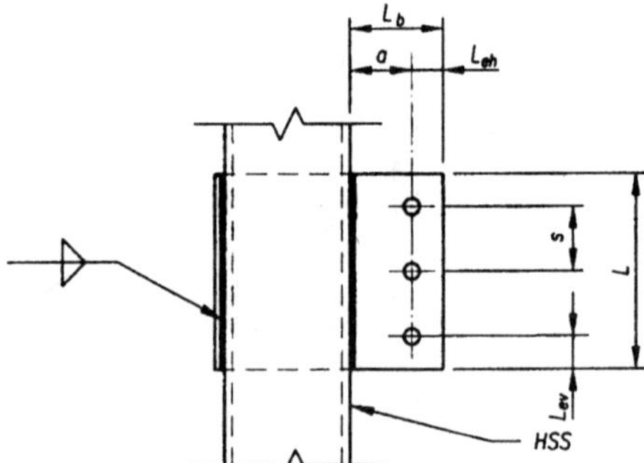

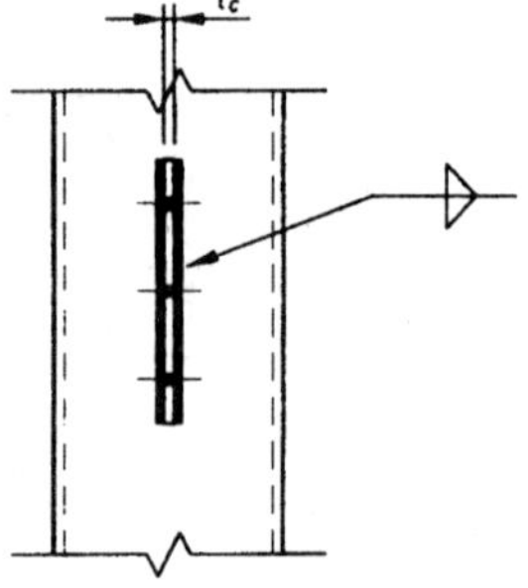

(e) Through Plate

A_b = bolt area
a = distance from weld line to bolt line
C = constants from tables for eccentric shear on welds or bolts
D = $\frac{1}{16}$-in. of weld size
d_b = bolt diameter
d_h = hole diameter
e = shear eccentricity as defined for connection
F_{EXX} = weld strength
F_u = HSS tensile strength
F_{uc} = connection material tensile strength
F_y = HSS yield stress
F_{yb} = beam material yield stress
F_{yc} = connection material yield stress

F_v = bolt shear strength
L = length of connection in the direction of loading
L_b = width of projecting element of connection
L_c = width of angle leg against column or $b_f/2$ for WT shape
L_{eh} = horizontal edge distance
L_{ev} = vertical edge distance
L_s = shear rupture length
L_t = tension rupture length
n = number of bolts
s = bolt spacing
t = HSS thickness
t_c = angle, plate or tee stem thickness

Fig. 4-1 (cont.).

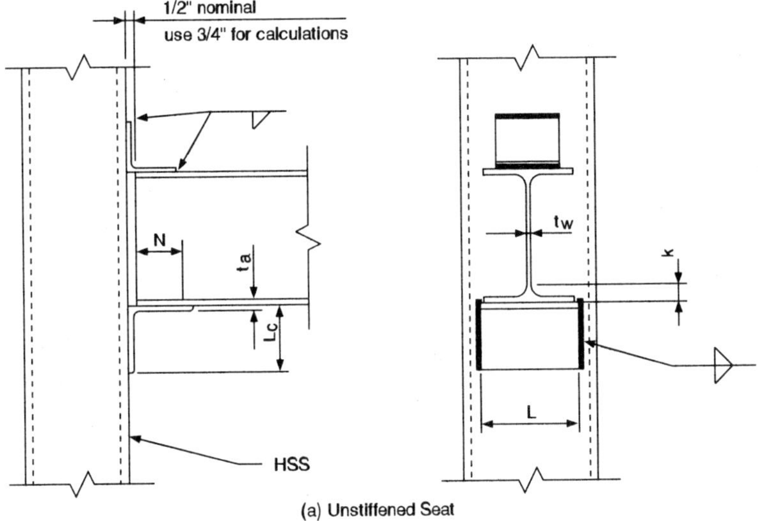

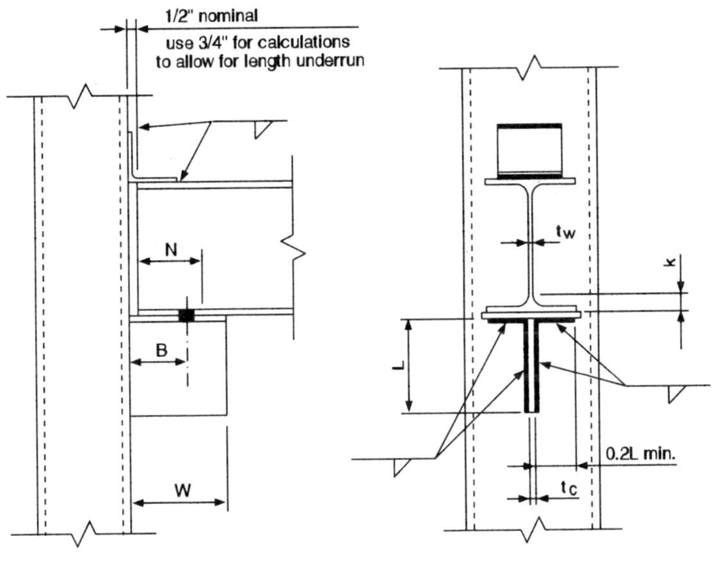

$$b_{\max} = \frac{W}{2} \geq 2\text{-}5/8 \text{ in.}$$

$$B_{\max} = W / 2 \geq 2\tfrac{5}{8}\text{-in.}$$

$$t_c \text{ is greater of } (F_{yb} / F_{yc})t_w,$$

$$2D / 16 \text{ for } F_{yc} = 36 \text{ ksi, or}$$

$$1.5D / 16 \text{ for } F_{yc} = 50 \text{ ksi}$$

$$W = \text{seat length}$$

Fig. 4-2. Notation for seated connections.

Shear with eccentricity by ultimate analysis

$$\phi R_n = CA_b F_v \qquad \qquad 4\text{-}2$$

where

$A_b = \pi d_b^2 / 4$
F_v = shear strength for threads -N or -X
C from Table 8-18 (AISC, 1994) includes $\phi = 0.75$ and n

   eccentricity $e = a$ for tee connections

   eccentricity varies for single plates and through-plates

      for flexible support and standard holes

$$e = |(n-1) - a| \geq a$$

      for flexible support and short-slotted holes

$$e = |2n/3 - a| \geq a$$

      for rigid support and standard holes

$$e = |(n-1) - a|$$

      for rigid support and short-slotted holes

$$e = |2n/3 - a|$$

Connecting Material

Bearing strength at bolt holes (no eccentricity) $L_{ev} \geq 1.5 d_b$ and $s \geq 3 d_b$

$$\phi R_n = \phi n(2.4 d_b k t_c F_{uc}) \qquad \qquad 4\text{-}3$$

where

$\phi = 0.75$
$k = 2$ for double angles
$k = 1$ for others

Bearing strength at bolt holes (with eccentricity)

$$\phi R_n = \phi C(2.4 d_b t_c F_{uc}) \qquad \qquad 4\text{-}4$$

Gross section shear yield

$$\phi R_n = \phi L(k t_c)(0.6 F_{yc}) \qquad \qquad 4\text{-}5$$

where

$\phi = 0.90$
L = width of unstiffened seat and length of others
$k = 2$ for double angles
$k = 1$ for others

Net section shear rupture

$$\phi R_n = \phi[L - n(d_h + \tfrac{1}{16})](k t_c)(0.6 F_{uc}) \qquad \qquad 4\text{-}6$$

where

$\phi = 0.75$

$k = 2$ for double angles
$k = 1$ for others

Flexural yielding

$$\phi R_n = \phi F_{yc} \frac{t_c L^2}{6a}$$

4-7

where

$\phi = 0.90$

$$\phi R_n = \phi F_{yc} \frac{L t_c^2}{4e} \leq \phi(0.6 F_{yc})L t_c$$

4-8a

where

$\phi = 0.9$
L = angle length
t_c = angle thickness

$$e = \frac{N}{2} + \tfrac{3}{4} - t_c - r = \frac{N}{2} + \tfrac{3}{8} - t_c$$

N = largest required bearing length determined from Equations 4-8b, 4-8c, 4-8d, and 4-8e (cannot exceed width of angle leg minus $\tfrac{3}{4}$-in.)
t_c = angle thickness
r = angle fillet radius taken as $\tfrac{3}{8}$-in.

for local web yielding,

$$N_{\min} = \frac{R_u}{F_{yw} t_w} - 2.5k = \frac{R_u - \phi R_1}{\phi R_2}$$

4-8b

$$N_{\min} = k$$

4-8c

for web crippling when $N/d \leq 0.2$,

$$N_{\min} = \frac{d}{3}\left[\frac{R_u t_f}{0.75(68 t_w^3 \sqrt{F_{yw}})} - \left(\frac{t_f}{t_w}\right)^{1.5} \right] = \frac{R_u - \phi_r R_3}{\phi_r R_4}$$

4-8d

for web crippling when $N/d > 0.2$,

$$N_{\min} = \frac{d}{4}\left[\frac{R_u t_f}{0.75(68 t_w^3 \sqrt{F_{yw}})} - \left(\frac{t_f}{t_w}\right)^{1.5} + 0.2 \right] = \frac{R_u - \phi_r R_5}{\phi_r R_6}$$

4-8e

where

d = beam depth, in.

In the above equations, ϕR_1, ϕR_2, $\phi_r R_3$, $\phi_r R_4$, $\phi_r R_5$, and $\phi_r R_6$ are constants tabulated in the factored uniform load tables in LRFD Manual Part 4.

Flexural rupture

$$\phi R_n = \phi F_{uc} \frac{t_c}{6a}\left[L^2 - \frac{s^2 n(n^2 - 1)(d_h + \tfrac{1}{16})}{L} \right]$$

4-9

where

$$\phi = 0.75$$

Block shear rupture

$$\phi = 0.75$$

$$L_s = s(n-1) + L_{ev} - (n-0.5)(d_h + \tfrac{1}{16})$$

$$L_t = L_{eh} - 0.5(d_h + \tfrac{1}{16})$$

For $0.6L_s \geq L_t$

$$\phi R_n = \phi(kt_c)[0.6F_{uc}L_s + F_{yc}L_{eh}] \qquad \text{4-10a}$$

For $0.6L_s < L_t$

$$\phi R_n = \phi(kt_c)\left\{F_{uc}L_t + 0.6F_{yc}[s(n-1) + L_{ev}]\right\} \qquad \text{4-10b}$$

where

    $k = 2$ for double angles
    $k = 1$ for all others

Rotational ductility

$$t_c \leq \frac{d_b}{2} + \tfrac{1}{16}\text{-in.} \qquad \text{4-11}$$

Buckling

$$t_c \geq \frac{L}{64} \geq \tfrac{1}{4}\text{-in.} \qquad \text{4-12}$$

Welds

Solve for D based on controlling load. For through-plate connection, multiply controlling load by $(H + e) / H$. Assume $F_{EXX} = 70$ ksi.

Direct shear (no eccentricity)

$$\phi R_n = \phi(0.6F_{EXX})\frac{D}{16\sqrt{2}}(2L) = 2.784DL \qquad \text{4-13}$$

where

$$\phi = 0.75$$

Shear with eccentricity by vector analysis

$$\phi R_n = \phi(0.6F_{EXX})\frac{D}{16\sqrt{2}}(2L) \Big/ \sqrt{1 + \frac{12.96L_c^2}{L^2}}$$

$$= \frac{2.784DL}{\sqrt{1 + \dfrac{12.96L_c^2}{L^2}}} \qquad \text{4-14}$$

where

    $\phi = 0.75$
    L_c = width of column leg of angle

$$\phi R_n = \phi(0.6F_{EXX}) \frac{D}{16\sqrt{2}} (2L) \Big/ \sqrt{1 + \frac{20.25e_w^2}{L^2}}$$

$$= \frac{2.784DL}{\sqrt{1 + \frac{20.25e_w^2}{L^2}}} \tag{4-15}$$

where

$\phi = 0.75$

e_w = eccentricity from column face to shear resultant

$L = L_c$ for unstiffened seat angle

$$\phi R_n = \phi(0.6F_{EXX}) \frac{D}{16\sqrt{2}} (2.4L) \Big/ \sqrt{16e^2 + L^2}$$

For $\phi = 0.75$, $e = 0.8W$, and each horizontal weld length = $0.2L$

$$= \frac{3.34DL}{\sqrt{1 + \frac{10.2W^2}{L^2}}} \tag{4-16}$$

Shear with eccentricity by ultimate analysis

$$\phi R_n = CDL \tag{4-17}$$

C from AISC LRFD Manual Table 8-44 for single angle welded at toe and bottom

$$e_x = L_c - 0.5L_c^2/(L + L_c)$$

C from AISC LRFD Manual Table 8-38 for single plates

Ductility

$$D \ge 16(0.0158) \frac{F_{yc} t_f^2}{b} \left(\frac{b^2}{L^2} + 2 \right) \le 16(0.75t_c) \tag{4-18}$$

where

t_f = flange thickness of tee connector

$b = (b_f - k_1)/2$ of tee connector

HSS Wall

Shear at weld (Use maximum effective weld size in Equations 4-13 through 4-17 as applicable.)

$$(D_{eff})_{max} = \frac{16\sqrt{2}(0.9F_y)}{(0.75F_{EXX})} t \tag{4-19}$$

Punching shear

$$0.9F_{yc} t_c \le 2(0.75)(0.6F_u)t$$

$$t_c \le \frac{F_u}{F_{yc}} t \tag{4-20}$$

Yield line

$$\phi R_n = \phi k \frac{t^2 F_y L}{4e}$$

4-21a

where

$\phi = 0.9$
$e = 0.8W$
$k = f[gh + m + n]$ 4-21b

$$f = \frac{1}{B - 0.2L}$$

4-21c

$$g = \left(1 + 0.661\frac{B}{L}\right)$$

4-21d

$$h = \sqrt{(B - 0.4L)(7B + 0.4L)}$$

4-21e

$$m = \frac{B(B - 0.4L)}{4L}$$

4-21f

$$n = 2L + 2.65B$$

4-21g

Beam Web

Bearing strength at bolt holes, $s \geq 3d_b$ and beam end distance $L_{eh} \geq 1\frac{1}{2}$-in.

Use Equation 4-3 or 4-4 as applicable with $t_c = t_w$

Local web yielding

$$\phi R_n = \phi(2.5k + N)F_{yw}t_w$$

4-22

where

$\phi = 1.0$

Web crippling

For $N/d \leq 0.2$,

$$\phi R_n = \phi 68 t_w^2 \left[1 + 3\left(\frac{N}{d}\right)\left(\frac{t_w}{t_f}\right)^{1.5}\right]\sqrt{\frac{F_{yw}t_f}{t_w}}$$

4-23a

For $N/d > 0.2$,

$$\phi R_n = \phi 68 t_w^2 \left[1 + \left(\frac{4N}{d} - 0.2\right)\left(\frac{t_w}{t_f}\right)^{1.5}\right]\sqrt{\frac{F_{yw}t_f}{t_w}}$$

4-23b

where

$\phi = 0.75$

TEE CONNECTIONS

Since load tables are not provided for tee connections, this type of connection will be used as an example of design using the limit state equations.

EXAMPLE 4.1—Tee Connection (Figure 4-1C)

Design a connection for a W16×50 beam to an HSS8×8×¼ column using a WT5×22.5 for a factored shear R_u = 37 kips. Both the WT and the beam are A572 Gr50 material. Use ¾-in. diameter A325-N bolts in standard holes with a bolt spacing s of 3 in., vertical edge distance L_{ev} of 1¼-in. and distance a of 3 in. from face of HSS to bolt line.

Section Properties

HSS8×8×¼	t = 0.233 in.	B = 8 in.	
	F_y = 46 ksi	F_u = 58 ksi	
W16×50	t_w = 0.380 in.	d = 16.26 in.	t_f = 0.630 in.
	F_y = 50 ksi	F_u = 65 ksi	T = 13⅝-in.
WT5×22.5	$t_c = t_w$ = 0.350 in.	d = 5.050 in.	t_f = 0.620 in.
	F_y = 50 ksi	F_u = 65 ksi	b_f = 8.020 in.
	k_1 = ¹¹⁄₁₆-in.		

Check limit on WT stem thickness, (Equation 4-11)

$$t_{c,\,max} = \frac{d_b}{2} + \frac{1}{16}$$

$$= \frac{3/4}{2} + \frac{1}{16}$$

$$= 0.438 \text{ in.} > 0.350 \text{ in.} \quad \textbf{o.k.}$$

Determine number of bolts (L_b = d of the WT)

$e = a = 3$ in.

shear per bolt

ϕr_v = 15.9 kips, (LRFD Manual Table 8-11 or Equation 4-2)

bearing per bolt on WT stem

L_{ev} = 1¼-in. > 1.5(3/4) = 1.125 in., s = 3 in. > 3(3/4) = 2.25 in.

ϕr_v = 87.8(0.350) = 30.7 kips, (LRFD Manual Table 8-13 or Equation 4-4)

Shear per bolt controls and

C_{reqd} = 37/15.9 = 2.33, (Equation 4-2)

From LRFD Manual Table 8-18 for $\theta = 0°$, s = 3 in., e = 3.00 in.

select $n = 4$, C = 2.81 > 2.33

Use 4 bolts

Determine length of WT and check associated limit states

Erection stability

$L_{min} = T/2$ = 13.625/2 = 6.8 in.

Bolt spacing and edge distances

$$L = 3(3) + 2(1.25) = 11.5 \text{ in. } (< T)$$

Try $L = 11.5$ in.

Check shear yield on stem, (Equation 4-5)

$$\phi R_n = 0.9(11.5)(0.350)(0.6 \times 50) = 109 \text{ kips} > 37 \text{ kips } \mathbf{o.k.}$$

Check shear rupture on stem, (Equation 4-6)

$$\phi R_n = 0.75[11.5 - 4(0.875)](0.350)(0.6 \times 65)$$

$$= 81.9 \text{ kips} > 37 \text{ kips } \mathbf{o.k.}$$

Check flexural yield of stem (Equation 4-7)

$$a = 3 \text{ in.}$$

$$\phi R_n = 0.9(50) \frac{0.350(11.5)^2}{6(3.00)}$$

$$= 116 \text{ kips} > 37 \text{ kips } \mathbf{o.k.}$$

Check flexural rupture of WT stem (Equation 4-9)

$$\phi R_n = 0.75(65)\frac{0.350}{6(3.00)}\left[11.5^2 - \frac{3^2(4)(4^2 - 1)(0.875)}{11.5} \right]$$

$$= 86.4 \text{ kips} > 37 \text{ kips } \mathbf{o.k.}$$

Check block shear rupture of WT stem (Equation 4-10)

$$L_s = 3(4 - 1) + 1.25 - (4 - 0.5)(0.875) = 7.19 \text{ in.}$$

$$L_{eh} = 5.050 - 3.00 = 2.05 \text{ in.}$$

$$L_t = 2.05 - 0.5(0.875) = 1.6125 \text{ in.}$$

$L_s > L_t$, use (Equation 4-10a)

$$\phi R_n = 0.75(0.350)[0.6(65)7.19 + 50(2.05)]$$

$$= 101 \text{ kips} > 37 \text{ kips } \mathbf{o.k.}$$

Beam web

$$t_w > t_c$$

0.380 in. > 0.350 in.

Beam web is satisfactory for bearing.

Determine the weld size

Since the flange width of the WT is slightly larger than the width of the HSS, a flare bevel groove weld is required. Taking the outside radius as $2(0.25) = 0.5$ in. and using AISC LRFD Specification Table J2.2, the effective throat thickness of the weld is $E = 5/16(0.5) = 0.156$ in. The equivalent fillet weld that provides the same throat dimension is

$$(D / 16)(1 / \sqrt{2}) = 0.156 \text{ or } D = 3.53$$

Check weld ductility (Equation 4-18)

$$b = [8.020 - 2(^{11}\!/_{16})] / 2 = 3.32 \text{ in.}$$

$$D_{\min} = 16(0.0158)\frac{50(0.620)^2}{3.32}\left[\frac{3.32^2}{11.5^2} + 2\right] \le 16(0.75)0.350$$

$$= 3.05 \le 4.20$$

$$D_{\min} = 3.05 < 3.53 \quad \textbf{o.k.}$$

Check shear in HSS at weld (Equation 4-19)

$$(D_{eff})_{\max} = \frac{16\sqrt{2}\ [0.9(46)]}{(0.75(70))}\ 0.233 = 4.16$$

$4.16 > 3.53$, therefore weld controls

Check weld shear (Equation 4-13)

$$\phi R_n = 2.784(3.53)11.5 = 113 \text{ kips} > 37 \text{ kips} \quad \textbf{o.k.}$$

For ease of welding an alternate design that uses a WT with a narrower flange to permit fillet welding is desirable.

EXAMPLE 4.2—Tee Connection with Fillet Welds

Redesign the connection in Example 4-1 with fillet welds on the flat width of the HSS.

Select the WT

Maximum flange width assuming ¼-in. welds and HSS corner radius equal to 1.5 times the nominal thickness $(1.5)(0.250) = 0.375$ in.

$$b_f < 8 - 2(0.375) - 2(^1\!/_4) = 6.75 \text{ in.}$$

Since all of the WT stem design strengths in Example 4-1 were more than 1.7 times the required strength 37 kips, a stem thickness greater than $0.350/1.7 = 0.206$ in. should be satisfactory.

Try a WT5×11

$t_c = t_w = 0.240$ in.	$d = 5.085$ in.	$t_f = 0.360$ in.
$F_y = 50$ ksi	$F_u = 65$ ksi	$b_f = 5.750$ in.
$k_1 = \tfrac{1}{2}$-in.		

Check that bearing at bolt holes does not control

Bearing per bolt

$$\phi R_n = 87.8(0.240) = 21.1 \text{ kips, (LRFD Manual Table 8-13 or Equation 4-4)}$$

Bearing is greater than bolt shear strength of 15.9 kips, so there is no change in bolt calculations if the distance from the HSS to the bolts is kept at 3.00 in., L_{eh} becomes $5.085 - 3.00 = 2.085$ in.

Use 4 bolts and WT length, $L = 11.5$ in.

Check WT stem limit states

The limit states in Example 4-1 are reduced by the ratio of the stem thicknesses,

0.240/0.350 = 0.686. The critical limit state in the WT stem was 86.4 kips for flexural rupture. Check that it is still satisfactory.

ϕR_n = 0.686(86.4) = 59.3 kips > 37 kips   **o.k.**

Determine the fillet weld size

Minimum size (LRFD Specification Table J2.4)

$\frac{3}{16}$-in. (D = 3) for welding to flange

Weld ductility (Equation 4-18)

b = (5.750 − 2(1/2))/2 = 2.38 in.

$$D_{min} = 16(0.0158)\frac{50 \times 0.360^2}{2.38}\left[\frac{2.38^2}{11.5^2} + 2\right] \le 16(0.75(0.240))$$

$$= 1.41 \le 2.88$$

$$D_{min} = 1.41$$

Shear in HSS at weld (Equation 4-19)

$$D_{eff\,max} = \frac{16\sqrt{2}\,[0.9(46)]}{(0.75(70))}\,0.233 = 4.16$$

Weld shear for ϕR_n = 37 kips (Equation 4-13)

$$D = 37/(2.784 \times 11.5) = 1.16$$

Minimum size controls, use $\frac{3}{16}$-in. fillet welds.

DOUBLE-ANGLE CONNECTIONS

Tables 4-4 are design aids for double-angle connections. The tables show the compatible sizes of W-beams for the various connection configurations. Based on maximum beam web thickness, maximum weld size, maximum HSS corner radius and 4-in. outstanding angle legs, double-angle connections may be used with any HSS having a width greater than or equal to 12 inches. If 3-in. outstanding angle legs are used for connections with six bolts or less, HSS with widths of 10-in. are acceptable for obtaining welds on the flat of the side. For smaller web thicknesses, welds and corner radii, it may be possible to fit the connection on widths of 10 inches if the outstanding angle legs are 4 inches and on widths of 8 inches for outstanding angle legs of 3 inches. However, these dimensions must be verified for a particular case.

Design strengths are tabulated for angle materials with F_y = 36 ksi and F_u = 58 ksi and with F_y = 50 ksi and F_u = 65 ksi. All values, including slip-critical bolt design strengths, are to be compared to factored loads. The design strengths are the lowest of the limit states from Equation 4-1 for the bolts and Equations 4-3, 4-5, 4-6, 4-8, and 4-10 for the angle material. Values are tabulated for two through twelve bolt rows of $\frac{3}{4}$-in., $\frac{7}{8}$-in. and 1-in. diameter A325 and A490 bolts at 3-in. spacing. For calculation purposes the horizontal and vertical angle edge distances are taken as $1\frac{1}{4}$-in.

The tables also include the bolt bearing limit state per inch of thickness of the beam web from Equation 4-3 for the ultimate strengths, F_{ub} of 58 and 65 ksi.

Design strengths of the welds are based on Equation 4-14 with 4-in. angle legs connected to the HSS. For connections with six or less bolts, the width of the legs may be reduced to 3 in. to permit attachment to smaller HSS. In this case the tabulated weld design strength will be conservative. Shorter leg length may restrict the rotational flexibility of the connection. Alternatively, a flare-bevel-welded detail can be used with an HSS that is narrower than the width of the angle legs. The minimum angle thickness must be the weld size plus $\frac{1}{16}$-in. but not less than the angle thickness determined for the angle and bolt limit states.

The minimum HSS thicknesses associated with the weld strengths are from Equation 4-19. If the HSS thickness is less than the minimum tabulated value, the weld strength must be reduced proportionally.

EXAMPLE 4.3—Double-Angle Connection

Use Table 4-4 to design a double-angle connection for a W36×230 beam of A572 Gr 50 steel to an HSS14×14×½ column for a factored reaction of 225 kips. Use $\frac{3}{4}$-in. diameter A325-N bolts in standard holes and A36 material for the angles.

Section Properties

HSS14×14×½	$t = 0.465$ in.	$B = 14$ in.
	$F_y = 46$ ksi	$F_u = 58$ ksi
W36×230	$t_w = 0.760$ in.	
	$F_y = 50$ ksi	$F_u = 65$ ksi

Design bolts and angles

From Table 4-4, for $\frac{3}{4}$-in. diameter A325-N bolts and angle material with $F_{yc} = 36$ ksi and $F_{uc} = 58$ ksi, select eight rows of bolts and $\frac{5}{16}$-in. angle thickness.

$\phi R_n = 254$ kips > 225 kips   **o.k.**

The W36 is in the acceptable group for angle length between T and $\frac{1}{2}T$. B of the HSS is greater than 12 inches, the minimum acceptable width for 4-in. OSL angles.

Check the beam web bearing

From the same Table with $F_{ub} = 65$ ksi

$\phi R_n = 702(0.760) = 534$ kips > 225 kips   **o.k.**

Determine the weld size

Also from the Table, a $\frac{5}{16}$-in. weld size provides $\phi R_n = 279$ kips > 225 kips.   **o.k.**

$t_{min} = 0.28$ in. < 0.465 in.   **o.k.**

Check the minimum angle thickness

Angle thickness must be greater than weld size plus $\frac{1}{16}$-in.

$t_{c\,min} = \frac{5}{16} + \frac{1}{16} = \frac{3}{8}$-in.

The angle thickness must be increased to $\frac{3}{8}$-in. to accommodate the welded legs of the double angle connection.

For Example 4.4, see page 4-86.

$F_y = 36$ ksi

$F_u = 58$ ksi

Table 4-4.
Double-Angle Connections

¾-in. Bolts 12 Rows W44	Bolt and Angle Design Strength, kips					
	ASTM Desig.	Thread Cond.	Hole Type	Angle Thickness, in.		
				$\frac{5}{16}$	$\frac{3}{8}$	$\frac{1}{2}$
	A325	N	—	382	382	382
		X	—	408	477	477
		SC Class A	STD	251	251	251
			OVS	213	213	213
			SSLT	213	213	213
		SC Class B	STD	380	380	380
			OVS	323	323	323
			SSLT	323	323	323
	A490	N	—	408	477	477
		X	—	408	489	596
		SC Class A	STD	313	313	313
			OVS	266	266	266
			SSLT	266	266	266
		SC Class B	STD	408	475	475
			OVS	383	403	403
			SSLT	403	403	403

Beam Web Design Strength per Inch Thickness, kips/in.

For $F_{ub} = 58$ ksi: 940 kips/in.
For $F_{ub} = 65$ ksi: 1,053 kips/in.

Welds: $F_{EXX} = 70$ ksi and HSS $F_y = 46$ ksi

Weld Size, in.	ϕR_n, kips	Min. HSS Thickness, in.
$\frac{3}{8}$	550	0.34
$\frac{5}{16}$	458	0.28
$\frac{1}{4}$	366	0.22

Notes:
STD = Standard holes   N = Threads included
OVS = Oversized holes   X = Threads excluded
SSLT = Short-slotted holes transverse   SC = Slip critical
     to direction of load

F_y = 50 ksi
F_u = 65 ksi

Table 4-4 (cont.). Double-Angle Connections

¾-in Bolts 12 Rows W44	Bolt and Angle Design Strength, kips					
	ASTM Desig.	Thread Cond.	Hole Type	Angle Thickness, in. ⁵⁄₁₆	⅜	½
	A325	N	—	382	382	382
		X	—	457	477	477
		SC Class A	STD	251	251	251
			OVS	213	213	213
			SSLT	213	213	213
		SC Class B	STD	380	380	380
			OVS	323	323	323
			SSLT	323	323	323
	A490	N	—	457	477	477
		X	—	457	548	596
		SC Class A	STD	313	313	313
			OVS	266	266	266
			SSLT	266	266	266
		SC Class B	STD	457	475	475
			OVS	403	403	403
			SSLT	403	403	403

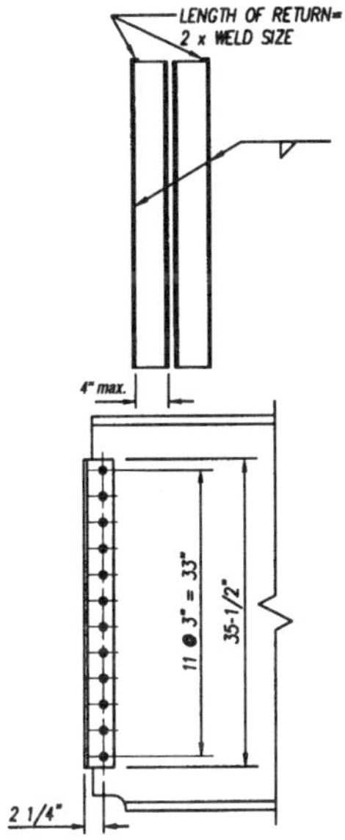

Beam Web Design Strength per Inch Thickness, kips/in.

For F_{ub} = 58 ksi: 940 kips/in.
For F_{ub} = 65 ksi: 1,053 kips/in.

Welds: F_{EXX} = 70 ksi and HSS F_y = 46 ksi

Weld Size, in.	ϕR_n, kips	Min. HSS Thickness, in.
⅜	550	0.34
⁵⁄₁₆	458	0.28
¼	366	0.22

Notes:
STD = Standard holes
OVS = Oversized holes
SSLT = Short-slotted holes transverse to direction of load
N = Threads included
X = Threads excluded
SC = Slip critical

$F_y = 36$ ksi
$F_u = 58$ ksi

Table 4-4 (cont.). Double-Angle Connections

¾-in. Bolts

11 Rows

W44, 40

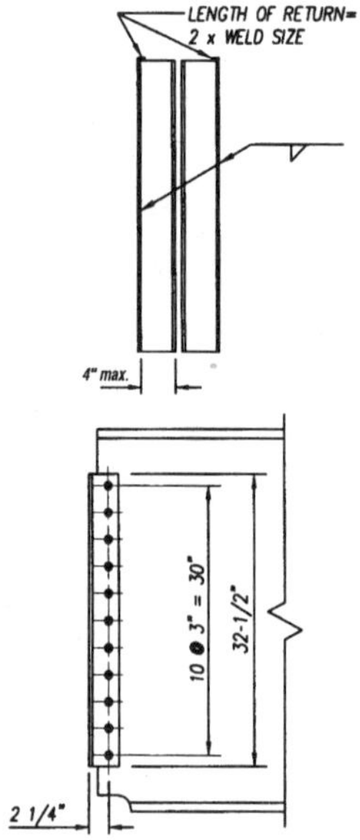

Bolt and Angle Design Strength, kips					
ASTM Desig.	**Thread Cond.**	**Hole Type**	**Angle Thickness, in.**		
			5⁄16	**3⁄8**	**1⁄2**
A325	N	—	350	350	350
	X	—	373	437	437
	SC Class A	STD	230	230	230
		OVS	195	195	195
		SSLT	195	195	195
	SC Class B	STD	348	348	348
		OVS	296	296	296
		SSLT	296	296	296
A490	N	—	373	437	437
	X	—	373	448	547
	SC Class A	STD	287	287	287
		OVS	244	244	244
		SSLT	244	244	244
	SC Class B	STD	373	435	435
		OVS	351	370	370
		SSLT	370	370	370

Beam Web Design Strength per Inch Thickness, kips/in.

For $F_{ub} = 58$ ksi: 861 kips/in.
For $F_{ub} = 65$ ksi: 965 kips/in.

Welds: $F_{EXX} = 70$ ksi and HSS $F_y = 46$ ksi

Weld Size, in.	ϕR_n, kips	Min. HSS Thickness, in.
3⁄8	496	0.34
5⁄16	414	0.28
1⁄4	331	0.22

Notes:
STD = Standard holes
OVS = Oversized holes
SSLT = Short-slotted holes transverse to direction of load
N = Threads included
X = Threads excluded
SC = Slip critical

				$F_y = 50$ ksi
				$F_u = 65$ ksi

Table 4-4 (cont.).
Double-Angle Connections

¾-in. Bolts	Bolt and Angle Design Strength, kips					
11 Rows	ASTM	Thread	Hole	Angle Thickness, in.		
W44, 40	Desig.	Cond.	Type	$\frac{5}{16}$	$\frac{3}{8}$	$\frac{1}{2}$
	A325	N	—	350	350	350
		X	—	418	437	437
		SC Class A	STD	230	230	230
			OVS	195	195	195
			SSLT	195	195	195
		SC Class B	STD	348	348	348
			OVS	296	296	296
			SSLT	296	296	296
	A490	N	—	418	437	437
		X	—	418	502	547
		SC Class A	STD	287	287	287
			OVS	244	244	244
			SSLT	244	244	244
		SC Class B	STD	418	435	435
			OVS	370	370	370
			SSLT	370	370	370

Beam Web Design Strength per Inch Thickness, kips/in.

For $F_{ub} = 58$ ksi: 861 kips/in.
For $F_{ub} = 65$ ksi: 965 kips/in.

Welds: $F_{EXX} = 70$ ksi and HSS $F_y = 46$ ksi

Weld Size, in.	ϕR_n, kips	Min. HSS Thickness, in.
$\frac{3}{8}$	496	0.34
$\frac{5}{16}$	414	0.28
$\frac{1}{4}$	331	0.22

Notes:
STD = Standard holes   N = Threads included
OVS = Oversized holes   X = Threads excluded
SSLT = Short-slotted holes transverse   SC = Slip critical
     to direction of load

| F_y = 36 ksi |
| F_u = 58 ksi |

Table 4-4 (cont.).
Double-Angle Connections

¾-in. Bolt 10 Rows W44, 40, 36	Bolt and Angle Design Strength, kips					
	ASTM Desig.	Thread Cond.	Hole Type	Angle Thickness, in.		
				5/16	3/8	1/2
	A325	N	—	318	318	318
		X	—	338	398	398
		SC Class A	STD	209	209	209
			OVS	178	178	178
			SSLT	178	178	178
		SC Class B	STD	316	316	316
			OVS	269	269	269
			SSLT	269	269	269
	A490	N	—	338	398	398
		X	—	338	406	497
		SC Class A	STD	261	261	261
			OVS	222	222	222
			SSLT	222	222	222
		SC Class B	STD	338	396	396
			OVS	318	336	336
			SSLT	336	336	336

Beam Web Design Strength per Inch Thickness, kips/in.

For F_{ub} = 58 ksi: 783 kips/in.
For F_{ub} = 65 ksi: 878 kips/in.

Welds: F_{EXX} = 70 ksi and HSS F_y = 46 ksi

Weld Size, in.	ϕR_n, kips	Min. HSS Thickness, in.
3/8	443	0.34
5/16	369	0.28
1/4	295	0.22

Notes:
STD = Standard holes   N = Threads included
OVS = Oversized holes   X = Threads excluded
SSLT = Short-slotted holes transverse   SC = Slip critical
     to direction of load

$F_y = 50$ ksi

$F_u = 65$ ksi

Table 4-4 (cont.).
Double-Angle Connections

¾-in. Bolts / 10 Rows / W44, 40, 36	Bolt and Angle Design Strength, kips					
	ASTM Desig.	Thread Cond.	Hole Type	Angle Thickness, in.		
				$^{5}/_{16}$	$^{3}/_{8}$	$^{1}/_{2}$
	A325	N	—	318	318	318
		X	—	379	398	398
		SC Class A	STD	209	209	209
			OVS	178	178	178
			SSLT	178	178	178
		SC Class B	STD	316	316	316
			OVS	269	269	269
			SSLT	269	269	269
	A490	N	—	379	398	398
		X	—	379	455	497
		SC Class A	STD	261	261	261
			OVS	222	222	222
			SSLT	222	222	222
		SC Class B	STD	379	396	396
			OVS	336	336	336
			SSLT	336	336	336

Beam Web Design Strength per Inch Thickness, kips/in.

For $F_{ub} = 58$ ksi: 783 kips/in.
For $F_{ub} = 65$ ksi: 878 kips/in.

Welds: $F_{EXX} = 70$ ksi and HSS $F_y = 46$ ksi

Weld Size, in.	ϕR_n, kips	Min. HSS Thickness, in.
$^{3}/_{8}$	443	0.34
$^{5}/_{16}$	369	0.28
$^{1}/_{4}$	295	0.22

Notes:
STD = Standard holes   N = Threads included
OVS = Oversized holes   X = Threads excluded
SSLT = Short-slotted holes transverse   SC = Slip critical
       to direction of load

| F_y = 36 ksi |
| F_u = 58 ksi |

Table 4-4 (cont.).
Double-Angle Connections

¾-in. Bolts	Bolt and Angle Design Strength, kips					
9 Rows	ASTM	Thread	Hole	Angle Thickness, in.		
W44, 40, 36, 33	Desig.	Cond.	Type	⁵⁄₁₆	³⁄₈	½
	A325	N	—	286	286	286
		X	—	304	358	358
		SC Class A	STD	188	188	188
			OVS	160	160	160
			SSLT	160	160	160
		SC Class B	STD	285	285	285
			OVS	242	242	242
			SSLT	242	242	242
	A490	N	—	304	358	358
		X	—	304	365	447
		SC Class A	STD	235	235	235
			OVS	200	200	200
			SSLT	200	200	200
		SC Class B	STD	304	356	356
			OVS	285	303	303
			SSLT	303	303	303

Beam Web Design Strength per Inch Thickness, kips/in.

For F_{ub} = 58 ksi: 705 kips/in.
For F_{ub} = 65 ksi: 790 kips/in.

Welds: F_{EXX} = 70 ksi and HSS F_y = 46 ksi

Weld Size, in.	ϕR_n, kips	Min. HSS Thickness, in.
³⁄₈	389	0.34
⁵⁄₁₆	324	0.28
¼	259	0.22

Notes:
STD = Standard holes
OVS = Oversized holes
SSLT = Short-slotted holes transverse
     to direction of load

N = Threads included
X = Threads excluded
SC = Slip critical

	F_y = 50 ksi
	F_u = 65 ksi

Table 4-4 (cont.).
Double-Angle Connections

¾-in. Bolts	Bolt and Angle Design Strength, kips					
9 Rows	ASTM Desig.	Thread Cond.	Hole Type	Angle Thickness, in.		
W44, 40, 36, 33				⁵⁄₁₆	³⁄₈	½
	A325	N	—	286	286	286
		X	—	340	358	358
		SC Class A	STD	188	188	188
			OVS	160	160	160
			SSLT	160	160	160
		SC Class B	STD	285	285	285
			OVS	242	242	242
			SSLT	242	242	242
	A490	N	—	340	358	358
		X	—	340	409	447
		SC Class A	STD	235	235	235
			OVS	200	200	200
			SSLT	200	200	200
		SC Class B	STD	340	356	356
			OVS	303	303	303
			SSLT	303	303	303

Beam Web Design Strength per Inch Thickness, kips/in.

For F_{ub} = 58 ksi: 705 kips/in.
For F_{ub} = 65 ksi: 790 kips/in.

Welds: F_{EXX} = 70 ksi and HSS F_y = 46 ksi

Weld Size, in.	ϕR_n, kips	Min. HSS Thickness, in.
³⁄₈	389	0.34
⁵⁄₁₆	324	0.28
¼	259	0.22

Notes:
STD = Standard holes
OVS = Oversized holes
SSLT = Short-slotted holes transverse
    to direction of load

N = Threads included
X = Threads excluded
SC = Slip critical

F_y = 36 ksi

F_u = 58 ksi

Table 4-4 (cont.).
Double-Angle Connections

¾-in. Bolts 8 Rows W44, 40, 36, 33, 30	Bolt and Angle Design Strength, kips					
	ASTM Desig.	Thread Cond.	Hole Type	Angle Thickness, in.		
				⁵⁄₁₆	³⁄₈	½
	A325	N	—	254	254	254
		X	—	269	318	318
		SC Class A	STD	167	167	167
			OVS	142	142	142
			SSLT	142	142	142
		SC Class B	STD	253	253	253
			OVS	215	215	215
			SSLT	215	215	215
	A490	N	—	269	318	318
		X	—	269	323	398
		SC Class A	STD	209	209	209
			OVS	178	178	178
			SSLT	178	178	178
		SC Class B	STD	269	316	316
			OVS	253	269	269
			SSLT	269	269	269

Beam Web Design Strength per Inch Thickness, kips/in.

For F_{ub} = 58 ksi: 626 kips/in.
For F_{ub} = 65 ksi: 702 kips/in.

Welds: F_{EXX} = 70 ksi and HSS F_y = 46 ksi

Weld Size, in.	ϕR_n, kips	Min. HSS Thickness, in.
³⁄₈	335	0.34
⁵⁄₁₆	279	0.28
¼	223	0.22

Notes:
STD = Standard holes   N = Threads included
OVS = Oversized holes   X = Threads excluded
SSLT = Short-slotted holes transverse   SC = Slip critical
     to direction of load

	$F_y = 50$ ksi
	$F_u = 65$ ksi

Table 4-4 (cont.).
Double-Angle Connections

¾-in. Bolts 8 Rows W44, 40, 36, 33, 30	Bolt and Angle Design Strength, kips					
	ASTM Desig.	Thread Cond.	Hole Type	Angle Thickness, in.		
				$\frac{5}{16}$	$\frac{3}{8}$	$\frac{1}{2}$
	A325	N	—	254	254	254
		X	—	302	318	318
		SC Class A	STD	167	167	167
			OVS	142	142	142
			SSLT	142	142	142
		SC Class B	STD	253	253	253
			OVS	215	215	215
			SSLT	215	215	215
	A490	N	—	302	318	318
		X	—	302	362	398
		SC Class A	STD	209	209	209
			OVS	178	178	178
			SSLT	178	178	178
		SC Class B	STD	302	316	316
			OVS	269	269	269
			SSLT	269	269	269

Beam Web Design Strength per Inch Thickness, kips/in.

For $F_{ub} = 58$ ksi: 626 kips/in.
For $F_{ub} = 65$ ksi: 702 kips/in.

Welds: $F_{EXX} = 70$ ksi and HSS $F_y = 46$ ksi

Weld Size, in.	ϕR_n, kips	Min. HSS Thickness, in.
$\frac{3}{8}$	335	0.34
$\frac{5}{16}$	279	0.28
$\frac{1}{4}$	223	0.22

Notes:
STD = Standard holes
OVS = Oversized holes
SSLT = Short-slotted holes transverse to direction of load

N = Threads included
X = Threads excluded
SC = Slip critical

F_y = 36 ksi
F_u = 58 ksi

Table 4-4 (cont.).
Double-Angle Connections

¾-in. Bolts 7 Rows W44, 40, 36, 33, 30, 27, 24 S24	Bolt and Angle Design Strength, kips					
	ASTM Desig.	Thread Cond.	Hole Type	Angle Thickness, in.		
				⁵⁄₁₆	⅜	½
	A325	N	—	223	223	223
		X	—	234	278	278
		SC Class A	STD	146	146	146
			OVS	124	124	124
			SSLT	124	124	124
		SC Class B	STD	221	221	221
			OVS	188	188	188
			SSLT	188	188	188
	A490	N	—	234	278	278
		X	—	234	281	348
		SC Class A	STD	183	183	183
			OVS	155	155	155
			SSLT	155	155	155
		SC Class B	STD	234	277	277
			OVS	220	235	235
			SSLT	234	235	235

Beam Web Design Strength per Inch Thickness, kips/in.

For F_{ub} = 58 ksi: 548 kips/in.
For F_{ub} = 65 ksi: 614 kips/in.

Welds: F_{EXX} = 70 ksi and HSS F_y = 46 ksi

Weld Size, in.	ϕR_n, kips	Min. HSS Thickness, in.
⅜	280	0.34
⁵⁄₁₆	234	0.28
¼	187	0.22

Notes:
STD = Standard holes
OVS = Oversized holes
SSLT = Short-slotted holes transverse
     to direction of load

N = Threads included
X = Threads excluded
SC = Slip critical

$F_y = 50$ ksi

$F_u = 65$ ksi

Table 4-4 (cont.).
Double-Angle Connections

¾-in. Bolts	Bolt and Angle Design Strength, kips					
7 Rows	**ASTM Desig.**	**Thread Cond.**	**Hole Type**	**Angle Thickness, in.**		
W44, 40, 36, 33, 30, 27, 24 S24				**⁵⁄₁₆**	**³⁄₈**	**½**
	A325	N	—	223	223	223
		X	—	263	278	278
		SC Class A	STD	146	146	146
			OVS	124	124	124
			SSLT	124	124	124
		SC Class B	STD	221	221	221
			OVS	188	188	188
			SSLT	188	188	188
	A490	N	—	263	278	278
		X	—	263	315	348
		SC Class A	STD	183	183	183
			OVS	155	155	155
			SSLT	155	155	155
		SC Class B	STD	263	277	277
			OVS	235	235	235
			SSLT	235	235	235

Beam Web Design Strength per Inch Thickness, kips/in.

For $F_{ub} = 58$ ksi: 548 kips/in.
For $F_{ub} = 65$ ksi: 614 kips/in.

Welds: $F_{EXX} = 70$ ksi and HSS $F_y = 46$ ksi

Weld Size, in.	ϕR_n, kips	Min. HSS Thickness, in.
³⁄₈	280	0.34
⁵⁄₁₆	234	0.28
¼	187	0.22

Notes:
STD = Standard holes　　　　N = Threads included
OVS = Oversized holes　　　　X = Threads excluded
SSLT = Short-slotted holes transverse　　SC = Slip critical
　　　　to direction of load

F_y = 36 ksi
F_u = 58 ksi

Table 4-4 (cont.).
Double-Angle Connections

¾-in. Bolts 6 Rows W40, 36, 33, 30, 27, 24, 21 S24	Bolt and Angle Design Strength, kips					
	ASTM Desig.	Thread Cond.	Hole Type	Angle Thickness, in.		
				5⁄16	3⁄8	1⁄2
	A325	N	—	191	191	191
		X	—	200	239	239
		SC Class A	STD	125	125	125
			OVS	107	107	107
			SSLT	107	107	107
		SC Class B	STD	190	190	190
			OVS	161	161	161
			SSLT	161	161	161
	A490	N	—	200	239	239
		X	—	200	240	298
		SC Class A	STD	157	157	157
			OVS	133	133	133
			SSLT	133	133	133
		SC Class B	STD	200	237	237
			OVS	188	202	202
			SSLT	200	202	202

Beam Web Design Strength per Inch Thickness, kips/in.

For F_{ub} = 58 ksi: 470 kips/in.
For F_{ub} = 65 ksi: 527 kips/in.

Welds: F_{EXX} = 70 ksi and HSS F_y = 46 ksi

Weld Size, in.	ϕR_n, kips	Min. HSS Thickness, in.
3⁄8	226	0.34
5⁄16	188	0.28
1⁄4	150	0.22

Notes:
STD = Standard holes
OVS = Oversized holes
SSLT = Short-slotted holes transverse
     to direction of load

N = Threads included
X = Threads excluded
SC = Slip critical

$F_y = 50$ ksi

$F_u = 65$ ksi

Table 4-4 (cont.).
Double-Angle Connections

³⁄₄-in. Bolts 6 Rows W40, 36, 33, 30, 27, 24, 21 S24	Bolt and Angle Design Strength, kips					
	ASTM Desig.	Thread Cond.	Hole Type	Angle Thickness, in.		
				$5/16$	$3/8$	$1/2$
	A325	N	—	191	191	191
		X	—	224	239	239
		SC Class A	STD	125	125	125
			OVS	107	107	107
			SSLT	107	107	107
		SC Class B	STD	190	190	190
			OVS	161	161	161
			SSLT	161	161	161
	A490	N	—	224	239	239
		X	—	224	269	298
		SC Class A	STD	157	157	157
			OVS	133	133	133
			SSLT	133	133	133
		SC Class B	STD	224	237	237
			OVS	202	202	202
			SSLT	202	202	202

Beam Web Design Strength per Inch Thickness, kips/in.

For $F_{ub} = 58$ ksi: 470 kips/in.
For $F_{ub} = 65$ ksi: 527 kips/in.

Welds: $F_{EXX} = 70$ ksi and HSS $F_y = 46$ ksi

Weld Size, in.	ϕR_n, kips	Min. HSS Thickness, in.
$3/8$	226	0.34
$5/16$	188	0.28
$1/4$	150	0.22

Notes:
STD = Standard holes
OVS = Oversized holes
SSLT = Short-slotted holes transverse
      to direction of load

N = Threads included
X = Threads excluded
SC = Slip critical

F_y = 36 ksi
F_u = 58 ksi

Table 4-4 (cont.).
Double-Angle Connections

¾-in. Bolts	Bolt and Angle Design Strength, kips					
5 Rows	**ASTM Desig.**	**Thread Cond.**	**Hole Type**	**Angle Thickness, in.**		
				⁵⁄₁₆	³⁄₈	½
W30, 27, 24, 21, 18 **S24, 20, 18** **MC18**	A325	N	—	159	159	159
		X	—	165	198	199
		SC Class A	STD	104	104	104
			OVS	88.8	88.8	88.8
			SSLT	88.8	88.8	88.8
		SC Class B	STD	158	158	158
			OVS	134	134	134
			SSLT	134	134	134
	A490	N	—	165	198	199
		X	—	165	198	249
		SC Class A	STD	131	131	131
			OVS	111	111	111
			SSLT	111	111	111
		SC Class B	STD	165	198	198
			OVS	155	168	168
			SSLT	165	168	168

Beam Web Design Strength per Inch Thickness, kips/in.

For F_{ub} = 58 ksi: 392 kips/in.
For F_{ub} = 65 ksi: 439 kips/in.

Welds: F_{EXX} = 70 ksi and HSS F_y = 46 ksi

Weld Size, in.	ϕR_n, kips	Min. HSS Thickness, in.
³⁄₈	172	0.34
⁵⁄₁₆	143	0.28
¼	115	0.22

Notes:
STD = Standard holes
OVS = Oversized holes
SSLT = Short-slotted holes transverse to direction of load

N = Threads included
X = Threads excluded
SC = Slip critical

F_y = 50 ksi
F_u = 65 ksi

Table 4-4 (cont.).
Double-Angle Connections

$\frac{3}{4}$-in. Bolts	Bolt and Angle Design Strength, kips					
5 Rows	**ASTM Desig.**	**Thread Cond.**	**Hole Type**	**Angle Thickness, in.**		
				$\frac{5}{16}$	$\frac{3}{8}$	$\frac{1}{2}$
W30, 27, 24, 21, 18 / S24, 20, 18 / MC18	A325	N	—	159	159	159
		X	—	185	199	199
		SC Class A	STD	104	104	104
			OVS	88.8	88.8	88.8
			SSLT	88.8	88.8	88.8
		SC Class B	STD	158	158	158
			OVS	134	134	134
			SSLT	134	134	134
	A490	N	—	185	199	199
		X	—	185	222	249
		SC Class A	STD	131	131	131
			OVS	111	111	111
			SSLT	111	111	111
		SC Class B	STD	185	198	198
			OVS	168	168	168
			SSLT	168	168	168

Beam Web Design Strength per Inch Thickness, kips/in.

For F_{ub} = 58 ksi: 392 kips/in.
For F_{ub} = 65 ksi: 439 kips/in.

Welds: F_{EXX} = 70 ksi and HSS F_y = 46 ksi

Weld Size, in.	ϕR_n, kips	Min. HSS Thickness, in.
$\frac{3}{8}$	172	0.34
$\frac{5}{16}$	143	0.28
$\frac{1}{4}$	115	0.22

Notes:
STD = Standard holes   N = Threads included
OVS = Oversized holes   X = Threads excluded
SSLT = Short-slotted holes transverse   SC = Slip critical
      to direction of load

| F_y = 36 ksi |
| F_u = 58 ksi |

Table 4-4 (cont.).
Double-Angle Connections

¾-in. Bolts 4 Rows W24, 21, 18, 16 S24, 20, 18, 15 C15 MC18	Bolt and Angle Design Strength, kips					
	ASTM Desig.	Thread Cond.	Hole Type	Angle Thickness, in.		
				5/16	3/8	1/2
	A325	N	—	127	127	127
		X	—	131	157	159
		SC Class A	STD	83.5	83.5	83.5
			OVS	71.0	71.0	71.0
			SSLT	71.0	71.0	71.0
		SC Class B	STD	127	127	127
			OVS	108	108	108
			SSLT	108	108	108
	A490	N	—	131	157	159
		X	—	131	157	199
		SC Class A	STD	104	104	104
			OVS	88.8	88.8	88.8
			SSLT	88.8	88.8	88.8
		SC Class B	STD	131	157	158
			OVS	122	134	134
			SSLT	131	134	134

Beam Web Design Strength per Inch Thickness, kips/in.

For F_{ub} = 58 ksi: 313 kips/in.
For F_{ub} = 65 ksi: 351 kips/in.

Welds: F_{EXX} = 70 ksi and HSS F_y = 46 ksi

Weld Size, in.	ϕR_n, kips	Min. HSS Thickness, in.
3/8	120	0.34
5/16	100	0.28
1/4	79.9	0.22

Notes:
STD = Standard holes   N = Threads included
OVS = Oversized holes   X = Threads excluded
SSLT = Short-slotted holes transverse   SC = Slip critical
        to direction of load

$F_y = 50$ ksi

$F_u = 65$ ksi

Table 4-4 (cont.).
Double-Angle Connections

¾-in. Bolts 4 Rows W24, 21, 18, 16 S24, 20, 18, 15 C15 MC18	Bolt and Angle Design Strength, kips					
	ASTM Desig.	Thread Cond.	Hole Type	Angle Thickness, in.		
				$\frac{5}{16}$	$\frac{3}{8}$	$\frac{1}{2}$
	A325	N	—	127	127	127
		X	—	146	159	159
		SC Class A	STD	83.5	83.5	83.5
			OVS	71.0	71.0	71.0
			SSLT	71.0	71.0	71.0
		SC Class B	STD	127	127	127
			OVS	108	108	108
			SSLT	108	108	108
	A490	N	—	146	159	159
		X	—	146	176	199
		SC Class A	STD	104	104	104
			OVS	88.8	88.8	88.8
			SSLT	88.8	88.8	88.8
		SC Class B	STD	146	158	158
			OVS	134	134	134
			SSLT	134	134	134

Beam Web Design Strength per Inch Thickness, kips/in.

For $F_{ub} = 58$ ksi: 313 kips/in.
For $F_{ub} = 65$ ksi: 351 kips/in.

Welds: $F_{EXX} = 70$ ksi and HSS $F_y = 46$ ksi

Weld Size, in.	ϕR_n, kips	Min. HSS Thickness, in.
$\frac{3}{8}$	120	0.34
$\frac{5}{16}$	100	0.28
$\frac{1}{4}$	79.9	0.22

Notes:
STD = Standard holes
OVS = Oversized holes
SSLT = Short-slotted holes transverse
     to direction of load

N = Threads included
X = Threads excluded
SC = Slip critical

F_y = 36 ksi
F_u = 58 ksi

Table 4-4 (cont.).
Double-Angle Connections

¾-in. Bolts	Bolt and Angle Design Strength, kips					
3 Rows	**ASTM Desig.**	**Thread Cond.**	**Hole Type**	**Angle Thickness, in.**		
				⁵⁄₁₆	⅜	½
W18, 16, 14, 12, 10* **S18, 15, 12** **C15, 12** **MC18, 13, 12**	A325	N	—	95.4	95.4	95.4
		X	—	95.8	115	119
*Limited to W10×12, 15, 17, 19, 22, 26, 30.		SC Class A	STD	62.6	62.6	62.6
			OVS	53.3	53.3	53.3
			SSLT	53.3	53.3	53.3
		SC Class B	STD	94.9	94.9	94.9
			OVS	80.7	80.7	80.7
			SSLT	80.7	80.7	80.7
	A490	N	—	95.8	115	119
		X	—	95.8	115	149
		SC Class A	STD	78.3	78.3	78.3
			OVS	66.6	66.6	66.6
			SSLT	66.6	66.6	66.6
		SC Class B	STD	95.8	115	119
			OVS	89.7	101	101
			SSLT	95.8	101	101

Beam Web Design Strength per Inch Thickness, kips/in.

For F_{ub} = 58 ksi: 235 kips/in.
For F_{ub} = 65 ksi: 263 kips/in.

Welds: F_{EXX} = 70 ksi and HSS F_y = 46 ksi

Weld Size, in.	ϕR_n, kips	Min. HSS Thickness, in.
⅜	72.2	0.34
⁵⁄₁₆	60.1	0.28
¼	48.1	0.22

Notes:
STD = Standard holes   N = Threads included
OVS = Oversized holes   X = Threads excluded
SSLT = Short-slotted holes transverse SC = Slip critical
   to direction of load

$F_y = 50$ ksi
$F_u = 65$ ksi

Table 4-4 (cont.).
Double-Angle Connections

¾-in. Bolts 3 Rows **W18, 16, 14, 12, 10*** **S18, 15, 12** **C15, 12** **MC18, 13, 12** *Limited to W10×12, 15, 17, 19, 22, 26, 30.	Bolt and Angle Design Strength, kips					
	ASTM Desig.	Thread Cond.	Hole Type	Angle Thickness, in.		
				⁵⁄₁₆	³⁄₈	½
	A325	N	—	95.4	95.4	95.4
		X	—	107	119	119
		SC Class A	STD	62.6	62.6	62.6
			OVS	53.3	53.3	53.3
			SSLT	53.3	53.3	53.3
		SC Class B	STD	94.9	94.9	94.9
			OVS	80.7	80.7	80.7
			SSLT	80.7	80.7	80.7
	A490	N	—	107	119	119
		X	—	107	129	149
		SC Class A	STD	78.3	78.3	78.3
			OVS	66.6	66.6	66.6
			SSLT	66.6	66.6	66.6
		SC Class B	STD	107	119	119
			OVS	101	101	101
			SSLT	101	101	101

Beam Web Design Strength per Inch Thickness, kips/in.

For $F_{ub} = 58$ ksi: 235 kips/in.
For $F_{ub} = 65$ ksi: 263 kips/in.

Welds: $F_{EXX} = 70$ ksi and HSS $F_y = 46$ ksi

Weld Size, in.	ϕR_n, kips	Min. HSS Thickness, in.
³⁄₈	72.2	0.34
⁵⁄₁₆	60.1	0.28
¼	48.1	0.22

Notes:
STD = Standard holes   N = Threads included
OVS = Oversized holes   X = Threads excluded
SSLT = Short-slotted holes transverse   SC = Slip critical
         to direction of load

F_y = 36 ksi
F_u = 58 ksi

Table 4-4 (cont.).
Double-Angle Connections

¾-in. Bolts	Bolt and Angle Design Strength, kips					
2 Rows	**ASTM Desig.**	**Thread Cond.**	**Hole Type**	**Angle Thickness, in.**		
W12, 10, 8 / S12, 10, 8 / C12, 10, 9, 8 / MC13, 12, 10, 9, 8				$^5/_{16}$	$^3/_8$	$^1/_2$
	A325	N	—	61.2	63.6	63.6
		X	—	61.2	73.4	79.5
		SC Class A	STD	41.8	41.8	41.8
			OVS	35.5	35.5	35.5
			SSLT	35.5	35.5	35.5
		SC Class B	STD	61.2	63.3	63.3
			OVS	53.8	53.8	53.8
			SSLT	53.8	53.8	53.8
	A490	N	—	61.2	73.4	79.5
		X	—	61.2	73.4	97.9
		SC Class A	STD	52.2	52.2	52.2
			OVS	44.4	44.4	44.4
			SSLT	44.4	44.4	44.4
		SC Class B	STD	61.2	73.4	79.1
			OVS	57.1	67.2	67.2
			SSLT	61.2	67.2	67.2

Beam Web Design Strength per Inch Thickness, kips/in.

For F_{ub} = 58 ksi: 157 kips/in.
For F_{ub} = 65 ksi: 176 kips/in.

Welds: F_{EXX} = 70 ksi and HSS F_y = 46 ksi

Weld Size, in.	ϕR_n, kips	Min. HSS Thickness, in.
$^3/_8$	32.8	0.34
$^5/_{16}$	27.3	0.28
$^1/_4$	21.9	0.22

Notes:
STD = Standard holes   N = Threads included
OVS = Oversized holes   X = Threads excluded
SSLT = Short-slotted holes transverse   SC = Slip critical
     to direction of load

	$F_y = 50$ ksi
	$F_u = 65$ ksi

Table 4-4 (cont.).
Double-Angle Connections

¾-in. Bolts 2 Rows **W12, 10, 8** **S12, 10, 8** **C12, 10, 9, 8** **MC13, 12, 10, 9, 8**	Bolt and Angle Design Strength, kips					
	ASTM Desig.	Thread Cond.	Hole Type	Angle Thickness, in.		
				$\frac{5}{16}$	$\frac{3}{8}$	$\frac{1}{2}$
	A325	N	—	63.6	63.6	63.6
		X	—	68.6	79.5	79.5
		SC Class A	STD	41.8	41.8	41.8
			OVS	35.5	35.5	35.5
			SSLT	35.5	35.5	35.5
		SC Class B	STD	63.3	63.3	63.3
			OVS	53.8	53.8	53.8
			SSLT	53.8	53.8	53.8
	A490	N	—	68.6	79.5	79.5
		X	—	68.6	82.3	99.4
		SC Class A	STD	52.2	52.2	52.2
			OVS	44.4	44.4	44.4
			SSLT	44.4	44.4	44.4
		SC Class B	STD	68.6	79.1	79.1
			OVS	64.0	67.2	67.2
			SSLT	67.2	67.2	67.2

Beam Web Design Strength per Inch Thickness, kips/in.

For $F_{ub} = 58$ ksi: 157 kips/in.
For $F_{ub} = 65$ ksi: 176 kips/in.

Welds: $F_{EXX} = 70$ ksi and HSS $F_y = 46$ ksi

Weld Size, in.	ϕR_n, kips	Min. HSS Thickness, in.
$\frac{3}{8}$	32.8	0.34
$\frac{5}{16}$	27.3	0.28
$\frac{1}{4}$	21.9	0.22

Notes:
STD = Standard holes
OVS = Oversized holes
SSLT = Short-slotted holes transverse
     to direction of load

N   = Threads included
X   = Threads excluded
SC = Slip critical

$F_y = 36$ ksi

$F_u = 58$ ksi

Table 4-4 (cont.).
Double-Angle Connections

⁷⁄₈-in. Bolts	Bolt and Angle Design Strength, kips					
12 Rows	ASTM Desig.	Thread Cond.	Hole Type	Angle Thickness, in.		
W44				$\frac{5}{16}$	$\frac{3}{8}$	$\frac{1}{2}$
	A325	N	—	383	460	520
		X	—	383	460	613
		SC Class A	STD	349	349	349
			OVS	297	297	297
			SSLT	297	297	297
		SC Class B	STD	383	460	520
			OVS	358	429	450
			SSLT	383	450	450
	A490	N	—	383	460	613
		X	—	383	460	613
		SC Class A	STD	383	439	439
			OVS	358	373	373
			SSLT	373	373	373
		SC Class B	STD	383	460	613
			OVS	358	429	565
			SSLT	383	460	565

Beam Web Design Strength per Inch Thickness, kips/in.

For $F_{ub} = 58$ ksi: 1,096 kips/in.
For $F_{ub} = 65$ ksi: 1,229 kips/in.

Welds: $F_{EXX} = 70$ ksi and HSS $F_y = 46$ ksi

Weld Size, in.	ϕR_n, kips	Min. HSS Thickness, in.
$\frac{3}{8}$	550	0.34
$\frac{5}{16}$	458	0.28
$\frac{1}{4}$	366	0.22

Notes:
STD = Standard holes
OVS = Oversized holes
SSLT = Short-slotted holes transverse
    to direction of load

N = Threads included
X = Threads excluded
SC = Slip critical

	$F_y = 50$ ksi
	$F_u = 65$ ksi

Table 4-4 (cont.).
Double-Angle Connections

7/8-in. Bolts	Bolt and Angle Design Strength, kips					
12 Rows	ASTM Desig.	Thread Cond.	Hole Type	Angle Thickness, in.		
W44				5/16	3/8	1/2
	A325	N	—	430	516	520
		X	—	430	516	649
		SC Class A	STD	349	349	349
			OVS	297	297	297
			SSLT	297	297	297
		SC Class B	STD	430	516	520
			OVS	401	450	450
			SSLT	430	450	450
	A490	N	—	430	516	649
		X	—	430	516	687
		SC Class A	STD	430	439	439
			OVS	373	373	373
			SSLT	373	373	373
		SC Class B	STD	430	516	649
			OVS	401	481	565
			SSLT	430	516	565

Beam Web Design Strength per Inch Thickness, kips/in.

For $F_{ub} = 58$ ksi: 1,096 kips/in.
For $F_{ub} = 65$ ksi: 1,229 kips/in.

Welds: $F_{EXX} = 70$ ksi and HSS $F_y = 46$ ksi

Weld Size, in.	ϕR_n, kips	Min. HSS Thickness, in.
3/8	550	0.34
5/16	458	0.28
1/4	366	0.22

Notes:
STD = Standard holes
OVS = Oversized holes
SSLT = Short-slotted holes transverse
    to direction of load

N = Threads included
X = Threads excluded
SC = Slip critical

$F_y = 36$ ksi
$F_u = 58$ ksi

Table 4-4 (cont.).
Double-Angle Connections

$\frac{7}{8}$-in. Bolts	Bolt and Angle Design Strength, kips					
11 Rows	ASTM Desig.	Thread Cond.	Hole Type	Angle Thickness, in.		
W44, 40				$\frac{5}{16}$	$\frac{3}{8}$	$\frac{1}{2}$
	A325	N	—	351	421	476
		X	—	351	421	561
		SC Class A	STD	320	320	320
			OVS	272	272	272
			SSLT	272	272	272
		SC Class B	STD	351	421	476
			OVS	327	393	412
			SSLT	351	412	412
	A490	N	—	351	421	561
		X	—	351	421	561
		SC Class A	STD	351	402	402
			OVS	327	342	342
			SSLT	342	342	342
		SC Class B	STD	351	421	561
			OVS	327	393	518
			SSLT	351	421	518

Beam Web Design Strength per Inch Thickness, kips/in.

For $F_{ub} = 58$ ksi: 1,005 kips/in.
For $F_{ub} = 65$ ksi: 1,126 kips/in.

Welds: $F_{EXX} = 70$ ksi and HSS $F_y = 46$ ksi

Weld Size, in.	ϕR_n, kips	Min. HSS Thickness, in.
$\frac{3}{8}$	496	0.34
$\frac{5}{16}$	414	0.28
$\frac{1}{4}$	331	0.22

Notes:
STD = Standard holes   N = Threads included
OVS = Oversized holes   X = Threads excluded
SSLT = Short-slotted holes transverse   SC = Slip critical
       to direction of load

$F_y = 50$ ksi

$F_u = 65$ ksi

Table 4-4 (cont.).
Double-Angle Connections

⁷⁄₈-in. Bolts 11 Rows W44, 40	Bolt and Angle Design Strength, kips					
	ASTM Desig.	Thread Cond.	Hole Type	Angle Thickness, in.		
				$5/16$	$3/8$	$1/2$
	A325	N	—	393	472	476
		X	—	393	472	595
		SC Class A	STD	320	320	320
			OVS	272	272	272
			SSLT	272	272	272
		SC Class B	STD	393	472	476
			OVS	367	412	412
			SSLT	393	412	412
	A490	N	—	393	472	595
		X	—	393	472	629
		SC Class A	STD	393	402	402
			OVS	342	342	342
			SSLT	342	342	342
		SC Class B	STD	393	472	595
			OVS	367	440	518
			SSLT	393	472	518

Beam Web Design Strength per Inch Thickness, kips/in.

For $F_{ub} = 58$ ksi: 1,005 kips/in.
For $F_{ub} = 65$ ksi: 1,126 kips/in.

Welds: $F_{EXX} = 70$ ksi and HSS $F_y = 46$ ksi

Weld Size, in.	ϕR_n, kips	Min. HSS Thickness, in.
$3/8$	496	0.34
$5/16$	414	0.28
$1/4$	331	0.22

Notes:
STD = Standard holes   N = Threads included
OVS = Oversized holes   X = Threads excluded
SSLT = Short-slotted holes transverse   SC = Slip critical
     to direction of load

F_y = 36 ksi
F_u = 58 ksi

Table 4-4 (cont.).
Double-Angle Connections

| 7/8-in. Bolts | Bolt and Angle Design Strength, kips | | | | | |
| 10 Rows | ASTM Desig. | Thread Cond. | Hole Type | Angle Thickness, in. | | |
W44, 40, 36				5/16	3/8	1/2
	A325	N	—	318	382	433
		X	—	318	382	509
		SC Class A	STD	291	291	291
			OVS	247	247	247
			SSLT	247	247	247
		SC Class B	STD	318	382	433
			OVS	297	356	375
			SSLT	318	375	375
	A490	N	—	318	382	509
		X	—	318	382	509
		SC Class A	STD	318	365	365
			OVS	297	311	311
			SSLT	311	311	311
		SC Class B	STD	318	382	509
			OVS	297	356	471
			SSLT	318	382	471

Beam Web Design Strength per Inch Thickness, kips/in.

For F_{ub} = 58 ksi: 914 kips/in.
For F_{ub} = 65 ksi: 1,024 kips/in.

Welds: F_{EXX} = 70 ksi and HSS F_y = 46 ksi

Weld Size, in.	ϕR_n, kips	Min. HSS Thickness, in.
3/8	443	0.34
5/16	369	0.28
1/4	295	0.22

Notes:
STD = Standard holes
OVS = Oversized holes
SSLT = Short-slotted holes transverse
    to direction of load

N = Threads included
X = Threads excluded
SC = Slip critical

$F_y = 50$ ksi
$F_u = 65$ ksi

Table 4-4 (cont.).
Double-Angle Connections

⅞-in. Bolts 10 Rows W44, 40, 36	Bolt and Angle Design Strength, kips					
	ASTM Desig.	Thread Cond.	Hole Type	Angle Thickness, in.		
				⁵⁄₁₆	⅜	½
	A325	N	—	356	428	433
		X	—	356	428	541
		SC Class A	STD	291	291	291
			OVS	247	247	247
			SSLT	247	247	247
		SC Class B	STD	356	428	433
			OVS	333	375	375
			SSLT	356	375	375
	A490	N	—	356	428	541
		X	—	356	428	570
		SC Class A	STD	356	365	365
			OVS	311	311	311
			SSLT	311	311	311
		SC Class B	STD	356	428	541
			OVS	333	399	471
			SSLT	356	428	471

Beam Web Design Strength per Inch Thickness, kips/in.

For $F_{ub} = 58$ ksi: 914 kips/in.
For $F_{ub} = 65$ ksi: 1,024 kips/in.

Welds: $F_{EXX} = 70$ ksi and HSS $F_y = 46$ ksi

Weld Size, in.	ϕR_n, kips	Min. HSS Thickness, in.
⅜	443	0.34
⁵⁄₁₆	369	0.28
¼	295	0.22

Notes:
STD = Standard holes   N = Threads included
OVS = Oversized holes   X = Threads excluded
SSLT = Short-slotted holes transverse   SC = Slip critical
     to direction of load

| F_y = 36 ksi |
| F_u = 58 ksi |

Table 4-4 (cont.).
Double-Angle Connections

| 7/8-in. Bolts | Bolt and Angle Design Strength, kips | | | | | |
| 9 Rows | ASTM Desig. | Thread Cond. | Hole Type | Angle Thickness, in. | | |
W44, 40, 36, 33				5/16	3/8	1/2
	A325	N	—	285	343	390
		X	—	285	343	457
		SC Class A	STD	262	262	262
			OVS	223	223	223
			SSLT	223	223	223
		SC Class B	STD	285	343	390
			OVS	266	320	337
			SSLT	285	337	337
	A490	N	—	285	343	457
		X	—	285	343	457
		SC Class A	STD	285	329	329
			OVS	266	280	280
			SSLT	280	280	280
		SC Class B	STD	285	343	457
			OVS	266	320	424
			SSLT	285	343	424

Beam Web Design Strength per Inch Thickness, kips/in.

For F_{ub} = 58 ksi: 822 kips/in.
For F_{ub} = 65 ksi: 921 kips/in.

Welds: F_{EXX} = 70 ksi and HSS F_y = 46 ksi

Weld Size, in.	ϕR_n, kips	Min. HSS Thickness, in.
3/8	389	0.34
5/16	324	0.28
1/4	259	0.22

Notes:
STD = Standard holes
OVS = Oversized holes
SSLT = Short-slotted holes transverse to direction of load

N = Threads included
X = Threads excluded
SC = Slip critical

$F_y = 50$ ksi
$F_u = 65$ ksi

Table 4-4 (cont.).
Double-Angle Connections

⅞-in. Bolts	**Bolt and Angle Design Strength, kips**					
9 Rows	**ASTM Desig.**	**Thread Cond.**	**Hole Type**	**Angle Thickness, in.**		
W44, 40, 36, 33				**⁵⁄₁₆**	**³⁄₈**	**½**
	A325	N	—	320	384	390
		X	—	320	384	487
		SC Class A	STD	262	262	262
			OVS	223	223	223
			SSLT	223	223	223
		SC Class B	STD	320	384	390
			OVS	299	337	337
			SSLT	320	337	337
	A490	N	—	320	384	487
		X	—	320	384	512
		SC Class A	STD	320	329	329
			OVS	280	280	280
			SSLT	280	280	280
		SC Class B	STD	320	384	487
			OVS	299	358	424
			SSLT	320	384	424

Beam Web Design Strength per Inch Thickness, kips/in.

For $F_{ub} = 58$ ksi: 822 kips/in.
For $F_{ub} = 65$ ksi: 921 kips/in.

Welds: $F_{EXX} = 70$ ksi and HSS $F_y = 46$ ksi

Weld Size, in.	**ϕR_n, kips**	**Min. HSS Thickness, in.**
³⁄₈	389	0.34
⁵⁄₁₆	324	0.28
¼	259	0.22

Notes:
STD = Standard holes
OVS = Oversized holes
SSLT = Short-slotted holes transverse
    to direction of load

N = Threads included
X = Threads excluded
SC = Slip critical

$F_y = 36$ ksi
$F_u = 58$ ksi

Table 4-4 (cont.).
Double-Angle Connections

⁷⁄₈-in. Bolts 8 Rows W44, 40, 36, 33, 30	Bolt and Angle Design Strength, kips					
	ASTM Desig.	Thread Cond.	Hole Type	Angle Thickness, in.		
				$^5/_{16}$	$^3/_8$	$^1/_2$
	A325	N	—	253	303	346
		X	—	253	303	405
		SC Class A	STD	233	233	233
			OVS	198	198	198
			SSLT	198	198	198
		SC Class B	STD	253	303	346
			OVS	236	283	300
			SSLT	253	300	300
	A490	N	—	253	303	405
		X	—	253	303	405
		SC Class A	STD	253	292	292
			OVS	236	249	249
			SSLT	249	249	249
		SC Class B	STD	253	303	405
			OVS	236	283	377
			SSLT	253	303	377

Beam Web Design Strength per Inch Thickness, kips/in.

For $F_{ub} = 58$ ksi: 731 kips/in.
For $F_{ub} = 65$ ksi: 819 kips/in.

Welds: $F_{EXX} = 70$ ksi and HSS $F_y = 46$ ksi

Weld Size, in.	ϕR_n, kips	Min. HSS Thickness, in.
$^3/_8$	335	0.34
$^5/_{16}$	279	0.28
$^1/_4$	223	0.22

Notes:
STD = Standard holes
OVS = Oversized holes
SSLT = Short-slotted holes transverse to direction of load

N = Threads included
X = Threads excluded
SC = Slip critical

	F_y = 50 ksi
	F_u = 65 ksi

Table 4-4 (cont.).
Double-Angle Connections

⅞-in. Bolts	Bolt and Angle Design Strength, kips					
8 Rows	**ASTM Desig.**	**Thread Cond.**	**Hole Type**	**Angle Thickness, in.**		
W44, 40, 36, 33, 30				$^5/_{16}$	$^3/_8$	$^1/_2$
	A325	N	—	283	340	346
		X	—	283	340	433
		SC Class A	STD	233	233	233
			OVS	198	198	198
			SSLT	198	198	198
		SC Class B	STD	283	340	346
			OVS	264	300	300
			SSLT	283	300	300
	A490	N	—	283	340	433
		X	—	283	340	453
		SC Class A	STD	283	292	292
			OVS	249	249	249
			SSLT	249	249	249
		SC Class B	STD	283	340	433
			OVS	264	317	377
			SSLT	283	340	377

Beam Web Design Strength per Inch Thickness, kips/in.

For F_{ub} = 58 ksi: 731 kips/in.
For F_{ub} = 65 ksi: 819 kips/in.

Welds: F_{EXX} = 70 ksi and HSS F_y = 46 ksi

Weld Size, in.	ϕR_n, kips	Min. HSS Thickness, in.
$^3/_8$	335	0.34
$^5/_{16}$	279	0.28
$^1/_4$	223	0.22

Notes:
STD = Standard holes   N = Threads included
OVS = Oversized holes   X = Threads excluded
SSLT = Short-slotted holes transverse   SC = Slip critical
     to direction of load

| F_y = 36 ksi |
| F_u = 58 ksi |

Table 4-4 (cont.).
Double-Angle Connections

$^7/_8$-in. Bolts 7 Rows W44, 40, 36, 33, 30, 27, 24 S24	Bolt and Angle Design Strength, kips					
	ASTM Desig.	Thread Cond.	Hole Type	Angle Thickness, in.		
				$^5/_{16}$	$^3/_8$	$^1/_2$
	A325	N	—	220	264	303
		X	—	220	264	352
		SC Class A	STD	204	204	204
			OVS	173	173	173
			SSLT	173	173	173
		SC Class B	STD	220	264	303
			OVS	205	246	262
			SSLT	220	262	262
	A490	N	—	220	264	352
		X	—	220	264	352
		SC Class A	STD	220	256	256
			OVS	205	217	217
			SSLT	217	217	217
		SC Class B	STD	220	264	352
			OVS	205	246	329
			SSLT	220	264	329

Beam Web Design Strength per Inch Thickness, kips/in.

For F_{ub} = 58 ksi: 639 kips/in.
For F_{ub} = 65 ksi: 717 kips/in.

Welds: F_{EXX} = 70 ksi and HSS F_y = 46 ksi

Weld Size, in.	ϕR_n, kips	Min. HSS Thickness, in.
$^3/_8$	280	0.34
$^5/_{16}$	234	0.28
$^1/_4$	187	0.22

Notes:
STD = Standard holes
OVS = Oversized holes
SSLT = Short-slotted holes transverse to direction of load

N = Threads included
X = Threads excluded
SC = Slip critical

$F_y = 50$ ksi

$F_u = 65$ ksi

Table 4-4 (cont.).
Double-Angle Connections

⅞-in. Bolts 7 Rows W44, 40, 36, 33, 30, 27, 24 S24	Bolt and Angle Design Strength, kips					
	ASTM Desig.	Thread Cond.	Hole Type	Angle Thickness, in.		
				$\frac{5}{16}$	$\frac{3}{8}$	$\frac{1}{2}$
	A325	N	—	247	296	303
		X	—	247	296	379
		SC Class A	STD	204	204	204
			OVS	173	173	173
			SSLT	173	173	173
		SC Class B	STD	247	296	303
			OVS	230	262	262
			SSLT	247	262	262
	A490	N	—	247	296	379
		X	—	247	296	395
		SC Class A	STD	247	256	256
			OVS	217	217	217
			SSLT	217	217	217
		SC Class B	STD	247	296	379
			OVS	230	276	329
			SSLT	247	296	329

Beam Web Design Strength per Inch Thickness, kips/in.

For $F_{ub} = 58$ ksi: 639 kips/in.
For $F_{ub} = 65$ ksi: 717 kips/in.

Welds: $F_{EXX} = 70$ ksi and HSS $F_y = 46$ ksi

Weld Size, in.	ϕR_n, kips	Min. HSS Thickness, in.
$\frac{3}{8}$	280	0.34
$\frac{5}{16}$	234	0.28
$\frac{1}{4}$	187	0.22

Notes:
STD = Standard holes
OVS = Oversized holes
SSLT = Short-slotted holes transverse
    to direction of load

N = Threads included
X = Threads excluded
SC = Slip critical

F_y = 36 ksi
F_u = 58 ksi

Table 4-4 (cont.).
Double-Angle Connections

⅞-in. Bolts	**Bolt and Angle Design Strength, kips**					
6 Rows	**ASTM Desig.**	**Thread Cond.**	**Hole Type**	**Angle Thickness, in.**		
W40, 36, 33, 30, 27, 24, 21 S24				$^5/_{16}$	$^3/_8$	$^1/_2$
	A325	N	—	188	225	260
		X	—	188	225	300
		SC Class A	STD	175	175	175
			OVS	148	148	148
			SSLT	148	148	148
		SC Class B	STD	188	225	260
			OVS	175	210	225
			SSLT	188	225	225
	A490	N	—	188	225	300
		X	—	188	225	300
		SC Class A	STD	188	219	219
			OVS	175	186	186
			SSLT	186	186	186
		SC Class B	STD	188	225	300
			OVS	175	210	280
			SSLT	188	225	282

Beam Web Design Strength per Inch Thickness, kips/in.

For F_{ub} = 58 ksi: 548 kips/in.
For F_{ub} = 65 ksi: 614 kips/in.

Welds: F_{EXX} = 70 ksi and HSS F_y = 46 ksi

Weld Size, in.	**ϕR_n, kips**	**Min. HSS Thickness, in.**
$^3/_8$	226	0.34
$^5/_{16}$	188	0.28
$^1/_4$	150	0.22

Notes:
STD = Standard holes   N = Threads included
OVS = Oversized holes   X = Threads excluded
SSLT = Short-slotted holes transverse   SC = Slip critical
      to direction of load

F_y = 50 ksi	
F_u = 65 ksi	

Table 4-4 (cont.).
Double-Angle Connections

⁷⁄₈-in. Bolts		Bolt and Angle Design Strength, kips					
6 Rows		ASTM Desig.	Thread Cond.	Hole Type	Angle Thickness, in.		
W40, 36, 33, 30, 27, 24, 21 S24					$^5⁄_{16}$	$^3⁄_8$	$^1⁄_2$
		A325	N	—	210	252	260
			X	—	210	252	325
			SC Class A	STD	175	175	175
				OVS	148	148	148
				SSLT	148	148	148
			SC Class B	STD	210	252	260
				OVS	196	225	225
				SSLT	210	225	225
		A490	N	—	210	252	325
			X	—	210	252	336
			SC Class A	STD	210	219	219
				OVS	186	186	186
				SSLT	186	186	186
			SC Class B	STD	210	252	325
				OVS	196	235	282
				SSLT	210	252	282

Beam Web Design Strength per Inch Thickness, kips/in.

For F_{ub} = 58 ksi: 548 kips/in.
For F_{ub} = 65 ksi: 614 kips/in.

Welds: F_{EXX} = 70 ksi and HSS F_y = 46 ksi

Weld Size, in.	ϕR_n, kips	Min. HSS Thickness, in.
$^3⁄_8$	226	0.34
$^5⁄_{16}$	188	0.28
$^1⁄_4$	150	0.22

Notes:
STD = Standard holes
OVS = Oversized holes
SSLT = Short-slotted holes transverse
    to direction of load

N = Threads included
X = Threads excluded
SC = Slip critical

| F_y = 36 ksi |
| F_u = 58 ksi |

Table 4-4 (cont.).
Double-Angle Connections

⁷⁄₈-in. Bolts	**Bolt and Angle Design Strength, kips**					
5 Rows	**ASTM Desig.**	**Thread Cond.**	**Hole Type**	**Angle Thickness, in.**		
				⁵⁄₁₆	**³⁄₈**	**¹⁄₂**
W30, 27, 24, 21, 18 **S24, 20, 18** **MC18**	A325	N	—	155	186	216
		X	—	155	186	248
		SC Class A	STD	145	145	145
			OVS	124	124	124
			SSLT	124	124	124
		SC Class B	STD	155	186	216
			OVS	144	173	187
			SSLT	155	186	187
	A490	N	—	155	186	248
		X	—	155	186	248
		SC Class A	STD	155	183	183
			OVS	144	155	155
			SSLT	155	155	155
		SC Class B	STD	155	186	248
			OVS	144	173	231
			SSLT	155	186	235

Beam Web Design Strength per Inch Thickness, kips/in.

For F_{ub} = 58 ksi: 457 kips/in.
For F_{ub} = 65 ksi: 512 kips/in.

Welds: F_{EXX} = 70 ksi and HSS F_y = 46 ksi

Weld Size, in.	**ϕR_n, kips**	**Min. HSS Thickness, in.**
³⁄₈	172	0.34
⁵⁄₁₆	143	0.28
¹⁄₄	115	0.22

Notes:
STD = Standard holes
OVS = Oversized holes
SSLT = Short-slotted holes transverse to direction of load

N = Threads included
X = Threads excluded
SC = Slip critical

	F_y = 50 ksi
	F_u = 65 ksi

Table 4-4 (cont.).
Double-Angle Connections

⁷⁄₈-in. Bolts **5 Rows** **W30, 27, 24, 21, 18** **S24, 20, 18** **MC18**	**Bolt and Angle Design Strength, kips**					
	ASTM Desig.	**Thread Cond.**	**Hole Type**	**Angle Thickness, in.**		
				⁵⁄₁₆	**³⁄₈**	**¹⁄₂**
	A325	N	—	174	208	216
		X	—	174	208	271
		SC Class A	STD	145	145	145
			OVS	124	124	124
			SSLT	124	124	124
		SC Class B	STD	174	208	216
			OVS	162	187	187
			SSLT	174	187	187
	A490	N	—	174	208	271
		X	—	174	208	278
		SC Class A	STD	174	183	183
			OVS	155	155	155
			SSLT	155	155	155
		SC Class B	STD	174	208	271
			OVS	162	194	235
			SSLT	174	208	235

Beam Web Design Strength per Inch Thickness, kips/in.

For F_{ub} = 58 ksi: 457 kips/in.
For F_{ub} = 65 ksi: 512 kips/in.

Welds: F_{EXX} = 70 ksi and HSS F_y = 46 ksi

Weld Size, in.	**ϕR_n, kips**	**Min. HSS Thickness, in.**
³⁄₈	172	0.34
⁵⁄₁₆	143	0.28
¹⁄₄	115	0.22

Notes:
STD = Standard holes
OVS = Oversized holes
SSLT = Short-slotted holes transverse
      to direction of load

N = Threads included
X = Threads excluded
SC = Slip critical

F_y = 36 ksi
F_u = 58 ksi

Table 4-4 (cont.).
Double-Angle Connections

7/8-in. Bolts	**Bolt and Angle Design Strength, kips**					
4 Rows	**ASTM Desig.**	**Thread Cond.**	**Hole Type**	**Angle Thickness, in.**		
W24, 21, 18, 16 **S24, 20, 18, 15** **C15** **MC18**				**5/16**	**3/8**	**1/2**
	A325	N	—	122	147	173
		X	—	122	147	196
		SC Class A	STD	116	116	116
			OVS	98.9	98.9	98.9
			SSLT	98.9	98.9	98.9
		SC Class B	STD	122	147	173
			OVS	114	137	150
			SSLT	122	147	150
	A490	N	—	122	147	196
		X	—	122	147	196
		SC Class A	STD	122	146	146
			OVS	114	124	124
			SSLT	122	124	124
		SC Class B	STD	122	147	196
			OVS	114	137	182
			SSLT	122	147	188

Beam Web Design Strength per Inch Thickness, kips/in.

For F_{ub} = 58 ksi: 365 kips/in.
For F_{ub} = 65 ksi: 410 kips/in.

Welds: F_{EXX} = 70 ksi and HSS F_y = 46 ksi

Weld Size, in.	**ϕR_n, kips**	**Min. HSS Thickness, in.**
3/8	120	0.34
5/16	100	0.28
1/4	79.9	0.22

Notes:
STD = Standard holes
OVS = Oversized holes
SSLT = Short-slotted holes transverse
    to direction of load

N = Threads included
X = Threads excluded
SC = Slip critical

				F_y = 50 ksi
				F_u = 65 ksi

Table 4-4 (cont.).
Double-Angle Connections

⁷⁄₈-in. Bolts	**Bolt and Angle Design Strength, kips**					
4 Rows	**ASTM Desig.**	**Thread Cond.**	**Hole Type**	**Angle Thickness, in.**		
W24, 21, 18, 16 **S24, 20, 18, 15** **C15** **MC18**				**⁵⁄₁₆**	**³⁄₈**	**¹⁄₂**
	A325	N	—	137	165	173
		X	—	137	165	216
		SC Class A	STD	116	116	116
			OVS	98.9	98.9	98.9
			SSLT	98.9	98.9	98.9
		SC Class B	STD	137	165	173
			OVS	128	150	150
			SSLT	137	150	150
	A490	N	—	137	165	216
		X	—	137	165	219
		SC Class A	STD	137	146	146
			OVS	124	124	124
			SSLT	124	124	124
		SC Class B	STD	137	165	216
			OVS	128	153	188
			SSLT	137	165	188

Beam Web Design Strength per Inch Thickness, kips/in.

For F_{ub} = 58 ksi: 365 kips/in.
For F_{ub} = 65 ksi: 410 kips/in.

Welds: F_{EXX} = 70 ksi and HSS F_y = 46 ksi

Weld Size, in.	**ϕR_n, kips**	**Min. HSS Thickness, in.**
³⁄₈	120	0.34
⁵⁄₁₆	100	0.28
¹⁄₄	79.9	0.22

Notes:
STD = Standard holes
OVS = Oversized holes
SSLT = Short-slotted holes transverse
    to direction of load

N = Threads included
X = Threads excluded
SC = Slip critical

F_y = 36 ksi
F_u = 58 ksi

Table 4-4 (cont.).
Double-Angle Connections

⅞-in. Bolts

3 Rows

W18, 16, 14, 12, 10*
S18, 15, 12
C15, 12
MC18, 13, 12

*Limited to W10×12, 15, 17, 19, 22, 26, 30

	Bolt and Angle Design Strength, kips					
ASTM Desig.	Thread Cond.	Hole Type	Angle Thickness, in.		
			5/16	3/8	1/2
A325	N	—	89.7	108	130
	X	—	89.7	108	144
	SC Class A	STD	87.3	87.3	87.3
		OVS	74.2	74.2	74.2
		SSLT	74.2	74.2	74.2
	SC Class B	STD	89.7	108	130
		OVS	83.4	100	112
		SSLT	89.7	108	112
A490	N	—	89.7	108	144
	X	—	89.7	108	144
	SC Class A	STD	89.7	108	110
		OVS	83.4	93.2	93.2
		SSLT	89.7	93.2	93.2
	SC Class B	STD	89.7	108	144
		OVS	83.4	100	133
		SSLT	89.7	108	141

Beam Web Design Strength per Inch Thickness, kips/in.

For F_{ub} = 58 ksi: 274 kips/in.
For F_{ub} = 65 ksi: 307 kips/in.

Welds: F_{EXX} = 70 ksi and HSS F_y = 46 ksi

Weld Size, in.	ϕR_n, kips	Min. HSS Thickness, in.
3/8	72.2	0.34
5/16	60.1	0.28
1/4	48.1	0.22

Notes:
STD = Standard holes
OVS = Oversized holes
SSLT = Short-slotted holes transverse to direction of load

N = Threads included
X = Threads excluded
SC = Slip critical

F_y = 50 ksi

F_u = 65 ksi

Table 4-4 (cont.).
Double-Angle Connections

$\frac{7}{8}$-in. Bolts 3 Rows	Bolt and Angle Design Strength, kips					
	ASTM Desig.	**Thread Cond.**	**Hole Type**	Angle Thickness, in.		
				$\frac{5}{16}$	$\frac{3}{8}$	$\frac{1}{2}$
W18, 16, 14, 12, 10*	A325	N	—	101	121	130
S18, 15, 12		X	—	101	121	161
C15, 12		SC Class A	STD	87.3	87.3	87.3
MC18, 13, 12			OVS	74.2	74.2	74.2
			SSLT	74.2	74.2	74.2
*Limited to W10×12, 15, 17, 19, 22, 26, 30		SC Class B	STD	101	121	130
			OVS	93.4	112	112
			SSLT	101	112	112
	A490	N	—	101	121	161
		X	—	101	121	161
		SC Class A	STD	101	110	110
			OVS	93.2	93.2	93.2
			SSLT	93.2	93.2	93.2
		SC Class B	STD	101	121	161
			OVS	93.4	112	141
			SSLT	101	121	141

Beam Web Design Strength per Inch Thickness, kips/in.

For F_{ub} = 58 ksi: 274 kips/in.
For F_{ub} = 65 ksi: 307 kips/in.

Welds: F_{EXX} = 70 ksi and HSS F_y = 46 ksi

Weld Size, in.	ϕR_n, kips	Min. HSS Thickness, in.
$\frac{3}{8}$	72.2	0.34
$\frac{5}{16}$	60.1	0.28
$\frac{1}{4}$	48.1	0.22

Notes:
STD = Standard holes
OVS = Oversized holes
SSLT = Short-slotted holes transverse to direction of load

N = Threads included
X = Threads excluded
SC = Slip critical

F_y = 36 ksi
F_u = 58 ksi

Table 4-4 (cont.).
Double-Angle Connections

$\frac{7}{8}$-in. Bolts 2 Rows W12, 10, 8 S12, 10, 8 C12, 10, 9, 8 MC13, 12, 10, 9, 8	Bolt and Angle Design Strength, kips					
	ASTM Desig.	Thread Cond.	Hole Type	Angle Thickness, in.		
				$\frac{5}{16}$	$\frac{3}{8}$	$\frac{1}{2}$
	A325	N	—	57.1	68.5	86.6
		X	—	57.1	68.5	91.4
		SC Class A	STD	57.1	58.2	58.2
			OVS	49.4	49.4	49.4
			SSLT	49.4	49.4	49.4
		SC Class B	STD	57.1	68.5	86.6
			OVS	52.9	63.4	74.9
			SSLT	57.1	68.5	74.9
	A490	N	—	57.1	68.5	91.4
		X	—	57.1	68.5	91.4
		SC Class A	STD	57.1	68.5	73.1
			OVS	52.9	62.1	62.1
			SSLT	57.1	62.1	62.1
		SC Class B	STD	57.1	68.5	91.4
			OVS	52.9	63.4	84.6
			SSLT	57.1	68.5	91.4

Beam Web Design Strength per Inch Thickness, kips/in.

For F_{ub} = 58 ksi: 183 kips/in.
For F_{ub} = 65 ksi: 205 kips/in.

Welds: F_{EXX} = 70 ksi and HSS F_y = 46 ksi

Weld Size, in.	ϕR_n, kips	Min. HSS Thickness, in.
$\frac{3}{8}$	32.8	0.34
$\frac{5}{16}$	27.3	0.28
$\frac{1}{4}$	21.9	0.22

Notes:
STD = Standard holes
OVS = Oversized holes
SSLT = Short-slotted holes transverse
     to direction of load

N = Threads included
X = Threads excluded
SC = Slip critical

F_y = 50 ksi
F_u = 65 ksi

Table 4-4 (cont.). Double-Angle Connections

⅞-in. Bolts

2 Rows

W12, 10, 8
S12, 10, 8
C12, 10, 9, 8
MC13, 12, 10, 9, 8

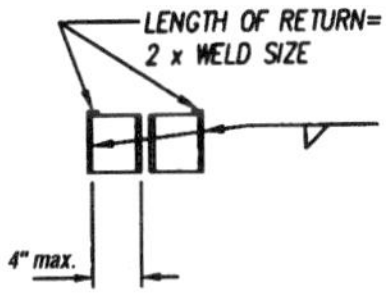

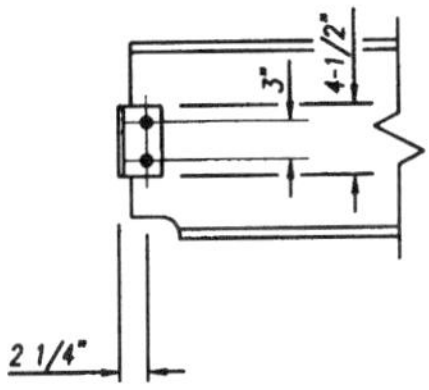

Bolt and Angle Design Strength, kips					
ASTM Desig.	Thread Cond.	Hole Type	Angle Thickness, in. ⁵⁄₁₆	⅜	½
A325	N	—	64.0	76.8	86.6
	X	—	64.0	76.8	102
	SC Class A	STD	58.2	58.2	58.2
		OVS	49.4	49.4	49.4
		SSLT	49.4	49.4	49.4
	SC Class B	STD	64.0	76.8	86.6
		OVS	59.2	71.1	74.9
		SSLT	64.0	74.9	74.9
A490	N	—	64.0	76.8	102
	X	—	64.0	76.8	102
	SC Class A	STD	64.0	73.1	73.1
		OVS	59.2	62.1	62.1
		SSLT	62.1	62.1	62.1
	SC Class B	STD	64.0	76.8	102
		OVS	59.2	71.1	94.1
		SSLT	64.0	76.8	94.1

Beam Web Design Strength per Inch Thickness, kips/in.

For F_{ub} = 58 ksi: 183 kips/in.
For F_{ub} = 65 ksi: 205 kips/in.

Welds: F_{EXX} = 70 ksi and HSS F_y = 46 ksi

Weld Size, in.	ϕR_n, kips	Min. HSS Thickness, in.
⅜	32.8	0.34
⁵⁄₁₆	27.3	0.28
¼	21.9	0.22

Notes:
STD = Standard holes
OVS = Oversized holes
SSLT = Short-slotted holes transverse to direction of load
N = Threads included
X = Threads excluded
SC = Slip critical

F_y = 36 ksi
F_u = 58 ksi

Table 4-4 (cont.).
Double-Angle Connections

1-in. Bolts	Bolt and Angle Design Strength, kips					
12 Rows	**ASTM Desig.**	**Thread Cond.**	**Hole Type**	**Angle Thickness, in.**		
W44				$5/16$	$3/8$	$1/2$
	A325	N	—	358	429	573
		X	—	358	429	573
		SC Class A	STD	358	429	456
			OVS	323	387	388
			SSLT	358	388	388
		SC Class B	STD	358	429	573
			OVS	323	387	516
			SSLT	358	429	573
	A490	N	—	358	429	573
		X	—	358	429	573
		SC Class A	STD	358	429	573
			OVS	323	387	487
			SSLT	358	429	487
		SC Class B	STD	358	429	573
			OVS	323	387	516
			SSLT	358	429	573

Beam Web Design Strength per Inch Thickness, kips/in.

For F_{ub} = 58 ksi: 1,253 kips/in.
For F_{ub} = 65 ksi: 1,404 kips/in.

Welds: F_{EXX} = 70 ksi and HSS F_y = 46 ksi

Weld Size, in.	ϕR_n, kips	Min. HSS Thickness, in.
$3/8$	550	0.34
$5/16$	458	0.28
$1/4$	366	0.22

Notes:
STD = Standard holes   N = Threads included
OVS = Oversized holes   X = Threads excluded
SSLT = Short-slotted holes transverse   SC = Slip critical
      to direction of load

	$F_y = 50$ ksi
	$F_u = 65$ ksi

Table 4-4 (cont.).
Double-Angle Connections

1-in. Bolts	Bolt and Angle Design Strength, kips					
12 Rows	ASTM Desig.	Thread Cond.	Hole Type	Angle Thickness, in.		
W44				$\frac{5}{16}$	$\frac{3}{8}$	$\frac{1}{2}$
	A325	N	—	401	481	642
		X	—	401	481	642
		SC Class A	STD	401	456	456
			OVS	362	388	388
			SSLT	388	388	388
		SC Class B	STD	401	481	642
			OVS	362	434	579
			SSLT	401	481	588
	A490	N	—	401	481	642
		X	—	401	481	642
		SC Class A	STD	401	481	573
			OVS	362	434	487
			SSLT	401	481	487
		SC Class B	STD	401	481	642
			OVS	362	434	579
			SSLT	401	481	642

Beam Web Design Strength per Inch Thickness, kips/in.

For $F_{ub} = 58$ ksi: 1,253 kips/in.
For $F_{ub} = 65$ ksi: 1,404 kips/in.

Welds: $F_{EXX} = 70$ ksi and HSS $F_y = 46$ ksi

Weld Size, in.	ϕR_n, kips	Min. HSS Thickness, in.
$\frac{3}{8}$	550	0.34
$\frac{5}{16}$	458	0.28
$\frac{1}{4}$	366	0.22

Notes:
STD = Standard holes   N = Threads included
OVS = Oversized holes   X = Threads excluded
SSLT = Short-slotted holes transverse   SC = Slip critical
      to direction of load

F_y = 36 ksi
F_u = 58 ksi

Table 4-4 (cont.).
Double-Angle Connections

1-in. Bolts	Bolt and Angle Design Strength, kips					
11 Rows	**ASTM Desig.**	**Thread Cond.**	**Hole Type**	**Angle Thickness, in.**		
W44, 40				$5/16$	$3/8$	$1/2$
	A325	N	—	327	393	524
		X	—	327	393	524
		SC Class A	STD	327	393	418
			OVS	295	354	356
			SSLT	327	356	356
		SC Class B	STD	327	393	524
			OVS	295	354	472
			SSLT	327	393	524
	A490	N	—	327	393	524
		X	—	327	393	524
		SC Class A	STD	327	393	524
			OVS	295	354	446
			SSLT	327	393	446
		SC Class B	STD	327	393	524
			OVS	295	354	472
			SSLT	327	393	524

Beam Web Design Strength per Inch Thickness, kips/in.

For F_{ub} = 58 ksi: 1,148 kips/in.
For F_{ub} = 65 ksi: 1,287 kips/in.

Welds: F_{EXX} = 70 ksi and HSS F_y = 46 ksi

Weld Size, in.	ϕR_n, kips	Min. HSS Thickness, in.
$3/8$	496	0.34
$5/16$	414	0.28
$1/4$	331	0.22

Notes:
STD = Standard holes
OVS = Oversized holes
SSLT = Short-slotted holes transverse to direction of load

N = Threads included
X = Threads excluded
SC = Slip critical

F_y = 50 ksi
F_u = 65 ksi

Table 4-4 (cont.).
Double-Angle Connections

1-in. Bolts	Bolt and Angle Design Strength, kips					
11 Rows	ASTM Desig.	Thread Cond.	Hole Type	Angle Thickness, in.		
W44, 40				$5/16$	$3/8$	$1/2$
	A325	N	—	367	440	587
		X	—	367	440	587
		SC Class A	STD	367	418	418
			OVS	331	356	356
			SSLT	356	356	356
		SC Class B	STD	367	440	587
			OVS	331	397	529
			SSLT	367	440	539
	A490	N	—	367	440	587
		X	—	367	440	587
		SC Class A	STD	367	440	525
			OVS	331	397	446
			SSLT	367	440	446
		SC Class B	STD	367	440	587
			OVS	331	397	529
			SSLT	367	440	587

Beam Web Design Strength per Inch Thickness, kips/in.

For F_{ub} = 58 ksi: 1,148 kips/in.
For F_{ub} = 65 ksi: 1,287 kips/in.

Welds: F_{EXX} = 70 ksi and HSS F_y = 46 ksi

Weld Size, in.	ϕR_n, kips	Min. HSS Thickness, in.
$3/8$	496	0.34
$5/16$	414	0.28
$1/4$	331	0.22

Notes:
STD = Standard holes   N = Threads included
OVS = Oversized holes   X = Threads excluded
SSLT = Short-slotted holes transverse   SC = Slip critical
        to direction of load

F_y = 36 ksi
F_u = 58 ksi

Table 4-4 (cont.).
Double-Angle Connections

1-in. Bolts	Bolt and Angle Design Strength, kips					
10 Rows	ASTM Desig.	Thread Cond.	Hole Type	Angle Thickness, in.		
W44, 40, 36				$^5/_{16}$	$^3/_8$	$^1/_2$
	A325	N	—	297	356	475
		X	—	297	356	475
		SC Class A	STD	297	356	380
			OVS	268	321	323
			SSLT	297	323	323
		SC Class B	STD	297	356	475
			OVS	268	321	428
			SSLT	297	356	475
	A490	N	—	297	356	475
		X	—	297	356	475
		SC Class A	STD	297	356	475
			OVS	268	321	406
			SSLT	297	356	406
		SC Class B	STD	297	356	475
			OVS	268	321	428
			SSLT	297	356	475

Beam Web Design Strength per Inch Thickness, kips/in.

For F_{ub} = 58 ksi: 1,044 kips/in.
For F_{ub} = 65 ksi: 1,170 kips/in.

Welds: F_{EXX} = 70 ksi and HSS F_y = 46 ksi

Weld Size, in.	ϕR_n, kips	Min. HSS Thickness, in.
$^3/_8$	443	0.34
$^5/_{16}$	369	0.28
$^1/_4$	295	0.22

Notes:
STD = Standard holes   N = Threads included
OVS = Oversized holes   X = Threads excluded
SSLT = Short-slotted holes transverse   SC = Slip critical
       to direction of load

				$F_y = 50$ ksi
				$F_u = 65$ ksi

Table 4-4 (cont.).
Double-Angle Connections

1-in. Bolts	Bolt and Angle Design Strength, kips					
10 Rows	ASTM Desig.	Thread Cond.	Hole Type	Angle Thickness, in.		
W44, 40, 36				$5/16$	$3/8$	$1/2$
	A325	N	—	333	399	532
		X	—	333	399	532
		SC Class A	STD	333	380	380
			OVS	300	323	323
			SSLT	323	323	323
		SC Class B	STD	333	399	532
			OVS	300	360	480
			SSLT	333	399	490
	A490	N	—	333	399	532
		X	—	333	399	532
		SC Class A	STD	333	399	477
			OVS	300	360	406
			SSLT	333	399	406
		SC Class B	STD	333	399	532
			OVS	300	360	480
			SSLT	333	399	532

Beam Web Design Strength per Inch Thickness, kips/in.

For $F_{ub} = 58$ ksi: 1,044 kips/in.
For $F_{ub} = 65$ ksi: 1,170 kips/in.

Welds: $F_{EXX} = 70$ ksi and HSS $F_y = 46$ ksi

Weld Size, in.	ϕR_n, kips	Min. HSS Thickness, in.
$3/8$	443	0.34
$5/16$	369	0.28
$1/4$	295	0.22

Notes:
STD = Standard holes   N = Threads included
OVS = Oversized holes   X = Threads excluded
SSLT = Short-slotted holes transverse   SC = Slip critical
      to direction of load

F_y = 36 ksi
F_u = 58 ksi

Table 4-4 (cont.).
Double-Angle Connections

1-in. Bolts	Bolt and Angle Design Strength, kips					
9 Rows	ASTM Desig.	Thread Cond.	Hole Type	Angle Thickness, in.		
W44, 40, 36, 33				$\frac{5}{16}$	$\frac{3}{8}$	$\frac{1}{2}$
	A325	N	—	266	320	426
		X	—	266	320	426
		SC Class A	STD	266	320	342
			OVS	240	288	291
			SSLT	266	291	291
		SC Class B	STD	266	320	426
			OVS	240	288	384
			SSLT	266	320	426
	A490	N	—	266	320	426
		X	—	266	320	426
		SC Class A	STD	266	320	426
			OVS	240	288	365
			SSLT	266	320	365
		SC Class B	STD	266	320	426
			OVS	240	288	384
			SSLT	266	320	426

Beam Web Design Strength per Inch Thickness, kips/in.

For F_{ub} = 58 ksi: 940 kips/in.
For F_{ub} = 65 ksi: 1,053 kips/in.

Welds: F_{EXX} = 70 ksi and HSS F_y = 46 ksi

Weld Size, in.	ϕR_n, kips	Min. HSS Thickness, in.
$\frac{3}{8}$	389	0.34
$\frac{5}{16}$	324	0.28
$\frac{1}{4}$	259	0.22

Notes:
STD = Standard holes
OVS = Oversized holes
SSLT = Short-slotted holes transverse to direction of load

N = Threads included
X = Threads excluded
SC = Slip critical

	F_y = 50 ksi
	F_u = 65 ksi

Table 4-4 (cont.).
Double-Angle Connections

1-in. Bolts 9 Rows W44, 40, 36, 33	Bolt and Angle Design Strength, kips					
	ASTM Desig.	Thread Cond.	Hole Type	Angle Thickness, in.		
				$5/16$	$3/8$	$1/2$
	A325	N	—	299	358	478
		X	—	299	358	478
		SC Class A	STD	299	342	342
			OVS	269	291	291
			SSLT	291	291	291
		SC Class B	STD	299	358	478
			OVS	269	323	430
			SSLT	299	358	441
	A490	N	—	299	358	478
		X	—	299	358	478
		SC Class A	STD	299	358	430
			OVS	269	323	365
			SSLT	299	358	365
		SC Class B	STD	299	358	478
			OVS	269	323	430
			SSLT	299	358	478

Beam Web Design Strength per Inch Thickness, kips/in.

For F_{ub} = 58 ksi: 940 kips/in.
For F_{ub} = 65 ksi: 1,053 kips/in.

Welds: F_{EXX} = 70 ksi and HSS F_y = 46 ksi

Weld Size, in.	ϕR_n, kips	Min. HSS Thickness, in.
$3/8$	389	0.34
$5/16$	324	0.28
$1/4$	259	0.22

Notes:
STD = Standard holes
OVS = Oversized holes
SSLT = Short-slotted holes transverse to direction of load

N = Threads included
X = Threads excluded
SC = Slip critical

$F_y = 36$ ksi
$F_u = 58$ ksi

Table 4-4 (cont.).
Double-Angle Connections

1-in. Bolts 8 Rows W44, 40, 36, 33, 30	Bolt and Angle Design Strength, kips					
	ASTM Desig.	Thread Cond.	Hole Type	Angle Thickness, in.		
				$\frac{5}{16}$	$\frac{3}{8}$	$\frac{1}{2}$
	A325	N	—	236	283	377
		X	—	236	283	377
		SC Class A	STD	236	283	304
			OVS	212	255	259
			SSLT	236	259	259
		SC Class B	STD	236	283	377
			OVS	212	255	340
			SSLT	236	283	377
	A490	N	—	236	283	377
		X	—	236	283	377
		SC Class A	STD	236	283	377
			OVS	212	255	325
			SSLT	236	283	325
		SC Class B	STD	236	283	377
			OVS	212	255	340
			SSLT	236	283	377

Beam Web Design Strength per Inch Thickness, kips/in.

For $F_{ub} = 58$ ksi: 835 kips/in.
For $F_{ub} = 65$ ksi: 936 kips/in.

Welds: $F_{EXX} = 70$ ksi and HSS $F_y = 46$ ksi

Weld Size, in.	ϕR_n, kips	Min. HSS Thickness, in.
$\frac{3}{8}$	335	0.34
$\frac{5}{16}$	279	0.28
$\frac{1}{4}$	223	0.22

Notes:
STD = Standard holes
OVS = Oversized holes
SSLT = Short-slotted holes transverse to direction of load

N = Threads included
X = Threads excluded
SC = Slip critical

	$F_y = 50$ ksi
	$F_u = 65$ ksi

Table 4-4 (cont.).
Double-Angle Connections

1-in. Bolts	Bolt and Angle Design Strength, kips					
8 Rows	ASTM Desig.	Thread Cond.	Hole Type	Angle Thickness, in.		
W44, 40, 36, 33, 30				$5/16$	$3/8$	$1/2$
	A325	N	—	264	317	423
		X	—	264	317	423
		SC Class A	STD	264	304	304
			OVS	238	259	259
			SSLT	259	259	259
		SC Class B	STD	264	317	423
			OVS	238	286	381
			SSLT	264	317	392
	A490	N	—	264	317	423
		X	—	264	317	423
		SC Class A	STD	264	317	382
			OVS	238	286	325
			SSLT	264	317	325
		SC Class B	STD	264	317	423
			OVS	238	286	381
			SSLT	264	317	423

Beam Web Design Strength per Inch Thickness, kips/in.

For $F_{ub} = 58$ ksi: 835 kips/in.
For $F_{ub} = 65$ ksi: 936 kips/in.

Welds: $F_{EXX} = 70$ ksi and HSS $F_y = 46$ ksi

Weld Size, in.	ϕR_n, kips	Min. HSS Thickness, in.
$3/8$	335	0.34
$5/16$	279	0.28
$1/4$	223	0.22

Notes:
STD = Standard holes
OVS = Oversized holes
SSLT = Short-slotted holes transverse to direction of load

N = Threads included
X = Threads excluded
SC = Slip critical

F_y = 36 ksi

F_u = 58 ksi

Table 4-4 (cont.).
Double-Angle Connections

1-in. Bolts	Bolt and Angle Design Strength, kips					
7 Rows	**ASTM Desig.**	**Thread Cond.**	**Hole Type**	**Angle Thickness, in.**		
W44, 40, 36, 33, 30, 27, 24 S24				$^{5}\!/_{16}$	$^{3}\!/_{8}$	$^{1}\!/_{2}$
	A325	N	—	205	246	329
		X	—	205	246	329
		SC Class A	STD	205	246	266
			OVS	185	222	226
			SSLT	205	226	226
		SC Class B	STD	205	246	329
			OVS	185	222	296
			SSLT	205	246	329
	A490	N	—	205	246	329
		X	—	205	246	329
		SC Class A	STD	205	246	329
			OVS	185	222	284
			SSLT	205	246	284
		SC Class B	STD	205	246	329
			OVS	185	222	296
			SSLT	205	246	329

Beam Web Design Strength per Inch Thickness, kips/in.

For F_{ub} = 58 ksi: 731 kips/in.
For F_{ub} = 65 ksi: 819 kips/in.

Welds: F_{EXX} = 70 ksi and HSS F_y = 46 ksi

Weld Size, in.	ϕR_n, kips	Min. HSS Thickness, in.
$^{3}\!/_{8}$	280	0.34
$^{5}\!/_{16}$	234	0.28
$^{1}\!/_{4}$	187	0.22

Notes:
STD = Standard holes
OVS = Oversized holes
SSLT = Short-slotted holes transverse
    to direction of load

N = Threads included
X = Threads excluded
SC = Slip critical

				$F_y = 50$ ksi
				$F_u = 65$ ksi

Table 4-4 (cont.).
Double-Angle Connections

1-in. Bolts	**Bolt and Angle Design Strength, kips**					
7 Rows	**ASTM Desig.**	**Thread Cond.**	**Hole Type**	**Angle Thickness, in.**		
W44, 40, 36, 33, 30, 27, 24 S24				$5/16$	$3/8$	$1/2$
	A325	N	—	230	276	368
		X	—	230	276	368
		SC Class A	STD	230	266	266
			OVS	207	226	226
			SSLT	226	226	226
		SC Class B	STD	230	276	368
			OVS	207	249	331
			SSLT	230	276	343
	A490	N	—	230	276	368
		X	—	230	276	368
		SC Class A	STD	230	276	334
			OVS	207	249	284
			SSLT	230	276	284
		SC Class B	STD	230	276	368
			OVS	207	249	331
			SSLT	230	276	368

Beam Web Design Strength per Inch Thickness, kips/in.

For $F_{ub} = 58$ ksi: 731 kips/in.
For $F_{ub} = 65$ ksi: 819 kips/in.

Welds: $F_{EXX} = 70$ ksi and HSS $F_y = 46$ ksi

Weld Size, in.	ϕR_n, **kips**	**Min. HSS Thickness, in.**
$3/8$	280	0.34
$5/16$	234	0.28
$1/4$	187	0.22

Notes:
STD = Standard holes
OVS = Oversized holes
SSLT = Short-slotted holes transverse
      to direction of load

N = Threads included
X = Threads excluded
SC = Slip critical

$F_y = 36$ ksi

$F_u = 58$ ksi

Table 4-4 (cont.).
Double-Angle Connections

1-in. Bolts 6 Rows W40, 36, 33, 30, 27, 24, 21 S24	Bolt and Angle Design Strength, kips					
	ASTM Desig.	Thread Cond.	Hole Type	Angle Thickness, in.		
				$\frac{5}{16}$	$\frac{3}{8}$	$\frac{1}{2}$
	A325	N	—	175	210	280
		X	—	175	210	280
		SC Class A	STD	175	210	228
			OVS	157	189	194
			SSLT	175	194	194
		SC Class B	STD	175	210	280
			OVS	157	189	252
			SSLT	175	210	280
	A490	N	—	175	210	280
		X	—	175	210	280
		SC Class A	STD	175	210	280
			OVS	157	189	243
			SSLT	175	210	243
		SC Class B	STD	175	210	280
			OVS	157	189	252
			SSLT	175	210	280

Beam Web Design Strength per Inch Thickness, kips/in.

For $F_{ub} = 58$ ksi: 626 kips/in.
For $F_{ub} = 65$ ksi: 702 kips/in.

Welds: $F_{EXX} = 70$ ksi and HSS $F_y = 46$ ksi

Weld Size, in.	ϕR_n, kips	Min. HSS Thickness, in.
$\frac{3}{8}$	226	0.34
$\frac{5}{16}$	188	0.28
$\frac{1}{4}$	150	0.22

Notes:
STD = Standard holes   N = Threads included
OVS = Oversized holes   X = Threads excluded
SSLT = Short-slotted holes transverse   SC = Slip critical
    to direction of load

$F_y = 50$ ksi

$F_u = 65$ ksi

Table 4-4 (cont.).
Double-Angle Connections

1-in. Bolts 6 Rows W40, 36, 33, 30,27, 24, 21 S24						

Bolt and Angle Design Strength, kips

ASTM Desig.	Thread Cond.	Hole Type	Angle Thickness, in.		
			$5/16$	$3/8$	$1/2$
A325	N	—	196	235	314
	X	—	196	235	314
	SC Class A	STD	196	228	228
		OVS	176	194	194
		SSLT	194	194	194
	SC Class B	STD	196	235	314
		OVS	176	211	282
		SSLT	196	235	294
A490	N	—	196	235	314
	X	—	196	235	314
	SC Class A	STD	196	235	286
		OVS	176	211	243
		SSLT	196	235	243
	SC Class B	STD	196	235	314
		OVS	176	211	282
		SSLT	196	235	314

Beam Web Design Strength per Inch Thickness, kips/in.

For $F_{ub} = 58$ ksi: 626 kips/in.
For $F_{ub} = 65$ ksi: 702 kips/in.

Welds: $F_{EXX} = 70$ ksi and HSS $F_y = 46$ ksi

Weld Size, in.	ϕR_n, kips	Min. HSS Thickness, in.
$3/8$	226	0.34
$5/16$	188	0.28
$1/4$	150	0.22

Notes:
STD = Standard holes
OVS = Oversized holes
SSLT = Short-slotted holes transverse
    to direction of load

N = Threads included
X = Threads excluded
SC = Slip critical

F_y = 36 ksi
F_u = 58 ksi

Table 4-4 (cont.).
Double-Angle Connections

1-in. Bolts	Bolt and Angle Design Strength, kips					
5 rows	ASTM Desig.	Thread Cond.	Hole Type	Angle Thickness, in.		
				$5/16$	$3/8$	$1/2$
W30, 27, 24, 21, 18 **S24, 20, 18** **MC18**	A325	N	—	144	173	231
		X	—	144	173	231
		SC Class A	STD	144	173	190
			OVS	130	156	162
			SSLT	144	162	162
		SC Class B	STD	144	173	231
			OVS	130	156	207
			SSLT	144	173	231
	A490	N	—	144	173	231
		X	—	144	173	231
		SC Class A	STD	144	173	231
			OVS	130	156	203
			SSLT	144	173	203
		SC Class B	STD	144	173	231
			OVS	130	156	207
			SSLT	144	173	231

Beam Web Design Strength per Inch Thickness, kips/in.

For F_{ub} = 58 ksi: 522 kips/in.
For F_{ub} = 65 ksi: 585 kips/in.

Welds: F_{EXX} = 70 ksi and HSS F_y = 46 ksi

Weld Size, in.	ϕR_n, kips	Min. HSS Thickness, in.
$3/8$	172	0.34
$5/16$	143	0.28
$1/4$	115	0.22

Notes:
STD = Standard holes   N = Threads included
OVS = Oversized holes   X = Threads excluded
SSLT = Short-slotted holes transverse   SC = Slip critical
      to direction of load

$F_y = 50$ ksi
$F_u = 65$ ksi

Table 4-4 (cont.).
Double-Angle Connections

1-in. Bolts	Bolt and Angle Design Strength, kips					
5 Rows	**ASTM Desig.**	**Thread Cond.**	**Hole Type**	**Angle Thickness, in.**		
				$\frac{5}{16}$	$\frac{3}{8}$	$\frac{1}{2}$
W30, 27, 24, 21, 18 **S24, 20, 18** **MC18**	A325	N	—	162	194	259
		X	—	162	194	259
		SC Class A	STD	162	190	190
			OVS	145	162	162
			SSLT	162	162	162
		SC Class B	STD	162	194	259
			OVS	145	174	233
			SSLT	162	194	245
	A490	N	—	162	194	259
		X	—	162	194	259
		SC Class A	STD	162	194	239
			OVS	145	174	203
			SSLT	162	194	203
		SC Class B	STD	162	194	259
			OVS	145	174	233
			SSLT	162	194	259

Beam Web Design Strength per Inch Thickness, kips/in.

For $F_{ub} = 58$ ksi: 522 kips/in.
For $F_{ub} = 65$ ksi: 585 kips/in.

Welds: $F_{EXX} = 70$ ksi and HSS $F_y = 46$ ksi

Weld Size, in.	ϕR_n, kips	Min. HSS Thickness, in.
$\frac{3}{8}$	172	0.34
$\frac{5}{16}$	143	0.28
$\frac{1}{4}$	115	0.22

Notes:
STD = Standard holes
OVS = Oversized holes
SSLT = Short-slotted holes transverse
      to direction of load

N = Threads included
X = Threads excluded
SC = Slip critical

$F_y = 36$ ksi
$F_u = 58$ ksi

Table 4-4 (cont.).
Double-Angle Connections

1-in. Bolts 4 Rows W24, 21, 18, 16 S24, 20, 18, 15 C15 MC18	Bolt and Angle Design Strength, kips					
	ASTM Desig.	Thread Cond.	Hole Type	Angle Thickness, in.		
				$\frac{5}{16}$	$\frac{3}{8}$	$\frac{1}{2}$
	A325	N	—	114	137	182
		X	—	114	137	182
		SC Class A	STD	114	137	152
			OVS	102	123	129
			SSLT	114	129	129
		SC Class B	STD	114	137	182
			OVS	102	123	163
			SSLT	114	137	182
	A490	N	—	114	137	182
		X	—	114	137	182
		SC Class A	STD	114	137	182
			OVS	102	123	162
			SSLT	114	137	162
		SC Class B	STD	114	137	182
			OVS	102	123	163
			SSLT	114	137	182

Beam Web Design Strength per Inch Thickness, kips/in.

For $F_{ub} = 58$ ksi: 418 kips/in.
For $F_{ub} = 65$ ksi: 468 kips/in.

Welds: $F_{EXX} = 70$ ksi and HSS $F_y = 46$ ksi

Weld Size, in.	ϕR_n, kips	Min. HSS Thickness, in.
$\frac{3}{8}$	120	0.34
$\frac{5}{16}$	100	0.28
$\frac{1}{4}$	79.9	0.22

Notes:
STD = Standard holes
OVS = Oversized holes
SSLT = Short-slotted holes transverse to direction of load

N = Threads included
X = Threads excluded
SC = Slip critical

	F_y = 50 ksi
	F_u = 65 ksi

Table 4-4 (cont.).
Double-Angle Connections

1-in. Bolts	Bolt and Angle Design Strength, kips					
4 Rows	ASTM Desig.	Thread Cond.	Hole Type	Angle Thickness, in.		
				$\frac{5}{16}$	$\frac{3}{8}$	$\frac{1}{2}$
W24, 21, 18, 16 S24, 20, 18, 15 C15 MC18	A325	N	—	128	153	204
		X	—	128	153	204
		SC Class A	STD	128	152	152
			OVS	114	129	129
			SSLT	128	129	129
		SC Class B	STD	128	153	204
			OVS	114	137	183
			SSLT	128	153	196
	A490	N	—	128	153	204
		X	—	128	153	204
		SC Class A	STD	128	153	191
			OVS	114	137	162
			SSLT	128	153	162
		SC Class B	STD	128	153	204
			OVS	114	137	183
			SSLT	128	153	204

Beam Web Design Strength per Inch Thickness, kips/in.

For F_{ub} = 58 ksi: 418 kips/in.
For F_{ub} = 65 ksi: 468 kips/in.

Welds: F_{EXX} = 70 ksi and HSS F_y = 46 ksi

Weld Size, in.	ϕR_n, kips	Min. HSS Thickness, in.
$\frac{3}{8}$	120	0.34
$\frac{5}{16}$	100	0.28
$\frac{1}{4}$	79.9	0.22

Notes:
STD = Standard holes
OVS = Oversized holes
SSLT = Short-slotted holes transverse to direction of load

N = Threads included
X = Threads excluded
SC = Slip critical

F_y = 36 ksi
F_u = 58 ksi

Table 4-4 (cont.).
Double-Angle Connections

<table>
<tr><td colspan="2" align="center">1-in. Bolts

3 Rows</td><td colspan="7" align="center">Bolt and Angle Design Strength, kips</td></tr>
<tr><td colspan="2" rowspan="2" align="center">W18, 16, 14, 12, 10*
S18, 15, 12
C15, 12
MC18, 13, 12</td><td rowspan="2" align="center">ASTM
Desig.</td><td rowspan="2" align="center">Thread
Cond.</td><td rowspan="2" align="center">Hole
Type</td><td colspan="3" align="center">Angle Thickness, in.</td></tr>
<tr><td align="center">5/16</td><td align="center">3/8</td><td align="center">1/2</td></tr>
<tr><td align="center" rowspan="8">A325</td><td align="center">N</td><td align="center">—</td><td align="center">83.4</td><td align="center">100</td><td align="center">133</td></tr>
<tr><td align="center">X</td><td align="center">—</td><td align="center">83.4</td><td align="center">100</td><td align="center">133</td></tr>
<tr><td align="center" rowspan="3">SC
Class A</td><td align="center">STD</td><td align="center">83.4</td><td align="center">100</td><td align="center">114</td></tr>
<tr><td align="center">OVS</td><td align="center">74.5</td><td align="center">89.5</td><td align="center">97.0</td></tr>
<tr><td align="center">SSLT</td><td align="center">83.4</td><td align="center">97.0</td><td align="center">97.0</td></tr>
<tr><td align="center" rowspan="3">SC
Class B</td><td align="center">STD</td><td align="center">83.4</td><td align="center">100</td><td align="center">133</td></tr>
<tr><td align="center">OVS</td><td align="center">74.5</td><td align="center">89.5</td><td align="center">119</td></tr>
<tr><td align="center">SSLT</td><td align="center">83.4</td><td align="center">100</td><td align="center">133</td></tr>
<tr><td colspan="2" align="center">*Limited to W10×12, 15,
17, 19, 22, 26, 30</td><td align="center" rowspan="8">A490</td><td align="center">N</td><td align="center">—</td><td align="center">83.4</td><td align="center">100</td><td align="center">133</td></tr>
<tr><td colspan="2" rowspan="7"></td><td align="center">X</td><td align="center">—</td><td align="center">83.4</td><td align="center">100</td><td align="center">133</td></tr>
<tr><td align="center" rowspan="3">SC
Class A</td><td align="center">STD</td><td align="center">83.4</td><td align="center">100</td><td align="center">133</td></tr>
<tr><td align="center">OVS</td><td align="center">74.5</td><td align="center">89.5</td><td align="center">119</td></tr>
<tr><td align="center">SSLT</td><td align="center">83.4</td><td align="center">100</td><td align="center">122</td></tr>
<tr><td align="center" rowspan="3">SC
Class B</td><td align="center">STD</td><td align="center">83.4</td><td align="center">100</td><td align="center">133</td></tr>
<tr><td align="center">OVS</td><td align="center">74.5</td><td align="center">89.5</td><td align="center">119</td></tr>
<tr><td align="center">SSLT</td><td align="center">83.4</td><td align="center">100</td><td align="center">133</td></tr>
</table>

Beam Web Design Strength per Inch Thickness, kips/in.

For F_{ub} = 58 ksi: 313 kips/in.
For F_{ub} = 65 ksi: 351 kips/in.

Welds: F_{EXX} = 70 ksi and HSS F_y = 46 ksi

Weld Size, in.	ϕR_n, kips	Min. HSS Thickness, in.
3/8	72.2	0.34
5/16	60.1	0.28
1/4	48.1	0.22

Notes:
STD = Standard holes   N = Threads included
OVS = Oversized holes   X = Threads excluded
SSLT = Short-slotted holes transverse   SC = Slip critical
　　　to direction of load

	F_y = 50 ksi
	F_u = 65 ksi

Table 4-4 (cont.).
Double-Angle Connections

1-in. Bolts	Bolt and Angle Design Strength, kips					
3 Rows	ASTM Desig.	Thread Cond.	Hole Type	Angle Thickness, in.		
				$^5/_{16}$	$^3/_8$	$^1/_2$
W18, 16, 14, 12, 10* **S18, 15, 12** **C15, 12** **C18, 13, 12**	A325	N	—	93.4	112	149
		X	—	93.4	112	149
		SC Class A	STD	93.4	112	114
*Limited to W10×12, 15, 17, 19, 22, 26, 30			OVS	83.5	97.0	97.0
			SSLT	93.4	97.0	97.0
		SC Class B	STD	93.4	112	149
			OVS	83.5	100	134
			SSLT	93.4	112	147
	A490	N	—	93.4	112	149
		X	—	93.4	112	149
		SC Class A	STD	93.4	112	143
			OVS	83.5	100	122
			SSLT	93.4	112	122
		SC Class B	STD	93.4	112	149
			OVS	83.5	100	134
			SSLT	93.4	112	149

Beam Web Design Strength per Inch Thickness, kips/in.

For F_{ub} = 58 ksi: 313 kips/in.
For F_{ub} = 65 ksi: 351 kips/in.

Welds: F_{EXX} = 70 ksi and HSS F_y = 46 ksi

Weld Size, in.	ϕR_n, kips	Min. HSS Thickness, in.
$^3/_8$	72.2	0.34
$^5/_{16}$	60.1	0.28
$^1/_4$	48.1	0.22

Notes:
STD = Standard holes   N = Threads included
OVS = Oversized holes   X = Threads excluded
SSLT = Short-slotted holes transverse   SC = Slip critical
        to direction of load

| $F_y = 36$ ksi |
| $F_u = 58$ ksi |

Table 4-4 (cont.).
Double-Angle Connections

1-in. Bolts	Bolt and Angle Design Strength, kips					
2 Rows	ASTM Desig.	Thread Cond.	Hole Type	Angle Thickness, in.		
				$\frac{5}{16}$	$\frac{3}{8}$	$\frac{1}{2}$
W12, 10, 8	A325	N	—	52.9	63.4	84.6
S12, 10, 8		X	—	52.9	63.4	84.6
C12, 10, 9, 8		SC Class A	STD	52.9	63.4	76.1
MC13, 12, 10, 9, 8			OVS	47.0	56.4	64.7
			SSLT	52.9	63.4	64.7
		SC Class B	STD	52.9	63.4	84.6
			OVS	47.0	56.4	75.2
			SSLT	52.9	63.4	84.6
	A490	N	—	52.9	63.4	84.6
		X	—	52.9	63.4	84.6
		SC Class A	STD	52.9	63.4	84.6
			OVS	47.0	56.4	75.2
			SSLT	52.9	63.4	81.1
		SC Class B	STD	52.9	63.4	84.6
			OVS	47.0	56.4	75.2
			SSLT	52.9	63.4	84.6

Beam Web Design Strength per Inch Thickness, kips/in.

For $F_{ub} = 58$ ksi: 209 kips/in.
For $F_{ub} = 65$ ksi: 234 kips/in.

Welds: $F_{EXX} = 70$ ksi and HSS $F_y = 46$ ksi

Weld Size, in.	ϕR_n, kips	Min. HSS Thickness, in.
$\frac{3}{8}$	32.8	0.34
$\frac{5}{16}$	27.3	0.28
$\frac{1}{4}$	21.9	0.22

Notes:
STD = Standard holes   N = Threads included
OVS = Oversized holes   X = Threads excluded
SSLT = Short-slotted holes transverse   SC = Slip critical
     to direction of load

F_y = 50 ksi
F_u = 65 ksi

Table 4-4 (cont.).
Double-Angle Connections

1-in. Bolts 2 Rows W12, 10, 8 S12, 10, 8 C12, 10, 9, 8 MC13, 12, 10, 9, 8	Bolt and Angle Design Strength, kips					
	ASTM Desig.	**Thread Cond.**	**Hole Type**	**Angle Thickness, in.**		
				$5/16$	$3/8$	$1/2$
	A325	N	—	59.2	71.1	94.8
		X	—	59.2	71.1	94.8
		SC Class A	STD	59.2	71.1	76.1
			OVS	52.7	63.2	64.7
			SSLT	59.2	64.7	64.7
		SC Class B	STD	59.2	71.1	94.8
			OVS	52.7	63.2	84.2
			SSLT	59.2	71.1	94.8
	A490	N	—	59.2	71.1	94.8
		X	—	59.2	71.1	94.8
		SC Class A	STD	59.2	71.1	94.8
			OVS	52.7	63.2	81.1
			SSLT	59.2	71.1	81.1
		SC Class B	STD	59.2	71.1	94.8
			OVS	52.7	63.2	84.2
			SSLT	59.2	71.1	94.8

Beam Web Design Strength per Inch Thickness, kips/in.

For F_{ub} = 58 ksi: 209 kips/in.
For F_{ub} = 65 ksi: 234 kips/in.

Welds: F_{EXX} = 70 ksi and HSS F_y = 46 ksi

Weld Size, in.	ϕR_n, kips	Min. HSS Thickness, in.
$3/8$	32.8	0.34
$5/16$	27.3	0.28
$1/4$	21.9	0.22

Notes:
STD = Standard holes   N = Threads included
OVS = Oversized holes   X = Threads excluded
SSLT = Short-slotted holes transverse   SC = Slip critical
       to direction of load

EXAMPLE 4.4—Double Angle Connection with Narrow HSS

Use Table 4-4 to design a double-angle connection for a W18×50 beam of A572 Gr 50 steel to an HSS8×8×¼ column for a factored reaction of 70 kips. Use ¾-in. diameter A325-N bolts in standard holes and A36 material for the angles.

Section properties

HSS8×8×¼	$t = 0.233$ in.	$B = 8$ in.
	$F_y = 46$ ksi	$F_u = 58$ ksi
W18×50	$t_w = 0.355$ in.	
	$F_y = 50$ ksi	$F_u = 65$ ksi

Design bolts and angles

From Table 4-4, for ¾-in. diameter A325-N bolts and angle material with $F_{yc} = 36$ ksi and $F_{uc} = 58$ ksi, select three rows of bolts and $\frac{5}{16}$-in. angle thickness.

$\phi R_n = 95.4$ kips > 70 kips **o.k.**

The W18 is in the acceptable group for angle length between T and $\frac{1}{2}T$.

Check beam web bearing

From the same Table with $F_{ub} = 65$ ksi

$\phi R_n = (263)(0.355) = 93.4$ kips > 70 kips **o.k.**

Determine the weld size

A $\frac{3}{8}$-in. weld size provides $\phi R_n = 72.2$ kips > 70 kips.

However, $t_{min} = 0.34$ in. > 0.233 in. and the weld strength is reduced.

$\phi R_n = (0.233/0.34)(72.2) = 49.5$ kips < 70 kips **n.g.**

Use a four bolt connection with ¼-in. welds that provides

$\phi R_n = 79.9$ kips with $t_{min} = 0.22$ in. < 0.233 in. **o.k.**

Check the minimum angle thickness

Select angle thickness equal to the weld size plus $\frac{1}{16}$-in.

$t_{c\,min} = \frac{1}{4} + \frac{1}{16} = \frac{5}{16}$-in. **o.k.**

Check fit of the double angles on the HSS

Flat width $b = 8 - 2(1.5)(0.233) = 7.30$ in.

Outside weld dimension with 3-in. angle legs

$= 0.355 + (2)(3) + (2)(5/16) = 6.98$ in. < 7.30 in. **o.k.**

As an alternative to using a 4-bolt connection, solve Equation 4-14 using $D \leq D_{eff}$ from Equation 4-19. This will also take advantage of the smaller eccentricity with the 3-in. angle leg.

$$D_{eff} = \frac{16\sqrt{2}\,[0.9(46)]}{0.75(70)}\,0.233 = 4.16$$

Use a $\frac{1}{4}$-in. weld ($D = 4$) to reduce the minimum angle thickness and try a 10-in. length in Equation 4-14.

$$\phi R_n = \frac{2.784(4)10}{\sqrt{1 + \dfrac{12.96 \times 3^2}{10^2}}}$$

$$= 75.6 \text{ kips} > 70 \text{ kips} \quad \textbf{o.k.}$$

Use a three bolt connection with L3×3½×⁵⁄₁₆ having a length of 10 in., so that $L_{ev} = 2$ in.

UNSTIFFENED SEATED CONNECTIONS

An unstiffened seated connection is made with a seat angle and a top angle, as illustrated in Figure 4-3. While the seat is assumed to carry the entire end reaction of the supported beam, the 4-in.-long 4×4-in. top angle must be placed as shown or in the optional side location for satisfactory performance and stability. To provide adequate flexibility for the connection, only the toe of the top angle is welded to the HSS. The thickness of the top angle is $\frac{1}{4}$-in. or greater to accommodate the minimum size fillet weld to the HSS or beam flange. Although there is no calculated horizontal shear force transfer between the beam flange and the seat angles, two $\frac{3}{4}$-in. diameter A325-N bolts should be used. Two bolts may also be used to connect the top angle to the beam flange, or a minimum size weld may be used across the toe of the top angle.

To have the welds on the flat of the HSS, 6-in. or 8-in. wide seats (angle length) may be used with the following HSS and are tabulated in Table 4-5. Seat widths other than 6 in. and 8 in. may be used.

Seat width (in.)	HSS width (in.)	Nominal HSS thickness (in.)
6	7	³⁄₁₆
	≥ 8	all
8	9	³⁄₁₆
	10	≤ ³⁄₈
	≥ 12	all

These sizes are based on an HSS corner radius of 1.5 times the nominal thickness and a weld size corresponding to the HSS wall shear strength from Equation 4-19.

Table 4-5 is a design aid for unstiffened seated connections. Seat design strengths are tabulated for angle material with $F_y = 36$ ksi and $F_u = 58$ ksi. This table will be conservative when used with higher strength angle material. Electrode strength is 70 ksi. All values are to be compared to factored loads.

The tabulated values are based on the limit states of Equations 4-5 and 4-8 for the angle material and beam web strength. Values are based on a nominal beam setback of $\frac{1}{2}$-in., which was increased to $\frac{3}{4}$-in. for calculation purposes to account for possible underrun in beam length.

Weld design strengths are tabulated using the vector analysis method of Equation 4-15 and a 70 ksi electrode material. The minimum HSS thicknesses associated with the weld strengths are from Equation 4-19. If the HSS thickness is less than the minimum tabulated value, the weld strength must be reduced proportionally.

Some common angle sizes with available ranges of thickness are indicated in Table 4-5. This is not intended to preclude the use of alternative angle sizes and thicknesses. The use of a longer outstanding angle leg than that indicated is permitted.

EXAMPLE 4.5—Unstiffened Seated Connection

Use Table 4-5 to design an unstiffened seated connection for a W21×62 beam of A572 Gr 50 steel to an HSS12×12×½ column for a factored reaction of 55 kips.

Section Properties

HSS12×12×½	$t = 0.465$ in.	$B = 12$ in.
	$F_y = 46$ ksi	$F_u = 58$ ksi
W21×62	$t_w = 0.400$ in.	$d = 20.99$ in.
	$F_y = 50$ ksi	$F_u = 65$ ksi

Design the seat angle and weld

An 8-in. angle length may be used with a 12-in. wide HSS.

Determine the required bearing length N_{req}. For local web yielding, from Equation 4-8b,

$$N_{min} = \frac{55 - 68.8}{20.0}$$

$$= -0.690 \text{ in.}$$

From Equation 4-8c,

$$N_{min} = 1\tfrac{3}{8}\text{-in.}$$

For web crippling when $N / d \leq 0.2$, from Equation 4-8d,

$$N_{min} = \frac{55 - 71.5}{5.36}$$

$$= -3.08 \text{ in.}$$

For web crippling when $N / d \leq 0.2$, from Equation 4-8e,

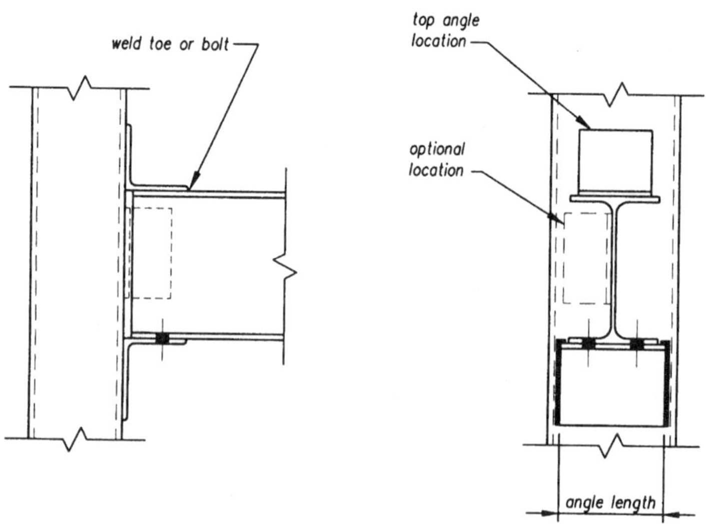

Fig. 4-3. Seated connection.

$$N_{min} = \frac{55 - 64.0}{7.15}$$

$$= -1.26 \text{ in.}$$

Note: generally, the value of N/d is not initially known and the larger value determined from Equations 4-8d and 4-8e can be used conservatively to determine the bearing length required for web crippling. If desired, however, an iterative approach can be used to determine which equation is appropriate.

For this beam and end reaction, the beam web strength exceeds the required strength (hence the negative bearing lengths) and the lower-bound bearing length controls ($N_{req} = k = 1\frac{3}{8}$-in.).

From Table 4-5 with $N_{req} = 1\frac{3}{8}$-in. and $L_a = 8$ in., an angle thickness of $\frac{5}{8}$-in. provides

$$\phi R_n = 57.9 \text{ kips} > 55 \text{ kips} \quad \textbf{o.k.}$$

An 8×4 angle with 4-in. leg outstanding and with $\frac{5}{16}$-in. fillet welds provides

$$\phi R_n = 66.8 \text{ kips} > 55 \text{ kips} \quad \textbf{o.k.}$$

Since t of the HSS is greater than t_{min} for the $\frac{5}{16}$-in. weld, no reduction in weld strength is required to account for shear in the HSS.

Connection to beam and top angle

Use a 4×4×$\frac{1}{4}$-in. top angle. Use the minimum weld of $\frac{3}{16}$-in. from LRFD Specification Table J2.4 across the toe of the angle for attachment to the HSS. Attach both the seat and top angles to the beam flanges with two $\frac{3}{4}$-in. diameter A325-N bolts.

STIFFENED SEATED CONNECTIONS

A stiffened seated connection is usually made with a seat plate and stiffening plate (or structural tee) and a top angle as illustrated in Figure 4-2b. While the seat is assumed to carry the entire end reaction of the supported beam, the 4-in.-long 4×4-in. top angle must be placed as shown or in the optional side location for satisfactory performance and stability. In order to provide adequate flexibility for the connection, only the toe of the top angle is welded to the HSS or beam flange. The thickness of the top angle is $\frac{1}{4}$-in. or greater to accommodate the minimum size fillet weld to the HSS. Although there is no calculated shear force transfer between the beam flange and the top angle, a minimum size fillet weld should be used across the toe of the top angle for stability. In lieu of welding, two $\frac{3}{4}$-in. diameter A325-N bolts may be used. The supported beam must be bolted to the seat plate with two high strength bolts having a strength of at least a $\frac{3}{4}$-in. diameter A325-N bolt to account for prying action caused by the rotation of the connection at ultimate load. Welding the beam to the seat plate is not recommended because welds lack the required strength and ductility. The centerline of the bolts should be located no more than the larger of $W/2$ or $2\frac{5}{8}$-in. from the HSS.

Tables 4-6 and 4-7 are design aids for stiffened seated connections. Table 4-6 is based on the yield line mechanism limit state for the HSS from Equation 4-21. However, some values for small L and large B have been reduced to meet the limit state for a line load with a width of $0.4L$ across the HSS from the AISC HSS Specification Section 8.1. The factored resistance of the connection is obtained by multiplying the tabulated value for a particular HSS width and stiffener length by the square of the HSS thickness and dividing by the width of the seat. For combinations of B and L that are not listed, the HSS does not have sufficient flat width to accommodate a weld to the seat that is $0.2L$ on each side of the stiffener. Since the required width also depends on the stiffener thickness and

Table 4-5
Unstiffened Seated Connections

Outstanding Angle Leg Design Strength, kips

Required Bearing Length N_{req}	Angle Length, in.										Min. Angle Leg
	6					8					
	Angle Thickness, in.										
in.	$3/8$	$1/2$	$5/8$	$3/4$	1	$3/8$	$1/2$	$5/8$	$3/4$	1	in.
$1/2$	27.3					36.5					
$9/16$	24.3					32.4					
$5/8$	21.9	58.3				29.2	77.8				
$11/16$	19.9	55.5				26.5	74.1				
$3/4$	18.2	48.6				24.3	64.8				
$13/16$	16.8	43.2				22.4	57.6				
$7/8$	15.6	38.9				20.8	51.8				
$15/16$	14.6	35.3				19.4	47.1				
1	13.7	32.4	72.9			18.2	43.2	97.2			
$1 1/16$	12.9	29.9	67.5			17.2	39.9	90.0			
$1 1/8$	12.2	27.8	60.8			16.2	37.0	81.0			
$1 3/16$	11.5	25.9	55.2			15.3	34.6	73.6			
$1 1/4$	10.9	24.3	50.6			14.6	32.4	67.5			
$1 5/16$	10.4	22.9	46.7			13.9	30.5	62.3			$3 1/2$
$1 3/8$	9.94	21.6	43.4	87.5		13.3	28.8	57.9	117		
$1 7/16$	9.51	20.5	40.5	79.5		12.7	27.3	54.0	106		
$1 1/2$	9.11	19.4	38.0	72.9		12.2	25.9	50.6	97.2		
$1 5/8$	8.41	17.7	33.8	62.5		11.2	23.6	45.0	83.3		
$1 3/4$	7.81	16.2	30.4	54.7		10.4	21.6	40.5	72.9		
$1 7/8$	7.29	15.0	27.6	48.6		9.72	19.9	36.8	64.8		
2	6.83	13.9	25.3	43.7	117	9.11	18.5	33.8	58.3	156	
$2 1/8$	6.43	13.0	23.4	39.8	111	8.58	17.3	31.2	53.0	148	
$2 1/4$	6.08	12.2	21.7	36.5	97.2	8.10	16.2	28.9	48.6	130	
$2 3/8$	5.76	11.4	20.3	33.6	86.4	7.67	15.2	27.0	44.9	115	
$2 1/2$	5.47	10.8	19.0	31.2	77.8	7.29	14.4	25.3	41.7	104	
$2 5/8$	5.21	10.2	17.9	29.2	70.7	6.94	13.6	23.8	38.9	94.3	
$2 3/4$	4.97	9.72	16.9	27.3	64.8	6.63	13.0	22.5	36.5	86.4	
$2 7/8$	4.75	9.26	16.0	25.7	59.8	6.34	12.3	21.3	34.3	79.8	
3	4.56	8.84	15.2	24.3	55.5	6.08	11.8	20.3	32.4	74.1	4
$3 1/8$	4.37	8.45	14.5	23.0	51.8	5.83	11.3	19.3	30.7	69.1	
$3 1/4$	4.21	8.10	13.8	21.9	48.6	5.61	10.8	18.4	29.2	64.8	

Weld (70 ksi) Design Strength, kips

| 70 ksi Weld Size, in. | Seat Angle Size (long leg vertical) | | | | |
	$4 \times 3 1/2$	$5 \times 3 1/2$	6×4	7×4	8×4
$1/4$	17.3	25.8	32.7	42.8	53.4
$5/16$	21.5	32.3	41.0	53.4	66.8
$3/8$	25.8	38.7	49.1	64.1	80.1
$7/16$	30.2	45.2	57.3	74.7	93.5
$1/2$	—	51.6	65.4	83.4	107
$5/8$	—	64.5	81.8	107	134
$11/16$	—	71.0	90.0	117	—

Available Angle Thickness, in.

	$4 \times 3 1/2$	$5 \times 3 1/2$	6×4	7×4	8×4
Minimum	$3/8$	$3/8$	$3/8$	$3/8$	$1/2$
Maximum	$1/2$	$3/4$	$3/4$	$3/4$	1

For tabulated values above the heavy line, shear yielding of the angle leg controls the design strength.

the HSS corner radius, the HSS width must be checked even when values are tabulated. Assuming a corner radius of 1.5 times the HSS thickness, $B \geq 0.4L + t_c + 3t$.

The values of ϕR_n in Table 4-7 are based on the limit state for the weld from Equation 4-16 with the electrode strength assumed to be 70 ksi. The minimum HSS thicknesses associated with the weld strengths are from Equation 4-19. If the HSS thickness is less than the minimum tabulated value, the weld strength must be reduced proportionally. All resistances obtained from Tables 4-6 and 4-7 are to be compared to factored loads.

The thickness of the horizontal seat plate or tee flange should not be less than $\frac{3}{8}$-in. Welds connecting the two plates should have a strength not less than the horizontal welds to the support under the seat.

The stiffener thickness may be conservatively determined as follows. When the stiffener has $F_y = 36$ ksi, the minimum stiffener thickness t_c for supported beams with unstiffened webs should not be less than t_w for beams with $F_y = 36$ ksi and not less than $1.4t_w$ for beams with $F_y = 50$ ksi. For stiffener material with $F_y = 50$ ksi or greater, the minimum plate thickness t_c for supported beams with unstiffened webs should be the beam web thickness t_w multiplied by the ratio of F_y of the beam material to F_y of the stiffener material. Additionally, the minimum stiffener thickness t_c should be at least $2w$ for stiffener material with $F_y = 36$ ksi or $1.5w$ for stiffener material with $F_y = 50$ ksi, where w is the weld size for 70 ksi electrodes.

The beam must be checked for local web yielding and web crippling in accordance with the limit states of Equations 22 and 23. This can be simplified by using the constants in the Factored Uniform Load Tables in Part 4 of the LRFD Manual.

$$\phi R_n = \phi R_1 + N(\phi R_2)$$

for $N/d \leq 0.2$

$$\phi R_n = \phi_r R_3 + N(\phi_r R_4)$$

for $N/d > 0.2$

$$\phi R_n = \phi_r R_5 + N(\phi_r R_6)$$

The nominal setback of $\frac{1}{2}$-in. should be assumed to be $\frac{3}{4}$-in. for calculation purposes to account for possible underrun in beam length.

EXAMPLE 4.6—Stiffened Seated Connection

Use Tables 4-6 and 4-7 to design an A36 steel stiffened seated connection for a W21×68 beam of A572 Gr 50 steel to an HSS12×12×½ column for a factored load of 125 kips. Use ¾-in. diameter A325-N bolts in standard holes to connect the beam to the seat plate. Use 70 ksi electrode welds to connect the stiffener, seat plate and top angle to the HSS.

Section Properties

HSS12×12×½	$t = 0.465$ in.	$B = 12$ in.	
	$F_y = 46$ ksi	$F_u = 58$ ksi	
W21×68	$t_w = 0.430$ in.	$d = 21.13$ in.	$t_f = 0.685$ in.
	$F_y = 50$ ksi	$F_u = 65$ ksi	

Determine stiffener width W required for web crippling and local web yielding

For web crippling, assume $N/d > 0.2$ and use constants $\phi_r R_5$ and $\phi_r R_6$ from the Factored Uniform Load Tables in LRFD Manual Part 4.

	Table 4-6.								
	Required Length for Stiffened Seated Connections								
	HSS Wall Strength Factor, $R_u W / t^2$, kips/in.								
	HSS Width B, in.								
L, in.	**5**	**5.5**	**6**	**7**	**8**	**9**	**10**	**12**	**14**
6	839	819	806	791	789	793	803	830	865
7	1033	997	972	940	925	920	922	940	968
8		1200	1159	1104	1074	1058	1052	1058	1078
9			1369	1287	1238	1208	1192	1183	1194
10			1604	1489	1417	1371	1343	1317	1317
11				1711	1612	1547	1505	1459	1446
12				1956	1825	1738	1679	1611	1583
13					2057	1944	1866	1772	1727
14					2310	2167	2067	1943	1880
15					2583	2407	2282	2125	2040
16						2664	2512	2318	2209
17						2941	2758	2522	2387
18							3020	2738	2574
19							3300	2967	2770
20							3596	3209	2977
21								3464	3194
22								3734	3422
23								4017	3661
24								4315	3912
25								4626	4175
26									4449
27									4736
28									5034
29									5344
30									5665
31									
32									

$$W_{\min} = \frac{R_u - \phi_r R_5}{\phi_r R_6} + \text{setback}$$

$$= \frac{125 - 75.8}{7.92} + \frac{3}{4}$$

$$= 6.96 \text{ in.}$$

For local web yielding, use constants ϕR_1 and ϕR_2 from the Factored Uniform Load Tables in LRFD Manual Part 4.

$$W_{\min} = \frac{R_u - \phi R_1}{\phi R_2} + \text{setback}$$

$$= \frac{125 - 77.3}{21.5} + \frac{3}{4}$$

$$= 2.97 \text{ in.}$$

The minimum stiffener width W for web crippling controls. Use $W = 7$ in.

	Table 4-6. (cont.) Required Length for Stiffened Seated Connections								
	HSS Wall Strength Factor, $R_u W / t^2$, kips/in.								
	HSS Width B, in.								
L in.	16	18	20	22	24	26	28	30	32
6	776	690	621	565	518	478	444	414	388
7	1003	939	845	768	704	650	604	564	528
8	1107	1141	1104	1004	920	849	789	736	690
9	1216	1245	1279	1270	1164	1075	998	932	873
10	1331	1354	1384	1417	1438	1327	1232	1150	1078
11	1451	1468	1492	1522	1556	1592	1491	1392	1305
12	1578	1587	1605	1630	1660	1694	1730	1656	1553
13	1710	1711	1723	1743	1768	1799	1832	1868	1822
14	1850	1840	1845	1859	1880	1907	1937	1970	2006
15	1995	1976	1972	1980	1996	2018	2045	2075	2109
16	2148	2116	2104	2105	2116	2133	2156	2183	2214
17	2308	2263	2241	2235	2239	2252	2270	2294	2321
18	2475	2417	2384	2369	2367	2374	2388	2408	2432
19	2650	2576	2532	2508	2498	2499	2509	2525	2545
20	2833	2742	2685	2651	2634	2629	2633	2644	2661
21	3024	2915	2844	2800	2774	2762	2761	2767	2780
22	3224	3095	3009	2954	2919	2900	2892	2894	2902
23	3433	3282	3180	3112	3068	3041	3027	3023	3027
24	3651	3476	3358	3277	3222	3187	3166	3156	3155
25	3878	3679	3541	3446	3381	3336	3308	3293	3286
26	4115	3889	3732	3622	3544	3491	3455	3432	3421
27	4362	4107	3929	3803	3713	3649	3605	3576	3558
28	4619	4334	4133	3990	3886	3812	3759	3723	3699
29	4887	4569	4543	4183	4065	3890	3917	3873	3843
30	5165	4813	4563	4382	4250	4152	4080	4028	3991
31	5454	5067	4790	4588	4440	4329	4247	4186	4142
32	5753	5329	5024	4801	4635	4511	4418	4348	4296

Check assumption

$N = 7 - \frac{3}{4} = 6.25$ in.

$N/d = 6.25/21.13 = 0.30 > 0.2$ as was assumed **o.k.**

Determine the stiffener length using wall strength factor

$R_u W / t^2 = 125 \times 7/(0.465)^2 = 4{,}047$

From Table 4-6 using the column for $B = 12$ in., select $L = 24$ in.

$4{,}315 > 4{,}047$ **o.k.**

Determine the weld strength requirements for the seat plate

From Table 4-7 with $W = 7$ in. and $L = 24$ in., a $\frac{1}{4}$-in. weld is adequate and is greater than the $\frac{3}{16}$-in. minimum size from LRFD Specification Table J2.4 for the HSS thickness.

293 kips > 125 kips **o.k.**

Table 4-7.
Stiffened Seated Connections

Weld Design Strength, kips

Width of Seat W, in.

L, in.	4 (70 ksi Weld Size, in.)				5 (70 ksi Weld Size, in.)				6 (70 ksi Weld Size, in.)			
	$\frac{1}{4}$	$\frac{5}{16}$	$\frac{3}{8}$	$\frac{7}{16}$	$\frac{5}{16}$	$\frac{3}{8}$	$\frac{7}{16}$	$\frac{1}{2}$	$\frac{5}{16}$	$\frac{3}{8}$	$\frac{7}{16}$	$\frac{1}{2}$
6	34.0	42.5	51.1	59.6	35.2	42.2	49.3	56.3	29.9	35.9	41.9	47.8
7	44.9	56.1	67.3	78.6	46.9	56.2	65.6	75.0	40.1	48.1	56.1	64.1
8	56.7	70.8	85.0	99.2	59.8	71.7	83.7	95.6	51.4	61.7	72.0	82.2
9	69.2	86.5	104	121	73.7	88.5	103	118	63.8	76.6	89.3	102
10	82.3	103	123	144	88.5	106	124	142	77.2	92.6	108	123
11	95.8	120	144	168	104	125	146	167	91.3	110	128	146
12	110	137	165	192	120	144	168	192	106	127	149	170
13	124	155	186	217	137	164	192	219	122	146	170	195
14	138	173	207	242	154	185	216	246	138	165	193	220
15	152	191	229	267	171	206	240	274	154	185	216	247
16	167	209	250	292	189	227	265	302	171	205	240	274
17	181	227	272	318	207	248	290	331	188	226	264	301
18	196	245	294	343	225	270	315	360	206	247	288	329
19	211	263	316	369	243	291	340	388	223	268	313	357
20	225	281	338	394	261	313	365	417	241	289	337	386
21	240	300	359	419	279	335	391	446	259	311	362	414
22	254	318	381	445	297	357	416	476	277	332	388	443
23	269	336	403	470	315	378	442	505	295	354	413	472
24	283	354	425	495	334	400	467	534	313	376	438	501
25	297	372	446	520	352	422	492	563	331	397	464	530
26	312	390	468	546	370	444	518	592	349	419	489	559
27	326	408	489	571	388	466	543	621	368	441	515	588

Weld Size, in.	Min. HSS Thickness, in.
$\frac{1}{4}$	0.22
$\frac{5}{16}$	0.28
$\frac{3}{8}$	0.34
$\frac{7}{16}$	0.39
$\frac{1}{2}$	0.45

Notes:

1. Values shown assume 70 ksi electrodes. For 60 ksi electrodes, multiply tabular values by 0.857, or enter table with 1.17 times the required strength R_u. For 80 ksi electrodes, multiply tabular values by 1.14, or enter table with 0.875 times the required strength R_u.
2. Tabulated values are valid for stiffeners with minimum thickness of

$$t_{c\,min} = \frac{F_{y\,beam}}{F_{y\,stiffener}} \times t_w$$

   but not less than $2w$ for stiffeners with $F_y = 36$ ksi nor $1.5w$ for stiffeners with $F_y = 50$ ksi. In the above, t_w is the thickness of the unstiffened supported beam web and w is the nominal weld size.
3. Tabulated values may be limited by shear yielding of or bearing on the stiffener; refer to LRFD Specification Sections F2.2 and J8, respectively.

Table 4-7 (cont.).
Stiffened Seated Connections

Weld Design Strength, kips

	Width of Seat W, in.											
	7				8				9			
L, in.	70 ksi Weld Size, in.				70 ksi Weld Size, in.				70 ksi Weld Size, in.			
	$\frac{1}{4}$	$\frac{5}{16}$	$\frac{3}{8}$	$\frac{7}{16}$	$\frac{5}{16}$	$\frac{3}{8}$	$\frac{7}{16}$	$\frac{1}{2}$	$\frac{5}{16}$	$\frac{3}{8}$	$\frac{7}{16}$	$\frac{1}{2}$
11	81.0	97.2	113	130	72.5	87.1	116	145	65.6	78.7	105	131
12	94.7	114	133	151	85.1	102	136	170	77.1	92.5	123	154
13	109	131	153	174	98.3	118	157	197	89.3	107	143	179
14	124	149	174	198	112	135	180	224	102	123	164	204
15	139	167	195	223	127	152	203	253	116	139	185	232
16	155	186	217	249	142	170	227	283	130	156	208	260
17	172	206	240	275	157	189	251	314	144	173	231	289
18	188	226	264	301	173	208	277	346	159	191	255	319
19	205	246	287	329	189	227	303	378	175	210	280	350
20	223	267	312	356	206	247	329	411	191	229	305	381
21	240	288	336	384	222	267	356	445	207	248	331	413
22	258	309	361	412	240	287	383	479	223	268	357	446
23	275	330	385	440	257	308	411	514	240	288	384	480
24	293	352	410	469	274	329	439	548	257	308	411	513
25	311	373	435	498	292	350	467	584	274	329	438	548
26	329	395	461	526	309	371	495	619	291	349	466	582
27	347	417	486	555	327	393	524	655	308	370	494	617
28	365	438	511	584	345	414	552	690	326	391	522	652
29	383	460	537	613	363	436	581	726	344	412	550	687
30	402	482	562	643	381	457	610	762	362	434	578	723
31	420	504	588	672	399	479	639	799	379	455	607	759
32	438	526	613	701	417	501	668	835	397	477	636	795

Weld Size, in.	Min. HSS Thickness, in.
$\frac{1}{4}$	0.22
$\frac{5}{16}$	0.28
$\frac{3}{8}$	0.34
$\frac{7}{16}$	0.39
$\frac{1}{2}$	0.45

Notes:
1. Values shown assume 70 ksi electrodes. For 60 ksi electrodes, multiply tabular values by 0.857, or enter table with 1.17 times the required strength R_u. For 80 ksi electrodes, multiply tabular values by 1.14, or enter table with 0.875 times the required strength R_u.
2. Tabulated values are valid for stiffeners with minimum thickness of
$$t_{c\,min} = \frac{F_{y\,beam}}{F_{y\,stiffener}} \times t_w$$
but not less than $2w$ for stiffeners with $F_y = 36$ ksi nor $1.5w$ for stiffeners with $F_y = 50$ ksi. In the above, t_w is the thickness of the unstiffened supported beam web and w is the nominal weld size.
3. Tabulated values may be limited by shear yielding of or bearing on the stiffener; refer to LRFD Specification Sections F2.2 and J8, respectively.

Since t of the HSS is greater than $t_{\min}$ for the ¼-in. weld, no reduction in weld strength to account for shear in the HSS is required.

The minimum length of the seat-plate-to-HSS weld on each side of the stiffener is $0.2L = 4.8$ in. This establishes the minimum weld between the seat plate and stiffener; use 5 inches of ¼-in. weld on each side of the stiffener.

Determine the stiffener plate thickness

To develop the stiffener-to-seat plate welds, the minimum stiffener thickness is

$$t_{c\,\min} = 2(0.25) = 0.50 \text{ in.}$$

For a stiffener with $F_y = 36$ ksi and a beam with $F_y = 50$ ksi, the minimum stiffener thickness is

$$t_{c\,\min} = 1.4t_w$$

$$= 1.4(0.430) = 0.602 \text{ in.}$$

The latter controls; use **PL**⅝-in. × 7-in. × 24-in. for the stiffener.

Check the HSS width

The minimum width is $0.4L + t_c + 3t$

$$B = 12 \text{ in.} > 0.4(24) + 0.625 + 3(0.465) = 11.6 \text{ in.} \quad \textbf{o.k.}$$

Determine the seat plate dimensions

To accommodate two ¾-in. diameter A325-N bolts on a 5½-in. gage connecting the beam flange to the seat plate, a width of 8 inches is required. To accommodate the seat-plate-to-HSS weld the required width is

$$0.4(24) + 0.625 = 10.2 \text{ in.}$$

Use **PL**⅜-in. × 7-in. × 10½-in. for the seat plate

Select the top angle, bolts and weld

The minimum weld size for the HSS thickness according to LRFD Specification Table J2.4 is 3⁄16-in. The angle thickness should be 1⁄16-in. larger.

Use **L**4×4×¼ with 3⁄16-in. fillet welds along the toes of the angle to the beam flange and HSS. Alternatively two ¾-in. diameter A325-N bolts may be used to connect the beam leg of the angle to the beam flange.

SINGLE-PLATE CONNECTIONS

A single-plate connection is made with a plate as illustrated in Figure 4-4. The plate is welded on both sides to the HSS and is bolted to the beam web.

As long as the HSS wall is not classified as a slender element, the local distortion caused by the single-plate connection will be insignificant in reducing the column strength of the HSS (Sherman, 1996). Therefore, single-plate connections may be used when $b/t \leq 253/\sqrt{F_y}$ [this is essentially the same as $1.4(E/F_y)^{0.5}$] or 37.3 for $F_y = 46$ ksi. Single-plate connections may also be used with round HSS as long as they are compact under axial load, $(D/t \leq 3300/F_y)$, or do not exceed the standard maximum A500 production limit of $D/t \leq 72$.

Tables 4-8 are design aids for single-plate connections with the plate material having $F_y = 36$ ksi and $F_u = 58$ ksi. The calculations are made for a 3-in. bolt spacing, 1½-in. vertical and horizontal edge distances and an a dimension from the weld to the bolt line of 3-in. It is conservative to use larger edge distances or an *a* between 3- and 2½-in. The tabulated values are valid for laterally supported beams, in steel and composite construction, all types of loading, snug-tightened and fully-tensioned bolts and for beams of all grades of steel. All values are for comparison with factored loads.

The tabulated values are based on the limit states from Equations 4-2, 4-4, 4-5, 4-6, and 4-10. When values are not listed for thin plates, the buckling limit state of Equation 4-12 has not been satisfied. Thicker plates that do not have tabulated values lack the rotational ductility specified in Equation 4-11. Values are tabulated for two through nine rows of ¾-in., ⅞-in. and 1-in. diameter A325 and A490 bolts.

Values are tabulated for rigid and flexible supports, which differ in the eccentricity used in Equation 4-2. Tests (Sherman and Ales, 1991) show that reaction eccentricities for single-plate connections to short rectangular HSS columns with large b/t are within the end connection, which is the definition used in Section 9 of the LRFD Manual for a rigid support. Tabulated values for flexible supports should be used only when the column itself is considered flexible at the connection.

Although the weld size can be determined from Equation 4-17, Section 9 of the LRFD Manual conservatively uses a weld size that is ¾ of the plate thickness. These weld sizes are tabulated in Table 4-9. The HSS limit states are Equations 4-19 and 4-20 for shear adjacent to the weld and punching shear through the thickness of the wall. With the weld size based on ¾ of the plate thickness, punching shear of the HSS will never control over shear in the HSS wall adjacent to the weld. Therefore, the minimum values for the HSS thickness in Table 4-9 are based on Equation 4-19. In the event that the plate used has greater strength than required, the minimum HSS thickness may be reduced by the ratio of the required load to the strength from Table 4-8, as long as the weld is still ¾ of the plate thickness. For large differences between tabulated and required capacities, or for *a* dimensions less than 3 inches, it may be advantageous to recalculate the strength and the weld requirements according to the limit states in Table 4-2.

Table 4-10 tabulates the strength of the beam web per inch of thickness. The values are based on the bolt bearing limit state from Equation 4-4.

(continued on page 4-107)

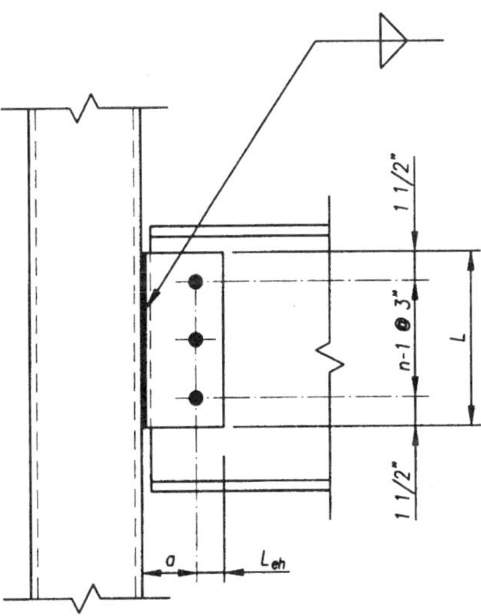

Fig. 4-4. Single-Plate connection.

¾-in. diameter bolts										
Table 4-8.										
Single-Plate Connections										
Bolt, Weld, and Single-Plate Design Strength, kips										

n	ASTM Desig.	Thread Cond.	Support Cond.	Hole Type	Plate Thickness, in.					
					$\frac{1}{4}$	$\frac{5}{16}$	$\frac{3}{8}$	$\frac{7}{16}$	$\frac{1}{2}$	$\frac{9}{16}$
9 ($L = 27$)	A325	N	Flexible	STD	—	—	—	115	—	—
				SSLT	—	—	—	130	—	—
			Rigid	STD	—	—	—	115	—	—
				SSLT	—	—	—	130	—	—
		X	Flexible	STD	—	—	—	144	—	—
				SSLT	—	—	—	162	—	—
			Rigid	STD	—	—	—	144	—	—
				SSLT	—	—	—	162	—	—
	A490	N	Flexible	STD	—	—	—	144	—	—
				SSLT	—	—	—	162	—	—
			Rigid	STD	—	—	—	144	—	—
				SSLT	—	—	—	162	—	—
		X	Flexible	STD	—	—	—	179	—	—
				SSLT	—	—	—	203	—	—
			Rigid	STD	—	—	—	179	—	—
				SSLT	—	—	—	203	—	—
8 ($L = 24$)	A325	N	Flexible	STD	—	—	106	106	—	—
				SSLT	—	—	113	113	—	—
			Rigid	STD	—	—	106	106	—	—
				SSLT	—	—	117	117	—	—
		X	Flexible	STD	—	—	132	132	—	—
				SSLT	—	—	142	142	—	—
			Rigid	STD	—	—	132	132	—	—
				SSLT	—	—	147	147	—	—
	A490	N	Flexible	STD	—	—	132	132	—	—
				SSLT	—	—	142	142	—	—
			Rigid	STD	—	—	132	132	—	—
				SSLT	—	—	147	147	—	—
		X	Flexible	STD	—	—	165	165	—	—
				SSLT	—	—	166	177	—	—
			Rigid	STD	—	—	165	165	—	—
				SSLT	—	—	166	183	—	—
7 ($L = 21$)	A325	N	Flexible	STD	—	96.4	96.4	96.4	—	—
				SSLT	—	96.4	96.4	96.4	—	—
			Rigid	STD	—	96.4	96.4	96.4	—	—
				SSLT	—	104	104	104	—	—
		X	Flexible	STD	—	120	120	120	—	—
				SSLT	—	120	120	120	—	—
			Rigid	STD	—	120	120	120	—	—
				SSLT	—	121	131	131	—	—
	A490	N	Flexible	STD	—	120	120	120	—	—
				SSLT	—	120	120	120	—	—
			Rigid	STD	—	120	120	120	—	—
				SSLT	—	121	131	131	—	—
		X	Flexible	STD	—	121	146	151	—	—
				SSLT	—	121	146	151	—	—
			Rigid	STD	—	121	146	151	—	—
				SSLT	—	121	146	163	—	—

STD = Standard holes
SSLT = Short-slotted holes transverse to direction of load
N = Threads included
X = Threads excluded

| | | | | | **3/4-in. diameter bolts** | | | | | |

Table 4-8 (cont.).
Single-Plate Connections

Bolt, Weld, and Single-Plate Design Strength, kips

n	ASTM Desig.	Thread Cond.	Support Cond.	Hole Type	Plate Thickness, in.					
					$1/4$	$5/16$	$3/8$	$7/16$	$1/2$	$9/16$
6 ($L = 18$)	A325	N	Flexible	STD	—	79.2	79.2	79.2	—	—
				SSLT	—	79.2	79.2	79.2	—	—
			Rigid	STD	—	86.7	86.7	86.7	—	—
				SSLT	—	91.1	91.1	91.1	—	—
		X	Flexible	STD	—	99.0	99.0	99.0	—	—
				SSLT	—	99.0	99.0	99.0	—	—
			Rigid	STD	—	104	108	108	—	—
				SSLT	—	104	114	114	—	—
	A490	N	Flexible	STD	—	99.0	99.0	99.0	—	—
				SSLT	—	99.0	99.0	99.0	—	—
			Rigid	STD	—	104	108	108	—	—
				SSLT	—	104	114	114	—	—
		X	Flexible	STD	—	104	124	124	—	—
				SSLT	—	104	124	124	—	—
			Rigid	STD	—	104	125	135	—	—
				SSLT	—	104	125	142	—	—
5 ($L = 15$)	A325	N	Flexible	STD	62.0	62.0	62.0	62.0	—	—
				SSLT	62.0	62.0	62.0	62.0	—	—
			Rigid	STD	69.3	74.8	74.8	74.8	—	—
				SSLT	69.3	77.9	77.9	77.9	—	—
		X	Flexible	STD	69.3	77.5	77.5	77.5	—	—
				SSLT	69.3	77.5	77.5	77.5	—	—
			Rigid	STD	69.3	86.7	93.4	93.4	—	—
				SSLT	69.3	86.7	97.4	97.4	—	—
	A490	N	Flexible	STD	69.3	77.5	77.5	77.5	—	—
				SSLT	69.3	77.5	77.5	77.5	—	—
			Rigid	STD	69.3	86.7	93.4	93.4	—	—
				SSLT	69.3	86.7	97.4	97.4	—	—
		X	Flexible	STD	69.3	86.7	96.9	96.9	—	—
				SSLT	69.3	86.7	96.9	96.9	—	—
			Rigid	STD	69.3	86.7	104	117	—	—
				SSLT	69.3	86.7	104	121	—	—
4 ($L = 12$)	A325	N	Flexible	STD	44.7	44.7	44.7	44.7	—	—
				SSLT	44.7	44.7	44.7	44.7	—	—
			Rigid	STD	55.5	63.6	63.6	63.6	—	—
				SSLT	55.5	61.9	61.9	61.9	—	—
		X	Flexible	STD	55.0	55.9	55.9	55.9	—	—
				SSLT	55.0	55.9	55.9	55.9	—	—
			Rigid	STD	55.5	69.3	79.5	79.5	—	—
				SSLT	55.5	69.3	77.3	77.3	—	—
	A490	N	Flexible	STD	55.0	55.9	55.9	55.9	—	—
				SSLT	55.0	55.9	55.9	55.9	—	—
			Rigid	STD	55.5	69.3	79.5	79.5	—	—
				SSLT	55.5	69.3	77.3	77.3	—	—
		X	Flexible	STD	55.0	68.8	69.8	69.8	—	—
				SSLT	55.0	68.8	69.8	69.8	—	—
			Rigid	STD	55.5	69.3	83.2	97.1	—	—
				SSLT	55.5	69.3	83.2	96.7	—	—

STD = Standard holes
SSLT = Short-slotted holes transverse to direction of load

N = Threads included
X = Threads excluded

¾-in. diameter bolts

Table 4-8 (cont.).
Single-Plate Connections

Bolt, Weld, and Single-Plate Design Strength, kips

n	ASTM Desig.	Thread Cond.	Support Cond.	Hole Type	Plate Thickness, in.					
					¼	⁵⁄₁₆	⅜	⁷⁄₁₆	½	⁹⁄₁₆
3 ($L = 9$)	A325	N	Flexible	STD	27.8	27.8	27.8	27.8	—	—
				SSLT	27.8	27.8	27.8	27.8	—	—
			Rigid	STD	41.6	41.7	41.7	41.7	—	—
				SSLT	41.6	41.7	41.7	41.7	—	—
		X	Flexible	STD	34.3	34.8	34.8	34.8	—	—
				SSLT	34.3	34.8	34.8	34.8	—	—
			Rigid	STD	41.6	52.0	52.1	52.1	—	—
				SSLT	41.6	52.0	52.1	52.1	—	—
	A490	N	Flexible	STD	34.3	34.8	34.8	34.8	—	—
				SSLT	34.3	34.8	34.8	34.8	—	—
			Rigid	STD	41.6	52.0	52.1	52.1	—	—
				SSLT	41.6	52.0	52.1	52.1	—	—
		X	Flexible	STD	34.3	42.8	43.5	43.5	—	—
				SSLT	34.3	42.8	43.5	43.5	—	—
			Rigid	STD	41.6	52.0	62.4	65.1	—	—
				SSLT	41.6	52.0	62.4	65.1	—	—
2 ($L = 6$)	A325	N	Flexible	STD	14.0	14.0	14.0	14.0	—	—
				SSLT	14.0	14.0	14.0	14.0	—	—
			Rigid	STD	18.8	18.8	18.8	18.8	—	—
				SSLT	21.0	21.0	21.0	21.0	—	—
		X	Flexible	STD	17.2	17.5	17.5	17.5	—	—
				SSLT	17.2	17.5	17.5	17.5	—	—
			Rigid	STD	23.1	23.5	23.5	23.5	—	—
				SSLT	25.8	26.2	26.2	26.2	—	—
	A490	N	Flexible	STD	17.2	17.5	17.5	17.5	—	—
				SSLT	17.2	17.5	17.5	17.5	—	—
			Rigid	STD	23.1	23.5	23.5	23.5	—	—
				SSLT	25.8	26.2	26.2	26.2	—	—
		X	Flexible	STD	17.2	21.5	21.9	21.9	—	—
				SSLT	17.2	21.5	21.9	21.9	—	—
			Rigid	STD	23.1	28.9	29.3	29.3	—	—
				SSLT	25.8	32.3	32.8	32.8	—	—

STD = Standard holes
SSLT = Short-slotted holes transverse to direction of load

N = Threads included
X = Threads excluded

| | | | | | 7/8-in. diameter bolts | | | | | |

Table 4-8 (cont.).
Single-Plate Connections

Bolt, Weld, and Single-Plate Design Strength, kips

	ASTM	Thread	Support	Hole	Plate Thickness, in.					
n	Desig.	Cond.	Cond.	Type	$\frac{1}{4}$	$\frac{5}{16}$	$\frac{3}{8}$	$\frac{7}{16}$	$\frac{1}{2}$	$\frac{9}{16}$
9 ($L = 27$)	A325	N	Flexible	STD	—	—	—	156	156	—
				SSLT	—	—	—	177	177	—
			Rigid	STD	—	—	—	156	156	—
				SSLT	—	—	—	177	177	—
		X	Flexible	STD	—	—	—	195	195	—
				SSLT	—	—	—	206	221	—
			Rigid	STD	—	—	—	195	195	—
				SSLT	—	—	—	206	221	—
	A490	N	Flexible	STD	—	—	—	195	195	—
				SSLT	—	—	—	206	221	—
			Rigid	STD	—	—	—	195	195	—
				SSLT	—	—	—	206	221	—
		X	Flexible	STD	—	—	—	206	235	—
				SSLT	—	—	—	206	235	—
			Rigid	STD	—	—	—	206	235	—
				SSLT	—	—	—	206	235	—
8 ($L = 24$)	A325	N	Flexible	STD	—	—	144	144	144	—
				SSLT	—	—	154	154	154	—
			Rigid	STD	—	—	144	144	144	—
				SSLT	—	—	157	160	160	—
		X	Flexible	STD	—	—	157	180	180	—
				SSLT	—	—	157	183	193	—
			Rigid	STD	—	—	157	180	180	—
				SSLT	—	—	157	183	200	—
	A490	N	Flexible	STD	—	—	157	180	180	—
				SSLT	—	—	157	183	193	—
			Rigid	STD	—	—	157	180	180	—
				SSLT	—	—	157	183	200	—
		X	Flexible	STD	—	—	157	183	209	—
				SSLT	—	—	157	183	209	—
			Rigid	STD	—	—	157	183	209	—
				SSLT	—	—	157	183	209	—
7 ($L = 21$)	A325	N	Flexible	STD	—	114	131	131	131	—
				SSLT	—	114	131	131	131	—
			Rigid	STD	—	114	131	131	131	—
				SSLT	—	114	137	142	142	—
		X	Flexible	STD	—	114	137	160	164	—
				SSLT	—	114	137	160	164	—
			Rigid	STD	—	114	137	160	164	—
				SSLT	—	114	137	160	178	—
	A490	N	Flexible	STD	—	114	137	160	164	—
				SSLT	—	114	137	160	164	—
			Rigid	STD	—	114	137	160	164	—
				SSLT	—	114	137	160	178	—
		X	Flexible	STD	—	114	137	160	183	—
				SSLT	—	114	137	160	183	—
			Rigid	STD	—	114	137	160	183	—
				SSLT	—	114	137	160	183	—

STD = Standard holes
SSLT = Short-slotted holes transverse to direction of load
N = Threads included
X = Threads excluded

7/8-in. diameter bolts

Table 4-8 (cont.).
Single-Plate Connections

Bolt, Weld, and Single-Plate Design Strength, kips

n	ASTM Desig.	Thread Cond.	Support Cond.	Hole Type	Plate Thickness, in.					
					$\frac{1}{4}$	$\frac{5}{16}$	$\frac{3}{8}$	$\frac{7}{16}$	$\frac{1}{2}$	$\frac{9}{16}$
6 (L = 18)	A325	N	Flexible	STD	—	97.9	108	108	108	—
				SSLT	—	97.9	108	108	108	—
			Rigid	STD	—	97.9	117	118	118	—
				SSLT	—	97.9	117	124	124	—
		X	Flexible	STD	—	97.9	117	135	135	—
				SSLT	—	97.9	117	135	135	—
			Rigid	STD	—	97.9	117	137	147	—
				SSLT	—	97.9	117	137	155	—
	A490	N	Flexible	STD	—	97.9	117	135	135	—
				SSLT	—	97.9	117	135	135	—
			Rigid	STD	—	97.9	117	137	147	—
				SSLT	—	97.9	117	137	155	—
		X	Flexible	STD	—	97.9	117	137	157	—
				SSLT	—	97.9	117	137	157	—
			Rigid	STD	—	97.9	117	137	157	—
				SSLT	—	97.9	117	137	157	—
5 (L = 15)	A325	N	Flexible	STD	65.3	81.6	84.4	84.4	84.4	—
				SSLT	65.3	81.6	84.4	84.4	84.4	—
			Rigid	STD	65.3	81.6	97.9	102	102	—
				SSLT	65.3	81.6	97.9	106	106	—
		X	Flexible	STD	65.3	81.6	97.9	106	106	—
				SSLT	65.3	81.6	97.9	106	106	—
			Rigid	STD	65.3	81.6	97.9	114	127	—
				SSLT	65.3	81.6	97.9	114	131	—
	A490	N	Flexible	STD	65.3	81.6	97.9	106	106	—
				SSLT	65.3	81.6	97.9	106	106	—
			Rigid	STD	65.3	81.6	97.9	114	127	—
				SSLT	65.3	81.6	97.9	114	131	—
		X	Flexible	STD	65.3	81.6	97.9	114	131	—
				SSLT	65.3	81.6	97.9	114	131	—
			Rigid	STD	65.3	81.6	97.9	114	131	—
				SSLT	65.3	81.6	97.9	114	131	—
4 (L = 12)	A325	N	Flexible	STD	52.2	60.8	60.8	60.8	60.8	—
				SSLT	52.2	60.8	60.8	60.8	60.8	—
			Rigid	STD	52.2	65.3	78.3	86.6	86.6	—
				SSLT	52.2	65.3	78.3	84.2	84.2	—
		X	Flexible	STD	52.2	65.3	76.0	76.0	76.0	—
				SSLT	52.2	65.3	76.0	76.0	76.0	—
			Rigid	STD	52.2	65.3	78.3	91.4	104	—
				SSLT	52.2	65.3	78.3	91.4	104	—
	A490	N	Flexible	STD	52.2	65.3	76.0	76.0	76.0	—
				SSLT	52.2	65.3	76.0	76.0	76.0	—
			Rigid	STD	52.2	65.3	78.3	91.4	104	—
				SSLT	52.2	65.3	78.3	91.4	104	—
		X	Flexible	STD	52.2	65.3	78.3	91.4	95.0	—
				SSLT	52.2	65.3	78.3	91.4	95.0	—
			Rigid	STD	52.2	65.3	78.3	91.4	104	—
				SSLT	52.2	65.3	78.3	91.4	104	—

STD = Standard holes
SSLT = Short-slotted holes transverse to direction of load
N = Threads included
X = Threads excluded

| | | | | | ⁷⁄₈-in. diameter bolts |

Table 4-8 (cont.).
Single-Plate Connections

Bolt, Weld, and Single-Plate Design Strength, kips

n	ASTM Desig.	Thread Cond.	Support Cond.	Hole Type	Plate Thickness, in.					
					$1/4$	$5/16$	$3/8$	$7/16$	$1/2$	$9/16$
3 ($L = 9$)	A325	N	Flexible	STD	37.9	37.9	37.9	37.9	37.9	—
				SSLT	37.9	37.9	37.9	37.9	37.9	—
			Rigid	STD	39.2	48.9	56.7	56.7	56.7	—
				SSLT	39.2	48.9	56.7	56.7	56.7	—
		X	Flexible	STD	39.2	47.4	47.4	47.4	47.4	—
				SSLT	39.2	47.4	47.4	47.4	47.4	—
			Rigid	STD	39.2	48.9	58.7	68.5	70.9	—
				SSLT	39.2	48.9	58.7	68.5	70.9	—
	A490	N	Flexible	STD	39.2	47.4	47.4	47.4	47.4	—
				SSLT	39.2	47.4	47.4	47.4	47.4	—
			Rigid	STD	39.2	48.9	58.7	68.5	70.9	—
				SSLT	39.2	48.9	58.7	68.5	70.9	—
		X	Flexible	STD	39.2	48.9	58.7	59.2	59.2	—
				SSLT	39.2	48.9	58.7	59.2	59.2	—
			Rigid	STD	39.2	48.9	58.7	68.5	78.3	—
				SSLT	39.2	48.9	58.7	68.5	78.3	—
2 ($L = 6$)	A325	N	Flexible	STD	19.0	19.0	19.0	19.0	19.0	—
				SSLT	19.0	19.0	19.0	19.0	19.0	—
			Rigid	STD	25.5	25.5	25.5	25.5	25.5	—
				SSLT	26.1	28.6	28.6	28.6	28.6	—
		X	Flexible	STD	20.1	23.8	23.8	23.8	23.8	—
				SSLT	20.1	23.8	23.8	23.8	23.8	—
			Rigid	STD	26.1	31.9	31.9	31.9	31.9	—
				SSLT	26.1	32.6	35.7	35.7	35.7	—
	A490	N	Flexible	STD	20.1	23.8	23.8	23.8	23.8	—
				SSLT	20.1	23.8	23.8	23.8	23.8	—
			Rigid	STD	26.1	31.9	31.9	31.9	31.9	—
				SSLT	26.1	32.6	35.7	35.7	35.7	—
		X	Flexible	STD	20.1	25.1	29.8	29.8	29.8	—
				SSLT	20.1	25.1	29.8	29.8	29.8	—
			Rigid	STD	26.1	32.6	39.2	39.9	39.9	—
				SSLT	26.1	32.6	39.2	44.6	44.6	—

STD = Standard holes
SSLT = Short-slotted holes transverse to direction of load

N = Threads included
X = Threads excluded

1-in. diameter bolts

Table 4-8 (cont.).
Single-Plate Connections

Bolt, Weld, and Single-Plate Design Strength, kips

n	ASTM Desig.	Thread Cond.	Support Cond.	Hole Type	Plate Thickness, in.					
					$\frac{1}{4}$	$\frac{5}{16}$	$\frac{3}{8}$	$\frac{7}{16}$	$\frac{1}{2}$	$\frac{9}{16}$
9 (L = 27)	A325	N	Flexible	STD	—	—	—	192	204	204
				SSLT	—	—	—	192	220	231
			Rigid	STD	—	—	—	192	204	204
				SSLT	—	—	—	192	220	231
		X	Flexible	STD	—	—	—	192	220	247
				SSLT	—	—	—	192	220	247
			Rigid	STD	—	—	—	192	220	247
				SSLT	—	—	—	192	220	247
	A490	N	Flexible	STD	—	—	—	192	220	247
				SSLT	—	—	—	192	220	247
			Rigid	STD	—	—	—	192	220	247
				SSLT	—	—	—	192	220	247
		X	Flexible	STD	—	—	—	192	220	247
				SSLT	—	—	—	192	220	247
			Rigid	STD	—	—	—	192	220	247
				SSLT	—	—	—	192	220	247
8 (L = 24)	A325	N	Flexible	STD	—	—	146	171	188	188
				SSLT	—	—	146	171	195	201
			Rigid	STD	—	—	146	171	188	188
				SSLT	—	—	146	171	195	209
		X	Flexible	STD	—	—	146	171	195	220
				SSLT	—	—	146	171	195	220
			Rigid	STD	—	—	146	171	195	220
				SSLT	—	—	146	171	195	220
	A490	N	Flexible	STD	—	—	146	171	195	220
				SSLT	—	—	146	171	195	220
			Rigid	STD	—	—	146	171	195	220
				SSLT	—	—	146	171	195	220
		X	Flexible	STD	—	—	146	171	195	220
				SSLT	—	—	146	171	195	220
			Rigid	STD	—	—	146	171	195	220
				SSLT	—	—	146	171	195	220
7 (L = 21)	A325	N	Flexible	STD	—	107	128	149	171	171
				SSLT	—	107	128	149	171	171
			Rigid	STD	—	107	128	149	171	171
				SSLT	—	107	128	149	171	186
		X	Flexible	STD	—	107	128	149	171	192
				SSLT	—	107	128	149	171	192
			Rigid	STD	—	107	128	149	171	192
				SSLT	—	107	128	149	171	192
	A490	N	Flexible	STD	—	107	128	149	171	192
				SSLT	—	107	128	149	171	192
			Rigid	STD	—	107	128	149	171	192
				SSLT	—	107	128	149	171	192
		X	Flexible	STD	—	107	128	149	171	192
				SSLT	—	107	128	149	171	192
			Rigid	STD	—	107	128	149	171	192
				SSLT	—	107	128	149	171	192

STD = Standard holes
SSLT = Short-slotted holes transverse to direction of load

N = Threads included
X = Threads excluded

				1-in. diameter bolts

Table 4-8 (cont.).
Single-Plate Connections

Bolt, Weld, and Single-Plate Design Strength, kips

n	ASTM Desig.	Thread Cond.	Support Cond.	Hole Type	Plate Thickness, in.					
					$\frac{1}{4}$	$\frac{5}{16}$	$\frac{3}{8}$	$\frac{7}{16}$	$\frac{1}{2}$	$\frac{9}{16}$
6 ($L = 18$)	A325	N	Flexible	STD	—	91.5	110	128	141	141
				SSLT	—	91.5	110	128	141	141
			Rigid	STD	—	91.5	110	128	146	154
				SSLT	—	91.5	110	128	146	162
		X	Flexible	STD	—	91.5	110	128	146	165
				SSLT	—	91.5	110	128	146	165
			Rigid	STD	—	91.5	110	128	146	165
				SSLT	—	91.5	110	128	146	165
	A490	N	Flexible	STD	—	91.5	110	128	146	165
				SSLT	—	91.5	110	128	146	165
			Rigid	STD	—	91.5	110	128	146	165
				SSLT	—	91.5	110	128	146	165
		X	Flexible	STD	—	91.5	110	128	146	165
				SSLT	—	91.5	110	128	146	165
			Rigid	STD	—	91.5	110	128	146	165
				SSLT	—	91.5	110	128	146	165
5 ($L = 15$)	A325	N	Flexible	STD	61.0	76.3	91.5	107	110	110
				SSLT	61.0	76.3	91.5	107	110	110
			Rigid	STD	61.0	76.3	91.5	107	122	133
				SSLT	61.0	76.3	91.5	107	122	137
		X	Flexible	STD	61.0	76.3	91.5	107	122	137
				SSLT	61.0	76.3	91.5	107	122	137
			Rigid	STD	61.0	76.3	91.5	107	122	137
				SSLT	61.0	76.3	91.5	107	122	137
	A490	N	Flexible	STD	61.0	76.3	91.5	107	122	137
				SSLT	61.0	76.3	91.5	107	122	137
			Rigid	STD	61.0	76.3	91.5	107	122	137
				SSLT	61.0	76.3	91.5	107	122	137
		X	Flexible	STD	61.0	76.3	91.5	107	122	137
				SSLT	61.0	76.3	91.5	107	122	137
			Rigid	STD	61.0	76.3	91.5	107	122	137
				SSLT	61.0	76.3	91.5	107	122	137
4 ($L = 12$)	A325	N	Flexible	STD	48.8	61.0	73.2	79.5	79.5	79.5
				SSLT	48.8	61.0	73.2	79.5	79.5	79.5
			Rigid	STD	48.8	61.0	73.2	85.4	97.6	110
				SSLT	48.8	61.0	73.2	85.4	97.6	110
		X	Flexible	STD	48.8	61.0	73.2	85.4	97.6	99.3
				SSLT	48.8	61.0	73.2	85.4	97.6	99.3
			Rigid	STD	48.8	61.0	73.2	85.4	97.6	110
				SSLT	48.8	61.0	73.2	85.4	97.6	110
	A490	N	Flexible	STD	48.8	61.0	73.2	85.4	97.6	99.3
				SSLT	48.8	61.0	73.2	85.4	97.6	99.3
			Rigid	STD	48.8	61.0	73.2	85.4	97.6	110
				SSLT	48.8	61.0	73.2	85.4	97.6	110
		X	Flexible	STD	48.8	61.0	73.2	85.4	97.6	110
				SSLT	48.8	61.0	73.2	85.4	97.6	110
			Rigid	STD	48.8	61.0	73.2	85.4	97.6	110
				SSLT	48.8	61.0	73.2	85.4	97.6	110

STD = Standard holes
SSLT = Short-slotted holes transverse to direction of load
N = Threads included
X = Threads excluded

1-in. diameter bolts

Table 4-8 (cont.).
Single-Plate Connections

Bolt, Weld, and Single-Plate Design Strength, kips

n	ASTM Desig.	Thread Cond.	Support Cond.	Hole Type	Plate Thickness, in.					
					$\frac{1}{4}$	$\frac{5}{16}$	$\frac{3}{8}$	$\frac{7}{16}$	$\frac{1}{2}$	$\frac{9}{16}$
3 ($L = 9$)	A325	N	Flexible	STD	36.6	45.8	49.5	49.5	49.5	49.5
				SSLT	36.6	45.8	49.5	49.5	49.5	49.5
			Rigid	STD	36.6	45.8	54.9	64.1	73.2	74.1
				SSLT	36.6	45.8	54.9	64.1	73.2	74.1
		X	Flexible	STD	36.6	45.8	54.9	61.9	61.9	61.9
				SSLT	36.6	45.8	54.9	61.9	61.9	61.9
			Rigid	STD	36.6	45.8	54.9	64.1	73.2	82.4
				SSLT	36.6	45.8	54.9	64.1	73.2	82.4
	A490	N	Flexible	STD	36.6	45.8	54.9	61.9	61.9	61.9
				SSLT	36.6	45.8	54.9	61.9	61.9	61.9
			Rigid	STD	36.6	45.8	54.9	64.1	73.2	82.4
				SSLT	36.6	45.8	54.9	64.1	73.2	82.4
		X	Flexible	STD	36.6	45.8	54.9	64.1	73.2	77.3
				SSLT	36.6	45.8	54.9	64.1	73.2	77.3
			Rigid	STD	36.6	45.8	54.9	64.1	73.2	82.4
				SSLT	36.6	45.8	54.9	64.1	73.2	82.4
2 ($L = 6$)	A325	N	Flexible	STD	23.0	24.9	24.9	24.9	24.9	24.9
				SSLT	23.0	24.9	24.9	24.9	24.9	24.9
			Rigid	STD	24.4	30.5	33.4	33.4	33.4	33.4
				SSLT	24.4	30.5	36.6	37.3	37.3	37.3
		X	Flexible	STD	23.0	28.7	31.1	31.1	31.1	31.1
				SSLT	23.0	28.7	31.1	31.1	31.1	31.1
			Rigid	STD	24.4	30.5	36.6	41.7	41.7	41.7
				SSLT	24.4	30.5	36.6	42.7	46.7	46.7
	A490	N	Flexible	STD	23.0	28.7	31.1	31.1	31.1	31.1
				SSLT	23.0	28.7	31.1	31.1	31.1	31.1
			Rigid	STD	24.4	30.5	36.6	41.7	41.7	41.7
				SSLT	24.4	30.5	36.6	42.7	46.7	46.7
		X	Flexible	STD	23.0	28.7	34.5	38.9	38.9	38.9
				SSLT	23.0	28.7	34.5	38.9	38.9	38.9
			Rigid	STD	24.4	30.5	36.6	42.7	48.8	52.1
				SSLT	24.4	30.5	36.6	42.7	48.8	54.9

STD = Standard holes
SSLT = Short-slotted holes transverse to direction of load

N = Threads included
X = Threads excluded

EXAMPLE 4.7—Single-Plate Connection to Rectangular HSS

Use Tables 4-8, 4-9 and 4-10 to design an A36 single-plate connection for a W18×35 beam of A572 Gr 50 steel to an A500 Grade C HSS6×6×3/8 column for a factored reaction of 40 kips. Use 3/4-in. A325-N bolts and 70 ksi weld electrode.

Section Properties

HSS6×6×3/8	$t = 0.349$ in.	$b/t = 14.2$
	$F_y = 50$ ksi	$F_u = 62$ ksi
W18×35	$t_w = 0.300$ in.	$T = 15\frac{1}{2}$-in.
	$F_y = 50$ ksi	$F_u = 65$ ksi

Check if single-plate connection is suitable

$(b/t)_{max} = 253/\sqrt{50} = 35.8$

HSS $b/t = 14.2 < 35.8$ **o.k.**

Design the bolts and single plate

From Table 4-8, for a rigid support, for 3/4-in. diameter A325-N bolts in standard holes, select 3 rows of bolts and a 1/4-in. plate thickness.

$\phi R_n = 41.6$ kips > 40 kips   **o.k.**

The 9-in. long plate is satisfactory.

$T/2 = 7.75$ in. < 9 in. $< T$

Determine the weld size

From Table 4-9 for a 1/4-in. plate, the weld size is 3/16-in.

t_{min} in Table 4-9 does not apply for A500 Grade C HSS

$$t_{min} = \frac{3[0.75 \times 70]}{16\sqrt{2}[0.9(50)]} = 0.15, \text{ (Equation 4-19)}$$

HSS $t = 0.349$ in. $> t_{min} = 0.15$ in.

No reduction in the weld strength is required.

Check the beam web

From Table 4-10 for three 3/4-in. diameter bolts in material with $F_u = 65$ ksi,

$\phi R_n = 263 \times 0.300 = 78.9$ kips > 40 kips   **o.k.**

EXAMPLE 4.8—Single Plate Connection to Round HSS

Repeat Example 4.7 using an HSS10.000×0.250 column with a yield strength of 42 ksi.

Section Properties

HSS10.000×0.250	$t = 0.233$ in.	$D/t = 42.9$
	$F_y = 42$ ksi	$F_u = 58$ ksi

Table 4-9. Welds for Single-Plate Connections				
Plate Thickness (in.)	Weld Size 70 ksi (in.)	Minimum HSS Thickness (in.)		Resistance ϕR_n (kips/in.)
		Rectangular (F_y = 46 ksi)	Round	
$\frac{1}{4}$	$\frac{3}{16}$	0.17	10.3 / F_y	8.35
$\frac{5}{16}$	$\frac{1}{4}$	0.22	12.9 / F_y	11.14
$\frac{3}{8}$	$\frac{5}{16}$	0.28	15.5 / F_y	13.92
$\frac{7}{16}$	$\frac{3}{8}$	0.34	18.0 / F_y	16.70
$\frac{1}{2}$	$\frac{3}{8}$	0.34	20.6 / F_y	16.70
$\frac{9}{16}$	$\frac{7}{16}$	0.39	23.2 / F_y	19.49

W18×35 $t_w = 0.300$ in. $T = 15\frac{1}{2}$ in.
          $F_y = 50$ ksi $F_u = 65$ ksi

Check if single-plate connection is suitable

$(D/t)_{\max} = 3{,}300/F_y \le 72$

$3{,}300/42 = 78.6$, therefore $(D/t)_{\max} = 72$

$D/t = 42.9 \le 72$ **o.k.**

Design the bolts and single plate

No change from Example 4.7. Use three $\frac{3}{4}$-in. diameter A325-N bolts in standard holes with a $\frac{1}{4}$-in. thick plate that is 9-in. long. The distance from the HSS to the bolt line is 3 inches and the horizontal edge distance is $1\frac{1}{2}$-in., making the plate width $4\frac{1}{2}$-in.

Determine the weld size

From Table 4-9 for a $\frac{1}{4}$-in. plate, the weld size is $\frac{3}{16}$-in. and $t_{\min} = 10.3 / F_y$

Since

HSS $t = 0.233$ in. $< t_{\min} = 10.3/42 = 0.245$ in.

Reevaluate the strength based on the shear of the HSS wall effective weld size (Equation 4-19).

$$D_{eff} = \frac{16\sqrt{2}\,(0.9)(42)}{0.75(70)}(0.233) = 3.80$$

Determine the weld eccentricity based on the bolt eccentricity from Equation 4-2.

bolt $e = \left| (3 - 1) - 3 \right| = \left| -1 \right| = 1$ in.

Since the value in the absolute braces is negative, the eccentricity is between the HSS and the bolt line.

weld eccentricity $= 3 - 1 = 2$ in.

Using $e_x = 2$ in., $a = e_x / l = 2/9 = 0.22$, in LRFD Manual Table 8-38 and $k = 0$ for an out-of-plane eccentricity by interpolation, $C = 2.58$

(Equation 4-17)

$\phi R_n = 2.58(3.80)9 = 88.1$ kips > 40 kips   **o.k.**

Check the beam web

No change from Example 4.7. The beam web is satisfactory.

THROUGH-PLATE CONNECTIONS

In the through-plate connection shown in Figure 4-1e, the front and rear faces of the HSS are slotted so that the plate can be passed completely through the HSS and welded to both faces. The plate acts as a reinforcement to the HSS walls and through-plate connections should be used when the HSS wall is classified as a slender element with $b/t > 253/\sqrt{F_y}$ or 37.3 for $F_y = 46$ ksi ($253/\sqrt{F_y}$ is the same as $1.4(E/F_y)^{0.5}$). However, a single plate connection is more economical and should be used if the HSS is neither slender nor inadequate for the punching shear rupture limit state in AISC HSS Specification Section 9.3.3.

Through-plate connections have the same limit states as single-plate connections and Table 4-8 may be used to determine the size and number of bolts and the plate thickness. All values are to be compared to factored loads. The welds, however, are subject to direct shear and may not have to be as large as those for single plate connections. For equilibrium of the forces in Fig. 4-5, the shear in the welds on the front face is $R_u(H + e_w)/H$, which should not exceed the resistance of the pair of welds, ϕR_n in Table 4-9. If the thickness of the HSS is less than the minimum in Table 4-9, the weld strength must be reduced proportionally. The eccentricity is $(n - 1)$ for n bolts in standard holes and $2n/3$ for short-slotted holes. Conservatively, the welds on the rear face may be the same size.

When a connection is made on both sides of the HSS with an extended through-plate, the portion of the plate inside the HSS is subject to a uniform bending moment. For long connections, this portion of the plate will buckle in a lateral-torsional mode prior to yielding, unless H is very small. Using a thicker plate to prevent lateral-torsional buckling would restrict the rotational flexibility of the connection. Therefore, it must be recognized that the plate may buckle and that the moment will be shared with the HSS wall in a complex manner. However, if the HSS is satisfactory when checked for the criteria of a single plate connection, the lateral-torsional buckling limit state is not a critical concern involving loss of strength.

EXAMPLE 4.9—Through-Plate Connection

Redesign the connection in Example 4.7 using an HSS6×4×⅛ with the connection to one of the 6-in. faces.

Section Properties

The width B and the wall slenderness b/t are referenced to the loaded face of the HSS. The b/t is tabulated as h/t in the section properties tables.

HSS6×4×⅛	$t = 0.116$ in.	$b/t = 48.7$	$H = 4$ in.
	$F_y = 46$ ksi	$F_u = 58$ ksi	

Table 4-10.
Beam Web Bearing Design Strength per Inch of Thickness (kips/in.)

Bolts		Beam F_u (ksi)		Bolts		Beam F_u (ksi)	
Number	Size (in.)	58	65	Number	Size (in.)	58	65
9	$\frac{3}{4}$	705	790	5	$\frac{3}{4}$	392	439
	$\frac{7}{8}$	822	921		$\frac{7}{8}$	457	512
	1	940	1,053		1	522	585
8	$\frac{3}{4}$	626	702	4	$\frac{3}{4}$	313	351
	$\frac{7}{8}$	731	819		$\frac{7}{8}$	365	410
	1	835	936		1	418	468
7	$\frac{3}{4}$	548	614	3	$\frac{3}{4}$	235	263
	$\frac{7}{8}$	639	717		$\frac{7}{8}$	274	307
	1	731	819		1	313	351
6	$\frac{3}{4}$	470	527	2	$\frac{3}{4}$	157	176
	$\frac{7}{8}$	548	614		$\frac{7}{8}$	183	205
	1	626	702		1	209	234

Check if a single-plate connection is suitable

HSS $b/t = 48.7 > 37.3$ **n.g.**

A through-plate connection should be used.

Design the bolts and plate

The selection of the bolts and plate are the same as in Example 4.7. Use 3 rows of bolts and a $\frac{1}{4}$-in. plate thickness. The plate is 9 inches long.

Determine the weld size

$e_w = 3 - 1 = 2$ in.

Weld force $V_f = 40(4 + 2)/4 = 60$ kips

Required weld resistance $= V_f/L = 60/9 = 6.67$ kips/in.

From Table 4-9, a $\frac{3}{16}$-in. weld is required but the resistance must be reduced because t of HSS is less than the minimum.

$\phi R_n = (0.116/0.17)8.53$

$= 5.82$ kips/in. < 6.67 kips/in.   **n.g.**

A longer weld is required.

$L \geq 60/5.82 = 10.3$ in.

Use a 10½-in. long plate and increase the vertical edge distances to 2¼-in.

Recheck the plate length

$L = 10\frac{1}{2}$-in. $< T = 15\frac{1}{2}$-in.   **o.k.**

Buckling with Equation 4-12.

$$t = \frac{1}{4}\text{-in.} > \frac{L}{64} = \frac{10.5}{64} = 0.164 \text{ in.}   \textbf{o.k.}$$

SINGLE-ANGLE CONNECTIONS

A single-angle connection is made with an angle on one side of the beam web. In order to provide adequate flexibility, the angle is shop welded to the HSS column along the toe and across the bottom of the angle with a return at the top per LRFD Specification Section J2.2b. Welding across the entire top must be avoided as it would inhibit flexibility.

A 4×3 angle is normally selected with the 3-in. leg shop welded to the HSS. A minimum angle thickness of ⅜-in. for ¾-in. and ⅞-in. diameter bolts, and ½-in. for 1-in. diameter bolts should be used. For fillet welding on the flat of the HSS side and keeping the center of the beam web in line with the center of the HSS, single angle connections may be used with the following HSS:

$B = 8$ inches and $t \leq \frac{1}{4}$-in.

$B = 9$ inches and $t \leq \frac{3}{8}$-in.

$B \geq 10$ inches and any nominal thickness

Alternatively, single angles can be welded to narrow HSS with a flare-bevel weld.

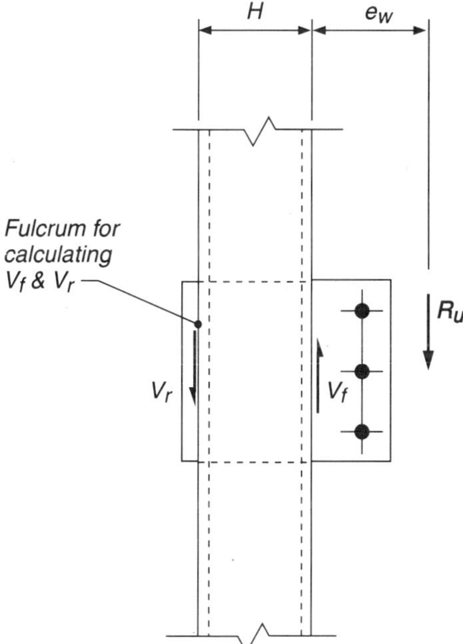

Fig. 4-5. Shear forces in a through-plate connection.

These sizes are based on an HSS corner radius of 1.5 times the nominal wall thickness and a weld size corresponding to the HSS wall shear strength from Equation 4-19.

Table 4-11 is the design aid for single-angle connections using 4×3×⅜ angle of material with F_y = 36 ksi. It includes a single row of from two to twelve A325-N bolts with ¾-in. and ⅞-in. diameters. No eccentricity is assumed for the bolts and tabulated values are based on the limit state from Equation 4-1. The limit states from Equations 4-3, 4-5, 4-6 and 4-10 do not control for the 4×3×⅜ angle. The tabulated values are to be compared to factored loads.

The tabulated values for the welds with a strength of 70 ksi are from Equation 4-17 with the eccentric reaction assumed at the middle of a ½-in. beam web. For half-web thicknesses less than ¼-in., the tabulated values are conservative. However, for larger half-web thicknesses that create larger eccentricities, the tabulated values should be reduced proportionally by up to eight percent for a half-web thickness of ½-in., or the weld strength should be recalculated. The limiting HSS thickness is from the shear limit state of Equation 4-19. If the HSS thickness is less than the minimum tabulated value, the weld strength must be reduced proportionally.

The limit state for the beam web is bolt bearing (Equation 4-4) and the tabulated values for resistance per inch of thickness in Table 4-10 apply.

EXAMPLE 4.10—Single-Angle Connection

Use Tables 4-11 and 4-10 to design a single-angle connection for a W16×50 beam of A572 Gr 50 steel to an HSS8×8×¼ column for a factored reaction of 55 kips. Use ¾-in. diameter A325-N bolts and 70 ksi weld electrode.

Section Properties

HSS8×8×¼	t = 0.233 in.	B = 8 in.	
	F_y = 46 ksi	F_u = 58 ksi	
W16×50	t_w = 0.380 in.	d = 16.26 in.	t_f = 0.630 in.
	F_y = 50 ksi	F_u = 65 ksi	T = 13⅝-in.

Design the bolts and single angle

Since the half-web dimension of the W16×58 is less than ¼-in., values in Table 4-11 may be used. Try a four-bolt single angle L4×3×⅜ with a length of 11½-in.

ϕR_n = 63.6 kips > 55 kips **o.k.**

$T/2$ = 6¹³⁄₁₆-in. < 11.5 in. < T **o.k.**

Determine the weld size

From Table 4-11 select a ³⁄₁₆-in. weld

ϕR_n = 56.6 kips > 55 kips **o.k.**

minimum HSS thickness = 0.17 in. < 0.233 in. (no reduction for shear in the HSS is required)

Check the beam web

From Table 4-10 for four ¾-in. diameter bolts, the resistance is 351 kips/in.

ϕR_n = 351 × 0.380 = 133 kips > 55 kips **o.k.**

Table 4-11.
Single-Angle Connections
L4×3×3/8, F_y = 36 ksi, F_u = 58 ksi

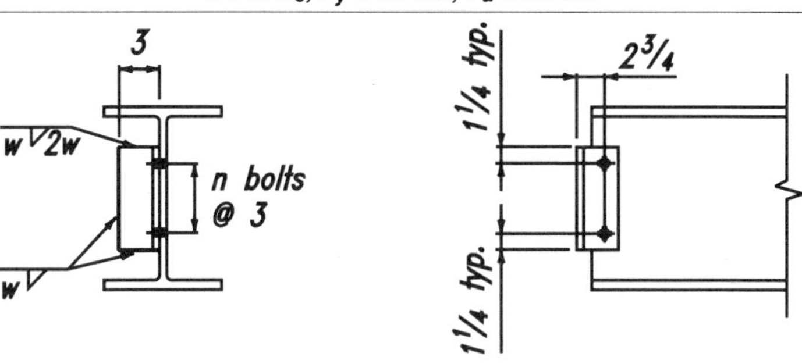

Number of Bolts in One Vertical Row	A325-N Bolt Shear Strength, kips		Angle Length in.	Weld (70 ksi)		Minimum HSS t, in.
	3/4-in.	7/8-in.		Design Strength, kips	Size W in.	
12	191	260	35½	270 216 162	5/16 1/4 3/16	0.28 0.22 0.17
11	175	238	32½	247 198 148	5/16 1/4 3/16	0.28 0.22 0.17
10	159	217	29½	227 182 136	5/16 1/4 3/16	0.28 0.22 0.17
9	143	195	26½	205 164 123	5/16 1/4 3/16	0.28 0.22 0.17
8	127	173	23½	185 150 111	5/16 1/4 3/16	0.28 0.22 0.17
7	111	152	20½	164 131 98.4	5/16 1/4 3/16	0.28 0.22 0.17
6	95.4	130	17½	141 113 84.5	5/16 1/4 3/16	0.28 0.22 0.17
5	79.5	108	14½	118 94.5 70.9	5/16 1/4 3/16	0.28 0.22 0.17
4	63.6	86.6	11½	94.3 75.4 56.6	5/16 1/4 3/16	0.28 0.22 0.17
3	47.7	64.9	8½	68.9 55.1 41.3	5/16 1/4 3/16	0.28 0.22 0.17
2	31.8	43.3	5½	42.1 33.7 25.2	5/16 1/4 3/16	0.28 0.22 0.17

Notes:

Gage in angle leg attached to beam web as well as leg width may be decreased. 3-in. welded leg may not be increased or decreased.

Tabulated weld design strengths are based on a 1/4-in. half web for the supported member. Smaller half webs will result in these values being conservative. For half webs over 1/4-in., weld values must be reduced proportionally to 8% for a 1/2-in. half web or recalculated.

SUGGESTED DETAILS

Details in Figures 4-6 and 4-7 are suggested treatments only and are not intended to limit the use of other connections not illustrated.

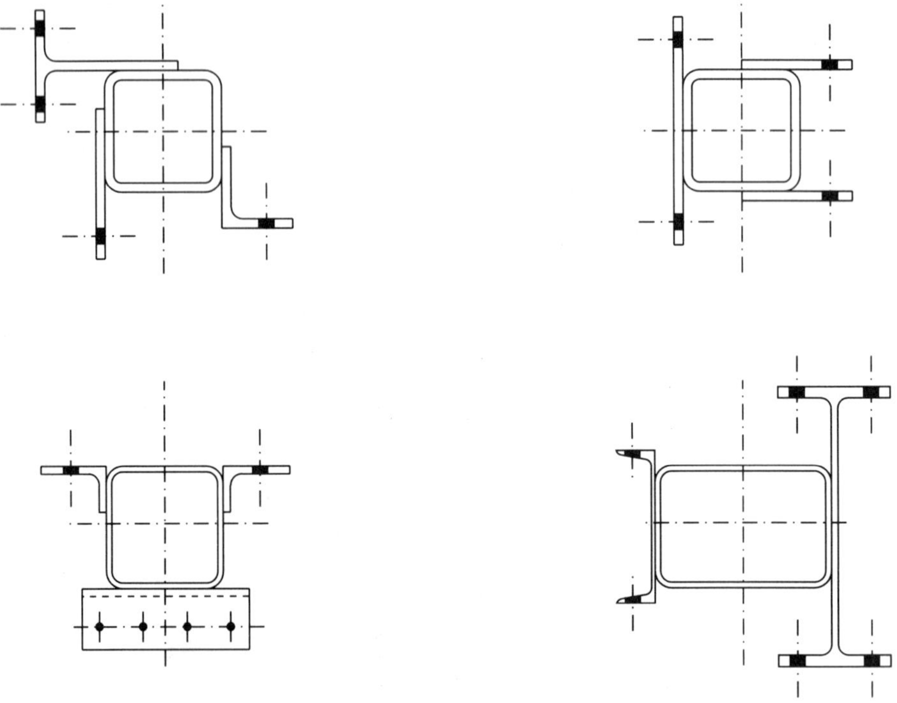

Fig. 4-6. Beam framing to HSS.

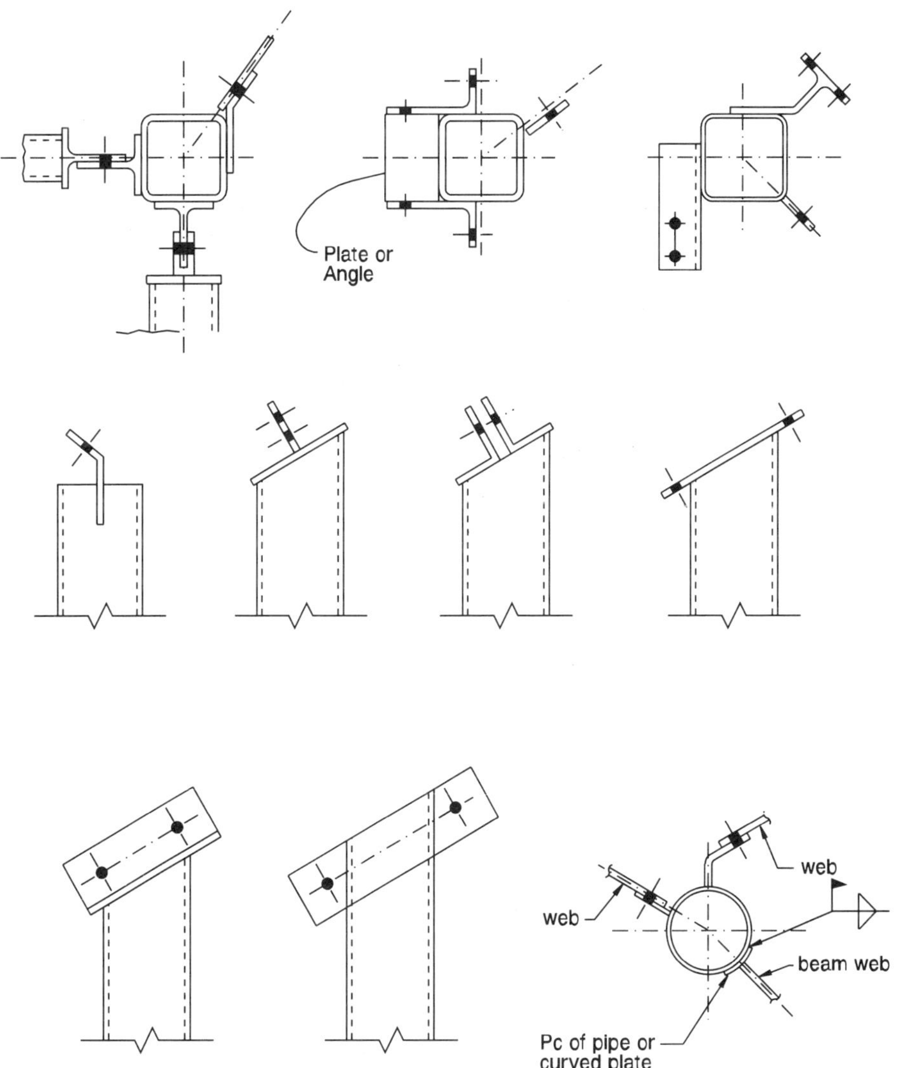

Fig. 4-7. Skewed and offset connections.

REFERENCES

Abolitz, A. L. and Warner, M. E., 1965, "Bending Under Seated Connections," *Engineering Journal*, AISC, Vol. 2, No. 1, pp. 1–5.

American Institute of Steel Construction, 1993, *Load and Resistance Design Specification for Structural Steel Buildings*, AISC, Chicago IL.

American Institute of Steel Construction, 1994, *Manual of Steel Construction LRFD, Vol. II, Connections*, 2nd ed., AISC, Chicago, IL.

Garret, J. H., Jr. and Brockenbrough, R. L., 1986, "Design Loads for Seated Beams in LRFS," *Engineering Journal*, AISC, Vol. 23, No. 2, pp. 84–88, AISC, Chicago, IL.

Salmon, C. G. and Johnson, J. E., 1995, *Steel Structures: Design and Behavior*, 4th ed., Harper-Collins, New York, NY.

Sherman, D. R. and Ales, J. M., 1991, "The Design of Shear Tabs With Tubular Columns," *Proceeding of the 1991 National Steel Construction Conference*, pp. 23.1–23.14, AISC, Chicago, IL.

Sherman, D. R., 1996, "Designing With Structural Tubing," *Engineering Journal*, AISC, Vol. 33, No. 3, pp. 101–109, AISC, Chicago, IL.

Sputo, T. and Ellifritt, D. S., 1991, "Proposed Design Criteria for Stiffened Seated Connections to Column Webs," *Proceeding of the 1991 National Steel Construction Conference*, pp. 8.1–8.26, AISC, Chicago, IL.

CHAPTER 5

MOMENT CONNECTIONS BETWEEN HSS SHAPES AND W SHAPES

OVERVIEW

Chapter 5 contains general information, design considerations, and examples for the design of round, square and rectangular HSS columns connected to W-shape beams. The information is based on generally accepted engineering principles, several references, the 1993 AISC LRFD Specification and the AISC HSS Specification.

Following is a detailed list of the connection types discussed.

MOMENT CONNECTIONS BETWEEN HSS AND W SHAPES

There is a wide range of concepts for transmitting moment between W-shapes and HSS. The best concept for a particular situation depends on the magnitude of moment that must be transmitted *to* an HSS, the magnitude of moment that must be transmitted *through* an HSS and the magnitude of the axial load in the HSS. The differences between the W-shape and the HSS make their flexural interconnection a challenging task.

CONTINUOUS BEAMS OVER HSS COLUMNS

If the principal load transfer is between W-shapes on each side of a joint, then it is best to have the W-shape be continuous, and to make the column a field connection to the flange of the W-shape. This is common for the case of the continuous roof beam where a 4-bolt cap plate is used on the column as illustrated in Figure 5-1.

Limit states for this type of connection are:

1. Bending with prying on the flange of the W shape.
2. Bending with prying on the cap plate of the HSS.
3. Strength of the beam web relative to yielding, crippling, and buckling.
4. Strength based on bolting to the cap plate.
5. Strength of the weld connecting the cap plate to the HSS.
6. Strength of the HSS wall relative to yielding and crippling.

The use of this connection alternative is illustrated in Example 5.1 on page 5-15.

HSS COLUMNS ABOVE AND BELOW CONTINUOUS BEAMS

Field connection to the flanges of the continuous beam can be used at joints where there is an HSS above and below. This situation is illustrated in Figure 5-2. If the column load is not high, stiffener plates may be used to transfer the axial load across the W-shape as shown in Figure 5-2a. If the axial load is higher it may be necessary to use a split HSS instead of plate stiffeners, as shown in Figure 5-2b. The width of the W-shape must be at least as wide as the HSS and preferably should be wider than the HSS for this detail to be used as shown. It may be necessary to use a rectangular HSS column in order to fit the HSS base plate on the beam flange. The design of the stiffeners or the split HSS can be accomplished using the LRFD Specification and is not treated herein. The moment

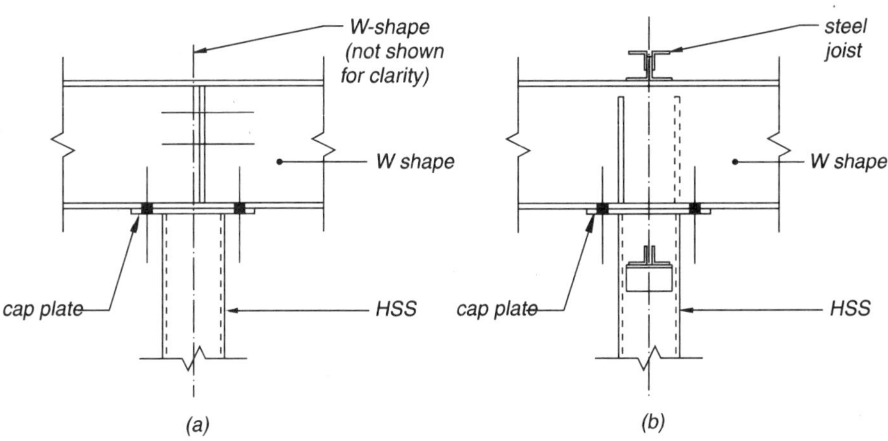

Fig. 5-1. Roof beam continuous over HSS column.

Table 5-1.

W Shapes, HP Shapes

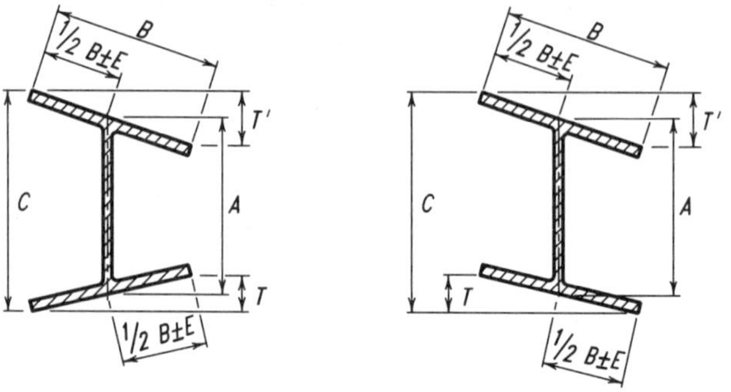

Permissible Variations in Cross Section

Section Nominal Size, in.	A, Depth, in.		B, Fig. Width, in.		T + T' Flanges, out of square, Max, in.	E [a], Web off Center, Max, in.	C, Max. Depth at any Cross-section over Theoretical Depth, in.
	Over Theoretical	Under Theoretical	Over Theoretical	Under Theoretical			
To 12, incl.	$\frac{1}{8}$	$\frac{1}{8}$	$\frac{1}{4}$	$\frac{3}{16}$	$\frac{1}{4}$	$\frac{3}{16}$	$\frac{1}{4}$
Over 12	$\frac{1}{8}$	$\frac{1}{8}$	$\frac{1}{4}$	$\frac{3}{16}$	$\frac{5}{16}$	$\frac{3}{16}$	$\frac{1}{4}$

Notes:
[a] Variation of $\frac{5}{16}$ in. max. for sections over 426 lb / ft.

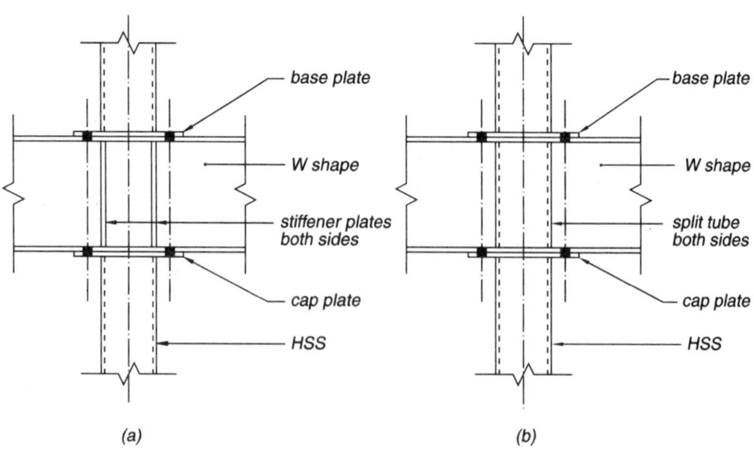

Fig. 5-2. HSS columns spliced to continuous beams.

transfer to the HSS is limited by the four bolts, the thickness of the W-shape flange, and the base and cap plate thicknesses.

Limit states are:

1. Bending with prying on the flange of the W shape.
2. Bending with prying on the cap plate of the lower HSS column.
3. Bending with prying on the upper HSS column base plate.
4. Strength based on bolting to the base and cap plates.
5. Strength of the weld connecting the base and cap plates to the HSS.

THROUGH-PLATE CONNECTIONS

If the required moment transfer to the column is larger than can be provided by the bolted base plate or cap plate, or if the HSS width is larger than that of the W-shape beam, a through-plate moment connection can be used as illustrated in Figure 5-3. It should be noted that through-plate connections are more difficult to erect than the continuous beam connected framing.

When moment connections are made using through-plates such as is shown in Figure

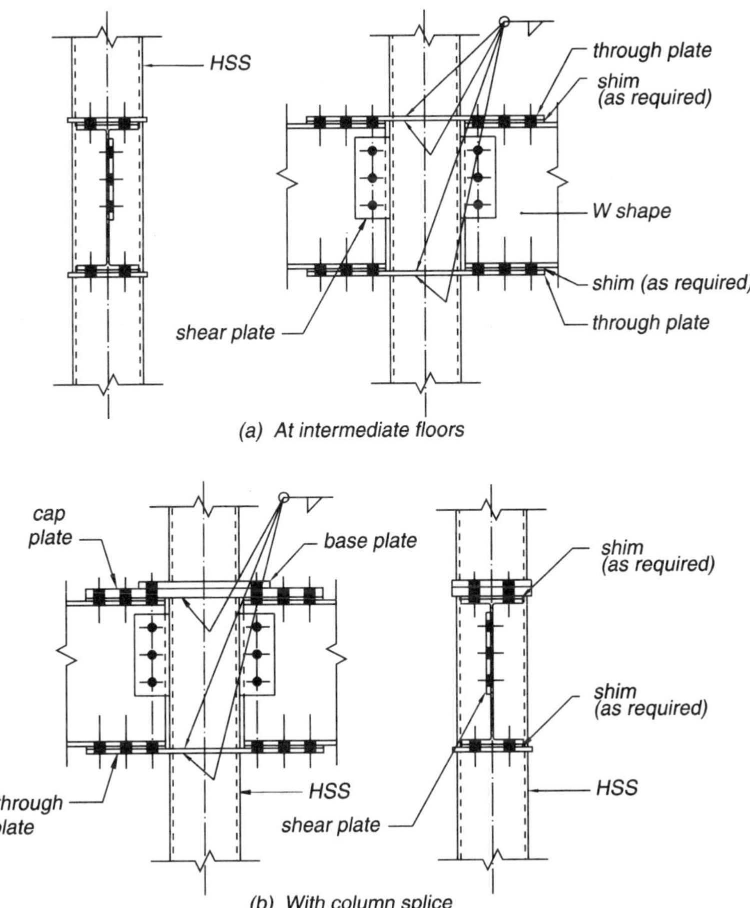

Fig. 5-3. Through-plate moment connection.

5-3 and similar figures in this chapter, the fabricator must allow adequate clearance between the through-plates and the W-shape so as to allow for the combined effects of mill, fabrication, and erection tolerances. Shown in Table 5-1 are the permissible mill tolerances for W-shape variations in depth and squareness (flange tip). ASTM A6 should be consulted for other sections such as S, M, or C shapes. Shimming in the field during erection with conventional or finger is the most commonly used method to fill the gap between the W-shape and the through-plate.

Limit states for the through-plate moment connections are:

1. Strength of the welds connecting the through-plates to the HSS.
2. Tension or compression of the through-plate.
3. Strength based on bolting to the through-plate.
4. Strength based on bolting to the beam flange.
5. Strength based on bolting to the cap plate.
6. Combined axial compression or tension and bending of the cap plate.
7. Bending with prying on the base plate.
8. Strength of the welds connecting the base and cap plates to the HSS.

Specific design considerations for through-plate moment connections are as follows:

1. In Figures 5-3a and 5-3b, the column moment transfer into the joint is limited by the fillet weld of the HSS to the through-plates. If necessary, a partial-joint-penetration (PJP) groove weld can be used to improve the connection strength or a complete-joint penetration (CJP) groove weld with backup bars could be used.
2. In Figure 5-3b an end plate (base plate) is employed to create a splice in the column. Bolt tension with prying on the base plate will determine its thickness and thus limit the moment that can be transferred to the upper HSS.
3. The cap plate, which is also a flange splice plate, should be at least the same thickness as the base plate so that moment transfer between the HSS columns need not rely on load transfer through the beam flanges. The cap plate may need to be thicker than the HSS base plate due to the combined effect of plate bending from the bolted base plate and plate tension or compression from the W-shape moment transfer.
4. The welding of the HSS to the cap and through-plate must be examined for both the HSS normal forces and the shear produced by the moment transfer from the W-shape.

The use of this connection alternative is illustrated in Example 5.2 on page 5-18.

CUT-OUT PLATE CONNECTIONS

An alternative to interrupting the HSS for the cap or through plate is to use a wider plate with a cut-out to slip around the HSS as illustrated in Figure 5-4. A shear plate can be placed on the front and rear of the HSS faces to provide simple connections for perpendicular beams. The cut-out plate can easily be extended on the near and far sides so that a moment splice is created about both horizontal axes through the joint. The perpendicular framing should ideally be of the same depth for this detail to work well or in the case of the simple connections, the perpendicular beams could be shallower than the space between the horizontal plates. The cut-out plates are shown as shop welded; however, they could be field welded.

The limit states are as follows:

1. Plate tensile and compressive strength at the sides of the HSS.
2. Shear strength of the HSS webs caused by opposing moments from lateral loads.
3. Strength based on bolting to the cut out plates.

For cut-out plate connections, the erection of the beams is more difficult than for continuous beam connections. The beams must be slipped between the two plates and against the single plate connection with shimming being required unless the upper plate is field welded in place.

The use of this connection alternative is illustrated in Example 5.3 on page 5-20.

WELDED TEE FLANGE CONNECTIONS

If the primary moment transfer is from a W-shape to an HSS, rather than through the HSS to another W-shape, a number of other connection concepts will work well. One of these is to use structural tee sections to transfer the force from the flanges of the W-shape to the walls of the HSS as is illustrated in Figure 5-5. The tees should be long enough so that a flare bevel groove (or single J-groove) weld with weld reinforcement can be used to connect the tee to the HSS. An alternative to using the tees to transfer the beam shear would be to use a single plate connection, if a deep enough plate can be fit between the flanges of the tees.

The limit states associated with the tee connection are as follows:

1. Strength of the flare-bevel groove welds attaching the tees to the HSS.
2. Local flange bending of the tee.
3. Local web yielding of the tee.
4. Local web yielding of the HSS.
5. Web crippling of the HSS.
6. Compression buckling of the side wall or web of the HSS.

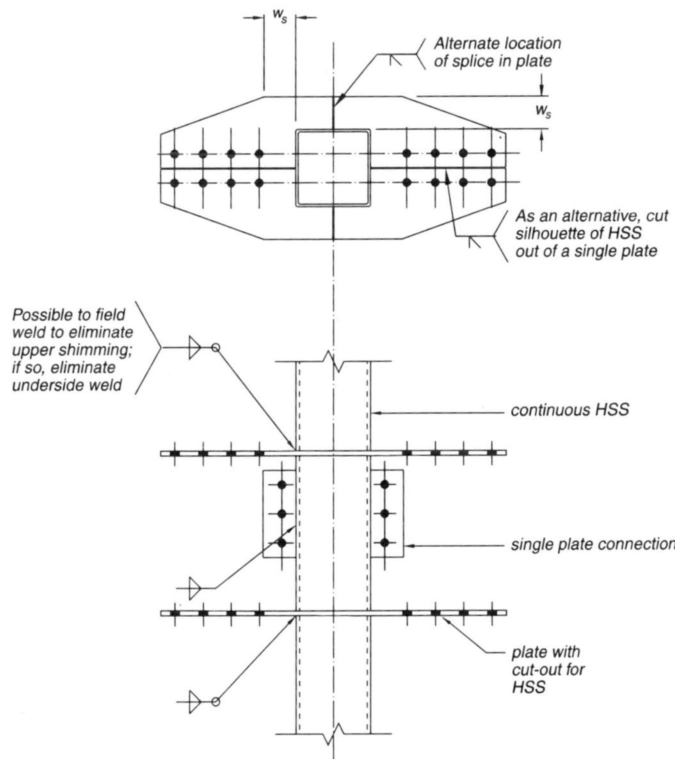

Fig. 5-4. Cut-out plate moment connection.

7. Strength based on bolting of the tee stem to the W-shape flange.

The use of this connection alternative is illutrated in Example 5.4 on page 5-22.

END PLATE CONNECTIONS

Another means of transferring the moment to the HSS is to use an end plate on the W-shape beam. This end plate is bolted either to an end plate welded to the face of the HSS as in Figure 5-6a or to angles welded to the sides of the HSS as is illustrated in Figure 5-6b. The projection of the end plate both above the W-shape and beyond the sides of the HSS may interfere with the construction of other building components.

For this connection to be practical, the beam flange width should be as large or larger than the HSS width. It is reasonable to consider only two bolts near each flange tip as effective in tension on the beam end plate. For the end plate connection the following limit states must be considered:

1. Bolt strength (combined tension and shear).
2. Bending with prying on the connection plate attached to the HSS.
3. Bending of the end plate on the W-shape beam.
4. Buckling of the HSS side wall.

Design Comments

1. If the two bolts are beyond the flange tip, bending at the flange tip must be checked

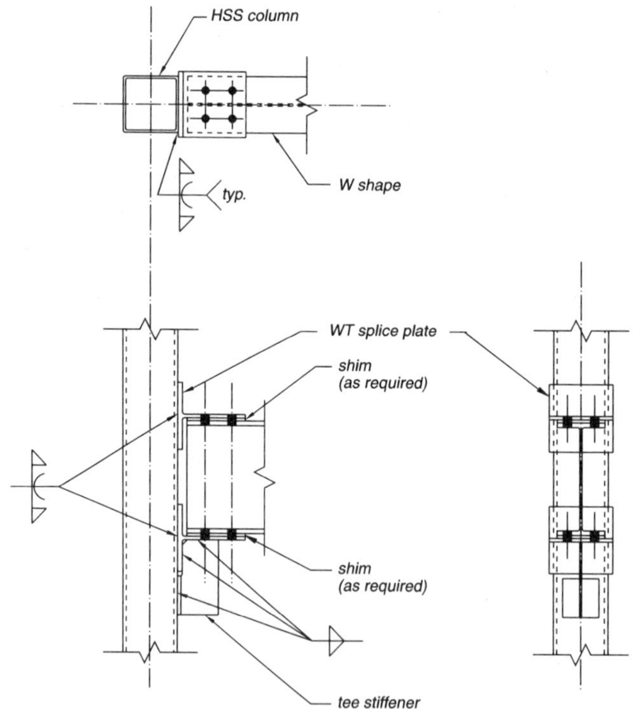

Note: A shear plate could be used in lieu of the vertical tee stiffener

Fig. 5-5. Welded tee flange connection to HSS column.

using the prying model, (limit state 3). There is no standard procedure for this situation. However, the edge of the flange can be assumed to behave like the bolt and the center of the two bolts could behave like the center of the web. Thus, the flange thickness plus twice the horizontal distance c from flange tip to the bolts could be used as the effective width for bending at the flange tip. See Figure 5-7a.

If the flange is wider than the HSS, the bolts may actually be above and below the flanges such that prying action will occur about the flange centerline and bending at the flange tip as discussed above will not occur. See Figure 5-7(b) where c is now the horizontal distance of the bolt line to the flange tip. In this case it is suggested that prying action be considered with the effective width p taken as $s/2 + c + t_f \leq s$. A situation may occur when c in Figure 5-7(b) is small relative to b where one could use c as negative in the computation of p to make an evaluation considering prying action.

2. The use of tee or end plate connections is limited by the amount of force that can be transferred into the HSS wall. Buckling of the HSS wall due to the flange force may make the wall partially ineffective for transmitting axial column load. The angles (see Figure 5-66) on the sides of the HSS stiffen the loaded wall, however the bolt tension acting on the angles will introduce some flexing of the angles which must be resisted by the HSS wall.

The use of this connection alternative is illustrated in Example 5.5 on page 5-24.

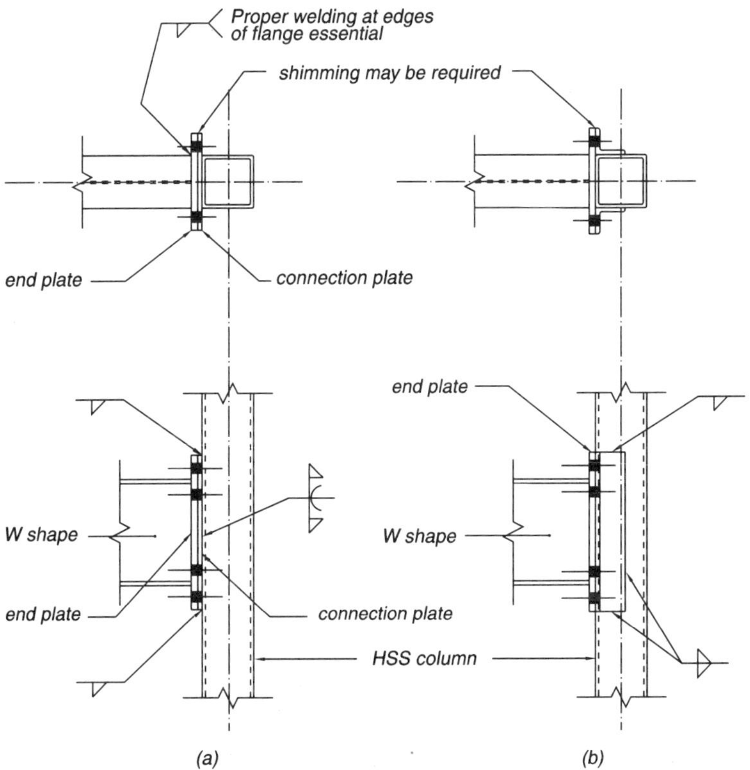

Fig. 5-6. End plate connections to HSS columns.

DIAPHRAGM PLATE CONNECTIONS

If the moment transferred by the W-shape to the HSS cannot be transmitted with the connection of the types shown in Figures 5-5 and 5-6, the use of diaphragm plates that transfer the flange loads to the sides of the HSS may be appropriate. This is illustrated in Figure 5-8. For this moment connection the limit states are those indicated for the cut-out plate connection plus a check of the weld transferring shear from the flange plate to the HSS wall.

DIRECT CONNECTIONS

It may be possible to accomplish the moment transfer to the HSS without having to use the WT splice plate, end plates or the diaphragm plates. Significant moment transfer can be achieved by attaching the W-shape directly to the face of the HSS either by welding as illustrated in Figure 5-9 or by bolting as illustrated in Figure 5-10. These connections are capable of developing the full flexural strength of the HSS. The full flexural strength of the W-shape, however, is seldom achievable.

The flexural strength for the welded W-shape as illustrated in Figure 5-9 is based on the strength of the respective flanges in tension and compression acting against the face of the HSS. This flange force can be considered to be the same as that of a plate with the dimensions of the flange.

Several limit states exist for the plate length (flange width) oriented perpendicular to the length of the HSS (Packer and Henderson, 1992) and (AISC, 1997).

These limit states are:

1. The AISC HSS Specification provides equations for the limit states of web crippling, punching shear and wall buckling.

   See Figure 5-11(a) for illustration of the effective bearing length to be used in the AISC equations.

2. Effective width of the flange: This is due to the non-uniform stress across the flange width, which can result in yielding at the tips of the flange. The flange design strength is a function of the local yielding strength of the face of the HSS and can be expressed as:

$$\phi R_n = \phi \frac{10 F_y t}{b_0 / t} b_1 \tag{5-1}$$

where

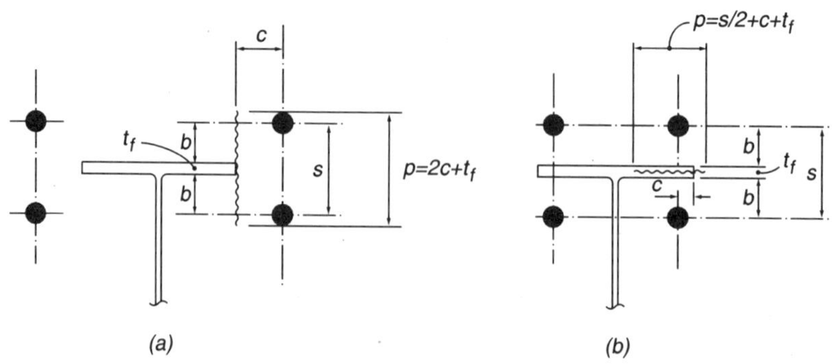

Fig. 5-7. End plate bolting near flange tip.

$\phi = 1.0$

$b_1 = b_f =$ the flange width of the beam. See Figure 5-9(b).

$b_o =$ the overall width of the loaded face of the HSS, either B or H

ϕR_n may not exceed the flange strength $\phi F_{ywf} t_f\, b_1$ where $\phi = 0.9$ and $F_{ywf} =$ the specified yield strength of the W-shape. Punching shear rupture, which is a possible limit state for HSS web to HSS chord members of trusses, will never govern for flange-to-HSS face connections since the effective width expression in Equation 5-1 will always control. It is also obvious by comparing the effective width expression with the AISC side wall failure expressions, that the effective width expression will always control over local web yielding, web crippling and punching shear.

3. Yielding of HSS face: This is the limit state for plates transversely positioned on the HSS face. This limit state is based on the expressions developed from yield line theory. If a full set of yield lines is developed as shown in Figure 5-9(b), the flange force can be expressed as:

$$\phi R_n = \phi F_y t^2 \left(\frac{2t_f}{b_o - b_1} + \frac{4}{\sqrt{1 - b_1/b_o}} \right) \tag{5-2}$$

where

$\phi = 0.9$

$0.95b < b_1 < b_o$

The flange width b_1, must be greater than or equal to 95 percent of the flat width $b = b_o - 3t$ to restrict the wall deformation and resulting moment rotation.

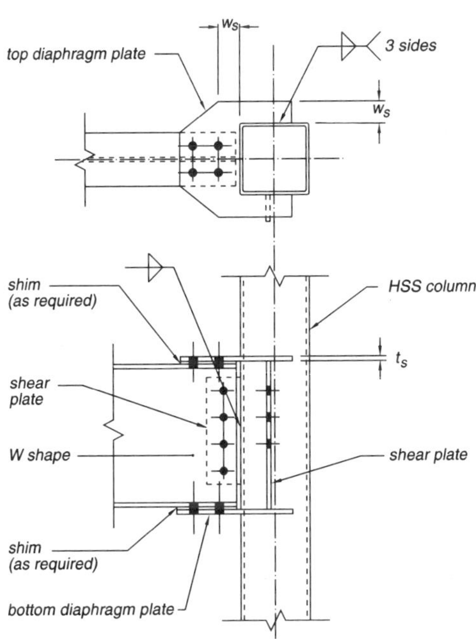

Note: A stiffened seat could also be used in lieu of the shear plate

Fig. 5-8. Diaphragm plate splice to exterior HSS column.

R_n determined from Equation 5-2 is always greater than R_n from Equation 5-1, thus this yield line expression need not be calculated.

However, if the yield lines across the HSS face above and below the flange (lines shown as dashed in Figure 5-9(b)) are ignored, ϕR_n becomes:

$$\phi R_n = \phi F_y t^2 \left(\frac{2t_f}{b_o - b_1} + \frac{3}{\sqrt{1 - b_1 / b_o}} \right) \tag{5-3}$$

where

$$\phi = 0.9 \text{ and } 0.95b < b_1 < b_o$$

Since these yield lines are difficult to develop as shown due to the flexibility of the face, reaching the R_n value in Equation 5-2 may require an extremely large deformation of the HSS face. This could reflect back to a significant inelastic deformation at service loads. Use of the smaller yield line expression (Equation 5-3) and the restriction on b_1 should reduce this possibility.

The flange force limit state can be determined by computation of Equation 5-1 for all values of b_1 / b_o unless the lower yield line limit state is used. In this case, the effective width expression (Equation 5-1) will control for larger values of b_1 / b_o and the yield line limit value will control for smaller values of b_1 / b_o.

Equations 5-1 to 5-3 should also be multiplied by a factor Q_f defined below which reflects the reduction in face capacity due to the presence of longitudinal compression stresses on the wall in question.

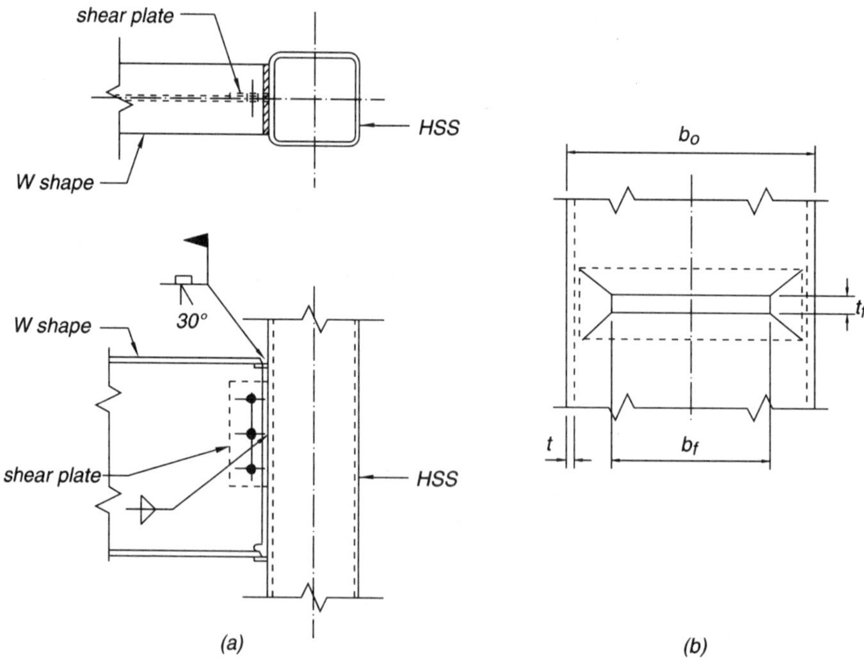

Fig. 5-9. Direct connections of W shape to HSS column by welding.

Q_f = 1.0 for HSS wall in longitudinal tension

    = $1 - 0.3f/F_y - 0.3(f/F_y)^2 \leq 1$ for HSS wall in compression

f = the magnitude of the sum of the maximum flexural plus axial stress on the wall of the HSS in question

4. Buckling of HSS walls: For compression forces on opposite faces of the HSS the AISC HSS Specification provides the equation:

$$\phi R_n = \phi 48 t^3 \sqrt{EF_y} / h$$

where

    $\phi = 0.90$

For the case of the end plate bolted to the face of the HSS as illustrated in

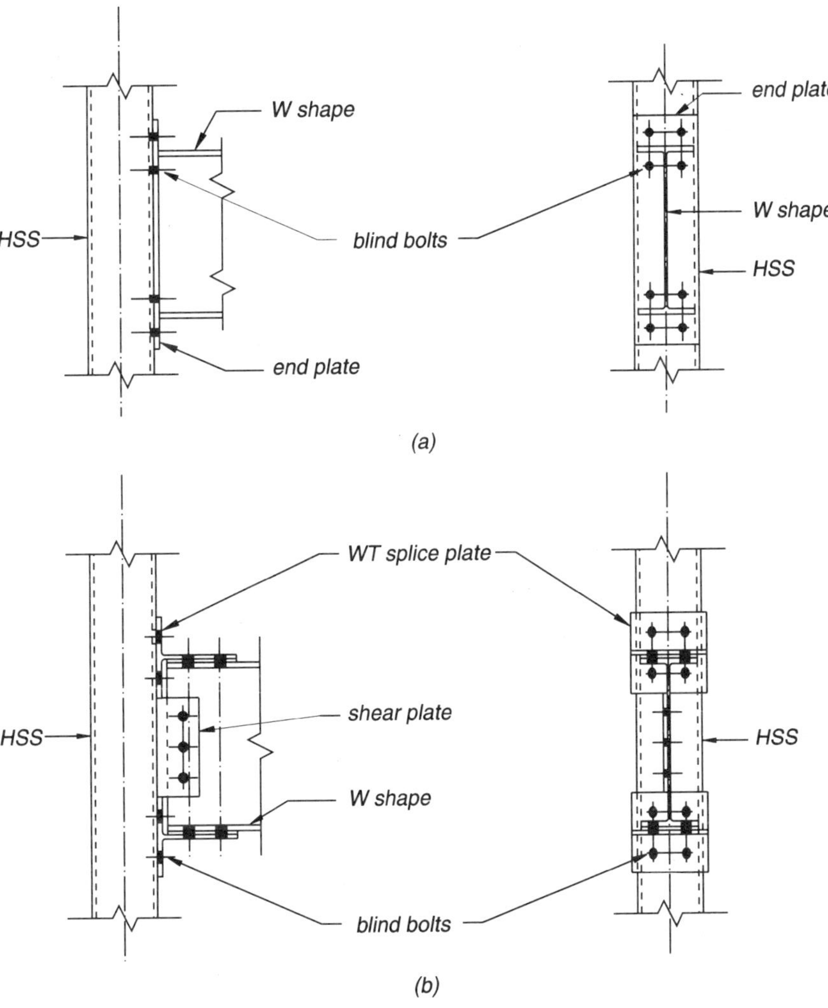

Fig. 5-10. Direct connection of W shape to HSS column by bolting.

Figure 5-10, the flexural strength will be limited by HSS wall bending at the tension bolts. On the compression side of the W-shape, the end plate, if at least 2.5 inches wider than the horizontal bolt spacing, will produce a less concentrated load on the HSS face and therefore not control the design.

The bolts will typically be blind bolts (See Chapter 3). The tension on a grouping of 4 bolts will produce a yield line as shown in Figure 5-12. Equations 5-2 and 5-3 are applicable for this case if t_f and b_1 are replaced by variables p and g respectively and account is taken of the reduction due to the bolt hole diameter d'. BCSA and SCI (1992) uses an expression very similar to Equation 5-3 to determine tensile design strength. However, the procedure in BCSA and SCI (1992) provides a more liberal design strength by using the HSS flat width b instead of the overall width b_o.

The recommended tensile design strength expression for the four-bolt case illustrated in Figure 5-12 is

$$\phi R_n = \phi F_y t^2 \left[\frac{2(p - d')}{b_o - g} + 3\sqrt{\frac{b_o - d'}{b_o - g}} \right] \tag{5-4}$$

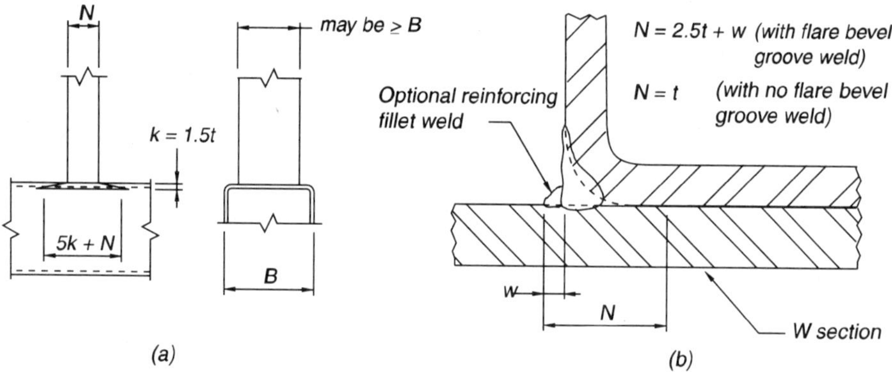

Application of Web Yielding Equations

Bearing width - HSS lateral load to W section

Fig. 5-11. N and k definitions.

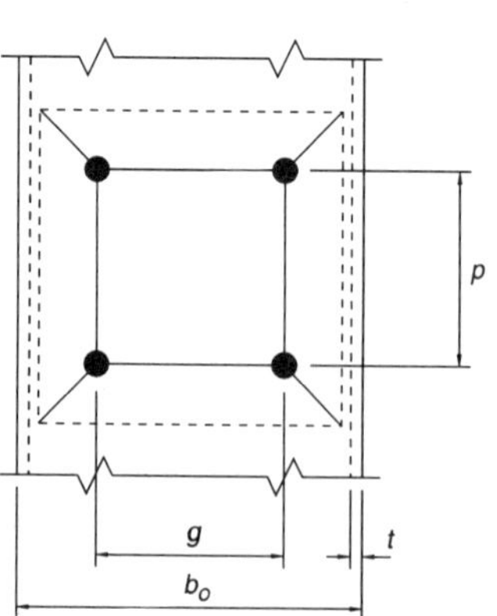

Fig. 5-12. Yield line for four bolt cluster.

where

$\phi = 1.0$

b_o = overall width of the loaded face of the HSS, either B or H

Since the bolts must be positioned some distance from the HSS corner, the tensile design strength determined by Equation 5-4 will invariably control. The flexural design strength will be determined by multiplying the tensile design strength by the vertical distance center-to-center of bolt groups at each flange.

The following examples illustrate the evaluation of the moment transfer for the various types of HSS to W-shape connections discussed.

Example 5.1—Continuous Beam over HSS Column

Investigate the connection strength of a two-span continuous W18×40 A36 beam supported by an HSS8×8×¼ center column using a detail similar to that in Figure 5-1b. The column is subjected to an axial load P_u = 45.0 kips, plus a moment M_u = 27.0 kips-ft. The bolts are ¾-in. diameter A325 with a gage of 3.5 inches. The bolts are 1.5 inches from the face of the HSS column. The cap plate is ½-in. thick.

Solution:

Bending with Prying on the Flange of the W-Shape

To evaluate the bolt tension resulting from the column loads, the axial load will be assumed to exist at the compression face of the column, see Figure Ex. 5-1.

From summation of forces and moments:

$$45 = C_u - T_u$$

$$27 \times 12 = 4C_u + 5.5T_u = 4(45 + T_u) + 5.5T_u$$

$$T_u = 15.16 \text{ kips on 2 bolts}$$

$$r_{ut} = \frac{15.16}{2} = 7.58 \text{ kips/bolt}$$

At this load, the size of the bolt is of little concern since ϕr_n using ¾-in. diameter A325 bolts is 29.8 kips.

Evaluate the required flange thickness considering prying per Part 11 of the AISC LRFD Manual (AISC, 1994). Due to the low level of load on the bolt, $\alpha' = 1.0$ indicating that full yield occurs at both the bolt and at the HSS wall.

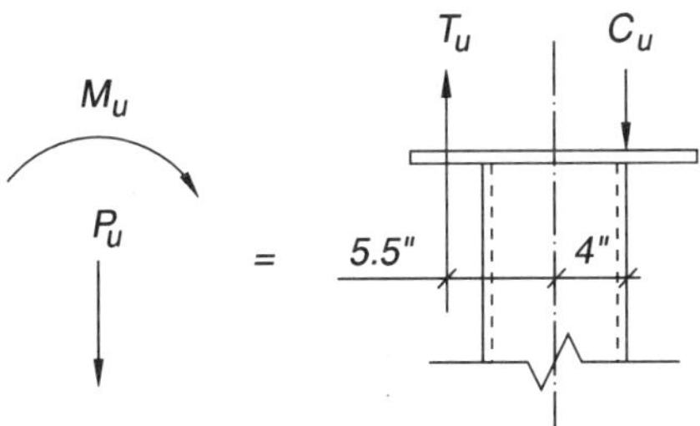

Figure Ex. 5-1. Evaluation of bolt tension.

$$\delta = 1 - d' / p = 1 - {}^{13}\!/_{16} \text{ hole size}/3.5 \text{ in. pitch} = 0.768$$

$$b = \frac{3.5 - 0.315}{2} = 1.59 \text{ in. (bolt centerline to face of web)}$$

$$b' = 1.59 - \frac{0.75}{2} = 1.22 \text{ in. (face of bolt to face of web)}$$

$$t_{req} = \sqrt{\frac{4.44 r_{ut} b'}{p F_y (1 + \delta \alpha')}} = \sqrt{\frac{4.44(7.58)(1.22)}{3.5(36)(1 + 0.768 \times 1)}} = 0.429 \text{ in.} < 0.525 \text{ in.} \quad \textbf{o.k.}$$

Bending with Prying on Cap Plate of HSS

In checking the cap plate, the value of $b' = b - d/2 + t$

where

b = the distance from the bolt centerline to the face of the HSS, in.
d = the bolt diameter, in.
t = the HSS wall thickness, in.

This reflects the recommended location of the inner plastic hinge line based on work done by Packer, Bruno and Birkemoe (1989).

Thus, $b' = 1.5 - 0.75/2 + 0.233 = 1.36$ in.

$\delta = 0.768$ since the pitch is 3.5 in. as before.

$$t_{req} = \sqrt{\frac{4.44(7.58)(1.36)}{3.5(36)(1.77)}} = 0.453 \text{ in.} < 0.5 \text{ in.} \quad \textbf{o.k.}$$

Strength of the Beam Relative to Web Yielding, Crippling and Buckling

Determine next whether the web of the W18 can satisfactorily transfer the compression force C_u without a stiffener directly over the HSS wall. $C_u = 45 + 15.16 = 60.2$ kips. Considering a 45° spread of load from the HSS wall use $N = 0.233 + 2(0.5) = 1.23$ in. Check Equations K1-2, K1-4 and K1-8 in the AISC LRFD Specification.

Web yielding resistance with $k = 1.19$ in. and $t_w = .315$ in. for the W18×40:

$$\phi = 1.0$$

$$\phi R_n = \phi(5k + N)F_{yw}t_w \tag{K1-2}$$

$$= 1.0[5(1.19) + 1.23](36)(0.315) = 81.4 \text{ kips} > 60.2 \text{ kips} \quad \textbf{o.k.}$$

Web crippling resistance of the W18×40 with $t_f = 0.525$ in.:

$$\phi = 0.75$$

$$\phi R_n = \phi 135 t_w^2 \left[1 + 3 \left(\frac{N}{d} \right) \left(\frac{t_w}{t_f} \right)^{1.5} \right] \sqrt{F_{yw} t_f / t_w} \tag{K1-4}$$

$$= 0.75(135)(0.315)^2 \left[1 + 3 \left(\frac{1.23}{18} \right) \left(\frac{0.315}{0.525} \right)^{1.5} \right] \sqrt{36(.525 / .315)}$$

$$= 85.2 \text{ kips} > 60.2 \text{ kips} \quad \textbf{o.k.}$$

Web buckling resistance for the W18×40, with h (the clear web height minus the corner radii) being 15.5 in.:

$$\phi = 0.9$$

$$\phi R_n = \phi 4{,}100 t_w^3 \sqrt{F_{yw}} \, / \, h \qquad \text{(K1-8)}$$

$$= 0.9(4{,}100)(0.315)^3 \sqrt{36} \, / \, 15.5 = 44.6 \text{ kips} < 60.2 \text{ kips} \quad \textbf{n.g.}$$

Place stiffeners above each wall as illustrated in Figure 5-1(b).

Strength Based on Bolting and Welding

The maximum acceptable bolt tension may be calculated from the equation on p. 11-10 of AISC LRFD Manual (AISC, 1994):

$$t_{req} = \sqrt{\frac{4.44 r_{ut} b'}{p F_y (1 + \delta \alpha')}}$$

Setting t_{req} = the cap plate thickness t_p and rearranging

$$r_{ut} = \frac{p F_y (1 + \delta \alpha') t_p^2}{4.44 b'}$$

$$= \frac{3.5(36)(1.768)(.5)^2}{4.44(1.375)} = 9.12 \text{ kips} < 29.8 \text{ kips} \quad \therefore \textbf{o.k.}$$

The bolt design strength (including prying) could be increased by increasing the cap plate thickness. The bolt design strength (including prying), as limited by the beam, could be increased by reducing the bolt gage. A 3-in. gage is a minimum based on the web and corner radius for this section. A wide flange section with a thicker flange or with a 50 ksi yield strength will permit a larger increase in bolt tension.

The weld of the cap plate to the HSS can be checked by evaluating $r_{ut} / p = 9.12/3.5 = 2.61$ kips/in. which can be satisfied with a $\frac{3}{16}$-in. fillet weld with $\phi r_n = 4.45$ kips/in.

HSS wall strength

In order to demonstrate the calculation for a concentrated load delivered from the W-shape web, ignore the fact that buckling of the W18 web required the use of stiffeners and evaluate wall yielding without stiffeners. Wall yielding can be evaluated using the AISC HSS Specification, with the bearing width, N, taken as twice the k of the W section, and "k" taken as the cap plate thickness.

$$\phi R_n = \phi(5k + N) F_{yw} t_w \qquad \text{(K1-2)}$$

$$= 1.0[5(0.50) + 2\,(1.19)](46)(0.233) = 52.3 \text{ kips} < C_u = 60.2 \text{ kips}$$

Since C_u is greater than 52.3 kips, the cap plate thickness could be increased or stiffeners added to the W-shape to increase N, thus increasing the HSS wall yielding strength.

Next check wall crippling considering the W18×40 without stiffeners. Using the AISC HSS Specification expression for wall crippling with $N = 2.38$ in. from above, cap plate thickness $t_1 = 0.50$ in., $t = 0.233$ in. and $B = 8$ in.

$$\phi R_n = \phi 0.80 t^2 \left[1 + 3\left(\frac{N}{B/2}\right)\left(\frac{t}{t_1}\right)^{1.5} \right] \sqrt{E F_y (t_1 / t)}$$

$$= 0.75(0.80)(0.233)^2 \left[1 + 3\left(\frac{2.38}{4}\right)\left(\frac{0.233}{0.50}\right)^{1.5} \right] \sqrt{29{,}000(36)\left(\frac{0.50}{0.233}\right)}$$

$$= 76.4 \text{ kips} > C_u = 60.2 \text{ kips}$$

Since the W18×40 in this example requires web stiffeners to resist web buckling, HSS wall yielding and crippling will not occur, as N is significantly larger than the 2.38 in. used above.

Example 5.2—Through Plate Moment Connection

Design a connection similar to that illustrated in Figure 5-3(b) where the beams are W12×45 sections and the columns are HSS10×10×⅜. The cap and the through plate are required to develop a moment of 120 kip-ft. from the W12×45. In addition, consider the moment on the upper HSS as producing an ultimate tensile bolt force of 18 kips at the base plate for each of the base plate bolts. Determine plate thicknesses.

Use ¾-in. diameter A325 bolts with a 5½-in. gage and with 13 inches between bolt lines. Use an 11 in. × 16 in. base plate.

Solution:

Bending with prying on the base plate

Packer and Henderson (1992) indicate that the pitch used should generally be no more than 4 to 5 bolt diameters. Considering this limitation and the limit suggested in the AISC LRFD Manual (AISC, 1994), prying action, which restricts p to the gage (which is just slightly more than twice the value of b), in this case, it is suggested that p, the length of plate tributary to each bolt, be limited to $(2)(1.5 + 0.375) = 3.75$ in. which is twice the distance from the bolt centerline to the yield location.

$$\delta = 1 - d'/p = 1 - 0.8125/3.75 = 0.783$$

$$b' = 1.5 - 0.75/2 + 0.349 = 1.47 \text{ in.}$$

$$a' = 1.5 + 0.75/2 = 1.875 \text{ in.}$$

$$\rho = b'/a' = 0.78$$

$$\beta = \frac{1}{\rho}\left(\frac{\phi r_n}{r_{ut}} - 1\right) = \frac{1}{0.78}\left(\frac{29.8}{18}\right) > 1$$

$$\therefore \ \alpha' = 1$$

$$t_{req} = \sqrt{\frac{4.44(18)(1.47)}{3.75(36)[1 + 0.783(1)]}} = 0.699 \text{ in.}$$

$$t_c = \sqrt{\frac{4.44(29.8)(1.47)}{3.75(36)}} = 1.20 \text{ in. for full bolt strength with no prying}$$

Try a ¾-in. A36 base plate. Per AISC LRFD Manual (AISC, 1994),

$$\alpha' = \frac{1}{0.783(1 + 0.78)}\left[\left(\frac{1.20}{0.75}\right)^2 - 1\right] = 1.12$$

$$q_u = 29.8\left[0.783(1.12)(0.78)\left(\frac{0.75}{1.20}\right)^2\right] = 7.96 \text{ kip prying force}$$

$r_{ut} + q_u = 18 + 7.96 = 26.0 < 29.8$ kips (the tensile bolt strength)

Use a ¾-in. A36 base plate thickness.

Combined axial and bending strength of the cap plate

The interaction of flexure and tension is limited by Equations H1-1a and H1-1b in the Specifications (AISC, 1994)

For $\dfrac{P_u}{\phi P_n} \geq 0.2$

$$\frac{P_u}{\phi P_n} + \frac{8}{9}\left(\frac{M_{ux}}{\phi_b M_{nx}} + \frac{M_{uy}}{\phi_b M_{ny}}\right) \leq 1.0 \tag{H1-1a}$$

For $\dfrac{P_u}{\phi P_n} < 0.2$

$$\frac{P_u}{2\phi P_n} + \left(\frac{M_{ux}}{\phi_b M_{nx}} + \frac{M_{uy}}{\phi_b M_{ny}}\right) \leq 1.0 \tag{H1-1b}$$

The cap plate is required to take an axial force of:

$$P_u = \frac{M_u}{d} = \frac{120(12)}{12.06} = 119 \text{ kips}$$

Note that the beam depth is used because the cap plate thickness is as yet unknown. If desired, the moment can be recalculated after the cap plate thickness is known.

Try a ¾-in. thick A36 cap plate.

$\phi P_n = \phi_t F_y A_g$ or $\phi P_n = \phi_t F_u A_e$

$\phi P_n = (0.9)(36)(0.75)(11) = 267$ kips $< (0.75)(58)(0.75)[11 - 2(0.875)] = 302$ kips

$$\frac{P_u}{\phi P_n} = \frac{119}{267} = 0.45 \geq 0.20$$

Therefore Eq. H1-1a applies.

In the interaction equation the ratio of $r_{ut}/\phi r_n$ equals the ratio of $M_u/\phi M_n$ from the bolt tension.

$$\phi r_n = \phi\,\frac{t^2 p F_y(1 + \delta\alpha')}{4b'} = 0.9\,\frac{(0.75)^2(3.75)(36)[1 + 0.783(1)]}{4(1.47)} = 20.7 \text{ kips}$$

Applying Eq. H1-1a:

$$\frac{119}{267} + \frac{8}{9}\left(\frac{18}{20.7}\right) = 1.22 > 1 \quad \textbf{n.g.}$$

Try 1 in. × 11 in. plate.

$\phi P_n = 356$ kips

$\phi r_n = 36.8$ kips

$$\frac{119}{356} + \frac{8}{9}\left(\frac{18}{36.8}\right) = 0.769 < 1 \quad \textbf{o.k.}$$

Use 1-in. thick A36 cap plate

Based on the yield and fracture calculation for the through plate,

Use a ¾-in. thick A36 through plate.

Example 5.3—Cut-Out Plate Connection

Consider a W12×45 beam and an HSS10×10×⅜ column with the cut-out plate as illustrated in Figure 5-4. Determine the required size of the plate assuming that the full beam moment is required to be transferred across the HSS. Consider both the plate and the beam to have a yield stress of 50 ksi and an ultimate strength of 65 ksi.

Solution:

Plate Tensile and Compressive Strength Along the Sides of the HSS

The design strength of the W12×45 equals:

$$\phi M_n = \phi Z F_y$$

$$\phi M_n = 0.9(64.7)(50) = 2910 \text{ kip-in.}$$

The flange force equals ϕM_n divided by the centroidal distance between the cut-out plates.

Try ¾-in. plates

$$P_u = 2910/(12.06 + 0.75) = 227 \text{ kips}$$

The effective area is represented by the projecting width of the plate with the critical section occurring where the flat portion of the HSS begins.

Therefore, the required plate projection w_s on each side can be determined for the maximum compression force:

$$P_u = \phi F_{yp}(2w_s)t_p$$

$$227 = (0.85)(50)(2w_s)(0.75)$$

$$w_s = 3.56 \text{ in.}$$

Check the width/thickness ratio of the plate projection for compression capability.

From AISC LRFD Specification Table B5.1

$$w_s / t = 3.56/0.75 = 4.75 < 95 / \sqrt{50} = 13.44 \quad \textbf{o.k.}$$

The required cut-out plate width equals $10 + 2(3.56) = 17.1$ in.

Use a 0.75-in. × 17.5-in. cut-out plate.

Strength Based on Bolting to the Cut-Out Plates

Strength Based on Bolt Shear:

Try 8 - 1 in. diameter A325-X Bolts

From the AISC Manual the design shear strength equals 35.3 kips/bolt

Design strength for (8) bolts = (8)(35.3) = 282 kips > 2910/12.06 = 241 kips   **o.k.**

Since the cut out plate is thicker than the flange of the W12×45, check bolt bearing on the beam flange.

$$\phi r_n = \phi 2.4 dt F_u$$

$$= (0.75)2.4(1)(0.349)65$$

$$= 40.8 \text{ kips/bolt}$$

Strength with 8 bolts:

$$\phi R_n = (8)(40.8) = 326 \text{ kips} > 241 \text{ kips} \quad \textbf{o.k.}$$

Check tension yielding of flange plate:

$$\phi R_n = \phi F_y A_g$$

$$= (0.9)(50)(17.5)(0.75) = 591 \text{ kips} > 241 \text{ kips} \quad \textbf{o.k.}$$

Check tension rupture of flange plate:

$$\phi R_n = \phi F_u A_n$$

$$= (0.75)(65)[17.5 - 2(1 + \tfrac{1}{8})]0.75 = 558 \text{ kips} > 241 \text{ kips} \quad \textbf{o.k.}$$

Check block shear rupture:

There are two cases for which block shear must be checked. The first case involves the tear-out of the two blocks outside the two rows of bolt holes in the flange plate. The second case involves the tear-out of the block between the two rows of holes in the flange plate. Based on the cut-out plate geometry and the bolt gage it is obvious that the latter case will control.

When $F_u A_{nt} \geq 0.6 F_u A_{nv}$:

$$\phi R_n = \phi[0.6 F_y A_{gv} + F_u A_{nt}]$$

When $0.6 F_u A_{nv} > F_u A_{nt}$

$$\phi R_n = \phi[0.6 F_u A_{nv} + F_y A_{gt}]$$

where

ϕ = 0.75
A_{gv} = gross area subject to shear, in.^{2}
A_{gt} = gross area subject to tension, in.^{2}
A_{nv} = net area subjected to shear, in.^{2}
A_{nt} = net area subjected to tension, in.^{2}
$F_u A_{nt}$ = $65(5.5 - 1.125)0.75 = 213$ kips
$0.6 F_u A_{nv}$ = $0.6(65)[14 - 3.565 - 3.5(1.125)]0.75 = 190$ kips

Therefore,

$$\phi R_n = 0.75[0.6(50)(14) + 213] = 633 \text{ kips} > 241 \text{ kips} \quad \textbf{o.k.}$$

Example 5.4—Welded Tee Flange Connection

Determine the moment strength that can be transferred between the HSS column and the W-shape by the WT detail as illustrated in Figure 5-5. The column is an HSS8×8×¼ with ASTM A500 Gr. B properties. The beam is a W16×36 and the tees are WT8×25 10 in. long. The beam and WTs are A36 material. A flare-bevel groove weld is used to attach the WT8 to the HSS. The bolts to be used are ⅞-in. diameter A325 bolts.

Solution:

Local Flange Bending

The force that can be taken by the WT8 at each web of the HSS is limited to 50 percent of the ϕR_n given by Eq. K1-1 of the LRFD Specification since the load is applied near the end of the WT8. Based on the criterion indicated in Section K1.2 regarding local flange bending it is appropriate that one half of the length of the WT8 be at least $10t_f$ or $10(0.630) = 6.3$ in. Considering the attachment at both webs, a WT8×25, 12.6 inches long (2×6.3) can carry a load of:

$$\phi R_n = (0.5)\phi(6.25t_f^2 F_{yf}) \text{ per web} \qquad \text{(K1-1)}$$

$$= (0.5)(0.9)(6.25)(0.63)^2(36) = 40.2 \text{ kips/web}$$

$$= 80.4 \text{ kips total for a 12.6-in. length of WT}$$

Since the WT8×25 is only 10 in. long, the design strength should be reduced to $(10/12.6)(80.4) = 63.8$ kips

Local Web Yielding of the WT Web

Per AISC LRFD Specification Section K1.3:

$$\phi R_n = \phi(2.5k + N)F_{yw}t_w$$

For the WT:

$t_w = 0.380$ in., $k = 1.3125$ in., and $N = 2.5t = 2.5(0.233) = 0.582$ in. (See Fig. 5-11(b)).

Effective web length $= 2.5(1.3125) + 0.582 = 3.86$ in. $< 10/2 = 5$ in.

$\phi R_n = 1.0(3.86)(36)(0.380) = 52.8$ kips/HSS web.

$= 106$ kips total

Local Web Yielding of the HSS Walls

Per AISC LRFD Specification Section K1.3:

For the HSS:

$t = 0.233$ in., $k = 1.5(0.233) + 0.63 = 0.98$ in., and $N = 0.38$ in.

$5k + N = 5(0.98) + 0.38 = 5.28$ in. < 7.07 in. flange width

$\phi R_n = 1.0[5.28](46)(0.233) = 56.6$ kips at each HSS sidewall

$= 113$ kips total

If $5k + N$ is less than the length of the flare bevel weld length (i.e. flange width), the tee flange is considered a flange on the HSS. Thus, for this situation, the use of Eq. K1-3 with N equal to the tee web thickness and with k equal to the flange thickness plus the

HSS corner radius of $1.5t$ is considered more appropriate than the HSS web yielding expressions for this situation.

Web Crippling of the HSS Walls

For the HSS the web crippling strength is:

$$\phi R_n = \phi 1.6t^2 \left[1 + 3\frac{N}{h} \right] \sqrt{EF_y}$$

where

$\phi = 0.75$
$t = 0.233$ in.; h = flat width = $8 - 3(0.233) = 7.30$ in.; Use $N = 2(1.312) = 2.624$ in.
  ($1.312 = k$ for WT)

$\phi R_n = 0.75\{1.6(.233)^2[1 + 3(2.624/7.30)]\sqrt{(29,000)46}\}$

  = 156.4 kips total for both walls

Compression Buckling of the Web

A conservative value for compression buckling of the web can be obtained from AISC HSS Specification Equation 8.1-6

$$\phi R_n = \phi 48t^3 \sqrt{EF_y} / h$$

where

$\phi = 0.9$
$t = 0.233$ in.
$h = 7.30$ in.

$\phi R_n = 0.9(48)(0.233)^3\sqrt{29,000(46)} / 7.30$

  = 86.5 kips total

This value is conservative since the compression force is applied to only one face of the HSS and not on opposite faces.

Based on the above calculations the limiting force on the WT is 63.8 kips, based on local flange bending.

Strength of the Weld Connecting the WT to the HSS

The flare-bevel groove weld (Single J-groove weld on p. 8-153 of LRFD Manual) provides an effective throat area of $\frac{5}{8}$ of the HSS wall thickness. Therefore:

$$\phi R_n = \phi(0.6F_{EXX})(E)L$$

where

$\phi = 0.8$
$F_{EXX} = 70$ ksi
E = the weld effective throat = $\frac{5}{8}t$
t = HSS wall thickness, in.
L = length of weld, in.

$\phi R_n = 0.8(0.6)(70)(0.625)(0.233)7.07 = 34.6$ kips/web

  = 69.2 kips total > 63.8 kip strength of the WT8×25.

Note that if the weld is reinforced with a fillet weld, the effective throat could be increased by 0.25 times the wall thickness per AWS D1.1 which would increase the weld strength by 40 percent.

Strength Based on Bolt Shear

From the AISC Manual, four $\frac{7}{8}$-in. diameter A325-N bolts provide a design strength of 86.4 kips.

Strength Based on Bolt Bearing

Bearing strength exceeds the shear strength of the bolts. Therefore, bearing strength is **o.k.**

Yielding and Fracture of the WT Web

Yielding:

$$\phi P_n = \phi_t F_y A_g$$

$$\phi P_n = (0.9)(36)(0.380)(10) = 123 \text{ kips}$$

Fracture:

$$\phi P_n = \phi_t F_u A_e$$

$$\phi P_n = (0.75)(58)(0.380)[10 - 2(0.875 + 0.125)] = 132 \text{ kips}$$

Thus, the flexural strength using the WT8×25 would be 63.8(15.86 + 0.380 + 0.125) = 1040 kip-in. considering an allowance for a $\frac{1}{8}$-in. shim.

The flexural strength can be increased by selecting a Tee with a thicker flange such as a WT8×28.5 with $t_f = 0.715$ in., in which case

$$\phi R_n = \frac{L_t}{10 t_f(2)} (\phi 6.25 t_f^2 F_{yt}) = \phi 0.3125 L_t t_f F_{yt} \leq \phi 6.25 t_f^2 F_{yt}$$

$$= 0.9(0.3125)(10)(.715)(36) = 72.4 \text{ kips} < 0.9(6.25)(.715)^2(36) = 104 \text{ kips}$$

or use a WT8×25 with a 50 ksi specified yield stress which would increase the flexural strength to (50/36) 63.8 = 88.6 kips.

In this case the bolts would govern the design strength provided that the fillet is added to the flare-bevel groove weld.

Example 5.5—End Plate Connection

Consider a W16×36 beam which has a 7-in. wide flange attached to an HSS7×7×$\frac{5}{16}$ column using an end plate connection as illustrated in Figure 5-6(a). Determine the flexural strength that can be transmitted by a $\frac{3}{4}$-in. end plate and a $\frac{1}{2}$-in. connection plate with eight $\frac{3}{4}$-in. diameter A325 bolts placed as shown in Figure 5-6(a). The beam and plates are A36 material.

Solution:

The bolts should be placed as close to the HSS section as possible to be most effective. Use 9.5 in. gage to obtain a $1\frac{1}{4}$-in. distance to the face of the HSS shape. The upper set of bolts should be centered on the top flange and the lower set centered on the bottom flange of the W16. Use vertical spacings of 3 in., 12.5 in., and 3 in. for the bolts.

Strength of the ½-in. connection plate and its attachment to the HSS column.

Bending with prying

Evaluate r_{ut} per bolt considering prying per Part 11 of the LRFD Manual. The pitch p should be 3 in. with hole size $d' = {}^{13}\!/_{16}$ in., making $\delta = 1 - 0.8125/3 = 0.729$.

By subtracting half the bolt diameter and adding the HSS wall thickness to the distance from the bolt to the face of HSS, $b' = 1.25 - 0.75/2 + 0.291 = 1.166$ in.

The HSS wall thickness is added as discussed earlier since the connection plate is expected to behave in the same manner as the end plates in Example 5.1. Since yielding will occur at the bolt for this plate thickness, set $\alpha' = 1$. Therefore,

$$r_{ut} = \frac{pF_y(1 + \delta\alpha')}{4.44b'}\,t_p^2$$

$$r_{ut} = \frac{3(36)(1 + 0.729(1))}{4.44(1.166)}(0.5)^2 = 9.02 \text{ kips/bolt}$$

Strength of the Flare-Bevel Weld

The flare-bevel groove weld has a strength of:

$$\phi R_n = \phi(0.6F_{EXX})(E)L$$

where

$$
\begin{aligned}
\phi &= 0.8 \\
F_{EXX} &= 70 \text{ ksi} \\
E &= \text{the weld effective throat} = {}^5\!/_8 t \\
t &= \text{HSS wall thickness, in.} \\
L &= \text{length of weld, in.}
\end{aligned}
$$

$$\phi R_n / L = 0.8(0.6)(70)(0.625)(0.291) = 6.11 \text{ kips/in.}$$

Using the 3-in. pitch as the effective weld length, the weld strength per bolt is $3(6.11) = 18.33$ kips which is much greater than 9.02 kips.

Strength of the ¾-in. end plate

Often with this type of connection the width of the wide flange shape will be less, equal to or only slightly greater than that of the HSS, which places the bolts beyond the ends of the flange and not above and below the flange as would be desired. It is not appropriate to use the prying expression for the end plate since the yield line that would develop along

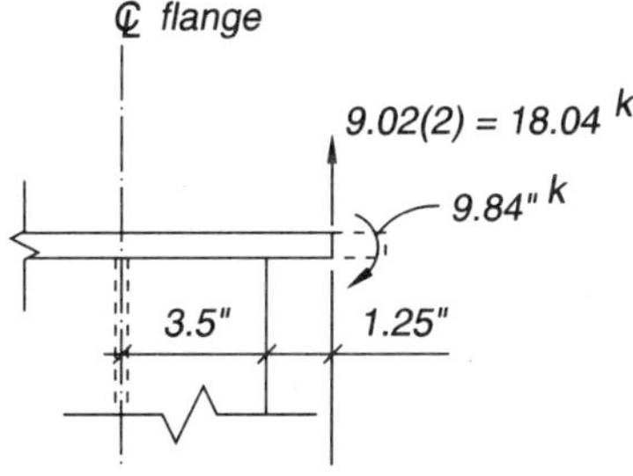

Figure Ex. 5-2.

a "web" (M_{u1} in Figure 11-2 of LRFD Manual) cannot be expected to develop to the same degree at the tip of the flange.

Therefore, it is suggested to limit the bending moment at the tip of the flange to the plastic moment over a width equal to the flange thickness plus twice the distance from the center of the bolt to the flange tip as discussed earlier (see Figure 5-7a). Place on the end plate the moment and load found above for the connection plate to determine the moment at the flange tip.

Applied moment:

$$M_{u2} = \frac{(3 - 0.8125)}{4}(0.5)^2(2 \text{ bolts})36 = 9.84 \text{ kip-in.}$$

$M_{utip} = 9.02(2 \text{ bolts})(1.25) - 9.84 = 12.7 \text{ kip-in.}$

Thus, based on statics and the bolt load from analysis of the ½-in. connection plate, the moment at the tip of the flange is 12.7 kip-in.

Design Strength:

$$\phi M_n = \phi F_y Z$$

Use an effective width at flange tip = $2(1.25) + 0.43 = 2.93$ in. (See Figure 5-7a)

$\phi M_n = 0.9(36)(2.93)(0.75)^2/4 = 13.3$ kip-in. > 12.7 kip-in.   **o.k.**

Since ϕM_n is greater than M_{utip} the ½-in. connection plate to the HSS controls the connection strength. The total design strength for the 4 bolts equals $(4)(9.02) = 36.1$ kips per flange. The design strength equals $36.1(12.5 + 3) = 559$ kip-in. for the connection based on the end plates and the bolting. Had the 1.25 inches been smaller, the value of M_{utip} would reduce and could even become negative. The end plate selected should never be less than the connection plate thickness.

Applied Force:

The length of flange required to transfer the 18.0 kip force at each flange tip can be obtained from the equation:

$$\phi R_n = \phi F_y tL$$

From which:

$$L = \frac{18.0}{0.9(36)(0.430)} = 1.29 \text{ in.}$$

The length of ⁵⁄₁₆-in. fillet weld using an E70 electrode to resist the applied load equals:

$$\phi R_n = \phi(0.6 F_{EXX})(0.707)L$$

From which:

$$L = \frac{18.0}{0.75(0.6)(70)(0.707)(0.3125)} = 2.59 \text{ in.}$$

Thus, weld on the tip of the flange plus 1.01 inches on each side of the flange ($t_f = 0.575$) is required which seems reasonable as well. For practical simplicity, use a ⁵⁄₁₆-in. fillet weld all around the flanges including the tips.

Had angles been employed in lieu of the connection plate for this example (as illustrated in Figure 5-6(b)), an L5×3×½ would be used. In computing the tensile strength, there is insufficient research to establish b'. A check should be made to establish that the

HSS wall can resist the moment developed at the angle corner from prying. With the angles, the bolt gage would have to be increased by about 1 inch, which would place the bolts ½-in. further from the flange tips. A b' value 0.5 inches larger than that used for the connection plate may be reasonable.

Example 5.6—Directly Welded Connection of W Shape to HSS Column

Determine the flexural strength that can be transferred to an HSS 8×8×⅜ column (A500 Gr. B) by using the welded connection shown in Figure 5-9(a). The W sections are A36 material.

1. A W14×43, or
2. A W16×36

Solution:

1. For the W14×43

   Effective width limit

   $$R_n = \frac{10F_y t}{B/t} b_1 \leq F_{y1} t_1 b_1$$

   where

   $\phi = 1.0$

   $b_1 = B = 8$ in., $t = 0.349$ in.

   $$\phi R_n = \frac{1.0(10)(46)(0.349)}{8/0.349}(8) = 56.0 \text{ kips} < (0.9)36(0.530)(8) = 137 \text{ kips}$$

   As noted earlier, this ϕR_n controls over the expression for local web yielding, web crippling and punching shear.

   Therefore, the design moment, $\phi M_n = (1.0)(13.66 - 0.53)(56.0) = 735$ kip-in.

2. For the W16×36

   Effective width limit

   Check $b_1 > 0.95b$; 6.985 in. > 0.95[8 − 3(0.349)] = 6.61 in.

   Flange width is okay.

   Per Equation 5.1

   $$\phi R_n = \frac{(1.0)(10)(46)(0.349)}{8/0.349}(6.985) = 48.9 \text{ kips}$$

   Face yielding limit

   Per Equation 5.3

   $$\phi R_n = (1.0)(46)(0.349)^2 \left[\frac{2(0.430)}{(8 - 6.985)} + \frac{3}{\sqrt{1 - 6.985/8}} \right] = 51.9 \text{ kips}$$

   For this section the effective width limit controls and

$$\phi M_n = (1.0)(15.86 - 0.43)(48.9) = 755 \text{ kip-in.}$$

Both values of ϕM_n are less than the plastic moment of the HSS8×8×⅜ which equals $(0.9)46(29.4) = 1,220$ kip-in.

Example 5.7—Bolted HSS Face Connection

Determine the flexural strength that can be transferred to an **HSS8×8×⅜** column (A500 Gr. B) by a **W18×35** using one of the bolted connections illustrated in Figure 5-10. Allowing for the tee stem thicknesses and shim spacing, the center-to-center distance of the tees will be 18³⁄₁₆-in. Consider use of nominal ¾-in. diameter blind bolts as shown in Chapter 3 at a pitch of 3.5 inches. The ¾-in. diameter bolt requires a ²⁷⁄₃₂-in. hole. Considering a possible maximum $3t$ bend radius for the HSS wall the gage should be no more than 4.75 inches to insure that the formed bolt head will bear on the flat portion of the wall. The largest gage will produce the greatest tensile design strength.

Face Yielding

Using Equation 5-4

$$\phi R_n = \phi F_y\, t^2 \left[\frac{2(p - d')}{b_o - g} + 3\sqrt{\frac{b_o - d'}{b_o - g}}\, \right]$$

$$\phi R_n = (1.0)(46)(0.349)^2 \left[\frac{2(3.5 - 0.844)}{8 - 4.75} + (3)\sqrt{\frac{8 - 0.844}{8 - 4.75}}\, \right] = 34.1 \text{ kips}$$

With an 18³⁄₁₆-in. spacing between the centers of the two bolt groups, there will be a design flexural strength of $\phi M_n = (1.0)(18.1875)(34.1) = 620$ kip-in. for the connection using the bolted end plate. This is less than achieved in Example 6 even though the beam used is deeper. This is due to a lower yield line strength.

SUGGESTED DETAILS

Beam Framing
Moment Connections

Details on these pages are suggested treatments only and are not intended to limit the use of other connections not illustrated.

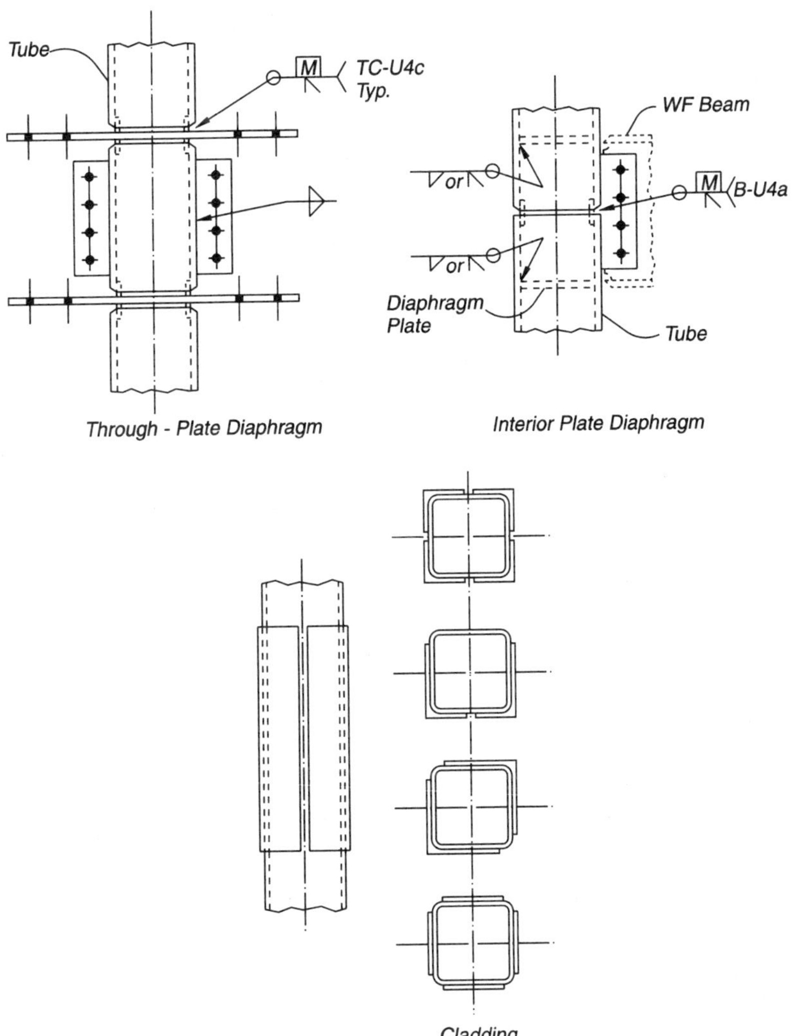

HSS COLUMN REINFORCEMENT

Figure 5-13.

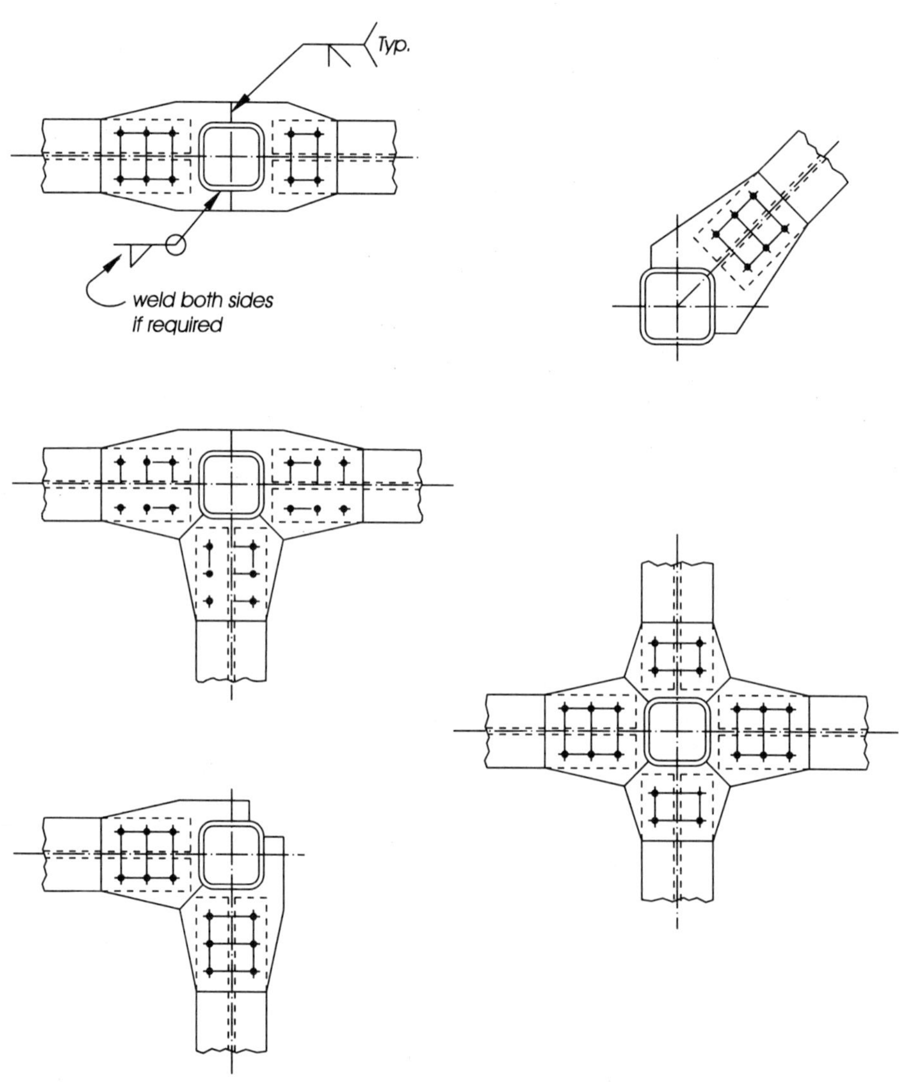

Note: Shear connections not shown for clarity.

Figure 5-14.

REFERENCES

American Institute of Steel Construction, 1994, *Manual of Steel Construction, Load and Resistance Factor Design, Vols. I and II*, 2nd ed., AISC, Chicago, IL.

Connection Guide for Hollow Structural Sections, Bull Moose Tube Company, 1994, Elkhart, IN.

Packer, J. A., Bruno, L., and Birkemoe, P. C., 1989, "Limit analysis of bolted RHS flange plate joints," *Journal of Structural Engineering*, American Society of Civil Engineers 115(9): 2226–2242.

Packer, J. A., and Henderson, J. E., 1992, *Design Guide for Hollow Structural Section Connections*, Canadian Institute of Steel Construction.

BCSA and SCI, 1992, *Joints in Simple Construction*, BCSA, London, and SCI, Ascot, England.

American Institute of Steel Construction, 1997, *Specification for the Design of Hollow Structural Sections*, AISC, Chicago, IL.

CHAPTER 6

TENSION AND COMPRESSION CONNECTIONS

OVERVIEW

Chapter 6 contains general information, design considerations, examples, and design aids for the design of tension and compression connections for round, square and rectangular hollow structural sections. The information is based on several references, but primarily on the 1993 AISC LRFD Specification.

Following is a detailed list of the topics addressed.

WELDED TEE END CONNECTIONS (AXIAL TENSION)

Welded tee end connections, illustrated in Figure 6-1 are usually composed of the HSS and a structural tee welded to the end of the HSS. In lieu of the structural tee a fabricated tee section is often used.

The design strength of HSS with welded tee ends can be obtained from the various limit states for the member.

For tension loads the limit states are:

1. Strength of the weld connecting the tee flange to the HSS.
2. Strength of the weld connecting the stem to the cap plate.
3. Strength of the HSS wall.
4. Shear strength of the tee flange.
5. Strength based on bolting to the tee stem.

Strength of the Weld Connecting the Tee Flange to the HSS

$$\phi R_n = \phi F_w A_w$$

where

$\phi \quad = 0.75$

$F_w \quad =$ nominal weld strength, ksi

$\quad \quad = 0.60 F_{EXX} (1.0 + 0.50 \sin^{1.5}\theta)$

$F_{EXX} =$ electrode classification number, i.e., minimum specified strength, ksi

$\theta \quad =$ angle of loading measured from the weld longitudinal axis, degrees

$\quad \quad = 90°$

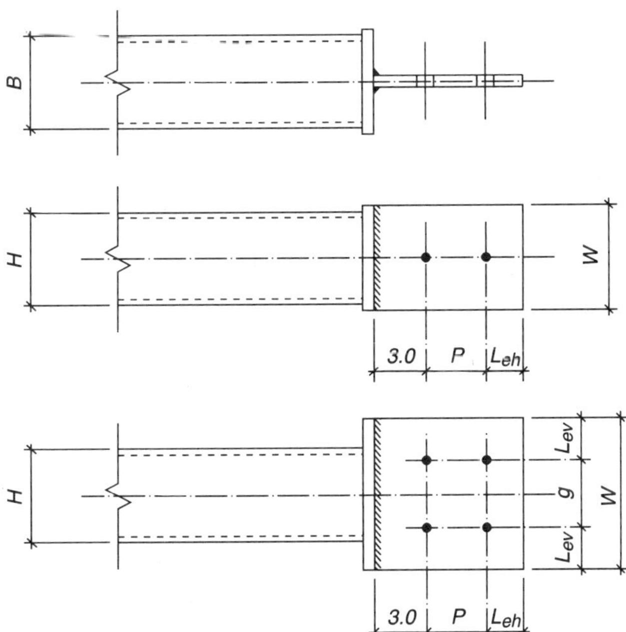

Fig. 6-1. Typical welded tee end.

A_w = effective area of weld throat, in.^{2} (see Figure 6-2)

    = $2(5t_1 + N)(0.707)W_w$

N = bearing length of load on the tee flange, in.

    = $t_2 + 2W_s$

t_1 = tee flange thickness, in.

t_2 = tee stem thickness, in.

W_s = weld size (tee flange to stem for built-up tee), in.

W_w = weld size (HSS to tee flange), in.

Strength of the Weld Connecting the Stem to the Tee Flange

$$\phi R_n = \phi F_w A_w$$

where

ϕ = 0.75

F_w = nominal weld strength, ksi

    = $0.60F_{EXX}(1.0 + 0.50\sin^{1.5}\theta)$

F_{EXX} = electrode classification number, i.e., minimum specified strength, ksi

θ = angle of loading measured from the weld longitudinal axis, degrees

    = 90°

A_w = effective area of weld throat, in.^{2}

    = $0.707(2W_s)W$

W_s = weld size (tee flange to stem), in.

W = width of tee flange, in.

    $\geq$ HSS depth H plus one inch to allow for fillet welds

    $\geq (H + 1)$

H = depth of HSS section, in.

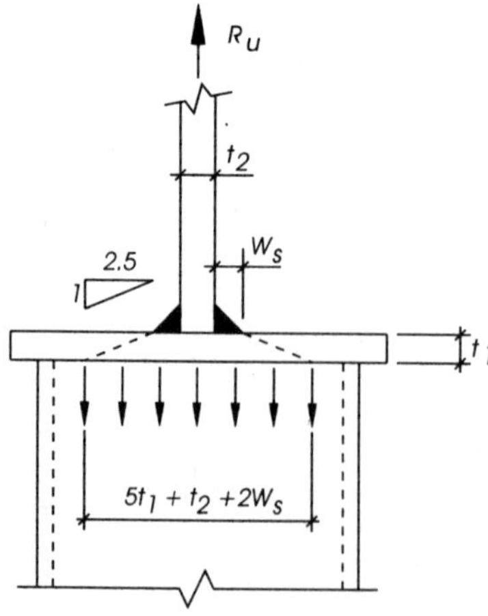

Fig. 6-2. Effective width.

Strength of the HSS Wall

$$\phi R_n = \phi F_y A_e$$

$$\le \phi 2 B F_y t$$

where

ϕ = 1.0
F_y = specified minimum yield stress of the HSS, ksi
A_e = effective area of HSS wall, in.^{2}
   = $2(5t_1 + N)t$
N = bearing length of load on the tee flange, in.
   = $t_2 + 2W_s$
B = width of HSS wall loaded by tee stem, in.
t = HSS wall thickness, in.
t_1 = tee flange thickness, in.
t_2 = tee stem thickness, in.
W_s = weld size (tee flange to stem), in.

Shear Strength of the Tee Flange

$$\phi R_n = \phi 2[(W)(t_1)(0.6 F_{y1}) + tuF_y]$$

where

ϕ = 0.75
t_1 = thickness of tee flange, in.
F_{y1} = specified minimum yield stress of the tee flange, ksi
F_y = specified minimum yield stress of the HSS, ksi
t = HSS wall thickness, in.
u = effective width of HSS wall, in.
   = $t_2 + 2W_s$

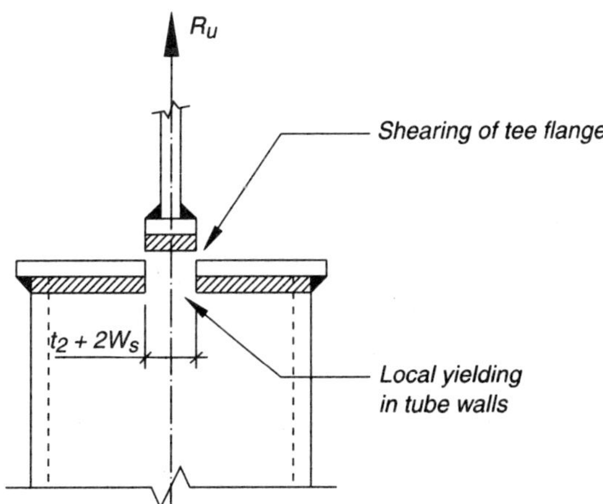

Fig. 6-3. Shear failure in the tee flange.

t_2 = tee stem plate thickness, in.
W = width of tee flange, in.
    $\geq (H + 1)$
W_s = weld size (tee flange to stem), in.
H = depth of HSS section, in.

Note: For HSS sections the stem is taken as oriented parallel to the long side of the section.

Strength Based on Bolting to the Tee Stem

Design Tensile Strength of the Tee Stem:

(a) For yielding in the gross section:

$$\phi = 0.90$$

$$R_n = F_{y2}A_g$$

(b) For rupture in the net section:

$$\phi = 0.75$$

$$R_n = F_{u2}A_e$$

where

A_e = net area $\leq 0.85A_g$, in.^{2}
A_g = gross area of the tee stem, in.^{2}
A_n = net area of tee stem, in.^{2}
    = $[W - n_r(d_h + \frac{1}{16})]t_2$
F_{y2} = specified minimum yield stress of the tee stem, ksi
F_{u2} = specified minimum tensile strength of the tee stem, ksi
R_n = nominal axial strength, kips
W = width of tee stem (same as cap plate), in.
n_r = number of rows of bolts
d_h = diameter of bolt holes, in.
t_2 = tee stem plate thickness, in.

(c) Block Shear Rupture Strength

The shear rupture term along a shear failure path in the tee stem is

$$\phi(0.6F_{u2}A_{nv})$$

where

ϕ = 0.75
F_{u2} = specified minimum tensile strength of the tee stem, ksi
A_{nv} = net area subject to shear, in.^{2}
    = $2[L_{eh} + (n - 1)p - (n - 0.5)(d_h + \frac{1}{16})]t_2$
L_{eh} = edge distance, in.
n = number of lines of bolts
p = spacing of lines of bolts (pitch), in.
d_h = diameter of bolt holes, in.
t_2 = tee stem plate thickness, in.

The tension rupture term along a tension path in the tee stem is

$$\phi(F_{u2}A_{nt})$$

where

ϕ = 0.75

F_{u2} = specified minimum tensile strength of the tee stem, ksi

A_{nt} = net area subject to tension, in.^{2}

= $[(n_r - 1)g - (n_r - 1)(d_h + \frac{1}{16})]\, t_2$

n_r = number of rows of bolts

g = spacing of rows of bolts (gage), in.

d_h = diameter of bolt holes, in.

t_2 = tee stem plate thickness, in.

The design block shear rupture strength is determined by the sum of the shear strength on a failure path(s) and the tensile strength on a perpendicular segment. When ultimate rupture strength on the net section is used to determine the resistance on one segment, yielding on the gross section is used on the perpendicular segment. The block shear rupture design strength, ϕR_n, shall be determined as follows:

(1) When $F_{u2}A_{nt} \geq 0.6F_{u2}A_{nv}$:

$$\phi R_n = \phi[0.6F_{y2}A_{gv} + F_{u2}A_{nt}], \text{ kips}$$

(2) When $0.6F_{u2}A_{nv} > F_{u2}A_{nt}$:

$$\phi R_n = \phi[0.6F_{u2}A_{nv} + F_{y2}A_{gt}], \text{ kips}$$

where

ϕ = 0.75

A_{gv} = gross area subject to shear, in.^{2}

= $2[L_{eh} + (n - 1)p]t_2$

A_{gt} = gross area subject to tension, in.^{2}

= $[(n_r - 1)g]t_2$

A_{nv} = net area subjected to shear, in.^{2}

A_{nt} = net area subjected to tension, in.^{2}

F_{y2} = specified minimum yield stress of the tee stem, ksi

F_{u2} = specified minimum tensile strength of the tee stem, ksi

t_2 = tee stem plate thickness, in.

Design Strength of the Bolts:

(a) Bolt Shear

See AISC Specification Section J3, and the AISC LRFD Manual Volume II, Part 8

(b) Bolt Bearing

See AISC Specification Section J3, and the AISC LRFD Manual Volume II, Part 8

EXAMPLE 6.1—Welded Tee End Connection (Axial Tension)

Determine the design axial tensile strength of the HSS8×6×$\frac{5}{16}$ shown.

Solution 1: Using Equations from this section

1. Strength of the weld connecting the tee flange to the HSS.

$$\phi R_n = \phi F_w A_w$$
$$\phi = 0.75$$
$$F_w = 0.60 F_{EXX}(1.0 + 0.50 \text{ in. } \sin^{1.5}\theta)$$
$$= 0.60(70)(1.5)$$
$$= 63.0 \text{ ksi}$$
$$A_w = 2(5t_1 + t_2 + 2W_s)(0.707)W_w$$
$$= 2[5(0.375) + 0.375 + 2(5/16)](0.707)(5/16)$$
$$= 1.27 \text{ in.}^2$$
$$\phi R_n = 0.75(63.0)(1.27)$$
$$= 60.0 \text{ kips}$$

2. Strength of the weld connecting the stem to the tee flange

$$\phi R_n = \phi F_w A_w$$
$$\phi = 0.75$$
$$F_w = 0.60 F_{EXX}(1.0 + 0.50\sin^{1.5}\theta)$$
$$= (0.60)(70)(1.5)$$
$$= 63.0 \text{ ksi}$$
$$W = H + 1 = 8 + 1 = 9.0 \text{ in.}$$
$$L_{ev} = (9.0 - 3.0)/2 = 3.0 \text{ in.} \geq 1.5 \,(0.75) = 1.125 \text{ in.}$$
$$A_w = (.707)(2)(5/16)(9.0) = 3.98 \text{ in.}^2$$
$$\phi R_n = 0.75(63.0)(3.98)$$
$$= 188 \text{ kips}$$

3. Strength of the HSS Wall

$$\phi R_n = \phi F_y A_e$$
$$\phi = 1.0$$
$$F_y = 46 \text{ ksi}$$

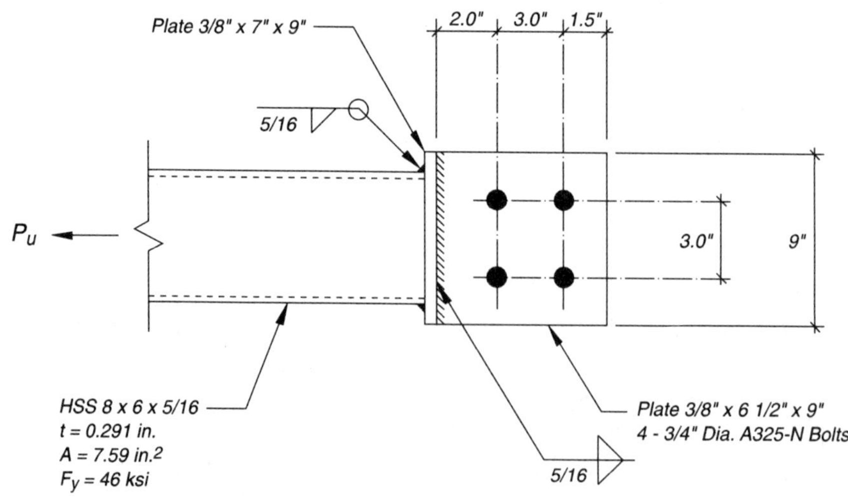

Figure Ex. 6.1.

$$A_e = 2(5t_1 + t_2 + 2W_s)t \leq 2(B)(t)$$
$$= 2[5(0.375) + 0.375 + 2(5/16)](0.291)$$
$$= 1.67 \text{ in.}^2 < 2(6.0)(0.291) = 3.49 \text{ in.}^2$$
$$\phi R_n = 1.0(46)(1.67)$$
$$= 76.8 \text{ kips}$$

4. Shear strength of the tee flange

$$\phi R_n = \phi 2[(W)(t_1)(0.6F_{y1}) + tuF_y]$$
$$\phi = 0.75$$
$$u = t_2 + 2W_s$$
$$= 0.375 + 2(5/16)$$
$$= 1.00 \text{ in.}$$
$$\phi R_n = (0.75)(2)[(9.0)(0.375)0.6(36.0) + (0.291)(1.00)(46.0)]$$
$$= 129 \text{ kips}$$

5. Strength Based on Bolting to the Tee Stem

Tensile strength of the tee stem:

(a) Yielding in the gross section

$$\phi_t = 0.90$$
$$R_n = F_{y2}A_g$$
$$A_g = (0.375)(9.0) = 3.38 \text{ in.}^2$$
$$\phi R_n = 0.9(36)(3.38)$$
$$= 110 \text{ kips}$$

(b) Rupture in the net section

$$\phi_t = 0.75$$
$$R_n = F_{u2}A_e$$
$$A_n = [9.0 - (2)(3/4 + 1/8)](0.375) = 2.72 \text{ in.}^2$$
$$\leq 0.85A_g = 2.87 \text{ in.}^2$$

Therefore,

$$A_e = 2.72 \text{ in.}^2$$
$$\phi R_n = (0.75)(58)(2.72)$$
$$= 118 \text{ kips}$$

(c) Block Shear Rupture Strength

$$A_{nv} = (2)[1.5 + (2-1)3 - (2 - 0.5)(3/4 + 1/8)](0.375)$$
$$= 2.39 \text{ in.}^2$$
$$A_{nt} = [(2 - 1)3 - (2 - 1)(3/4 + 1/8)](0.375)$$
$$= 0.797 \text{ in.}^2$$

$$0.6F_{u2}A_{nv} > F_{u2}A_{nt}$$

$$83.2 \text{ kips} > 46.2 \text{ kips}$$

Therefore,

$$\phi R_n = \phi[0.6F_{u2}A_{nv} + F_{y2}A_{gt}]$$
$$= 0.75[83.2 + (36)[(2 - 1)3](.375)]$$

$$= 0.75(83.2 + 40.5)$$
$$= 92.8 \text{ kips}$$

Design Strength of the Bolts

(a) Bolt shear:

$$\phi R_n = \phi(F_v A_b)n$$

From the AISC LRFD Manual Table 8-11

$$\phi R_n = (4)(15.9)$$
$$= 63.6 \text{ kips}$$

(b) Bolt bearing:

$$\phi \quad = 0.75$$
$$L_{eh} \geq 1.5d_b = 1.5(.75) = 1.125 \text{ in.}$$
$$s \quad \geq 3.0d_b = 3.0(.75) = 2.25 \text{ in.}$$
$$R_n \quad = (2.4d_b t_2 F_{u2})n$$
$$= (2.4)(0.75)(0.375)(58)(4)$$
$$= 157 \text{ kips}$$
$$\phi R_n = 0.75(157)$$
$$= 118 \text{ kips}$$

Results: The connection strength in tension is governed by the strength of the weld connecting the tee flange to the HSS.

$$\phi R_n = 60.0 \text{ kips}$$

Solution 2: Using the Tables

Using Table 6-1, select the weld sizes, plate thickness, and number of bolts required to resist a factored tension load of 60 kips.

From Table 6-1, 25 percent Member Design Strength, $\phi R_n = 78.6$ kips > 60.0 kips **o.k.**

Weld, W_w $\quad = \frac{3}{8}$-in.
Tee Flange, $t_1 = \frac{1}{2}$-in.
Weld, W_s $\quad = \frac{3}{16}$-in.
Tee Stem, t_2 $\quad = \frac{3}{8}$-in.
Bolts, n $\quad = (6) \frac{3}{4}$-in. A325-N

WELDED TEE END CONNECTION (AXIAL COMPRESSION)

For compression loads the limit states are:

1. Strength of the weld connecting the tee flange to the HSS.
2. Strength of the weld connecting the stem to the tee flange.
3. Strength of the HSS wall.
4. Strength based on buckling of the tee stem.
5. Strength based on bolting to the tee stem.

Strength of the Weld Connecting the Tee Flange to the HSS

$$\phi R_n = \phi F_w A_w$$

where

ϕ = 0.75
F_w = nominal weld strength, ksi
$\quad = 0.60 F_{EXX}(1.0 + 0.50\sin^{1.5}\theta)$
F_{EXX} = electrode classification number, i.e., minimum specified strength, ksi
θ = angle of loading measured from the weld longitudinal axis, degrees
$\quad = 90°$
A_w = effective area of weld throat, in.^{2}
$\quad = 2(5t_1 + N)(0.707)W_w$
N = bearing length of load on the cap plate, in.
$\quad = t_2 + 2W_s$
t_1 = tee flange thickness, in.
t_2 = tee stem thickness, in.
W_s = weld size (tee flange to stem), in.
W_w = weld size (HSS to tee flange), in.

Strength of the Weld Connecting the Stem to the Tee Flange

$$\phi R_n = \phi F_w A_w$$

where

ϕ = 0.75
F_w = nominal weld strength, ksi
$\quad = 0.60 F_{EXX}(1.0 + 0.50\sin^{1.5}\theta)$
F_{EXX} = electrode classification number, i.e., minimum specified strength, ksi
θ = angle of loading measured from the weld longitudinal axis, degrees
$\quad = 90°$
A_w = effective area of weld throat, in.^{2}
$\quad = 0.707(2W_s)W$
W_s = weld size (tee flange to stem), in.
W = width of tee flange, in.
$\quad \geq (H + 1)$
H = depth of HSS section, in.

Strength of the HSS Wall
(a) Local Yielding

$$\phi R_n = \phi F_y A_e$$

$$\leq \phi 2 B F_y t$$

where

ϕ = 1.0
F_y = specified minimum yield stress of the HSS, ksi
A_e = effective area of HSS wall, in.^{2}
$\quad = 2(5t_1 + N)t$
N = bearing length of load on the tee flange, in.
$\quad = t_2 + 2W_s$
B = width of HSS wall loaded by tee stem, in.
t = HSS wall thickness, in.
t_1 = tee flange thickness, in.

t_2 = tee stem thickness, in.

W_s = weld size (tee flange to stem), in.

(b) Local Crippling

$$\phi R_n = \phi 1.6 t^2 \left[1 + 3 \left(\frac{N}{B/2} \right) \left(\frac{t}{t_1} \right)^{1.5} \right] \sqrt{\frac{E F_y t_1}{t}}$$

where

ϕ = 0.75

N = bearing length of load on the tee flange, in.

   = $t_2 + 2W_s$

B = width of HSS wall loaded by tee stem, in.

t = HSS wall thickness, in.

t_1 = tee flange thickness, in.

t_2 = tee stem thickness, in.

W_s = weld size (tee flange to stem), in.

E = modulus of elasticity, ksi

F_y = minimum specified yield stress of the HSS, ksi

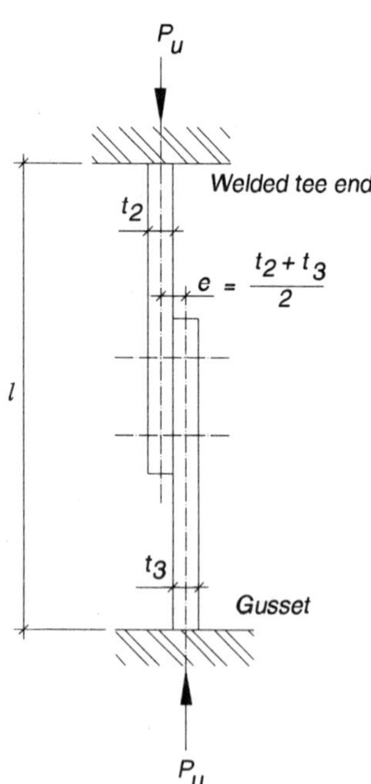

Fig. 6-4. Tee stem/gusset compression.

Strength Based on Buckling of the Tee Stem

The nominal compression strength of the tee stem/gusset connection presented below is based on simple column buckling procedures. The procedure assumes that both ends of the connection are fixed and can sway laterally.

For $\dfrac{P_u}{\phi_c P_n} \geq 0.2$

$$\frac{P_u}{\phi_c P_n} + \frac{8}{9}\left(\frac{M_u}{\phi_b M_n}\right) \leq 1.0$$

For $\dfrac{P_u}{\phi_c P_n} < 0.2$

$$\frac{P_u}{2\phi_c P_n} + \frac{M_u}{\phi_b M_n} \leq 1.0$$

where

ϕ_c = 0.85

ϕ_b = 0.90

P_u = required compressive strength, kips

P_n = nominal compressive strength, kips

    = $A_g F_{cr}$

A_g = gross area of member, in.^{2}

F_{cr} = critical compressive stress, ksi

    = $(0.658^{\lambda_c^2})F_{y2}$ for $\lambda_c \leq 1.5$

    = $\left[\dfrac{0.877}{\lambda_c^2}\right]F_{y2}$ for $\lambda_c > 1.5$

λ_c = column slenderness parameter

$$= \frac{Kl}{r\pi}\sqrt{\frac{F_{y2}}{E}}$$

F_{y2} = specified minimum yield stress of the tee stem, ksi

E = modulus of elasticity, ksi

K = effective length factor

    = 1.2

l = laterally unbraced length of plates, in.

r = governing radius of gyration, in.

    = $t_2 / \sqrt{12}$

M_u = required flexural strength, kip-in.

    = $P_u e / 2$

e = eccentricity of plate centerlines, in.

    = $(t_2 + t_3) / 2$

M_n = nominal flexural strength of the stem, kip-in.

    = $F_{y2} Z$

Z = plastic section modulus, in.^{3}

    = $W t_2^2 / 4$

W = width of tee stem (same as tee flange), in.

t_2 = tee stem plate thickness, in.

t_3 = gusset plate thickness, in.

therefore

For $\dfrac{P_u}{\phi_c P_n} \geq 0.2$

$$P_u \leq \cfrac{1}{\dfrac{1}{\phi_c P_n} + \dfrac{4}{9}\dfrac{e}{\phi_b M_n}} = \phi R_n$$

For $\dfrac{P_u}{\phi_c P_n} < 0.2$

$$P_u \leq \cfrac{2}{\dfrac{1}{\phi_c P_n} + \dfrac{e}{\phi_b M_n}} = \phi R_n$$

Strength Based on Bolting:

(a) Bolt Shear
See AISC Specification Section J3, and the AISC LRFD Manual Volume II, Part 8

(b) Bolt Bearing
See AISC Specification Section J3, and the AISC LRFD Manual Volume II, Part 8

EXAMPLE 6.2—Welded Tee End (Axial Compression)

Determine the design axial compressive strength of the HSS8×6×$^5/_{16}$ shown.

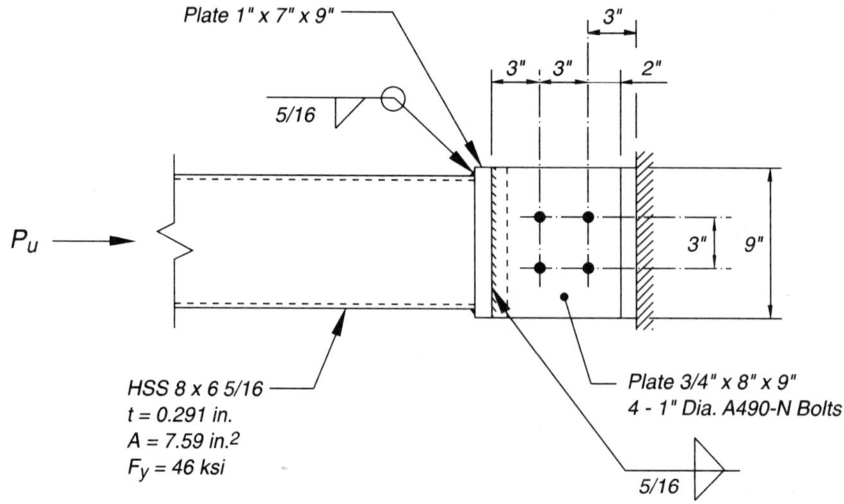

Figure Ex. 6-2.

Solution 1: Using Equations from this section.

1. Strength of the weld connecting the tee flange to the HSS

$\phi R_n = \phi F_w A_w$

$\phi \quad = 0.75$

$F_w \quad = 0.60F_{EXX}\,(1.0 + 0.5\sin^{1.5}\theta)$
$\qquad = 0.60(70)(1.5)$
$\qquad = 63.0 \text{ ksi}$

$A_w \quad = 2(5t_1 + t_2 + 2W_s)(0.707)W_w$
$\qquad = 2[5(1.0) + 0.75 + 2(5/16)](0.707)(5/16)$
$\qquad = 2.82 \text{ in.}^2$

$\phi R_n = 0.75(63.0)(2.82)$
$\qquad = 133 \text{ kips}$

2. Strength of the weld connecting the stem to the tee flange

$\phi R_n = \phi F_w A_w$

$\phi \quad = 0.75$

$F_w \quad = 0.60F_{EXX}\,(1.0 + 0.5\sin^{1.5}\theta)$
$\qquad = 0.60(70)(1.5)$
$\qquad = 63.0 \text{ ksi}$

$A_w \quad = 0.707(2W_s)W$
$W \quad = H + 1 = 8 + 1 = 9.0 \text{ in.}$
$L_{ev} \quad = (9.0 - 3.0)/2 = 3.0 \text{ in.} \geq 1.5(1.0) = 1.5 \text{ in.}$
$A_w \quad = 0.707(2)(5/16)(9.0)$
$\qquad = 3.98 \text{ in.}^2$

$\phi R_n = 0.75(63.0)(3.98)$
$\qquad = 188 \text{ kips}$

3. Strength of the HSS wall

(a) Local Yielding

$\phi R_n = \phi F_y A_e$

$\phi \quad = 1.0$

$F_y \quad = 46.0 \text{ ksi}$

$A_e \quad = 2(5t_1 + t_2 + 2W_s)t \leq 2(B)(t)$
$\qquad = 2[5(1.0) + 0.75 + 2(5/16)](.291)$
$\qquad = 3.71 \text{ in.}^2 > 2(6)(0.291) = 3.49 \text{ in.}^2$

$\phi R_n = 1.0(46.0)(3.49)$
$\qquad = 161 \text{ kips}$

(b) Web Crippling

$$\phi R_n = \phi 1.6t^2\left[1 + 3\left(\frac{N}{B/2}\right)\left(\frac{t}{t_1}\right)^{1.5}\right]\sqrt{\frac{EF_y t_1}{t}}$$

$\phi \quad = 0.75$
$t \quad = 0.291 \text{ in.}$
$N \quad = 0.75 + 2(5/16) = 1.38 \text{ in.}$
$B \quad = 6 \text{ in.}$
$t_1 \quad = 0.75 \text{ in.}$

$$E = 29{,}000 \text{ ksi}$$

$$F_y = 46 \text{ ksi}$$

$$\phi R_n = 0.75(1.6)(0.291)^2 \left[1 + 3\left(\frac{1.38}{6/2}\right)\left(\frac{0.291}{0.75}\right)^{1.5} \right] \sqrt{\frac{29{,}000(46)(0.75)}{0.291}}$$

$$= 252 \text{ kips}$$

4. Strength based on buckling of the tee stem

$$\phi_c P_n = \phi_c F_{cr} A_g$$

$$\phi_c = 0.85$$

$$A_g = 0.75(9) = 6.75 \text{ in.}^2$$

$$\lambda_c = \frac{Kl}{r\pi}\sqrt{\frac{F_{y2}}{E}}$$

$$= \frac{1.2(9)\sqrt{12}}{0.75\pi}\sqrt{\frac{36}{29{,}000}}$$

$$= 0.559 < 1.5$$

$$F_{cr} = (0.658^{\lambda_c^2})F_{y2}$$

$$= (0.658^{(0.559)^2})36$$

$$= 31.6 \text{ ksi}$$

$$\phi_c P_n = 0.85(31.6)(6.75)$$

$$= 181 \text{ kips}$$

$$\phi_b M_n = \phi_b F_{y2} Z$$

$$\phi_b = 0.90$$

$$F_{y2} = 36 \text{ ksi}$$

$$Z = 9.0(0.75)^2/4$$

$$= 1.27 \text{ in.}^3$$

$$\phi_b M_n = 0.90(36.0)(1.27)$$

$$= 41.1 \text{ kip-in.}$$

$$e = t_2 = 0.75 \text{ in.}$$

$$R_u \leq \cfrac{1}{\cfrac{1}{\phi_c P_n} + \cfrac{4}{9}\cfrac{e}{\phi_b M_n}} = \phi R_n$$

$$= \cfrac{1}{\cfrac{1}{181} + \cfrac{4}{9}\cfrac{(.75)}{41.1}}$$

$$= 73.3 \text{ kips } (> 0.2\phi_c P_n = 0.2\,(181) = 36.2 \text{ kips})$$

therefore

$$\phi R_n = 73.3 \text{ kips}$$

5. Strength based on bolting

a. Bolt shear:

$$\phi R_n = \phi(F_v A_b)n$$

From the AISC LRFD Manual Table 8-11

$$\phi R_n = (4)(35.3)$$
$$= 141 \text{ kips}$$

b. Bolt bearing:

$$\phi = 0.75$$
$$L_{eh} \geq 1.5d_b = 1.5(1.0) = 1.5 \text{ in.}$$
$$s \geq 3.0d_b = 3.0(1.0) = 3.0 \text{ in.}$$
$$R_n = (2.4d_b t_2 F_{u2})n$$
$$= 2.4(1.0)(0.75)(58)(4)$$
$$= 418 \text{ kips}$$
$$\phi R_n = 0.75 \ (418)$$
$$= 314 \text{ kips}$$

Results: The connection strength in compression is governed by the buckling strength of the tee stem.

$$\phi R_n = 73.3 \text{ kips}$$

Solution 2: Using the Tables

Using Table 6-2, select the weld sizes, plate thickness, and number of bolts required to resist a factored tension load of 73.3 kips.

From Table 6-2, 50 percent Member Design Strength, $\phi R_n = 116 \text{ kips} > 73.3 \text{ kips}$ **o.k.**

Weld, W_w $= \frac{7}{16}$-in.
Tee Flange, $t_1 = \frac{5}{8}$-in.
Weld, W_s $= \frac{5}{16}$-in.
Tee Stem, t_2 $= 1\frac{1}{4}$-in.
Bolts, n $= (6)$ 1 inch A325-N

SLOTTED HSS/GUSSET PLATE CONNECTION (AXIAL TENSION)

Shown in Figure 6-5 is a typical slotted HSS/gusset plate connection.

The design strength of HSS with slotted HSS/gusset plates can be obtained from the limit states for the member.

For tension loads the limit states are:

1. Strength based on member shear.
2. Strength of the weld connecting the gusset plate to the HSS.
3. Strength based on gusset plate shear.
4. Strength based on bolting to the gusset plate.

Strength Based on Member Shear

(a) Shear strength at welds

$$\phi R_n = \phi V_n, \text{ kips}$$

where

$$\phi = 0.9$$
$$V_n = 0.6F_y A_e, \text{ kips}$$
$$F_y = \text{specified minimum yield stress of the HSS, ksi}$$
$$A_e = \text{effective area of HSS wall, in.}^2$$
$$= 4L_w t$$
$$L_w = \text{length of weld to HSS, in. (As a "rule of thumb" } L_w \text{ should be equal to or greater}$$
$$\text{than the HSS depth)}$$

$\geq 1.0H$
t = HSS wall thickness, in.
H = depth of HSS section, in.

(b) Shear lag fracture in the HSS

$$\phi R_n = \phi F_u A_e$$

where

ϕ = 0.75
F_u = specified minimum tensile strength of the HSS, ksi
A_e = effective net area of the HSS, in.^{2}
$\quad = A_n U$
$A_n = A_g - 2t(t_1)$
$U = 1 - (\overline{x} / L_w) \leq 0.9$
$\overline{x}$ = D / π for round HSS with a single gusset
$\quad = \dfrac{B^2 + 2BH}{4(B + H)}$ for rectangular HSS with a single gusset
L_w = length of the weld to HSS, in.
$\quad \geq 1.0H$
D = outside diameter of HSS section, in.
B = width of HSS section, in.
H = depth of HSS section, in.
t = HSS wall thickness, in.
t_1 = gusset plate thickness, in.

Strength of the Weld Connecting the Gusset Plate to the HSS
$\phi R_n = \phi F_w A_w$, kips

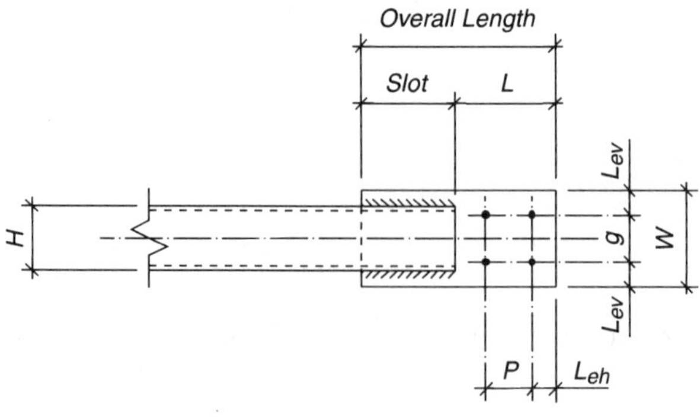

Fig. 6-5. Typical slotted HSS/gusset plate connection.

where

ϕ = 0.75

F_w = nominal weld strength, ksi

    = $0.60F_{EXX}(1.0 + 0.5\sin^{1.5}\theta)$

F_{EXX} = electrode classification number, i.e., minimum specified strength, ksi

θ = angle of loading measured from the weld longitudinal axis, degrees

    = $0°$

A_w = effective area of weld throat, in.^{2}

    = $4(0.707)W_e L_w$

W_e = effective weld size, in.

    = $W_w - \frac{1}{16}$

W_w = weld size, in.

L_w = length of weld to HSS, in.

    $\geq 1.0H$

H = depth of HSS section, in.

Strength Based on Gusset Plate Shear

The procedure below is based on the minimum thickness that matches the shear yielding strength of the gusset plate with the strength of the weld metal.

$$\phi V_n = 0.9(0.6F_{y1} t_1)$$
$$\phi V_w = 0.75(0.6F_{EXX} W_w)(0.707)2$$
$$\phi V_n \geq \phi V_w$$

The thickness of the gusset plate can be determined as follows:

$$t_1 \geq \frac{0.75(0.6F_{EXX} W_w)(0.707)2}{0.9(0.6F_{y1})} = \frac{1.18F_{EXX} W_w}{F_{y1}}$$

where

t_1 = gusset plate thickness, in.

F_{EXX} = electrode classification number, i.e., minimum specified strength, ksi

W_e = effective weld size, in.

    = $W_w - \frac{1}{16}$

W_w = weld size, in.

F_{y1} = specified minimum yield stress of the gusset plate, ksi

When the gusset plate thickness is less than the minimum, t_1, the strength of the gusset plate in shear is equal to the strength of the weld multiplied by the ratio of the actual thickness to t_1.

Strength Based on Bolting to the Gusset Plate

Design Tensile Strength of the Gusset Plate:

(a) For yielding in the gross section:

ϕ_t = 0.90

$R_n = F_{y1} A_g$, kips

(b) For rupture in the net section:

ϕ_t = 0.75

$$R_n = F_{u1}A_e, \text{ kips}$$

where

A_e = effective net area of the gusset plate, in.^{2}
    = minimum of A_n and $0.85A_g$
A_g = gross area of the gusset plate, in.^{2}
A_n = net area of gusset plate, in.^{2}
    = $[W - n_r(d_h + \frac{1}{16})]t_1$
F_{y1} = specified minimum yield stress of the gusset plate, ksi
F_{u1} = specified minimum tensile strength of the gusset plate, ksi
R_n = nominal axial strength, kips
W = width of gusset plate, in.
    $\geq (H + 1)$
H = depth of HSS section, in.
n_r = number of rows of bolts
d_h = diameter of bolt holes, in.
t_1 = gusset plate thickness, in.

(c) Block Shear Rupture Strength

The shear rupture term along a shear path in the gusset plate is

$$\phi(0.6F_{u1}A_{nv})$$

where

ϕ = 0.75
F_{u1} = specified minimum tensile strength of the gusset plate, ksi
A_{nv} = net area subject to shear, in.^{2}
    = $2[L_{eh} + (n - 1)p - (n - 0.5)(d_n + \frac{1}{16})]t_1$
L_{eh} = edge distance, in.
n = number of lines of bolts
d_n = diameter of bolt hole, in.
t_1 = gusset plate thickness, in.
p = spacing of lines of bolts (pitch), in.

The tension rupture term along a tension path in the gusset plate is

$$\phi(F_{u1}A_{nt})$$

where

ϕ = 0.75
F_{u1} = specified minimum tensile strength of the gusset plate, ksi
A_{nt} = net area subject to tension, in.^{2}
    = $[(n_r - 1)g - (n_r - 1)(d_h + \frac{1}{16})]t_1$
n_r = number of rows of bolts
g = spacing of rows of bolts, in.
d_h = diameter of bolt holes, in.
t_1 = gusset plate thickness, in.

The design block shear rupture strength is determined by the sum of the shear strength on a failure path(s) and the tensile strength on a perpendicular segment. When ultimate rupture strength on the net section is used to determine the resistance on one segment, yielding on the gross section is used on the perpendicular segment. The block shear rupture design strength, ϕR_n, shall be determined as follows:

(1) When $F_{u1}A_{nt} \geq 0.6F_{u1}A_{nv}$:

$$\phi R_n = \phi[0.6F_{y1}A_{gv} + F_{u1}A_{nt}], \text{ kips}$$

(2) When $0.6F_{u1}A_{nv} > F_{u1}A_{nt}$:

$$\phi R_n = \phi[0.6F_{u1}A_{nv} + F_{y1}A_{gt}], \text{ kips}$$

where

ϕ = 0.75
A_{gv} = gross area subject to shear, in.^{2}
  $= 2[L_{eh} + (n-1)p]t_1$
A_{gt} = gross area subject to tension, in.^{2}
  $= [(n_r - 1)g]t_1$
A_{nv} = net area subjected to shear, in.^{2}
A_{nt} = net area subjected to tension, in.^{2}
F_{y1} = stem specified yield stress, ksi
F_{u1} = stem specified minimum tensile strength, ksi

Design Strength of the Bolts:

(a) Bolt Shear
See AISC Specification Section J3, and the AISC Manual Volume II, Part 8

(b) Bolt Bearing
See AISC Specification Section J3, and the AISC Manual Volume II, Part 8

EXAMPLE 6.3—Slotted HSS/Gusset Plate Connection (Axial Tension)

For the HSS6.000×0.250 shown determine the design axial tensile strength.

Solution 1: Using Equations from this section.

1. Strength Based on Member Shear

(a) Shear strength at welds

$\phi R_n = \phi V_n$, kips
V_n = $0.6F_y A_e$, kips
ϕ = 0.9
V_n = 0.6(46)(4)(6)(0.233) = 154 kips
ϕR_n = 0.9(154)
  = 139 kips

(b) Shear lag fracture in the HSS

$\phi R_n = \phi F_u A_e$
ϕ = 0.75
F_u = 62 ksi
A_n = 4.22 − 2(0.233)(0.375)
  = 4.05 in.^{2}

$$\overline{x} \quad = 6/\pi = 1.91 \text{ in.}$$
$$U \quad = 1 - (1.91/6) = 0.682 < 0.9$$
$$\phi R_n = 0.75(62)[(4.05)(0.682)]$$
$$\quad = 128 \text{ kips}$$

2. Strength of the Weld

$$\phi R_n = \phi F_w A_w, \text{ kips}$$
$$\phi \quad = 0.75$$
$$F_w \quad = 0.60 F_{EXX}(1.0 + 0.5\sin^{1.5}\theta)$$
$$F_w \quad = 0.6(70)(1.0 + 0.0) = 42 \text{ ksi}$$
$$R_n \quad = 42(4)(0.707)(0.1875)(6) = 134 \text{ kips}$$
$$\phi R_n = 0.75(134)$$
$$\quad = 101 \text{ kips}$$

3. Strength Based on Gusset Plate Shear

(a) Shear strength at welds

$$t_1 \quad \geq \frac{1.18 F_{EXX} W_w}{F_{y1}}$$
$$t_1 \quad \geq 1.18(70)(0.1875)/36 = 0.430 \text{ in.}$$
$$\phi R_n = 101\left(\frac{0.375}{0.430}\right)$$
$$\quad = 88.1 \text{ kips}$$

4. Strength Based on Bolting

Tensile strength of the gusset plate

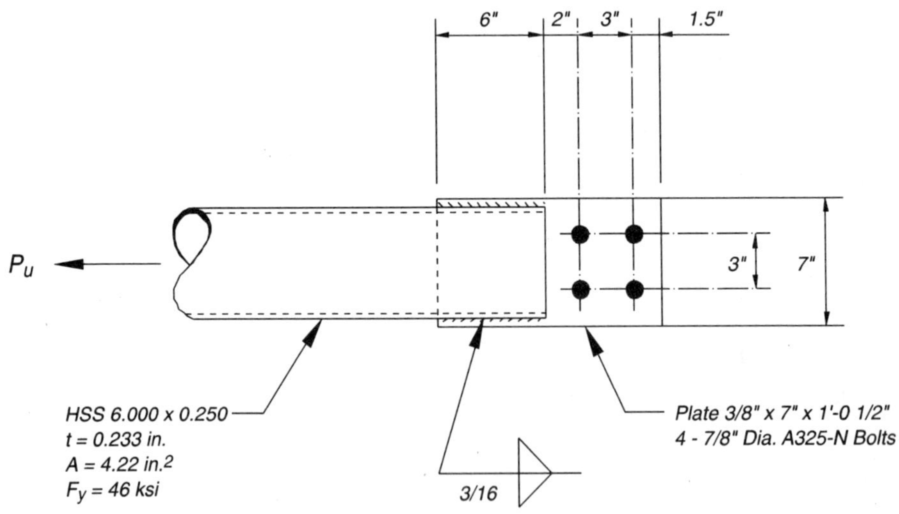

Figure Ex. 6.3.

(a) Yielding in the gross section

$$\phi = 0.9$$
$$R_n = F_{y1}A_g, \text{ kips}$$
$$R_n = 36(.375)(7) = 94.5 \text{ kips}$$
$$\phi R_n = 0.9(94.5)$$
$$= 85.1 \text{ kips}$$

(b) Rupture in the net section

$$\phi = 0.75$$
$$R_n = F_{u1}A_e, \text{ kips}$$
$$A_n = [7 - 2(\tfrac{7}{8} + \tfrac{1}{8})]0.375 = 1.88 \text{ in.}^2$$
$$0.85A_g = 0.85(.375)(7) = 2.23 \text{ in.}^2$$
$$A_e = 1.88 \text{ in.}^2$$
$$R_n = 58(1.88) = 109 \text{ kips}$$
$$\phi R_n = 0.75(109)$$
$$= 81.8 \text{ kips}$$

(c) Block Shear Rupture Strength

$$\phi = 0.75$$
$$A_{nv} = 2[L_{eh} + (n - 1)p - (n - 0.5)(d_b + \tfrac{1}{8})]t_1$$
$$A_{nv} = 2[1.5 + (2 - 1)3 - (2 - 0.5)(\tfrac{7}{8} + \tfrac{1}{8})] \, 0.375$$
$$= 2.25 \text{ in.}^2$$
$$A_{nt} = [(n_r - 1)g - (n_r - 1)(d_b + \tfrac{1}{8})]t_1$$
$$A_{nt} = [(2 - 1)3 - (2 - 1)(\tfrac{7}{8} + \tfrac{1}{8})]0.375$$
$$= 0.75 \text{ in.}^2$$

$$0.6F_{u1}A_{nv} > F_{u1}A_{nt}$$

$$78.3 \text{ kips} > 43.5 \text{ kips}$$

Therefore,

$$\phi R_n = 0.75[78.3 + 36(2 - 1)(3)(0.375)]$$
$$= 0.75(78.3 + 40.5)$$
$$= 89.1 \text{ kips}$$

Strength Based on Bolting:

(a) Bolt Shear

$$\phi R_n = (\phi F_v A_b)n$$

From the AISC LRFD Manual Table 8-11

$$\phi R_n = (21.6)(4)$$
$$= 86.4 \text{ kips}$$

(b) Bolt Bearing

$$\phi = 0.75$$
$$L_{eh} \geq 1.5d_b = 1.5(0.875) = 1.31 \text{ in.}$$
$$s \geq 3.0d_b = 3.0(0.875) = 2.62 \text{ in.}$$
$$R_n = (2.4d_b t_1 F_u)n$$

$$R_n = 2.4(0.875)(0.375)(58)(4)$$
$$= 183 \text{ kips}$$
$$\phi R_n = 137 \text{ kips}$$

Results: The connection strength in tension is governed by rupture in the net section of the gusset plate.

$$\phi R_n = 81.8 \text{ kips}$$

Solution 2: Using the Tables

Using Table 6-3, select the weld sizes, weld length, gusset plate thickness, and number of bolts required to resist a factored tension load of 75 kips.

From Table 6-3, 50 percent Member Design Strength, $\phi R_n = 79.8$ kips > 75 kips   **o.k.**

Weld, W_w $= \frac{1}{4}$-in.
Weld Length, $L_w = 6$ inches
Gusset Plate, t_1 $= \frac{1}{2}$-in.
Bolts, n $= (4) \frac{7}{8}$-in. A325-N

SLOTTED HSS/GUSSET PLATE CONNECTION (AXIAL COMPRESSION)

For compression loads the limit states are:

1. Strength based on member shear.
2. Strength of the weld connecting the gusset plate to the HSS.
3. Strength based on gusset plate shear.
4. Strength based on buckling of the gusset plate.
5. Strength based on bolting to the gusset plate.

Strength Based on Member Shear

$$\phi R_n = \phi V_n$$

where

$\phi = 0.9$
$V_n = 0.6 F_y A_e$, kips
F_y = specified minimum yield stress of the gusset plate, ksi
A_e = effective area of HSS wall, in.^{2}
$\quad = 4 L_w t$
L_w = length of weld to HSS, in.
$\quad \geq 1.0H$
t = HSS wall thickness, in.
H = depth of HSS section, in.

Strength of the Weld Connecting the Gusset Plate to the HSS

$$\phi R_n = \phi F_w A_w$$

where

$\phi   = 0.75$
$F_w $ = nominal weld strength, ksi
$\quad = 0.6 F_{EXX} (1.0 + 0.5 \sin^{1.5} \theta)$
F_{EXX} = electrode classification number, i.e., minimum specified strength, ksi
θ = angle of loading measured from the weld longitudinal axis, degrees

$\qquad = 0°$

$A_w\ \ =$ effective area of weld throat, in.^{2}

$\qquad = 4(0.707)W_e L_w$

$W_e\ \ =$ effective weld size, in.

$\qquad = W_w - \frac{1}{16}$

$W_w\ =$ weld size, in.

$L_w\ \ =$ length of weld to HSS, in.

$\qquad \geq 1.0H$

$H\ \ \ =$ depth of HSS section, in.

Strength Based on Gusset Plate Shear

The procedure below is based on the minimum thickness which matches the shear yielding strength of the gusset plate with the strength of the weld metal.

$$\phi V_n = 0.9(0.6 F_{y1} t_1)$$

$$\phi V_w = 0.75(0.6 F_{EXX} W_e)(0.707)2$$

$$\phi V_n \geq \phi V_w$$

therefore

$$t_1 \geq \frac{0.75(0.6 F_{EXX} W_e)(0.707)2}{(0.9(0.6 F_{y1}))} = \frac{1.18 F_{EXX} W_e}{F_{y1}}$$

where

$t_1\ \ \ \ =$ gusset plate thickness, in.

$F_{EXX} =$ electrode classification number, i.e., minimum specified strength, ksi

$W_e\ \ \ =$ effective weld size, in.

$\qquad = W_w - \frac{1}{16}$

$W_w\ \ =$ weld size, in.

$F_{y1} =$ specified minimum yield stress of the gusset plate, ksi

When the gusset plate thickness is less than the minimum, t_1, the strength of the gusset plate in shear is equal to the strength of the weld multiplied by the ratio of the actual thickness to t_1.

Strength Based on Buckling of the Gusset Plate

The nominal compression strength of the slotted HSS/gusset plate connection presented below is based on simple column buckling procedures. The procedure assumes that both ends of the connection are fixed and can sway laterally.

For $\quad \dfrac{P_u}{\phi_c P_n} \geq 0.2$

$$\frac{P_u}{\phi_c P_n} + \frac{8}{9}\left(\frac{M_u}{\phi_b M_n}\right) \leq 1.0$$

For $\quad \dfrac{P_u}{\phi_c P_n} < 0.2$

$$\frac{P_u}{2\phi_c P_n} + \frac{M_u}{\phi_b M_n} \leq 1.0$$

where

ϕ_c = 0.85

ϕ_b = 0.90

P_u = required compressive strength, kips

P_n = nominal compressive strength, kips

    = $A_g F_{cr}$

A_g = gross area of member, in.^{2}

F_{cr} = critical compressive stress, ksi

    = $(0.658^{\lambda_c^2})F_{y1}$, for $\lambda_c \leq 1.5$

    = $\left[\dfrac{0.877}{\lambda_c^2}\right] F_{y1}$, for $\lambda_c > 1.5$

λ_c = column slenderness parameter

$$= \frac{Kl}{r\pi}\sqrt{\frac{F_{y1}}{E}}$$

F_{y1} = specified minimum yield stress of the gusset plate, ksi

E = modulus of elasticity, ksi

K = effective length factor

    = 1.2

l = laterally unbraced length of plates, in.

r = governing radius of gyration, in.

    = $t_1 / \sqrt{12}$

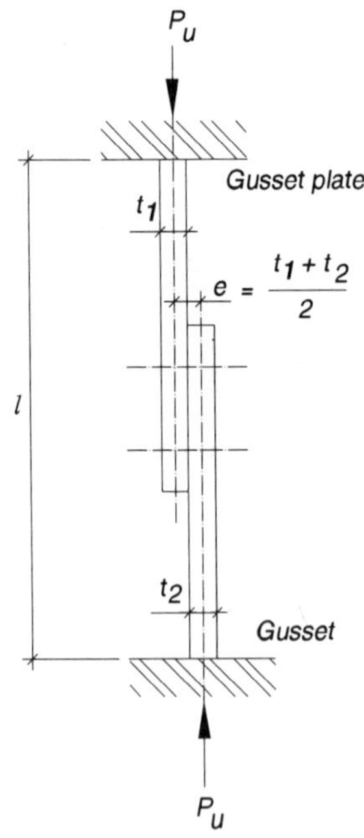

Fig. 6-6. Gusset plate/gusset compression.

M_u = required flexural strength, kip-in.

    = $P_u e / 2$

e = eccentricity of plate centerlines, in.

    = $(t_1 + t_2) / 2$

M_n = nominal flexural strength of the gusset plate, kip-in.

    = $F_{y1} Z$

Z = plastic section modulus, in.^{3}

    = $W t_1^2 / 4$

W = width of gusset plate, in.

t_1 = gusset plate thickness, in.

t_2 = gusset thickness, in.

therefore

For $\dfrac{P_u}{\phi_c P_n} \geq 0.2$

$$P_u \leq \frac{1}{\dfrac{1}{\phi_c P_n} + \dfrac{4}{9}\dfrac{e}{\phi_b M_n}} = \phi R_n$$

For $\dfrac{P_u}{\phi_c P_n} < 0.2$

$$P_u \leq \frac{2}{\dfrac{1}{\phi_c P_n} + \dfrac{e}{\phi_b M_n}} = \phi R_n$$

Strength Based on Bolting to the Gusset Plate

(a) Bolt Shear

See AISC Specification Section J3, and the AISC Manual Volume II, Part 8

(b) Bolt Bearing

See AISC Specification Section J3, and the AISC Manual Volume II, Part 8

EXAMPLE 6.4—Slotted HSS/Gusset Plate Connection (Axial Compression)

For the HSS6.000×0.250 shown, determine the design axial compressive strength.

Solution 1: Using Equations from this section.

1. Strength Based on Member Shear

    $\phi R_n = \phi V_n$

    V_n = $0.6 F_y A_e$

    ϕ = 0.9

    V_n = $0.6(46)(4)(6)(0.233) = 154$ kips

    ϕR_n = 0.9(154)

        = 139 kips

2. Strength of the Weld

    $\phi R_n = \phi F_w A_w$

    ϕ = 0.75

$$F_w = 0.60F_{EXX}(1.0 + 0.5\sin^{1.5}\theta)$$
$$F_w = 0.6(70)(1.0 + 0.0) = 42 \text{ ksi}$$
$$R_n = 42(4)(0.707)(.25)(6) = 178 \text{ kips}$$
$$\phi R_n = 0.75(178)$$
$$= 134 \text{ kips}$$

3. Strength Based on Gusset Plate Shear

$$t_1 \geq \frac{1.18F_{EXX}W_w}{F_{y1}}$$

$$t_1 \geq 1.18(70)(.25)/36 = 0.574 \text{ in.}$$

$$\phi R_n = 134\left(\frac{0.5}{0.574}\right)$$
$$= 117 \text{ kips}$$

4. Strength Based on Buckling of the Gusset Plate

$$\phi_c P_n = \phi_c F_{cr} A_g$$
$$\phi_c = 0.85$$
$$A_g = 0.5(7) = 3.50 \text{ in.}^2$$

$$\lambda_c = \frac{Kl}{r\pi}\sqrt{\frac{F_{y1}}{E}}$$

$$= \frac{1.2(12)\sqrt{12}}{0.50\pi}\sqrt{\frac{36}{29,000}}$$

$$= 1.12 < 1.5$$

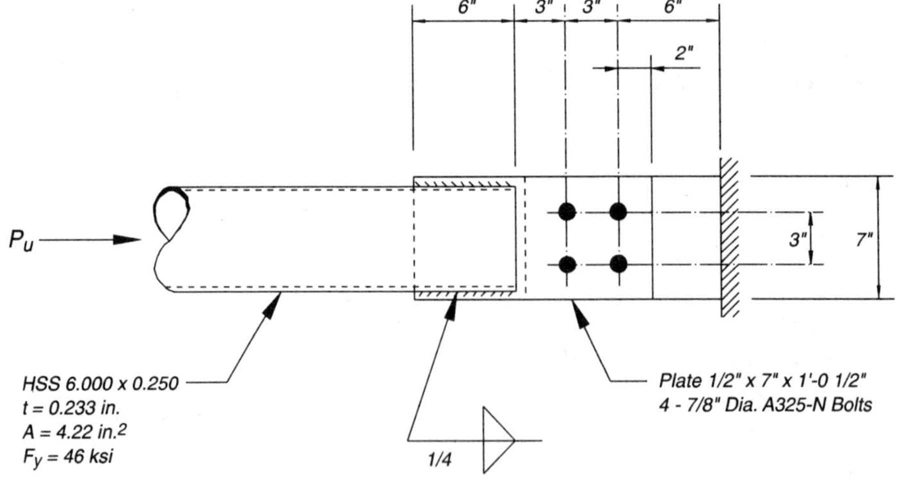

Figure Ex. 6.4.

$$F_{cr} = (0.658^{\lambda_c^2})F_{y1}$$
$$= (0.658^{(1.12)^2})\, 36$$
$$= 21.3 \text{ ksi}$$

$$\phi_c P_n = 0.85(21.3)(3.50)$$
$$= 63.4 \text{ kips}$$

$$\phi_b M_n = \phi_b F_{y1} Z$$

$$\phi_b = 0.9$$
$$F_{y1} = 36 \text{ ksi}$$
$$Z = 7.0(.5)^2/4$$
$$= 0.438 \text{ in.}^3$$

$$\phi_b M_n = 0.9(36.0)(0.438)$$
$$= 14.2 \text{ kip-in.}$$

$$e = t_1 = 0.5 \text{ in.}$$

$$P_u \leq \cfrac{1}{\cfrac{1}{\phi_c P_n} + \cfrac{4}{9}\cfrac{e}{\phi_b M_n}} = \phi R_n$$

$$= \cfrac{1}{\cfrac{1}{63.4} + \cfrac{4}{9}\cfrac{0.5}{14.2}}$$

$$= 31.8 \text{ kips} \ (> 0.2\phi_c P_n = 0.2(63.4) = 12.7 \text{ kips})$$

therefore,

$$\phi R_n = 31.8 \text{ kips}$$

5. Strength Based on Bolting to the Gusset Plate:

(a) Bolt Shear

$$\phi R_n = (\phi F_v A_b)n$$

From the AISC LRFD Manual Table 8-11

$$\phi R_n = (4)(21.6)$$
$$= 86.4 \text{ kips}$$

(b) Bolt Bearing

$$\phi = 0.75$$
$$L_{eh} \geq 1.5 d_b = 1.5(0.875) = 1.31 \text{ in.}$$
$$s \geq 3.0 d_b = 3.0(0.875) = 2.63 \text{ in.}$$
$$R_n = (2.4 d_b t_1 F_u)n$$
$$= 2.4(0.875)(0.5)(58)4$$
$$= 244 \text{ kips}$$
$$\phi R_n = 0.75(244)$$
$$= 183 \text{ kips}$$

Results: The connection strength is governed by the buckling strength of the gusset plate.
$$\phi R_n = 31.8 \text{ kips}$$

Solution 2: Using the Tables

Using Table 6-4, select the weld size, weld length, gusset plate thickness and number of bolts required to resist a factored tension load of 31.8 kips.

From Table 6-4, 25 percent Member Design Strength, $\phi R_n = 32.4$ kips > 31.8 kips **o.k.**

Weld, W_w	$= \frac{1}{4}$-in.
Weld Length, L_w	$= 6$ inches
Gusset Plate, t_1	$= \frac{1}{2}$-in.
Bolts, n	$= (2)$ $\frac{7}{8}$-in. A325-N

END PLATES (AXIAL TENSION)

The limit states for an HSS with end plates are:

1. Yielding of the end plate.
2. Tensile strength of the bolts.
3. Strength of the weld connecting the end plate to the HSS.

End Plates on Round Sections

The method provided below for the design of end plates for Round HSS is based on research by Igarashi et al (1985). This method permits prying action to occur at the limit state. The connection is designed based on the limit state of yielding of the end plate.

Yielding of the End Plate

The thickness of the end plate is determined from the equation (Packer, 1992):

$$t_1 \geq \sqrt{\frac{2P_u}{\phi F_{y1} \pi f_3}}$$

where

P_u = the factored tension load on the HSS, kips

ϕ = 0.90

F_{y1} = specified minimum yield stress of the end plate, ksi

and f_3 is as defined below.

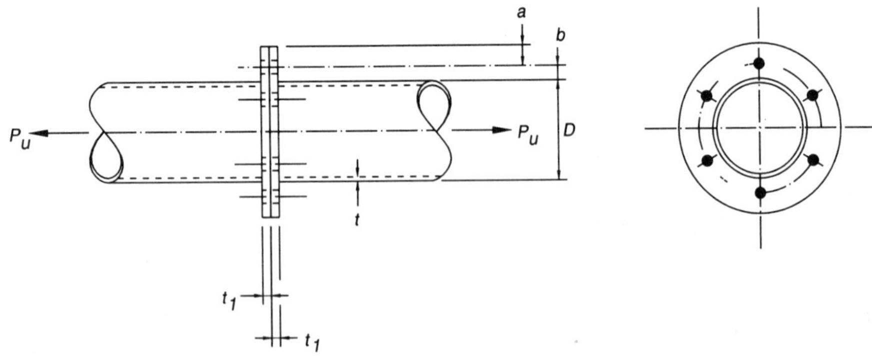

Fig. 6-7. End plate on round section.

Tensile Strength of the Bolts

The number of bolts required can be obtained from (Packer, 1992):

$$n \geq \frac{P_u}{\phi R_n}\left(1 - \frac{1}{f_3} + \frac{1}{f_3 \ln(r_1 / r_2)}\right)$$

where

P_u = the factored tension load on the HSS, kips

ϕR_n = bolt tension strength

$f_3 = \dfrac{1}{2k_1}(k_3 + \sqrt{k_3^2 - 4k_1})$

$k_1 = \ln\left(\dfrac{r_2}{r_3}\right)$

$k_3 = k_1 + 2$

$r_1 = \dfrac{D}{2} + 2b$

$r_2 = \dfrac{D}{2} + b$

$r_3 = \dfrac{D - t}{2}$

D = outer diameter of HSS, in.

t = HSS wall thickness, in.

b = distance from HSS wall to bolt line, in.

≥ 1.5 in.

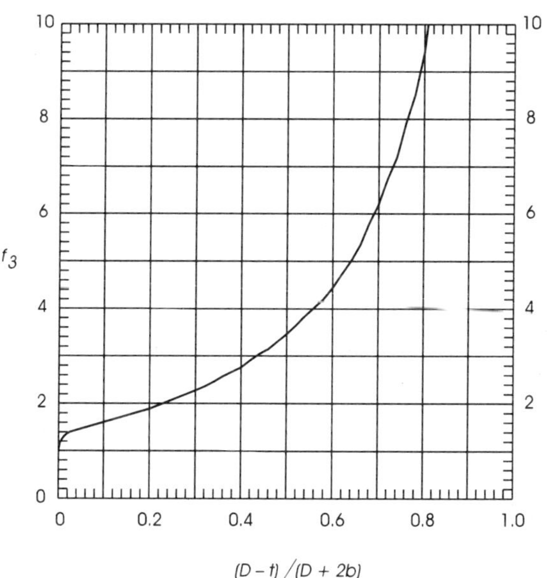

Fig. 6-8. Graph representing f_3.

Strength of the Weld Connecting the End Plate to the HSS

The weld size can be determined from:

$$W_w = \frac{P_u}{\phi F_w(0.707)\pi D}$$

where

P_u = the factored nominal tension load on the HSS, kips

ϕ = 0.75

F_w = nominal weld strength, ksi

      = $0.60 F_{EXX}(1.0 + 0.5\sin^{1.5}\theta)$

F_{EXX} = electrode classification number i.e., minimum specified strength, ksi

θ = angle of loading measured from the weld longitudinal axis, degrees

      = $90°$

D = outer diameter of the HSS, in.

To obtain minimum end plate thickness, b should be as small as practical, keeping in mind clearances for an impact wrench and minimum spacing requirements per AISC LRFD Section J3.3.

The method assumes that the flange is continuous and that the bolts are arranged symmetrically. The equation to determine the number of required bolts assumes that $a = b$.

The method is dependent on the HSS being welded to the flange plate in a manner which will fully develop the HSS yield strength. Shown in Figure 6-8 is a graph representing f_3.

EXAMPLE 6.5—Round End Plate (Axial Tension)

Determine the weld leg size, end plate thickness, and the number of ¾-in. ϕ A325 bolts required to resist a factored tension load of 100 kips on an HSS4.000×0.250 section. The end plate has a yield strength of 36 ksi.

Solution 1: Using Equations from this section.

1. Determine plate thickness required

ϕ = 0.90

F_{y1} = 36 ksi

$$r_2 = \frac{D}{2} + b$$

$$= \frac{4}{2} + 1.5$$

$$= 3.50 \text{ in.}$$

$$r_3 = \frac{D - t}{2}$$

$$= \frac{4 - .25}{2}$$

$$= 1.88 \text{ in.}$$

$$k_1 = \ln\left(\frac{r_2}{r_3}\right)$$

$$= \ln\left(\frac{3.50}{1.88}\right)$$

$$= 0.621$$

$$k_3 = k_1 + 2$$

$$= 2.62$$

$$f_3 = \frac{1}{2k_1}\left(k_3 + \sqrt{k_3^2 - 4k_1}\right)$$

$$= \frac{1}{2(0.621)}\left(2.62 + \sqrt{2.62^2 - 4(0.621)}\right)$$

$$= 3.79$$

$$t_1 \geq \sqrt{\frac{2P_u}{\phi F_{y1} \pi f_3}}$$

$$\geq \sqrt{\frac{2(100)}{0.9(36.0)\pi(3.79)}}$$

$$t_1 = 0.720 \text{ in.}$$

Use ¾-in. end plate

2. Determine number of bolts required

$$\phi R_n = 29.8 \text{ kips}$$

$$r_1 = \frac{D}{2} + 2b$$

$$= \frac{4}{2} + 2(1.5)$$

$$= 5.00$$

$$n = \frac{P_u}{\phi R_n}\left(1 - \frac{1}{f_3} + \frac{1}{f_3 \ln(r_1/r_2)}\right)$$

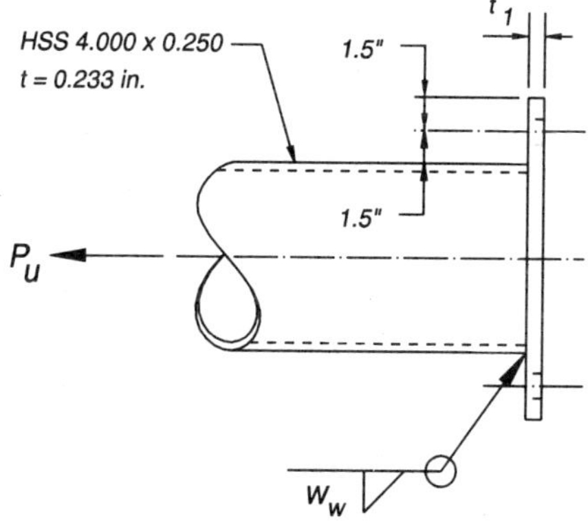

Figure Ex. 6.5.

$$= \frac{100}{29.8}\left(1 - \frac{1}{3.79} + \frac{1}{3.79\ln(5.00/3.50)}\right)$$

$$n \quad = 4.95$$

Use (5) ¾-in. diameter A325 bolts

3. Determine weld size required

$$\phi = 0.75$$
$$D = 4.0 \text{ in.}$$
$$F_w = 0.60F_{EXX}\,(1.0 + 0.5\sin^{1.5}(\theta))$$
$$= 0.60(70)(1.5)$$
$$= 63.0 \text{ ksi}$$

$$W_w = \frac{P_u}{\phi F_w(0.707)\pi D}$$

$$W_w = \frac{100}{0.75(63.0)(0.707)(\pi)(4.0)}$$

$$= 0.238 \text{ in.}$$

4. **Check minimum weld size requirements**
From AISC LRFD Table J2.4 for $t = $ ¾-in.

$$W_{wmin} = \text{¼-in.} > 0.238 \text{ in.}$$

Use ¼-in. weld leg size

5. **Check minimum bolt spacing requirements:**
The bolt spacing equals the circumference at the bolt location divided by the number of bolts.

$$s = \frac{2\pi r_2}{n}$$

where

$$r_2 = \frac{D}{2} + b = \frac{4}{2} + 1.5 = 3.5$$
$$n = 5$$
$$s = \frac{2\pi(3.5)}{5} = 4.40 \text{ in.} > 3.0d_b = 3.0(0.75) = 2.25 \quad \textbf{o.k.}$$

Results: Use ¼-in. weld with ¾-in. end plate and (5) ¾-in. diameter A325 bolts. Edge distance $a = b = 1.5$ in.

Solution 2: Using the Tables

Using Table 6-5, select the weld size, end plate thickness, number of bolts, and edge distance required to resist a factored tension load of 100 kips.
From Table 6-5, 100 percent Member Design Strength, $\phi R_n = 104$ kips > 100 kips   **o.k.**

Weld, W_w = ¼-in.
End Plate, t_1 = ¾-in.
Bolts, n = (6) ¾-in. A325
Edge Distance, b = 1½-in.

End Plates on Rectangular Sections

Two types of end plate configurations are commonly used for rectangular and square HSS. The most common is to bolt on all four sides of the HSS as illustrated in Fig. 6-9. The second configuration is to bolt only on two sides of the HSS as shown in Fig. 6-10.

End Plates Bolted on Four Sides of the HSS

Three limit states exist for the end plate bolted along four sides. These are:

1. Bending failure of the end plate.
2. Tensile strength of the bolts.
3. Strength of the weld connecting the end plate to the HSS.

The design procedure for the end plate bolted on all four sides is based on the AISC design procedure for Hanger Connections (LRFD Manual Part 11). For this design procedure to be applicable, bolts cannot be positioned beyond the corner of the HSS. See Fig. 6-11.

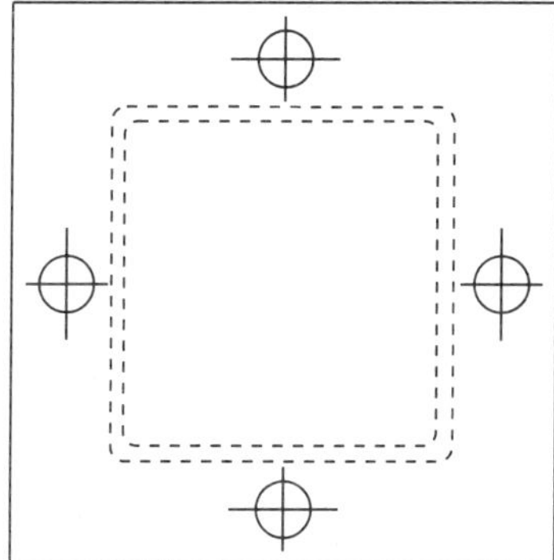
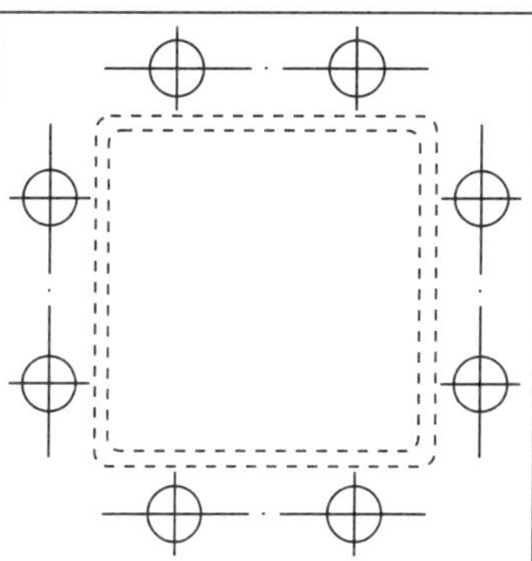

Fig. 6-9. End plate bolted on all four sides.

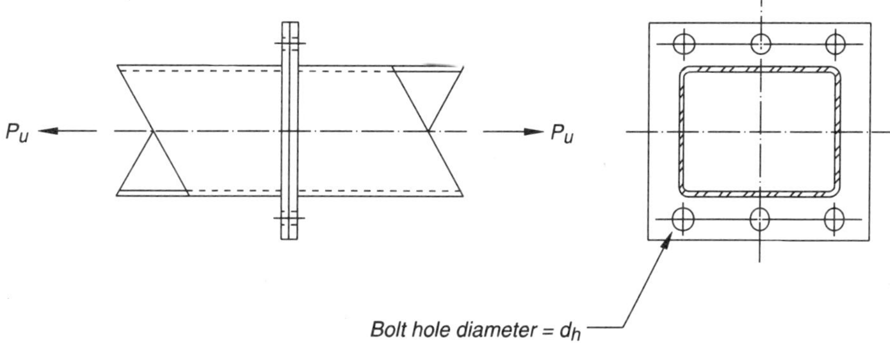

Fig. 6-10. End plate bolted on two sides.

This procedure checks limit states 1 and 2. The weld strength is checked independently. The procedure is outlined as follows:

First select the number and size of bolts required such that the design tensile strength of one bolt ϕr_n exceeds the factored tensile force per bolt r_{ut}.

With the preliminary number and size of bolts selected, determine the end plate thickness required. Given b, and the size of the bolts, calculate b', a', and ρ as

$$a' = \left(a + \frac{d_b}{2} \right)$$

$$b' = \left(b - \frac{d_b}{2} \right)$$

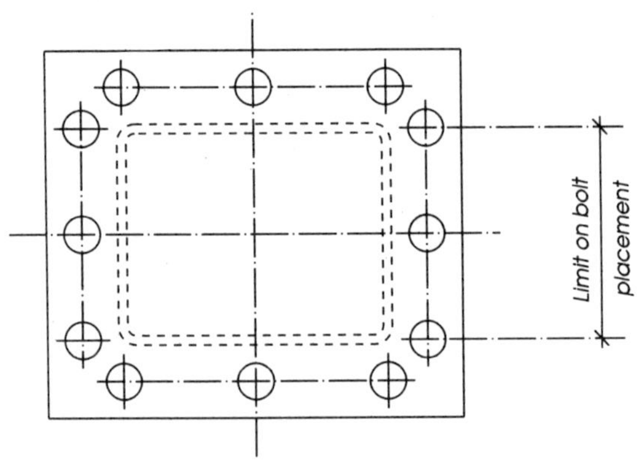

Fig. 6-11. End plate limit on bolt position.

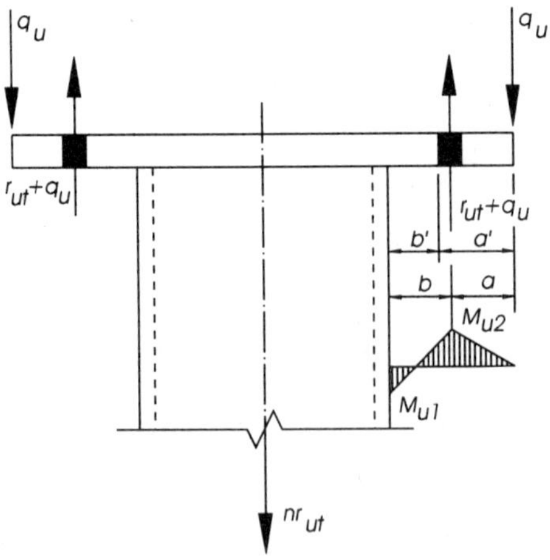

Fig. 6-12. End plate prying action.

$$\rho = \frac{b'}{a'}$$

In the above equations, *(a)* is the distance from the bolt centerline to the edge of the end plate for calculation purposes, *(a)* should not be taken to be greater than $1.25b$.

Next, calculate β as follows:

$$\beta = \frac{1}{\rho}\left(\frac{\phi r_n}{r_{ut}} - 1\right)$$

if $\beta \geq 1$, set $\alpha' = 1.0$

if $\beta < 1$, set $\alpha' = $ the lesser of 1.0 and

$$\frac{1}{\delta}\left(\frac{\beta}{1-\beta}\right)$$

where δ, the ratio of the net area at the bolt line to the gross area at the face of the HSS is

$$\delta = 1 - \frac{d'}{p}$$

and

$d' = $ width of bolt hole parallel to the HSS wall, in.
$p = $ length of end plate, parallel to the HSS wall, tributary to each bolt, in.

The required end plate thickness t_{req} may then be calculated as:

$$t_{req} = \sqrt{\frac{4.44 r_{ut}b'}{pF_{yl}(1 + \delta\alpha')}}$$

Note: Positioning the bolts as shown in Fig. 6-11 severely limits the capacity of the end plate since p must be taken as half of the bolt spacing.

The factored prying force per bolt q_u may be calculated from α as follows:

$$\alpha = \frac{1}{\delta}\left[\frac{r_{ut}}{\phi r_n}\left(\frac{t_c}{t_1}\right)^2 - 1\right] \geq 0$$

$$q_{ut} = \phi r_n\left[\delta\alpha\rho\left(\frac{t_1}{t_c}\right)^2\right]$$

and the factored force per bolt including prying action is $r_{ut} + q_u$. In the above equations, t_c, the end plate thickness required to develop the design strength of the bolt ϕr_n with no prying action, is calculated as:

$$t_c = \sqrt{\frac{4.44\phi r_n b'}{pF_{yl}}}$$

Analyzing a Connection for Prying Action

The foregoing procedure is somewhat simplified when analyzing a connection for prying action. As before, check that $r_{ut} \leq \phi r_n$. Then calculate α' as:

$$\alpha' = \frac{1}{\delta(1+\rho)}\left[\left(\frac{t_c}{t_1}\right)^2 - 1\right]$$

If $\alpha' < 0$, r_{ut} must be such that

$$r_{ut} \le \phi r_n$$

If $0 \le \alpha' \le 1$, r_{ut} must be such that

$$r_{ut} \le \phi r_n \left(\frac{t_1}{t_c}\right)^2 (1 + \delta\alpha')$$

If $\alpha' > 1$, r_{ut} must be such that

$$r_{ut} \le \phi r_n \left(\frac{t_1}{t_c}\right)^2 (1 + \delta)$$

If desired, the factored prying force per bolt q_u may be determined as before.

Strength of the Weld Connecting the End Plate to the HSS

$$W_w \ge \frac{P_u}{\phi F_w (0.707) L_w}$$

where

P_u = the factored nominal tension load on the HSS, kips
ϕ = 0.75
F_w = nominal weld strength, ksi
      = $0.60 F_{EXX} (1.0 + 0.50\sin^{1.5}\theta)$
F_{EXX} = electrode classification number, i.e., minimum specified strength, ksi
θ = angle of loading measured from the weld longitudinal axis, degrees
      = 90°
L_w = effective length of weld throat, in.^{2}
      = $2(B + H)$
B = width of HSS section, in.
H = depth of HSS section, in.

EXAMPLE 6.6—End Plate Bolted on all Four Sides

Determine the end plate thickness, size of A325 bolts, and weld leg size required to resist a factored tension load of 100 kips on the HSS4×4×¼ shown. The end plate has a yield strength of 36 ksi.

Solution 1: Using Equations from the preceding section.

Determine the size of the (4) A325 bolts:

$$r_{ut} = \frac{P_u}{n} = \frac{100}{4} = 25 \text{ kips}$$

Try ¾-in. diameter A325 bolts

$$\phi r_n = 29.8 \text{ kips}$$

Check trial selection:

$$a' = a + \frac{d_b}{2}$$

$$= 1.5 + \frac{\frac{3}{4}}{2}$$

$$= 1.88 \text{ in.}$$

$$b' = b - \frac{d_b}{2}$$

$$= 1.5 - \frac{\frac{3}{4}}{2}$$

$$= 1.12 \text{ in.}$$

$$\rho = \frac{b'}{a'}$$

$$= \frac{1.12}{1.88}$$

$$= 0.596$$

$$\beta = \frac{1}{\rho}\left(\frac{\phi r_n}{r_{ut}} - 1\right)$$

$$= \frac{1}{0.596}\left(\frac{29.8}{25} - 1\right)$$

$$= 0.322$$

$$\delta = 1 - \frac{d'}{p}$$

$$= 1 - \frac{(\frac{3}{4} + \frac{1}{8})}{4.0}$$

$$= 0.781$$

$$\alpha' = \frac{1}{\delta}\left(\frac{\beta}{1-\beta}\right)$$

$$\alpha' = \frac{1}{0.781}\left(\frac{0.322}{1-0.322}\right) \leq 1.0$$

$$= 0.608$$

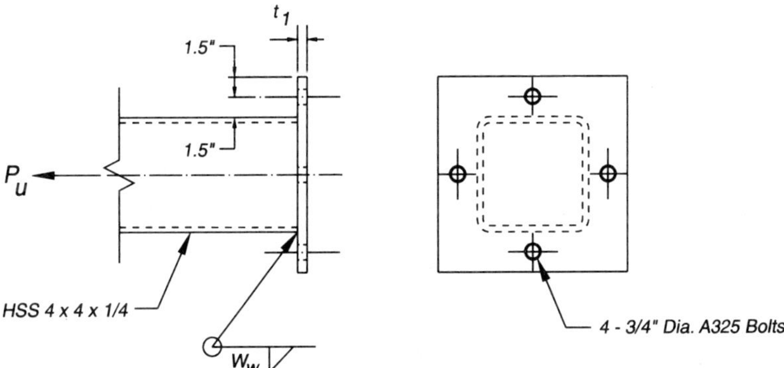

Figure Ex. 6.6.

Calculate required end plate thickness:

$$t_{req} = \sqrt{\frac{4.44 r_{ut} b'}{p F_{y1}(1 + \delta\alpha')}}$$

$$= \sqrt{\frac{4.44(25)(1.12)}{4(36)(1 + 0.781(0.608))}}$$

$$= 0.765 \text{ in.}$$

Use $\frac{7}{8}$-in. end plate, $t_1 > 0.765$, bolt check not required.

Use (4) $\frac{3}{4}$-in. diameter A325 bolts

Determine weld leg size:

$$\phi = 0.75$$
$$L_w = 4(4.0) = 16.0 \text{ in.}$$
$$F_w = 0.60 F_{EXX}(1.0 + 0.5\sin^{1.5}\theta)$$
$$= 0.60(70)(1.5)$$
$$= 63.0 \text{ ksi}$$
$$W_w \geq \frac{P_u}{\phi F_w(0.707)L_w}$$
$$\geq \frac{100}{0.75(63.0)(0.707)(16.0)} = 0.187 \text{ in.}$$

From AISC LRFD Table J2.4 for $t = 1$ in.

$$W_{min} = \frac{5}{16}\text{-in.} > 0.187 \text{ in.}$$

Use $\frac{5}{16}$-in. weld leg size

Results: Use $\frac{5}{16}$-in. weld with $\frac{7}{8}$-in. end plate and (4) $\frac{3}{4}$-in. diameter A325 bolts.

Solution 2: Using the Tables

Using Table 6-6, select the weld size, end plate thickness, and number of bolts required to resist a factored tension load of 100 kips.
From Table 6-6, 75 percent Member Design Strength, $\phi R_n = 105$ kips > 100 kips   **o.k.**

Weld, W_w $= \frac{5}{16}$-in.
End Plate, t_1 $= \frac{7}{8}$-in.
Bolts per side, $n = (2)\frac{3}{4}$-in. A325

End Plates Bolted on Two Sides of the HSS

Packer et al (1989) have shown that, by selecting specific connection parameters, one can fully develop the tensile resistance of the HSS by bolting along only two sides of the HSS. This form of connection lends itself to analysis as a 2-dimensional prying problem, but the application of traditional prying models developed for T-hangers was found to correlate poorly with the test results. One reason for this was that the location of the plastic hinge lines tended to form within the width of the HSS as shown in Fig. 6-13.

A modified T-stub design procedure was proposed (Birkemoe and Packer 1986) and verified against a set of possible failure mechanisms based on the observed failure modes. The design procedure involves a redefinition of various parameters in the T-hanger design method of Struik and de Back (1969). To reflect the observed location of the inner plastic hinge line, and also represent the connection behavior illustrated by the more complex analytical models, the distance b was adjusted to b' as shown on Fig. 6-13, where

$$b' = b - \frac{d_h}{2} + t$$

The term α has been used in Struik and de Back's T-hanger prying model to represent the ratio of the bending moment per unit plate width at the bolt line to the bending moment per unit plate width at the inner plastic hinge. For the limiting case of a rigid plate $\alpha = 0$, and for the limiting case of a flexible plate in double curvature with plastic hinges occurring both at the bolt line and the edge of the T-hanger web $\alpha = 1.0$. Hence, the term α in Struik and de Back's model was restricted to the range $0 \leq \alpha \leq 1.0$.

Thus, a suitable design method (Packer and Henderson, 1992) for this connection is as follows:

1. Estimate the number, grade and size of bolts required, knowing the applied tensile force P_u and allowing for some amount of prying. In general, the applied external load per bolt should be only 70 percent to 80 percent of the bolt tensile resistance in anticipation of bolt load amplification due to prying. Hence, determine a suitable connection arrangement. The bolt pitch p should generally be about 4 to 5 bolt diameters and the edge distance a about $1.25b$, which is the maximum allowed in calculations. Prying decreases as a is increased up to $1.25b$, beyond which there is no advantage. Then determine:

$$\delta = 1 - \frac{d_h}{p}$$

where

d_h = bolt hole diameter, in.

    = $d_b + \frac{1}{8}$-in.

d_b = diameter of bolt, in.

p = length of end plate tributary to each bolt, or bolt pitch, in.

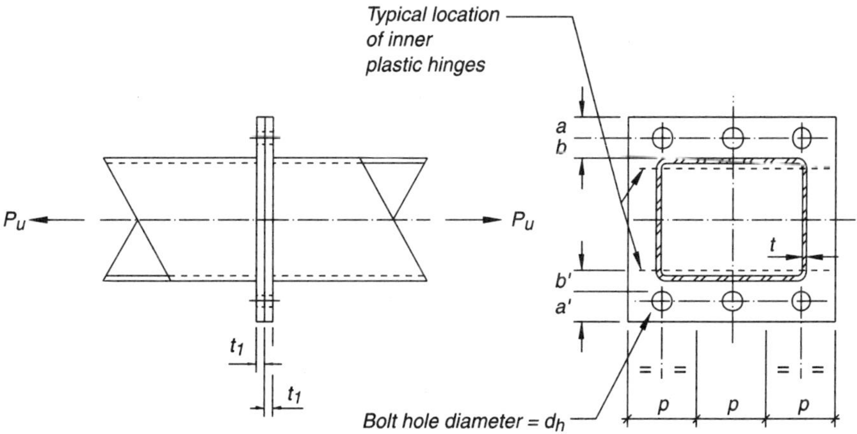

Fig. 6-13. End plate with bolts along two sides.

Also determine a trial end plate thickness t_1 from:

$$\sqrt{\frac{Kr_{ut}}{1+\delta}} \leq t_1 \leq \sqrt{Kr_{ut}}$$

where

$r_{ut} = \dfrac{P_u}{n}$, the external factored tensile load on one bolt, kips

$$K = \frac{4b'}{\phi F_{y1} p}$$

where

$\phi\ = 0.9$
$F_{y1} =$ specified minimum yield stress of the end plate, ksi

2. With the number, size and grade of bolts preselected, plus a trial end plate thickness, calculate the ratio α necessary for equilibrium from:

$$\alpha = \left(\frac{K\phi r_n}{t_1^2} - 1\right)\left(\frac{a + (d_h/2)}{\delta(a + b + t)}\right)$$

$\phi r_n\ =$ design tensile strength of one bolt, kips

3. Calculate the connection factored resistance ϕR_n set $\alpha = 0$ if $\alpha < 0$.

$$\phi R_n = \frac{t_1^2(1 + \delta\alpha)n}{K}$$

where n is the number of bolts. ϕR_n must be $\geq P_u$

The actual total bolt tension, including prying, can be determined from:

$$r_{ut} + q_u \approx r_{ut}\left(1 + \frac{b'}{a'}\left(\frac{\alpha}{1 + \alpha}\right)\right)$$

where

$r_{ut} + q_u =$ the total bolt tension, kips
$q_u \qquad =$ the prying force, kips
$a' \qquad =$ "effective a" plus $d_h/2$, in.
$\alpha \qquad = \left(\dfrac{Kr_{ut}}{t_1^2} - 1\right)$

"Effective a" is simply the "a" dimension (See Figure 6-13) up to a maximum of $1.25b$.

Note that this value of α is not necessarily the same as that from step 2 which was premised on the bolts being loaded to their full tensile resistance.

This design method should be restricted to the range of end plate thicknesses over which it has been validated experimentally and analytically (Packer et al. 1989; Birkemoe and Packer 1986), namely ½ to 1 in. It should be borne in mind that when a connection with bolts in tension is subject to repeated loads, the flange must be made thick enough and stiff enough so that deformation of the flange is virtually eliminated ($\alpha \leq 0$).

The method is dependent on the HSS being welded to the end plate in a manner which will fully develop the HSS yield strength.

This can be accomplished by sizing the weld to resist the factored tension load using only those welds parallel to the bolt lines. Since the resistance provided by the bolts will be concentrated along those HSS walls it is more economical to place the bolts along the long side of the HSS whenever possible. This will also allow for a greater number of bolts in the connection.

The weld leg size can be determined from the following procedure.

$$W_w \geq \frac{P_u}{\phi F_w(0.707)2H}$$

where

P_u = the factored nominal tension load on the HSS, kips

ϕ = 0.75

F_w = nominal weld strength, ksi

    = $0.60F_{EXX}(1.0 + 0.5\sin^{1.5}\theta)$

F_{EXX} = electrode classification number, i.e., minimum specified strength, ksi

θ = angle of loading measured from the weld longitudinal axis, degrees

    = 90°

H = depth of the HSS section, in.

Example 6.7—End Plate Bolted on Two Sides

Determine the weld leg size, end plate thickness, and the size and type of bolts required to resist a factored tension load of 100 kips for the HSS shown. The end plate has a yield strength of 36 ksi.

Solution 1: Using Equations from the preceding section.

1. Estimate the size and type of bolts required. Note that because of the depth of the HSS, the practical limit on the number of bolts is 6.

$$r_{ut} = \frac{P_u}{n} = \frac{100}{6} = 16.7 \text{ kips}$$

Amplifying the tension load to account for prying action:

$$r_{ut}/0.7 = 16.7/0.7 = 23.9 \text{ kips}$$

Try ¾-in. diameter A325 bolts

$$\phi r_n = 29.8 \text{ kips}$$

2. Determine a trial end plate thickness

ϕ = 0.9

F_{y1} = 36 ksi

p = $(8 - 2(3))/2 + 3/2 = 2.5$ in.

b' = $b - d_h/2 + t$

    = $1.5 - (¾ + ⅛)/2 + 0.233$

    = 1.30 in.

δ = $1 - d_h/p$

    = $1 - (¾ + ⅛)/2.5$

    = 0.65

$$K = \frac{4b'}{\phi F_{y1} p}$$

$$= \frac{4(1.30)}{0.9(36.0)(2.5)}$$

$$= 0.0642$$

$$t_{1min} = \sqrt{\frac{K r_{ut}}{1 + \delta}}$$

$$= \sqrt{\frac{0.0642(16.7)}{1 + 0.65}}$$

$$= 0.806 \text{ in.}$$

$$t_{1max} = \sqrt{K r_{ut}}$$

$$= \sqrt{0.0642(16.7)}$$

$$= 1.04 \text{ in.}$$

Try 1-in. end plate

3. Calculate α

$$\alpha = \left(\frac{K\phi r_n}{t_1^2} - 1\right)\left(\frac{a + (d_h/2)}{\delta(a + b + t)}\right)$$

$$= \left(\frac{0.0642(29.8)}{1.0^2} - 1\right)\left(\frac{1.5 + (0.875/2)}{0.65(1.5 + 1.5 + 0.233)}\right)$$

$$= 0.842$$

4. Calculate the factored resistance

$$\phi R_n = \frac{t_1^2(1 + \delta\alpha)n}{K}$$

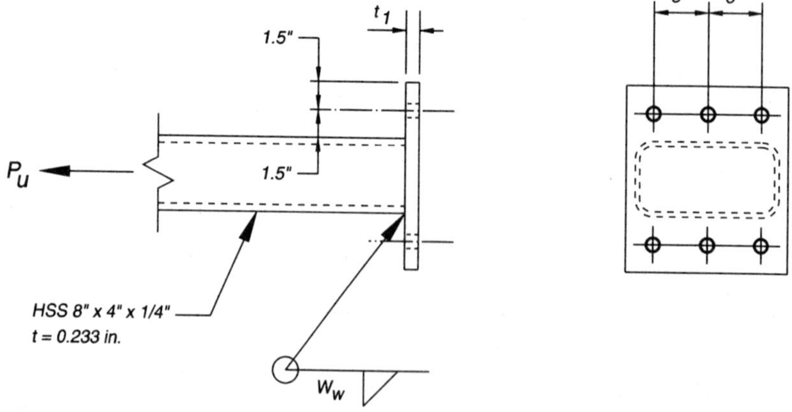

Figure Ex. 6.7.

$$= \frac{1.0^2(1 + 0.65(0.842))6}{0.0642}$$

$$= 145 \text{ kips}$$

$$\phi R_n = 145 \text{ kips} > P_u = 100 \text{ kips} \quad \textbf{o.k.}$$

Use 1-in. end plate

Use (6) ¾-in. diameter A325 bolts

5. Determine weld leg size required

$$\phi = 0.75$$
$$H = 8.0 \text{ in.}$$
$$F_w = 0.60F_{EXX}\ (1.0 + 0.5\sin^{1.5}\theta)$$
$$= 0.60(70)(1.5)$$
$$= 63.0 \text{ ksi}$$
$$W_w \geq \frac{P_u}{\phi F_w(0.707)2H}$$
$$\geq \frac{100}{0.75(63.0)(0.707)2(8.0)} = 0.187 \text{ in.}$$

From AISC LRFD Table J2.4 for $t = 1$ in.

$$W_{w\,min} = \tfrac{5}{16}\text{-in.} > 0.187 \text{ in.}$$

Use ⁵⁄₁₆-in. weld leg size

Results: Use ⁵⁄₁₆-in. weld with 1-in. end plate and (6) ¾-in. diameter A325 bolts.

Solution 2: Using the Tables

Using Table 6-7, select the weld size, end plate thickness, and number of bolts required to resist a factored load of 100 kips.
From Table 6-7, 50 percent Member Design Strength, $\phi R_n = 108$ kips > 100 kips   **o.k.**

Weld, W_w = ⁵⁄₁₆-in.
End Plate, t_1 = 1 in.
Bolts, n = (6) ¾-in. A325, (3) per side along the depth of HSS

END PLATES (AXIAL COMPRESSION)
End plates can be effectively used to transfer compressive loads from HSS to supports or to each other. See splice connections, Chapter 7. No special design requirements exist. Only the welds connecting the HSS to the end plate must be sized. Care must be taken to position the end plate squarely on the end of the HSS so that full contact between the end plate and the supporting element occurs.

FACE MOUNTED GUSSET PLATES
The design strength for face mounted gusset plates is determined from the limit states for the wall of the HSS and the gusset plate itself.
The limit states for the HSS and gusset plate are:

1. Shear strength of the HSS wall.
2. Yielding of the HSS wall.
3. Strength of the weld connecting the gusset to the HSS.
4. Strength based on buckling of the gusset plate (compression).
5. Strength based on bolting to the gusset plate.

Shear Strength of the HSS Wall

$$\phi R_n = \phi 0.6 F_y A_w$$

where

$\phi \;\; = 0.90$
$R_n = $ nominal shear strength, kips
$F_y \; = $ specified minimum yield stress of the HSS, ksi
$A_w = 2 L_w t$, in.^{2}
$L_w = $ length of weld to HSS, in.
$t \;\;\; = $ HSS wall thickness, in.

Yielding of the HSS Wall

For square and rectangular hollow sections:

$$\phi R_n = \phi \, \frac{F_y t^2}{(1 - t_1 / B)} \left(\frac{2N}{B} + 4\sqrt{1 - t_1 / B} \right) Q_f, \text{ kips}$$

where

$\phi \;\; = 1.0$
$B \; = $ width of hollow section face with gusset attached, in.
$t \;\; = $ HSS wall thickness, in.
$t_1 \; = $ thickness of gusset plate, in.
$N \; = $ length of gusset plate, in.
$F_y \; = $ specified minimum yield stress of the HSS, ksi
$Q_f = 1.0$ for tension in HSS
$\quad\;\; = 1 - 0.3(f / F_y) - 0.3(f / F_y)^2$ for compression in HSS

Note: in this equation the sign of f is positive (the magnitude of the compressive stress).

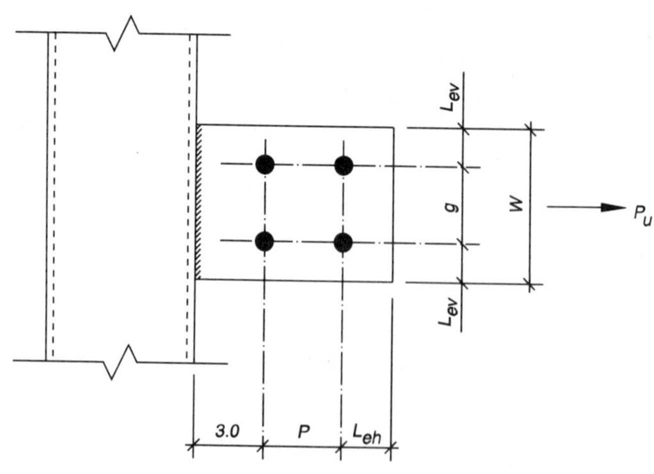

Fig. 6-14. Face mounted gusset.

For round hollow sections:

$$\phi R_n = \phi 5 F_y t^2 (1 + 0.25 N / D) Q_f, \text{ kips}$$

where

ϕ = 1.0
F_y = specified minimum yield stress of the HSS, ksi
t = HSS wall thickness, in.
N = length of gusset plate, in.
D = outer diameter of HSS, in.
Q_f = 1.0 for tension in HSS
    = $1 - 0.3(f / F_y) - 0.3(f / F_y)^2$ for compression in HSS

Note: in this equation the sign of f is positive (the magnitude of the compressive stress)

Strength of Weld Connecting the Gusset Plate to the HSS

$$\phi R_n = \phi F_w A_w$$

where

ϕ = 0.75
F_w = nominal weld strength, ksi
    = $0.60 F_{EXX} (1.0 + 0.50 \sin^{1.5}\theta)$
F_{EXX} = electrode classification number, i.e., minimum specified strength, ksi
θ = angle of loading measured from the weld longitudinal axis, degrees
    = 90°
A_w = effective area of weld throat, in.^{2}
    = $0.707(2)(L_w)W_w$
L_w = length of weld to HSS, in.
W_w = weld size (HSS to gusset plate), in.

Strength Based on Buckling of the Gusset Plate

The nominal compression strength of the gusset plate connection (see Fig. 6-6) is based on simple column buckling procedures. The procedure assumes that both ends of the connection are fixed and can sway laterally.

$$\text{For } \frac{P_u}{\phi_c P_n} \geq 0.2$$

$$\frac{P_u}{\phi_c P_n} + \frac{8}{9}\left(\frac{M_u}{\phi_b M_n}\right) \leq 1.0$$

$$\text{For } \frac{P_u}{\phi_c P_n} < 0.2$$

$$\frac{P_u}{2\phi_c P_n} + \frac{M_u}{\phi_b M_n} \leq 1.0$$

where

ϕ_c = 0.85
ϕ_b = 0.90
P_u = required compressive strength, kips

P_n = nominal compressive strength, kips

    = $A_g F_{cr}$

A_g = gross area of member, in.^{2}

F_{cr} = critical compressive stress, ksi

    = $(0.658^{\lambda_c^2})F_{y1}$, for $\lambda_c \leq 1.5$

    = $\left[\dfrac{0.877}{\lambda_c^2}\right] F_{y1}$, for $\lambda_c > 1.5$

λ_c = column slenderness parameter

$$= \frac{Kl}{r\pi}\sqrt{\frac{F_{y1}}{E}}$$

F_{y1} = specified minimum yield stress of the gusset plate, ksi

E = modulus of elasticity, ksi

K = effective length factor

    = 1.2

l = laterally unbraced length of plates, in.

r = governing radius of gyration, in.

    = $t_1 / \sqrt{12}$

M_u= required flexural strength, kip-in.

    = $P_u e / 2$

e = eccentricity of plate centerlines, in.

    = $(t_1 + t_2)/2$

M_n= nominal flexural strength of the gusset plate, kip-in.

    = $F_{y1} Z$

Z = plastic section modulus, in.^{3}

    = $W t_1^2 / 4$

W = width of gusset plate, in.

t_1 = gusset plate thickness, in.

t_2 = gusset thickness, in.

therefore

For $\dfrac{P_u}{\phi_c P_n} \geq 0.2$

$$P_u \leq \frac{1}{\dfrac{1}{\phi_c P_n} + \dfrac{4}{9}\dfrac{e}{\phi_b M_n}} = \phi R_n$$

For $\dfrac{P_u}{\phi_c P_n} < 0.2$

$$P_u \leq \frac{2}{\dfrac{1}{\phi_c P_n} + \dfrac{e}{\phi_b M_n}} = \phi R_n$$

Strength Based on Bolting to the Gusset Plate

Design Tensile Strength of the Gusset Plate:

(a) For yielding in the gross section:

$\phi_t = 0.90$

$R_n = F_{y1}A_g$, kips

(b) For rupture in the net section:

$\phi_t = 0.75$

$R_n = F_{u1}A_e$, kips

where

A_e = effective net area of the gusset plate, in.^{2}
 = minimum of A_n and $0.85A_g$
A_g = gross area of the gusset plate, in.^{2}
A_n = net area of gusset plate, in.^{2}
 = $[W - n_r(d_h + \frac{1}{16})]t_1$
F_{y1} = specified minimum yield stress of the gusset plate, ksi
F_{u1} = specified minimum tensile strength of the gusset plate, ksi
R_n = nominal axial strength, kips
W = width of gusset plate, in.
n_r = number of rows of bolts
d_h = diameter of bolt holes, in.
t_1 = gusset plate thickness, in.

Design Rupture Strength

(a) Shear Rupture Strength

The design strength for the limit state of rupture along a shear path in the gusset plate shall be taken as ϕR_n.

where

$\phi = 0.75$
$R_n = 0.6F_{u1}A_{nv}$, kips
F_{u1} = specified minimum tensile strength of the gusset plate, ksi
A_{nv} = net area subject to shear, in.^{2}
 = $2[L_{eh} + (n - 1)p - (n - 0.5)(d_h + \frac{1}{16})]t_1$
L_{eh} = edge distance, in.
n = number of lines of bolts
d_h = diameter of bolt holes, in.
t_1 = gusset plate thickness, in.
p = spacing of lines of bolts, in.

(b) Tension Rupture Strength

The design strength for the limit state of rupture along a tension path in the gusset plate shall be taken as ϕR_n

where

$\phi = 0.75$
$R_n = F_{u1}A_{nt}$, kips
F_{u1} = specified minimum tensile strength of the gusset plate, ksi
A_{nt} = net area subject to tension, in.^{2}
 = $[(n_r - 1)g - (n_r - 1)(d_h + \frac{1}{16})]t_1$
n_r = number of rows of bolts

g = spacing of rows of bolts, in.
d_h = diameter of bolt holes, in.
t_1 = gusset plate thickness, in.

(c) Block Shear Rupture Strength

Block shear is a limit state in which the resistance is determined by the sum of the shear strength on a failure path(s) and the tensile strength on a perpendicular segment. When ultimate rupture strength on the net section is used to determine the resistance on one segment, yielding on the gross section shall be used on the perpendicular segment. The block shear rupture design strength, ϕR_n, shall be determined as follows:

(1) When $F_{u1}A_{nt} \geq 0.6F_{u1}A_{nv}$:

$$\phi R_n = \phi[0.6F_{y1}A_{gv} + F_{u1}A_{nt}], \text{ kips}$$

(2) When $0.6F_{u1}A_{nv} > F_{u1}A_{nt}$:

$$\phi R_n = \phi[0.6F_{u1}A_{nv} + F_{y1}A_{gt}], \text{ kips}$$

where

ϕ = 0.75
A_{gv} = gross area subject to shear, in.^{2}
   = $2[L_{eh} + (n-1)p]t_1$
A_{gt} = gross area subject to tension, in.^{2}
   = $[(n_r - 1)g]t_1$
A_{nv} = net area subjected to shear, in.^{2}
A_{nt} = net area subjected to tension, in.^{2}
F_{y1} = specified minimum yield stress of the gusset plate, ksi
F_{u1} = specified minimum tensile strength of the gusset plate, ksi

Design Strength of the Bolts:

(a) Bolt Shear
See AISC Specification Section J3, and the AISC Manual Volume II, Part 8

(b) Bolt Bearing
See AISC Specification Section J3, and the AISC Manual Volume II, Part 8

EXAMPLE 6.8—Face Mounted Gusset Plate

Determine the design axial tensile and compressive strengths for the face mounted gusset plate shown.

Solution:

1. Shear Strength of the HSS Wall

$\phi R_n = \phi 0.6 F_y A_w$
ϕ = 0.90
F_y = 46.0 ksi
L_w = 7.0 in.
t = 0.291 in.
A_w = 2(7)(0.291)
   = 4.07 in.^{2}

$$\phi R_n = 0.9(0.6)(46.0)(4.07)$$
$$= 101 \text{ kips}$$

2. Yielding of the HSS Wall

$$\phi R_n = \phi \frac{F_y t^2}{(1 - t_1/B)} \left(\frac{2N}{B} + 4\sqrt{1 - t_1/B} \right) Q_f, \text{ kips}$$

F_y = 46.0 ksi
B = 7.0 in.
t = 0.291 in.
t_1 = 0.5 in.
N = 7.0 in.
Q_f = 1.0 (HSS in axial tension)

$$\phi R_n = \frac{46.0(0.291)^2}{(1 - 0.5/7.0)} \left[2\left(\frac{7.0}{7.0} \right) + 4\sqrt{1 - 0.5/7} \right] 1.0$$
$$= 24.6 \text{ kips}$$

3. Strength of the Weld Connecting the Gusset to the HSS

$\phi R_n = \phi F_w A_w$
ϕ = 0.75
F_{EXX} = 70 ksi
F_w = $0.6(70)(1.0 + 0.5 \sin^{1.5}(90°))$
  = 63.0 ksi
L_w = 7.0 in.
W_w = 0.25 in.
A_w = 0.707(2)(7.0)(0.25)
  = 2.47 in.^{2}
ϕR_n = 0.75(63.0)(2.47)
  = 117 kips

4. Strength Based on Buckling of the Gusset Plate

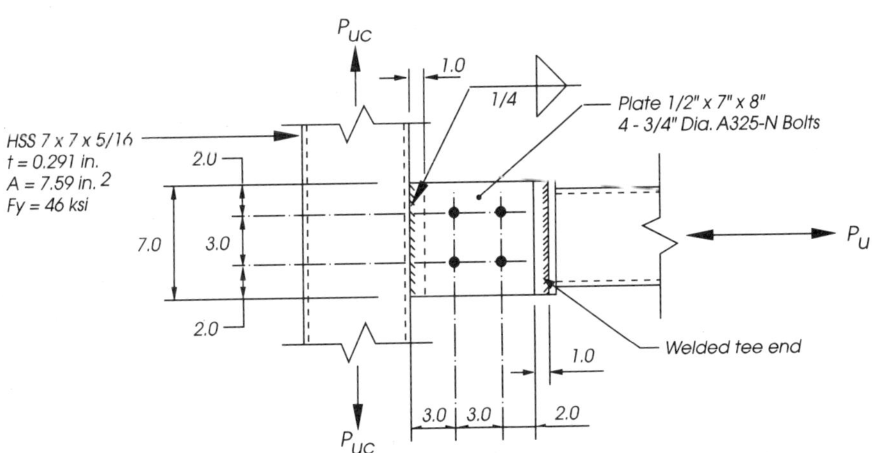

Figure Ex. 6.8.

$$\phi_c P_n = \phi_c F_{cr} A_g$$
$$\phi_c = 0.85$$
$$A_g = 0.5(7.0) = 3.50 \text{ in.}^2$$

$$\lambda_c = \frac{Kl}{r\pi}\sqrt{\frac{F_{y1}}{E}}$$

$$= \frac{1.2(9.0)\sqrt{12}}{0.50\pi}\sqrt{\frac{36.0}{29,000}}$$

$$= 0.839 < 1.5$$
$$F_{cr} = (0.658^{\lambda_c^2})F_{y1}$$
$$= (0.658^{(0.839)^2})(36.0)$$
$$= 26.8 \text{ ksi}$$
$$\phi_c P_n = 0.85(26.8)(3.50)$$
$$= 79.7 \text{ kips}$$
$$\phi_b M_n = \phi_b F_{y1} Z$$
$$\phi_b = 0.9$$
$$F_{y1} = 36.0 \text{ ksi}$$
$$Z = 7.0(0.5)^2/4$$
$$= 0.438 \text{ in.}^3$$
$$\phi_b M_n = 0.9(36.0)(0.438)$$
$$= 14.2 \text{ kip-in.}$$
$$e = t_1 = 0.5 \text{ in.}$$
$$P_u \leq \frac{1}{\dfrac{1}{\phi_c P_n} + \dfrac{4}{9}\dfrac{e}{\phi_b M_n}} = \phi R_n$$
$$= 35.5 \text{ kips} \ (> 0.2\phi_c P_n = 0.2(79.7) = 15.9 \text{ kips})$$

Therefore,

$$\phi R_n = 35.5 \text{ kips}$$

5. Strength Based on Bolting to the Gusset Plate

Tensile strength of the gusset plate

(a) Yielding in the gross section

$$R_n = F_{y1} A_g, \text{ kips}$$
$$\phi = 0.9$$
$$R_n = 36(0.5)(7.0) = 126 \text{ kips}$$
$$\phi R_n = 0.9(126)$$
$$= 113 \text{ kips}$$

(b) Rupture in the net section

$$R_n = F_{u1} A_e, \text{ kips}$$
$$\phi = 0.75$$
$$A_n = [7 - 2(\tfrac{3}{4} + \tfrac{1}{8})]0.5 = 2.63 \text{ in.}^2$$
$$0.85A_g = 0.85(0.5)(7.0) = 2.98 \text{ in.}^2$$
$$A_e = 2.63 \text{ in.}^2$$

$$R_n = 58(2.63) = 153 \text{ kips}$$
$$\phi R_n = 0.75(153)$$
$$= 115 \text{ kips}$$

(c) Block Shear Rupture Strength

$$\phi = 0.75$$
$$A_{nv} = 2[L_{eh} + (n-1)p - (n-0.5)(d_b + \tfrac{1}{8})]t_1$$
$$A_{nv} = 2[2 + (2-1)3 - (2-0.5)(\tfrac{3}{4} + \tfrac{1}{8})]0.5$$
$$= 3.69 \text{ in.}^2$$
$$A_{nt} = [(n_r - 1)g - (n_r - 1)(d_b + \tfrac{1}{8})]t_1$$
$$A_{nt} = [(2-1)3 - (2-1)(\tfrac{3}{4} + \tfrac{1}{8})]0.5$$
$$= 1.06 \text{ in.}^2$$

$$0.6F_{u1}A_{nv} > F_{u1}A_{nt}$$

$$128 \text{ kips} > 61.5 \text{ kips}$$

Therefore,

$$\phi R_n = 0.75[128 + 36(2-1)(3)(0.5)]$$
$$= 0.75(128 + 54.0)$$
$$= 137 \text{ kips}$$

Strength Based on Bolting

(a) Bolt Shear

$$\phi R_n = (\phi F_v A_b)n$$

From the AISC LRFD Manual Table 8-11

$$\phi R_n = (15.9)4$$
$$= 63.6 \text{ kips}$$

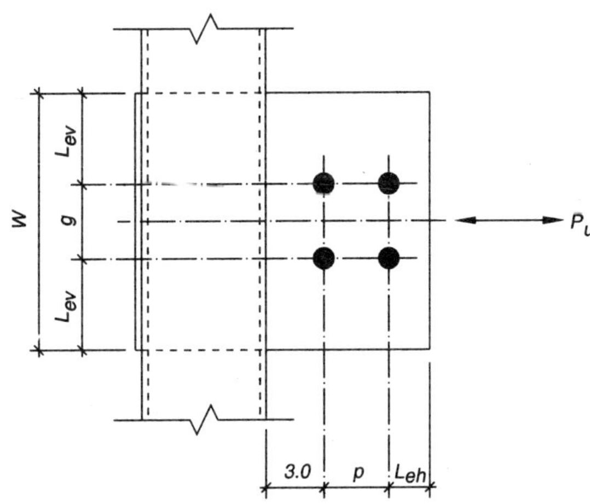

Fig. 6-15. Through plate connection.

(b) Bolt Bearing

$$R_n = (2.4 d_b t_1 F_u)n$$
$$\phi = 0.75$$
$$L_{eh} \geq 1.5 d_b = 1.5(0.75) = 1.12 \text{ in.}$$
$$s \geq 3.0 d_b = 3.0(0.75) = 2.25 \text{ in.}$$
$$R_n = 2.4(0.75)(0.5)(58)(4)$$
$$= 209 \text{ kips}$$
$$\phi R_n = 0.75(209)$$
$$= 157 \text{ kips}$$

Results: The connection strength in tension and compression is governed by yielding of the HSS wall.

$$\phi R_n = 24.6 \text{ kips}$$

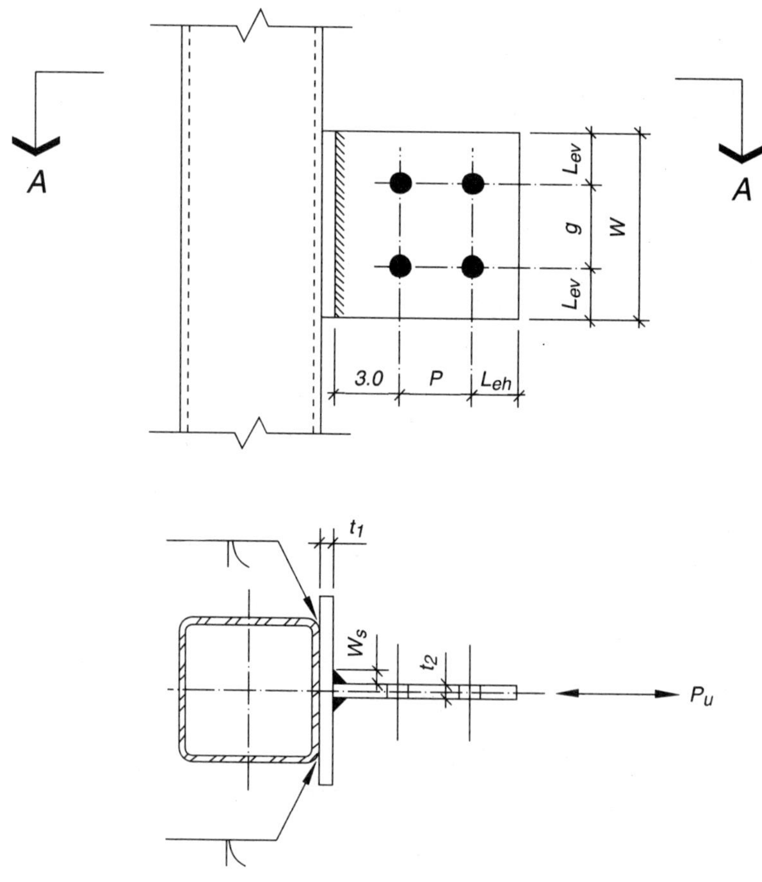

Fig. 6-16. Face mounted tee.

THROUGH PLATES

The limit states for an HSS with through plates are:

1. Shear strength of the HSS wall.
2. Yielding of the HSS wall.
3. Strength of weld connecting the through plate to the HSS.
4. Strength based on buckling of the through plate.
5. Strength based on bolting to the through plate.

The design strength for each of the limit states is the same as that for face mounted gusset plates (see page 6-47) with the exception that the design strengths of limit states 1, 2 and 3 are doubled.

FACE MOUNTED TEE

Face mounted tee connections, illustrated in Figure 6-16, are usually composed of the HSS and a structural tee welded to the wall of the HSS with the plate oriented along the axis of the HSS. In lieu of the structural tee a fabricated tee section is often used.

The design strength for a face mounted tee is determined from the limit states of the HSS and tee.

The limit states for the HSS and tee are:

1. Strength of the HSS side wall.
2. Strength of the weld connecting the HSS to the tee flange.
3. Yielding of the tee flange.
4. Shear yielding of the tee flange.
5. Strength of the weld connecting the tee flange to the tee stem.
6. Strength based on buckling of the tee stem.
7. Strength based on bolting to the tee stem.

Strength of the HSS Side Wall

(a) Local Yielding

$$\phi R_n = \phi F_y t\, (5k + N)$$

where

$\phi = 1.0$

k = outside corner radius of the HSS, in., which if not known may be taken as $1.5t$

t = HSS wall thickness, in.

N = length of the tee flange, in.

F_y = specified minimum yield stress of the HSS, ksi

(b) Web Crippling for Compressive Load Only

$$\phi R_n = \phi 1.6 t^2 \left[1 + 3\frac{N}{h} \right] \sqrt{EF_y}$$

where

$\phi = 0.75$

t = HSS wall thickness, in.

N = width of the tee flange, in.

h = flat width of the side wall of the HSS, in.

E = modulus of elasticity, ksi

F_y = minimum specified yield stress of the HSS, ksi

Strength of the Weld Connecting the HSS to the Face Plate

$$\phi R_n = \phi F_w A_w$$

where

ϕ $= 0.75$

F_w $=$ nominal weld strength, ksi

$\quad = 0.60 F_{EXX}$

$F_{EXX} =$ electrode classification number, i.e., minimum specified strength, ksi

A_w $=$ effective area of weld throat, in.^{2}

$\quad = {}^5\!/_{16} R (2W)$

R $=$ nominal outside corner radius, in.

$\quad = 2t$

t $=$ HSS wall thickness, in.

W $=$ width of the tee flange, in.

Yielding of the Tee Flange

$$\phi R_n = \phi F_{y1} W t_1^2 / B$$

where

ϕ $= 0.9$

F_{y1} $=$ specified minimum yield stress of the tee flange, ksi

W $=$ width of the tee flange, in.

t_1 $=$ tee flange thickness, in.

B $=$ width of the HSS wall face to which the tee flange is connected, in.

Shear Yielding of the Tee Flange

$$\phi R_n = \phi 0.6 F_{y1} W t_1$$

where

ϕ $= 0.75$

$F_{y1} =$ specified minimum yield stress of the tee flange, ksi

$W =$ width of the tee stem, in.

t_1 $=$ tee flange thickness, in.

Strength of the Weld Connecting the Tee Flange to the Tee Stem

$$\phi R_n = \phi F_w A_w$$

where

ϕ $= 0.75$

F_w $=$ nominal weld strength, ksi

$\quad = 0.60 F_{EXX} (1.0 + 0.5 \sin^{1.5}\theta)$

$F_{EXX} =$ electrode classification number, i.e., minimum specified strength, ksi

θ $=$ angle of loading measured from the weld longitudinal axis, degrees

$\quad = 90°$

A_w $=$ effective area of weld throat, in.^{2}

$\quad = 0.707 (2W_s) W$

W_s $=$ weld size (tee flange to tee stem), in.

W = width of the tee stem, in.

Strength Based on Buckling of the Tee Stem

The nominal compression strength of the tee stem/gusset connection (See Fig. 6-4) is based on simple column buckling procedures. The procedure assumes that both ends of the connection are fixed and can sway laterally.

For $\dfrac{P_u}{\phi_c P_n} \geq 0.2$

$$\frac{P_u}{\phi_c P_n} + \frac{8}{9}\left(\frac{M_u}{\phi_b M_n}\right) \leq 1.0$$

For $\dfrac{P_u}{\phi_c P_n} < 0.2$

$$\frac{P_u}{2\phi_c P_n} + \frac{M_u}{\phi_b M_n} \leq 1.0$$

where

$\phi_c = 0.85$

$\phi_b = 0.90$

P_u = required compressive strength, kips

P_n = nominal compressive strength, kips

    = $A_g F_{cr}$

A_g = gross area of member, in.^{2}

F_{cr} = critical compressive stress, ksi

    = $(0.658^{\lambda_c^2})F_{y2}$ for $\lambda_c \leq 1.5$

    = $\left[\dfrac{0.877}{\lambda_c^2}\right] F_{y2}$ for $\lambda_c > 1.5$

λ_c = column slenderness parameter

$$= \frac{Kl}{r\pi}\sqrt{\frac{F_{y2}}{E}}$$

F_{y2} = specified minimum yield stress of the tee stem, ksi

E = modulus of elasticity, ksi

K = effective length factor

    = 1.2

l = laterally unbraced length of plates, in.

r = governing radius of gyration, in.

    = $t_2 / \sqrt{12}$

M_u = required flexural strength, kip-in.

    = $P_u e / 2$

e = eccentricity of plate centerlines, in.

    = $(t_3 + t_2) / 2$

M_n = nominal flexural strength of the stem, kip-in.

    = $F_{y2} Z$

Z = plastic section modulus, in.^{3}

    = $W t_2^2 / 4$

t_2 = tee stem plate thickness, in.

t_3 = gusset plate thickness, in.
W = width of tee stem (same as cap plate), in.

Therefore

For $\dfrac{P_u}{\phi_c P_n} \geq 0.2$

$$P_u \leq \cfrac{1}{\dfrac{1}{\phi_c P_n} + \dfrac{4}{9}\dfrac{e}{\phi_b M_n}} = \phi R_n$$

For $\dfrac{P_u}{\phi_c P_n} < 0.2$

$$P_u \leq \cfrac{2}{\dfrac{1}{\phi_c P_n} + \dfrac{e}{\phi_b M_n}} = \phi R_n$$

Strength Based on Bolting to the Tee Stem

Design Tensile Strength of the Tee Stem:

(a) For yielding in the gross section:

$\phi_t = 0.90$
$R_n = F_{y2} A_g$, kips

(b) For rupture in the net section:

$\phi_t = 0.75$
$R_n = F_{u2} A_e$, kips

where

A_e = effective net area of the tee stem, in.^{2}
    = minimum of A_n and $0.85 A_g$
A_g = gross area of the tee stem, in.^{2}
A_n = net area of tee stem, in.^{2}
    = $[W - n_r(d_h + \frac{1}{16})]t_2$
F_{y2} = specified minimum yield stress of the tee stem, ksi
F_{u2} = specified minimum tensile strength of the tee stem, ksi
R_n = nominal axial strength, kips
W = width of tee stem (same as tee flange), in.
n_r = number of rows of bolts
d_h = diameter of bolt holes, in.
t_2= tee stem plate thickness, in.

Design Rupture Strength

(a) Shear Rupture Strength

The design strength for the limit state of rupture along a shear path in the tee stem shall be taken as ϕR_n

where

$\phi = 0.75$

$R_n = 0.6F_{u2}A_{nv}$, kips

F_{u2} = specified minimum tensile strength of the tee stem, ksi

A_{nv} = net area subject to shear, in.^{2}

$\quad = 2[L_{eh} + (n-1)p - (n-0.5)(d_h + \frac{1}{16})]t_2$

L_{eh} = edge distance, in.

$n\quad$ = number of lines of bolts

$p\quad$ = spacing of lines of bolts (pitch), in.

$d_h\quad$ = diameter of bolt holes, in.

$t_2\quad$ = tee stem plate thickness, in.

(b) Tension Rupture Strength

The design strength for the limit state of rupture along a tension path in the tee stem shall be taken as ϕR_n

where

$\phi\quad = 0.75$

$R_n\quad = F_{u2}A_{nt}$, kips

F_{u2} = specified minimum tensile strength of the tee stem, ksi

A_{nt} = net area subject to tension, in.^{2}

$\quad = [(n_r - 1)g - (n_r - 1)(d_h + \frac{1}{16})]t_2$

$n_r\quad$ = number of rows of bolts

$g\quad$ = spacing of rows of bolts, in.

$d_h\quad$ = diameter of bolt holes, in.

$t_2\quad$ = tee stem plate thickness, in.

(c) Block Shear Rupture Strength

Block shear is a limit state in which the resistance is determined by the sum of the shear strength on a failure path(s) and the tensile strength on a perpendicular segment. When ultimate rupture strength on the net section is used to determine the resistance on one segment, yielding on the gross section shall be used on the perpendicular segment. The block shear rupture design strength, ϕR_n, shall be determined as follows:

(1) When $F_{u2}A_{nt} \geq 0.6F_{u2}A_{nv}$:

$\quad \phi R_n = \phi[0.6F_{y2}A_{gv} + F_{u2}A_{nt}]$, kips

(2) When $0.6F_{u2}A_{nv} > F_{u2}A_{nt}$:

$\quad \phi R_n = \phi[0.6F_{u2}A_{nv} + F_{y2}A_{gt}]$, kips

where

$\phi\quad = 0.75$

A_{gv} = gross area subject to shear, in.^{2}

$\quad = 2[L_{eh} + (n-1)p]t_2$

A_{gt} = gross area subject to tension, in.^{2}

$\quad = [(n_r - 1)g]t_2$

A_{nv} = net area subjected to shear, in.^{2}

A_{nt} = net area subjected to tension, in.^{2}

F_{y2} = specified minimum yield stress of the tee stem, ksi

F_{u2} = specified minimum tensile strength of the tee stem, ksi

Design Strength of the Bolts:

(a) Bolt Shear
See AISC Specification Section J3, and the AISC Manual Volume II, Part 8

(b) Bolt Bearing
See AISC Specification Section J3, and the AISC Manual Volume II, Part 8

EXAMPLE 6.9—Face Mounted Tee

Determine the design tensile and compressive strengths for the face mounted tee shown.

Solution:

1. Strength of the HSS Side Wall

(a) Local Yielding

$$\phi R_n = \phi(5k + N)F_{yt}, \text{ kips}$$

$\phi \quad = 1.0$
$k \quad = 1.5(0.291) = 0.436 \text{ in.}$

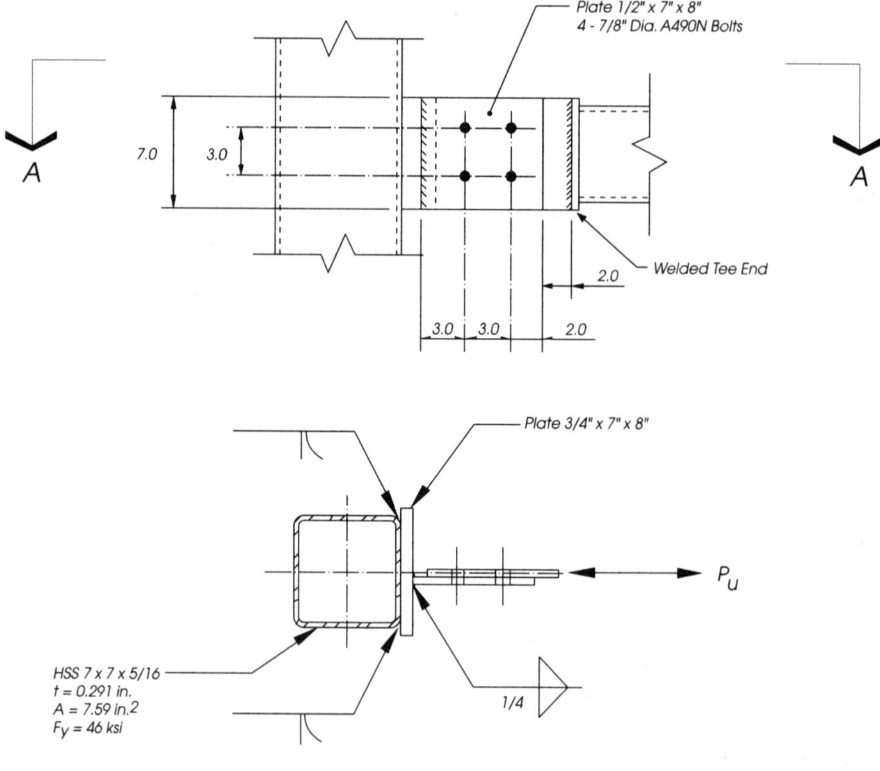

Section A-A

Figure Ex. 6-9.

N = 7.0 in.
F_y = 46.0 ksi
ϕR_n = 1.0(5(0.436) + 7.0)(46.0)(0.291)
    = 123 kips

(b) Web Crippling for Compressive Load Only

$$\phi R_n = \phi 1.6 t^2 \left[1 + 3\frac{N}{h} \right] \sqrt{EF_y}$$

ϕ = 0.75
t = 0.291 in.
N = 7.0 in.
h = $7.0 - 2(1.5)(0.291) = 6.13$ in.
E = 29,000 ksi
F_y = 46 ksi

$$\phi R_n = 0.75(1.6)(0.291)^2 \left[1 + 3\left(\frac{7.0}{6.13}\right) \right] \sqrt{29,000(46)}$$

    = 519 kips

2. Strength of the Weld Connecting the HSS to the Tee Flange

$\phi R_n = \phi F_w A_w$, kips
ϕ = 0.75
F_{EXX} = 70 ksi
F_w = 0.6(70)
    = 42.0 ksi
W = 7.0 in.
A_w = $\frac{5}{16}[2(0.291)][2(7.0)]$
    = 2.55 in.^{2}
ϕR_n = 0.75(42.0)(2.55)
    = 80.3 kips

3. Yielding of the Tee Flange

$\phi R_n = \phi F_{y1} W t_1^2 / B$, kips
ϕ = 0.9
F_{y1} = 36.0 ksi
W = 7.0 in.
t_1 = 0.75 in.
B = 7.0 in.
ϕR_n = $0.9(36.0)(7.0)(0.75)^2 / 7.0$
    = 18.2 kips

4. Shear Yielding of the Tee Flange

$\phi R_n = \phi 0.6 F_{y1} W t_1$, kips
ϕ = 0.75
F_{y1} = 36.0 ksi
W = 7.0 in.
t_1 = 0.75 in.
ϕR_n = 0.75(0.6)(36.0)(7.0)(0.75)
    = 85.0 kips

5. Strength of the Weld Connecting the Tee Flange to the Tee Stem

$$\phi R_n = \phi F_w A_w, \text{ kips}$$
$$\phi = 0.75$$
$$F_{EXX} = 70 \text{ ksi}$$
$$F_w = 0.6(70)(1.0 + 0.5 \sin^{1.5}(90°))$$
$$= 63.0 \text{ ksi}$$
$$W_s = 0.25 \text{ in.}$$
$$W = 7.0 \text{ in.}$$
$$A_w = 0.707(2(0.25))7.0$$
$$= 2.47 \text{ in.}^2$$
$$\phi R_n = 0.75(63.0)(2.47)$$
$$= 117 \text{ kips}$$

6. Strength Based on Buckling of the Tee Stem

$$\phi_c P_n = \phi_c F_{cr} A_g$$
$$\phi_c = 0.85$$
$$A_g = 0.5(7.0) = 3.5 \text{ in.}^2$$

$$\lambda_c = \frac{Kl}{r\pi}\sqrt{\frac{F_{y2}}{E}}$$

$$= \frac{1.2(10.0)\sqrt{12}}{0.5\pi}\sqrt{\frac{36}{29,000}}$$

$$= 0.932 < 1.5$$
$$F_{cr} = (0.658^{\lambda_c^2})F_{y2}$$
$$= (0.658^{(0.932)^2})36.0$$
$$= 25.0 \text{ ksi}$$
$$\phi_c P_n = 0.85(25.0)(3.5)$$
$$= 74.4 \text{ kips}$$
$$\phi_b M_n = \phi_b F_{y2} Z$$
$$\phi_b = 0.9$$
$$F_{y2} = 36.0 \text{ ksi}$$
$$Z = 7.0(0.5)^2 / 4$$
$$= 0.438 \text{ in.}^3$$
$$\phi_b M_n = 0.9(36.0)(0.438)$$
$$= 14.2 \text{ kip-in.}$$
$$e = t_2 = 0.5 \text{ in.}$$
$$P_u \leq \frac{1}{\dfrac{1}{\phi_c P_n} + \dfrac{4}{9}\dfrac{e}{\phi_b M_n}} = \phi R_n$$

$$= \frac{1}{\dfrac{1}{74.4} + \dfrac{4}{9}\dfrac{(0.5)}{14.2}}$$

$$= 34.4 \text{ kips } (> 0.2\phi_c P_n = 0.2(74.4) = 14.9 \text{ kips})$$

Therefore

$\phi R_n = 34.4$ kips

Strength Based on Bolting to the Tee Stem

Tensile strength of the tee stem

(a) Yielding in the gross section

$$R_n = F_{y2}A_g$$
$$\phi = 0.9$$
$$R_n = 36.0(0.5)(7.0) = 126 \text{ kips}$$
$$\phi R_n = 0.9(126)$$
$$= 113 \text{ kips}$$

(b) Rupture in the net section

$$R_n = F_{u2}A_e$$
$$\phi = 0.75$$
$$A_n = (7 - 2(\tfrac{7}{8} + \tfrac{1}{8}))0.5 = 2.50 \text{ in.}^2$$
$$0.85A_g = 0.85(0.5)(7.0) = 2.98 \text{ in.}^2$$
$$A_e = 2.50 \text{ in.}^2$$
$$R_n = 58(2.50)$$
$$= 145 \text{ kips}$$
$$\phi R_n = 0.75(145)$$
$$= 109 \text{ kips}$$

(c) Block Shear Rupture Strength

$$R_n = 0.6F_{u2}A_{nv}$$
$$\phi = 0.75$$
$$A_{nv} = 2[L_{eh} + (n - 1)p - (n - 0.5)(d_h + \tfrac{1}{16})]t_2$$
$$A_{nv} = 2(2 + (2 - 1)3 - (2 - 0.5)(\tfrac{5}{16} + \tfrac{1}{16}))0.5$$
$$= 3.5 \text{ in.}^2$$
$$A_{nt} = [(n_r - 1)g - (n_r - 1)(d_h + \tfrac{1}{16})]t_2$$
$$A_{nt} = ((2 - 1)3 - (2 - 1)(\tfrac{15}{16} + \tfrac{1}{16}))0.5$$
$$= 1.0 \text{ in.}^2$$

$$0.6F_{u2}A_{nv} > F_{u2}A_{nt}$$

$$122 \text{ kips} > 58 \text{ kips}$$

Therefore

$$\phi R_n = 0.75(122 + 36(2 - 1)(3)(0.5))$$
$$= 0.75(122 + 54)$$
$$= 132 \text{ kips}$$

Capacity Based on Bolting

(a) Bolt Shear

$$\phi R_n = \phi(F_v A_b)n$$

From the AISC LRFD Manual Table 8-11

$$\phi R_n = 4(27.1)$$
$$= 108 \text{ kips}$$

(b) Bolt Bearing

ϕ $\quad= 0.75$

L_{eh} $\geq 1.5 d_b = 1.5(0.875) = 1.31$ in.

s $\quad\geq 3.0 d_b = 3.0(0.875) = 2.62$ in.

R_n $\quad= 2.4(0.875)(0.5)(58)(4)$

$\quad\quad= 244$ kips

ϕR_n $= 0.75(244)$

$\quad\quad= 183$ kips

Results: The connection strength in tension and compression is governed by yielding of the tee flange

$\phi R_n = 18.2$ kips

WELDED TEE END TABLES

General Notes

The Welded Tee End Tables can be used for selecting the weld sizes, plate thicknesses, and the number of bolts required to develop the Connection Strength listed on the left side of the table. Table 6-1 is used for tension ties and Table 6-2 is used for compression struts. For tension-compression members, select the more conservative design from the two tables. The connection design strength is based on a percentage of the member design strength of the HSS. Tables are provided for 25 percent and 50 percent of the member design strength. For each HSS, the member design strength is computed using the specifications found in the AISC LRFD Manual, Second Edition. For the purposes of calculation, a maximum KL/r value of 60 was used to compute the design compressive strength of the HSS. The tables are presented for 2 rows of $\frac{3}{4}$-, $\frac{7}{8}$-, and 1-in. A325 bolts.

Material Specifications

The tables are presented for commonly used square, rectangular, and round HSS manufactured under ASTM A500 Grade B. All plates are ASTM A36 with $F_y = 36$ ksi and $F_u = 58$ ksi. The weld metal used has a minimum specified strength, $F_{EXX} = 70$ ksi. All bolts are ASTM A325 Specifications.

Design Procedure

All connection results are calculated following the limit state procedures outlined in Chapter 6, Welded Tee Ends. For connection strength calculations in Table 6-2, the length of the gusset plate is taken to be the same as the tee stem. The length of the tee stem is calculated by

$$L_e + (n/n_r - 1)p + L_e + 1$$

where

L_e $\;=$ edge distance

n_r $\;=$ number of rows of bolts

p $\;=$ bolt pitch

n $\;=$ number of $\frac{3}{4}$-in. bolts used in the connection

Also, the thickness of the gusset plate is assumed to be the same as the tee stem. For longer lengths or thinner gusset plates, the user is directed to AISC LRFD Manual Volume II, Part 11, Connections For Tension and Compression, to evaluate the strength of the gusset plate.

The value of n is determined by dividing the percent of member strength by ϕr_n, from AISC LRFD Table 8-11 for single shear, and rounding up to the next multiple of n_r.

The rounded value of n for 1 in. bolts is then used to calculate the minimum required thickness of the tee stem, t_2, based on bearing, yielding, rupture, and block shear rupture. For the struts, an iterative procedure based on buckling of the tee stem and gusset plate is also performed using the length as explained above. For thicknesses < 1 in. and > 1 in., t_2 is rounded up to the nearest $\frac{1}{8}$-in. and $\frac{1}{4}$-in., respectively.

The minimum required size of the fillet weld connecting the tee stem to the tee flange, W_s, is calculated and rounded up to the nearest $\frac{1}{16}$-in.

With the values of t_2 and W_s determined, the minimum required thickness of the tee flange, t_1, is calculated based on shear in the tee flange and yielding in the HSS wall. The value of t_1 is rounded up in the same manner as t_2.

The minimum required size of the weld connecting the tee flange to the HSS, W_w, is calculated and rounded upward to the nearest $\frac{1}{16}$-in. Both weld sizes, W_s and W_w, are checked for the minimum sizes shown in AISC LRFD Table J2.4 based on the computed plate thickness, and increased as required.

The values of W_w, t_1, W_s, t_2, and n are shown in the tables.

Design Parameters

The following values are held constant for calculations in Tables 6-1 and 6-2:

Plates: $F_y = 36$ ksi
$F_u = 58$ ksi

Welds: $F_{EXX} = 70$ ksi
$F_w = 63$ ksi

HSS: Square and Rectangular Round
$F_y = 46$ ksi $F_y = 42$ ksi
$F_u = 58$ ksi $F_u = 58$ ksi

Bolts: ASTM A325 $p = 3.0$ in
$g = 3.0$ in
$L_e = 1.5$ in

For different parameters, the user is directed to Welded Tee End Connections for evaluating the connection design strength.

25% of Member Design Strength								

Table 6-1.
Tension Ties
Welded Tee Ends—Square Sections

Designation			Connection Strength (kips)	Weld Size W_w (in.)	Tee Flange t_1 (in.)	Weld Size W_s (in.)	Tee Stem t_2 (in.)	No. of Bolts n		
H in.	B in.	t in.						$\frac{3}{4}$-in.	$\frac{7}{8}$-in.	1-in.
8	8	$\frac{5}{8}$	170	$\frac{5}{8}$	$\frac{5}{8}$	$\frac{5}{16}$	$\frac{5}{8}$	12	8	6
		$\frac{1}{2}$	140	$\frac{5}{8}$	$\frac{5}{8}$	$\frac{1}{4}$	$\frac{1}{2}$	10	8	6
		$\frac{3}{8}$	108	$\frac{1}{2}$	$\frac{5}{8}$	$\frac{1}{4}$	$\frac{1}{2}$	8	6	4
		$\frac{5}{16}$	90.7	$\frac{3}{8}$	$\frac{5}{8}$	$\frac{1}{4}$	$\frac{1}{2}$	6	6	4
		$\frac{1}{4}$	73.5	$\frac{5}{16}$	$\frac{5}{8}$	$\frac{1}{4}$	$\frac{1}{2}$	6	4	4
		$\frac{3}{16}$	55.6	$\frac{1}{4}$	$\frac{5}{8}$	$\frac{1}{4}$	$\frac{1}{2}$	4	4	2
7	7	$\frac{5}{8}$	145	$\frac{1}{2}$	$\frac{5}{8}$	$\frac{5}{16}$	$\frac{5}{8}$	10	8	6
		$\frac{1}{2}$	120	$\frac{9}{16}$	$\frac{1}{2}$	$\frac{1}{4}$	$\frac{1}{2}$	8	6	6
		$\frac{3}{8}$	92.8	$\frac{7}{16}$	$\frac{1}{2}$	$\frac{3}{16}$	$\frac{1}{2}$	6	6	4
		$\frac{5}{16}$	78.6	$\frac{3}{8}$	$\frac{1}{2}$	$\frac{3}{16}$	$\frac{3}{8}$	6	4	4
		$\frac{1}{4}$	63.9	$\frac{5}{16}$	$\frac{1}{2}$	$\frac{3}{16}$	$\frac{3}{8}$	6	4	4
		$\frac{3}{16}$	48.3	$\frac{1}{4}$	$\frac{1}{2}$	$\frac{3}{16}$	$\frac{3}{8}$	4	4	2
6	6	$\frac{5}{8}$	121	$\frac{7}{16}$	$\frac{5}{8}$	$\frac{5}{16}$	$\frac{5}{8}$	8	6	6
		$\frac{1}{2}$	101	$\frac{7}{16}$	$\frac{1}{2}$	$\frac{1}{4}$	$\frac{1}{2}$	8	6	4
		$\frac{3}{8}$	78.5	$\frac{7}{16}$	$\frac{1}{2}$	$\frac{3}{16}$	$\frac{1}{2}$	6	4	4
		$\frac{5}{16}$	66.6	$\frac{7}{16}$	$\frac{1}{2}$	$\frac{3}{16}$	$\frac{1}{2}$	6	4	4
		$\frac{1}{4}$	54.2	$\frac{1}{4}$	$\frac{1}{2}$	$\frac{3}{16}$	$\frac{1}{2}$	4	4	2
		$\frac{3}{16}$	41.2	$\frac{1}{4}$	$\frac{1}{2}$	$\frac{3}{16}$	$\frac{3}{8}$	4	2	2
		$\frac{1}{8}$	27.9	$\frac{3}{16}$	$\frac{1}{2}$	$\frac{3}{16}$	$\frac{3}{8}$	2	2	2
5	5	$\frac{1}{2}$	81.6	$\frac{3}{8}$	$\frac{1}{2}$	$\frac{1}{4}$	$\frac{1}{2}$	6	4	4
		$\frac{3}{8}$	64.0	$\frac{5}{16}$	$\frac{1}{2}$	$\frac{3}{16}$	$\frac{1}{2}$	6	4	4
		$\frac{5}{16}$	54.4	$\frac{5}{16}$	$\frac{1}{2}$	$\frac{3}{16}$	$\frac{1}{2}$	4	4	2
		$\frac{1}{4}$	44.5	$\frac{5}{16}$	$\frac{3}{8}$	$\frac{3}{16}$	$\frac{3}{8}$	4	4	2
		$\frac{3}{16}$	33.9	$\frac{1}{4}$	$\frac{3}{8}$	$\frac{3}{16}$	$\frac{3}{8}$	4	2	2
		$\frac{1}{8}$	23.1	$\frac{3}{16}$	$\frac{3}{8}$	$\frac{3}{16}$	$\frac{3}{8}$	2	2	2
$4\frac{1}{2}$	$4\frac{1}{2}$	$\frac{1}{2}$	71.9	$\frac{3}{8}$	$\frac{1}{2}$	$\frac{3}{16}$	$\frac{1}{2}$	6	4	4
		$\frac{3}{8}$	56.7	$\frac{3}{8}$	$\frac{3}{8}$	$\frac{3}{16}$	$\frac{3}{8}$	4	4	4
		$\frac{5}{16}$	48.4	$\frac{5}{16}$	$\frac{3}{8}$	$\frac{3}{16}$	$\frac{3}{8}$	4	4	2
		$\frac{1}{4}$	39.7	$\frac{1}{4}$	$\frac{3}{8}$	$\frac{3}{16}$	$\frac{3}{8}$	4	2	2
		$\frac{3}{16}$	30.3	$\frac{3}{16}$	$\frac{3}{8}$	$\frac{3}{16}$	$\frac{3}{8}$	2	2	2
		$\frac{1}{8}$	20.7	$\frac{3}{16}$	$\frac{3}{8}$	$\frac{3}{16}$	$\frac{3}{8}$	2	2	2

25% of Member Design Strength

Table 6-1.
Tension Ties
Welded Tee Ends—Square Sections

Designation			Con-nection Strength (kips)	Weld Size W_w (in.)	Tee Flange t_1 (in.)	Weld Size W_s (in.)	Tee Stem t_2 (in.)	No. of Bolts n		
H in.	B in.	t in.						$\frac{3}{4}$-in.	$\frac{7}{8}$-in.	1-in.
4	4	$\frac{1}{2}$	62.3	$\frac{5}{16}$	$\frac{1}{2}$	$\frac{3}{16}$	$\frac{1}{2}$	4	4	4
		$\frac{3}{8}$	49.5	$\frac{5}{16}$	$\frac{1}{2}$	$\frac{3}{16}$	$\frac{1}{2}$	4	4	2
		$\frac{5}{16}$	42.4	$\frac{5}{16}$	$\frac{3}{8}$	$\frac{3}{16}$	$\frac{3}{8}$	4	2	2
		$\frac{1}{4}$	34.9	$\frac{1}{4}$	$\frac{3}{8}$	$\frac{3}{16}$	$\frac{3}{8}$	4	2	2
		$\frac{3}{16}$	26.7	$\frac{3}{16}$	$\frac{3}{8}$	$\frac{3}{16}$	$\frac{3}{8}$	2	2	2
		$\frac{1}{8}$	18.3	$\frac{3}{16}$	$\frac{3}{8}$	$\frac{3}{16}$	$\frac{3}{8}$	2	2	2
$3\frac{1}{2}$	$3\frac{1}{2}$	$\frac{3}{8}$	42.3	$\frac{5}{16}$	$\frac{3}{8}$	$\frac{3}{16}$	$\frac{3}{8}$	4	2	2
		$\frac{5}{16}$	36.4	$\frac{1}{4}$	$\frac{3}{8}$	$\frac{3}{16}$	$\frac{3}{8}$	4	2	2
		$\frac{1}{4}$	30.1	$\frac{3}{16}$	$\frac{3}{8}$	$\frac{3}{16}$	$\frac{3}{8}$	2	2	2
		$\frac{3}{16}$	23.2	$\frac{3}{16}$	$\frac{3}{8}$	$\frac{3}{16}$	$\frac{3}{8}$	2	2	2
		$\frac{1}{8}$	15.9	$\frac{3}{16}$	$\frac{3}{8}$	$\frac{3}{16}$	$\frac{3}{8}$	2	2	2
3	3	$\frac{3}{8}$	35.1	$\frac{1}{4}$	$\frac{3}{8}$	$\frac{3}{16}$	$\frac{3}{8}$	4	2	2
		$\frac{5}{16}$	30.4	$\frac{3}{16}$	$\frac{3}{8}$	$\frac{3}{16}$	$\frac{3}{8}$	2	2	2
		$\frac{1}{4}$	25.3	$\frac{3}{16}$	$\frac{3}{8}$	$\frac{3}{16}$	$\frac{3}{8}$	2	2	2
		$\frac{3}{16}$	19.6	$\frac{3}{16}$	$\frac{3}{8}$	$\frac{3}{16}$	$\frac{3}{8}$	2	2	2
		$\frac{1}{8}$	13.5	$\frac{3}{16}$	$\frac{3}{8}$	$\frac{3}{16}$	$\frac{3}{8}$	2	2	2
$2\frac{1}{2}$	$2\frac{1}{2}$	$\frac{5}{16}$	24.3	$\frac{3}{16}$	$\frac{3}{8}$	$\frac{3}{16}$	$\frac{3}{8}$	2	2	2
		$\frac{1}{4}$	20.4	$\frac{3}{16}$	$\frac{3}{8}$	$\frac{3}{16}$	$\frac{3}{8}$	2	2	2
		$\frac{3}{16}$	15.9	$\frac{3}{16}$	$\frac{3}{8}$	$\frac{3}{16}$	$\frac{3}{8}$	2	2	2
		$\frac{1}{8}$	11.1	$\frac{3}{16}$	$\frac{3}{8}$	$\frac{3}{16}$	$\frac{3}{8}$	2	2	2
2	2	$\frac{1}{4}$	15.6	$\frac{3}{16}$	$\frac{3}{8}$	$\frac{3}{16}$	$\frac{3}{8}$	2	2	2
		$\frac{3}{16}$	12.3	$\frac{3}{16}$	$\frac{3}{8}$	$\frac{3}{16}$	$\frac{3}{8}$	2	2	2
		$\frac{1}{8}$	8.7	$\frac{3}{16}$	$\frac{3}{8}$	$\frac{3}{16}$	$\frac{3}{8}$	2	2	2

Table 6-1.
Tension Ties
Welded Tee Ends—Square Sections

Designation			Con-nection Strength (kips)	Weld Size W_w (in.)	Tee Flange t_1 (in.)	Weld Size W_s (in.)	Tee Stem t_2 (in.)	No. of Bolts n		
H in.	B in.	t in.						$\frac{3}{4}$-in.	$\frac{7}{8}$-in.	1-in.
8	8	$\frac{5}{8}$	339	$\frac{5}{8}$	$1\frac{1}{4}$	$\frac{5}{8}$	$1\frac{1}{4}$	22	16	12
		$\frac{1}{2}$	279	$\frac{5}{8}$	$1\frac{1}{4}$	$\frac{1}{2}$	1	18	14	10
		$\frac{3}{8}$	215	$\frac{7}{16}$	$1\frac{1}{4}$	$\frac{3}{8}$	$\frac{3}{4}$	14	10	8
		$\frac{5}{16}$	181	$\frac{3}{8}$	$1\frac{1}{4}$	$\frac{5}{16}$	$\frac{5}{8}$	12	10	8
		$\frac{1}{4}$	147	$\frac{5}{16}$	$1\frac{1}{4}$	$\frac{5}{16}$	$\frac{5}{8}$	10	8	6
		$\frac{3}{16}$	111	$\frac{5}{16}$	$1\frac{1}{4}$	$\frac{5}{16}$	$\frac{1}{2}$	8	6	4
7	7	$\frac{5}{8}$	290	$\frac{9}{16}$	$1\frac{1}{4}$	$\frac{9}{16}$	$1\frac{1}{4}$	20	14	12
		$\frac{1}{2}$	240	$\frac{9}{16}$	$1\frac{1}{4}$	$\frac{1}{2}$	1	16	12	10
		$\frac{3}{8}$	186	$\frac{1}{2}$	$1\frac{1}{4}$	$\frac{3}{8}$	$\frac{3}{4}$	12	10	8
		$\frac{5}{16}$	157	$\frac{3}{8}$	$1\frac{1}{4}$	$\frac{5}{16}$	$\frac{3}{4}$	10	8	6
		$\frac{1}{4}$	128	$\frac{5}{16}$	$1\frac{1}{4}$	$\frac{5}{16}$	$\frac{5}{8}$	10	6	6
		$\frac{3}{16}$	96.7	$\frac{5}{16}$	$1\frac{1}{4}$	$\frac{5}{16}$	$\frac{1}{2}$	8	6	4
6	6	$\frac{5}{8}$	242	$\frac{1}{2}$	$1\frac{1}{4}$	$\frac{9}{16}$	$1\frac{1}{4}$	16	12	10
		$\frac{1}{2}$	202	$\frac{1}{2}$	1	$\frac{7}{16}$	1	14	10	8
		$\frac{3}{8}$	157	$\frac{7}{16}$	1	$\frac{3}{8}$	$\frac{7}{8}$	10	8	6
		$\frac{5}{16}$	133	$\frac{7}{16}$	1	$\frac{5}{16}$	$\frac{3}{4}$	10	8	6
		$\frac{1}{4}$	108	$\frac{5}{16}$	1	$\frac{5}{16}$	$\frac{5}{8}$	8	6	4
		$\frac{3}{16}$	82.4	$\frac{5}{16}$	1	$\frac{5}{16}$	$\frac{1}{2}$	6	4	4
		$\frac{1}{8}$	55.9	$\frac{5}{16}$	1	$\frac{5}{16}$	$\frac{1}{2}$	4	4	2
5	5	$\frac{1}{2}$	163	$\frac{3}{8}$	1	$\frac{7}{16}$	1	12	8	6
		$\frac{3}{8}$	128	$\frac{3}{8}$	$\frac{7}{8}$	$\frac{3}{8}$	$\frac{7}{8}$	10	6	6
		$\frac{5}{16}$	109	$\frac{3}{8}$	$\frac{3}{4}$	$\frac{5}{16}$	$\frac{3}{4}$	8	6	4
		$\frac{1}{4}$	89.0	$\frac{3}{8}$	$\frac{3}{4}$	$\frac{1}{4}$	$\frac{5}{8}$	6	6	4
		$\frac{3}{16}$	67.9	$\frac{1}{4}$	$\frac{3}{4}$	$\frac{1}{4}$	$\frac{1}{2}$	6	4	4
		$\frac{1}{8}$	46.2	$\frac{1}{4}$	$\frac{3}{4}$	$\frac{1}{4}$	$\frac{3}{8}$	4	4	2
$4\frac{1}{2}$	$4\frac{1}{2}$	$\frac{1}{2}$	144	$\frac{3}{8}$	1	$\frac{3}{8}$	1	10	8	6
		$\frac{3}{8}$	113	$\frac{3}{8}$	$\frac{3}{4}$	$\frac{5}{16}$	$\frac{3}{4}$	8	6	6
		$\frac{5}{16}$	96.9	$\frac{3}{8}$	$\frac{3}{4}$	$\frac{1}{4}$	$\frac{5}{8}$	8	6	4
		$\frac{1}{4}$	79.5	$\frac{5}{16}$	$\frac{3}{4}$	$\frac{1}{4}$	$\frac{1}{2}$	6	4	4
		$\frac{3}{16}$	60.7	$\frac{1}{4}$	$\frac{3}{4}$	$\frac{1}{4}$	$\frac{3}{8}$	4	4	4
		$\frac{1}{8}$	41.4	$\frac{1}{4}$	$\frac{3}{4}$	$\frac{1}{4}$	$\frac{3}{8}$	4	2	2

Table 6-1.
Tension Ties
Welded Tee Ends—Square Sections

Designation			Con-nection Strength (kips)	Weld Size W_w (in.)	Tee Flange t_1 (in.)	Weld Size W_s (in.)	Tee Stem t_2 (in.)	No. of Bolts n		
H in.	B in.	t in.						$\frac{3}{4}$-in.	$\frac{7}{8}$-in.	1-in.
4	4	$\frac{1}{2}$	125	$\frac{3}{8}$	$\frac{7}{8}$	$\frac{5}{16}$	$\frac{7}{8}$	8	6	6
		$\frac{3}{8}$	98.9	$\frac{3}{8}$	$\frac{5}{8}$	$\frac{1}{4}$	$\frac{5}{8}$	8	6	4
		$\frac{5}{16}$	84.9	$\frac{5}{16}$	$\frac{5}{8}$	$\frac{1}{4}$	$\frac{5}{8}$	6	4	4
		$\frac{1}{4}$	69.8	$\frac{5}{16}$	$\frac{5}{8}$	$\frac{1}{4}$	$\frac{1}{2}$	6	4	4
		$\frac{3}{16}$	53.4	$\frac{1}{4}$	$\frac{5}{8}$	$\frac{1}{4}$	$\frac{1}{2}$	4	4	2
		$\frac{1}{8}$	36.6	$\frac{1}{4}$	$\frac{5}{8}$	$\frac{1}{4}$	$\frac{3}{8}$	4	2	2
$3\frac{1}{2}$	$3\frac{1}{2}$	$\frac{3}{8}$	84.7	$\frac{3}{8}$	$\frac{5}{8}$	$\frac{1}{4}$	$\frac{5}{8}$	6	4	4
		$\frac{5}{16}$	72.9	$\frac{3}{8}$	$\frac{1}{2}$	$\frac{3}{16}$	$\frac{1}{2}$	6	4	4
		$\frac{1}{4}$	60.2	$\frac{5}{16}$	$\frac{1}{2}$	$\frac{3}{16}$	$\frac{3}{8}$	4	4	4
		$\frac{3}{16}$	46.4	$\frac{1}{4}$	$\frac{1}{2}$	$\frac{3}{16}$	$\frac{3}{8}$	4	4	2
		$\frac{1}{8}$	31.9	$\frac{3}{16}$	$\frac{1}{2}$	$\frac{3}{16}$	$\frac{3}{8}$	4	2	2
3	3	$\frac{3}{8}$	70.2	$\frac{3}{8}$	$\frac{1}{2}$	$\frac{3}{16}$	$\frac{1}{2}$	6	4	4
		$\frac{5}{16}$	60.9	$\frac{3}{8}$	$\frac{1}{2}$	$\frac{3}{16}$	$\frac{1}{2}$	4	4	4
		$\frac{1}{4}$	50.5	$\frac{1}{4}$	$\frac{1}{2}$	$\frac{3}{16}$	$\frac{1}{2}$	4	4	2
		$\frac{3}{16}$	39.1	$\frac{1}{4}$	$\frac{1}{2}$	$\frac{3}{16}$	$\frac{3}{8}$	4	2	2
		$\frac{1}{8}$	26.9	$\frac{3}{16}$	$\frac{1}{2}$	$\frac{3}{16}$	$\frac{3}{8}$	2	2	2
$2\frac{1}{2}$	$2\frac{1}{2}$	$\frac{5}{16}$	48.6	$\frac{5}{16}$	$\frac{3}{8}$	$\frac{3}{16}$	$\frac{3}{8}$	4	4	2
		$\frac{1}{4}$	40.8	$\frac{1}{4}$	$\frac{3}{8}$	$\frac{3}{16}$	$\frac{3}{8}$	4	2	2
		$\frac{3}{16}$	31.9	$\frac{1}{4}$	$\frac{3}{8}$	$\frac{3}{16}$	$\frac{3}{8}$	4	2	2
		$\frac{1}{8}$	22.1	$\frac{3}{16}$	$\frac{3}{8}$	$\frac{3}{16}$	$\frac{3}{8}$	2	2	2
2	2	$\frac{1}{4}$	31.3	$\frac{3}{16}$	$\frac{3}{8}$	$\frac{3}{16}$	$\frac{3}{8}$	2	2	2
		$\frac{3}{16}$	24.6	$\frac{3}{16}$	$\frac{3}{8}$	$\frac{3}{16}$	$\frac{3}{8}$	2	2	2
		$\frac{1}{8}$	17.4	$\frac{3}{16}$	$\frac{3}{8}$	$\frac{3}{16}$	$\frac{3}{8}$	2	2	2

| 25% of Member Design Strength | | | | | | | | | |

Table 6-1.
Tension Ties
Welded Tee Ends—Rectangular Sections

Designation			Connection Strength (kips)	Weld Size W_w (in.)	Tee Flange t_1 (in.)	Weld Size W_s (in.)	Tee Stem t_2 (in.)	No. of Bolts n		
H in.	B in.	t in.						¾-in.	⅞-in.	1-in.
8	6	⅝	145	⅝	½	¼	½	10	8	6
		½	120	9/16	½	¼	½	8	6	6
		⅜	92.8	7/16	½	3/16	½	6	6	4
		5/16	78.6	⅜	½	3/16	⅜	6	4	4
		¼	63.9	5/16	½	3/16	⅜	6	4	4
		3/16	48.3	¼	½	3/16	⅜	4	4	2
8	4	⅝	121	9/16	½	¼	½	8	6	6
		½	101	½	½	3/16	½	8	6	4
		⅜	78.5	½	½	3/16	½	6	4	4
		5/16	66.6	7/16	½	3/16	½	6	4	4
		¼	54.2	¼	½	3/16	½	4	4	2
		3/16	41.2	¼	½	3/16	⅜	4	2	2
		⅛	27.9	3/16	½	3/16	⅜	2	2	2
8	3	½	91.2	7/16	½	3/16	½	6	6	4
		⅜	71.2	7/16	½	3/16	½	6	4	4
		5/16	60.5	⅜	½	3/16	½	4	4	4
		¼	49.4	¼	½	3/16	½	4	4	2
		3/16	37.6	¼	½	3/16	⅜	4	2	2
		⅛	25.5	3/16	½	3/16	⅜	2	2	2
8	2	⅜	64.0	7/16	½	3/16	½	6	4	4
		5/16	54.4	5/16	½	3/16	½	4	4	2
		¼	44.5	5/16	⅜	3/16	⅜	4	4	2
		3/16	33.9	¼	⅜	3/16	⅜	4	2	2
		⅛	23.1	3/16	⅜	3/16	⅜	2	2	2
6	4	½	81.6	⅜	½	3/16	½	6	4	4
		⅜	64.0	⅜	½	3/16	½	6	4	4
		5/16	54.4	5/16	½	3/16	½	4	4	2
		¼	44.5	5/16	⅜	3/16	⅜	4	4	2
		3/16	33.9	¼	⅜	3/16	⅜	4	2	2
		⅛	23.1	3/16	⅜	3/16	⅜	2	2	2

<table>
<tr><td colspan="12">25% of Member Design Strength</td></tr>
<tr><td colspan="12" align="center">Table 6-1.
Tension Ties
Welded Tee Ends—Rectangular Sections</td></tr>
</table>

Designation			Con-nection Strength (kips)	Weld Size W_w (in.)	Tee Flange t_1 (in.)	Weld Size W_s (in.)	Tee Stem t_2 (in.)	No. of Bolts n		
H in.	B in.	t in.						¾-in.	⅞-in.	1-in.
6	3	½	71.9	7/16	3/8	3/16	3/8	6	4	4
		3/8	56.7	3/8	3/8	3/16	3/8	4	4	4
		5/16	48.4	5/16	3/8	3/16	3/8	4	4	2
		¼	39.7	¼	3/8	3/16	3/8	4	2	2
		3/16	30.3	3/16	3/8	3/16	3/8	2	2	2
		⅛	20.7	3/16	3/8	3/16	3/8	2	2	2
6	2	3/8	49.5	5/16	½	3/16	½	4	4	2
		5/16	42.4	5/16	3/8	3/16	3/8	4	2	2
		¼	34.9	¼	3/8	3/16	3/8	4	2	2
		3/16	26.7	3/16	3/8	3/16	3/8	2	2	2
		⅛	18.3	3/16	3/8	3/16	3/8	2	2	2
5	3	½	62.3	5/16	½	3/16	½	4	4	4
		3/8	49.5	5/16	½	3/16	½	4	4	2
		5/16	42.4	5/16	3/8	3/16	3/8	4	2	2
		¼	34.9	¼	3/8	3/16	3/8	4	2	2
		3/16	26.7	3/16	3/8	3/16	3/8	2	2	2
		⅛	18.3	3/16	3/8	3/16	3/8	2	2	2
5	2	3/8	42.3	5/16	3/8	3/16	3/8	4	2	2
		5/16	36.4	¼	3/8	3/16	3/8	4	2	2
		¼	30.1	3/16	3/8	3/16	3/8	2	2	2
		3/16	23.2	3/16	3/8	3/16	3/8	2	2	2
		⅛	15.9	3/16	3/8	3/16	3/8	2	2	2
4	3	3/8	42.3	5/16	3/8	3/16	3/8	4	2	2
		5/16	36.4	¼	3/8	3/16	3/8	4	2	2
		¼	30.1	3/16	3/8	3/16	3/8	2	2	2
		3/16	23.2	3/16	3/8	3/16	3/8	2	2	2
		⅛	15.9	3/16	3/8	3/16	3/8	2	2	2
4	2	3/8	35.1	¼	3/8	3/16	3/8	4	2	2
		5/16	30.4	3/16	3/8	3/16	3/8	2	2	2
		¼	25.3	3/16	3/8	3/16	3/8	2	2	2
		3/16	19.6	3/16	3/8	3/16	3/8	2	2	2
		⅛	13.5	3/16	3/8	3/16	3/8	2	2	2

Table 6-1.
Tension Ties
Welded Tee Ends—Rectangular Sections

Designation			Connection Strength (kips)	Weld Size W_w (in.)	Tee Flange t_1 (in.)	Weld Size W_s (in.)	Tee Stem t_2 (in.)	No. of Bolts n		
H in.	B in.	t in.						¾-in.	⅞-in.	1-in.
8	6	5/8	290	5/8	1¼	1/2	1	20	14	12
		1/2	240	5/8	1¼	7/16	7/8	16	12	10
		3/8	186	7/16	1¼	5/16	3/4	12	10	8
		5/16	157	7/16	1¼	5/16	5/8	10	8	6
		1/4	128	3/8	1¼	5/16	1/2	10	6	6
		3/16	96.7	5/16	1¼	5/16	1/2	8	6	4
8	4	5/8	242	5/8	1	7/16	7/8	16	12	10
		1/2	202	5/8	1	3/8	3/4	14	10	8
		3/8	157	1/2	1	5/16	5/8	10	8	6
		5/16	133	3/8	1	5/16	1/2	10	8	6
		1/4	108	5/16	1	5/16	1/2	8	6	4
		3/16	82.4	5/16	1	5/16	3/8	6	4	4
		1/8	55.9	5/16	1	5/16	1/2	4	4	2
8	3	1/2	182	9/16	7/8	5/16	3/4	12	10	8
		3/8	142	1/2	7/8	5/16	1/2	10	8	6
		5/16	121	7/16	7/8	5/16	1/2	8	6	6
		1/4	98.7	3/8	7/8	5/16	1/2	8	6	4
		3/16	75.1	5/16	7/8	5/16	1/2	6	4	4
		1/8	50.9	5/16	7/8	5/16	1/2	4	4	2
8	2	3/8	128	1/2	3/4	1/4	1/2	10	6	6
		5/16	109	3/8	3/4	1/4	1/2	8	6	4
		1/4	89.0	5/16	3/4	1/4	1/2	6	6	4
		3/16	67.9	1/4	3/4	1/4	3/8	6	4	4
		1/8	46.2	1/4	3/4	1/4	3/8	4	4	2
6	4	1/2	163	1/2	7/8	3/8	7/8	12	8	6
		3/8	128	1/2	3/4	5/16	5/8	10	6	6
		5/16	109	7/16	3/4	1/4	5/8	8	6	4
		1/4	89.0	5/16	3/4	1/4	1/2	6	6	4
		3/16	67.9	1/4	3/4	1/4	3/8	6	4	4
		1/8	46.2	1/4	3/4	1/4	3/8	4	4	2

Table 6-1.
Tension Ties
Welded Tee Ends—Rectangular Sections

Designation			Con-nection Strength (kips)	Weld Size W_w (in.)	Tee Flange t_1 (in.)	Weld Size W_s (in.)	Tee Stem t_2 (in.)	No. of Bolts n		
H in.	B in.	t in.						¾-in.	⅞-in.	1-in.
6	3	½	144	⁷⁄₁₆	¾	⁵⁄₁₆	¾	10	8	6
		⅜	113	⁷⁄₁₆	¾	¼	⅝	8	6	6
		⁵⁄₁₆	96.9	⅜	¾	¼	½	8	6	4
		¼	79.5	⁵⁄₁₆	¾	¼	½	6	4	4
		³⁄₁₆	60.7	¼	¾	¼	⅜	4	4	4
		⅛	41.4	¼	¾	¼	⅜	4	2	2
6	2	⅜	98.9	⁷⁄₁₆	⅝	¼	½	8	6	4
		⁵⁄₁₆	84.9	⁷⁄₁₆	⅝	¼	½	6	4	4
		¼	69.8	⁵⁄₁₆	⅝	¼	½	6	4	4
		³⁄₁₆	53.4	¼	⅝	¼	½	4	4	2
		⅛	36.6	¼	⅝	¼	⅜	4	2	2
5	3	½	125	⅜	⅞	⁵⁄₁₆	⅞	8	6	6
		⅜	98.9	⅜	⅝	¼	⅝	8	6	4
		⁵⁄₁₆	84.9	⁵⁄₁₆	⅝	¼	⅝	6	4	4
		¼	69.8	⁵⁄₁₆	⅝	¼	½	6	4	4
		³⁄₁₆	53.4	¼	⅝	¼	½	4	4	2
		⅛	36.6	¼	⅝	¼	⅜	4	2	2
5	2	⅜	84.7	⅜	⅝	¼	⅝	6	4	4
		⁵⁄₁₆	72.9	⅜	½	³⁄₁₆	½	6	4	4
		¼	60.2	⁵⁄₁₆	½	³⁄₁₆	⅜	4	4	4
		³⁄₁₆	46.4	¼	½	³⁄₁₆	⅜	4	4	2
		⅛	31.9	³⁄₁₆	½	³⁄₁₆	⅜	4	2	2
4	3	⅜	84.7	⅜	⅝	¼	⅝	6	4	4
		⁵⁄₁₆	72.9	⅜	½	³⁄₁₆	½	6	4	4
		¼	60.2	⁵⁄₁₆	½	³⁄₁₆	⅜	4	4	4
		³⁄₁₆	46.4	¼	½	³⁄₁₆	⅜	4	4	2
		⅛	31.9	³⁄₁₆	½	³⁄₁₆	⅜	4	2	2
4	2	⅜	70.2	⅜	½	³⁄₁₆	½	6	4	4
		⁵⁄₁₆	60.9	⅜	½	³⁄₁₆	½	4	4	4
		¼	50.5	¼	½	³⁄₁₆	½	4	4	2
		³⁄₁₆	39.1	¼	½	³⁄₁₆	⅜	4	2	2
		⅛	26.9	³⁄₁₆	½	³⁄₁₆	⅜	2	2	2

25% of Member Design Strength		

Table 6-1.
Tension Ties
Welded Tee Ends—Round Sections

Designation		Con-nection Strength (kips)	Weld Size W_w (in.)	Tee Flange t_1 (in.)	Weld Size W_s (in.)	Tee Stem t_2 (in.)	No. of Bolts n		
D in.	t in.						¾-in.	⅞-in.	1-in.
8.625	0.500	112	½	½	³⁄₁₆	½	8	6	4
	0.375	85.7	⁷⁄₁₆	½	³⁄₁₆	½	6	4	4
	0.322	73.7	⅜	½	³⁄₁₆	⅜	6	4	4
	0.250	58.0	⁵⁄₁₆	½	³⁄₁₆	⅜	4	4	4
	0.188	43.7	¼	½	³⁄₁₆	⅜	4	4	2
7.500	0.500	97.3	⁷⁄₁₆	½	³⁄₁₆	½	8	6	4
	0.375	74.1	⁷⁄₁₆	½	³⁄₁₆	⅜	6	4	4
	0.312	62.3	⁵⁄₁₆	½	³⁄₁₆	⅜	4	4	4
	0.250	50.3	¼	½	³⁄₁₆	½	4	4	2
	0.188	37.8	³⁄₁₆	½	³⁄₁₆	⅜	4	2	2
7.000	0.500	90.2	⁷⁄₁₆	½	³⁄₁₆	½	6	6	4
	0.375	68.9	⁷⁄₁₆	½	³⁄₁₆	⅜	6	4	4
	0.312	57.9	⅜	½	³⁄₁₆	⅜	4	4	4
	0.250	46.8	⁵⁄₁₆	½	³⁄₁₆	⅜	4	4	2
	0.188	35.2	¼	½	³⁄₁₆	⅜	4	2	2
	0.125	23.7	³⁄₁₆	½	³⁄₁₆	⅜	2	2	2
6.875	0.500	88.5	⁷⁄₁₆	½	³⁄₁₆	½	6	6	4
	0.375	67.7	⁷⁄₁₆	⅜	³⁄₁₆	⅜	6	4	4
	0.312	56.9	⅜	⅜	³⁄₁₆	⅜	4	4	4
	0.250	45.9	⁵⁄₁₆	⅜	³⁄₁₆	⅜	4	4	2
	0.188	34.6	¼	⅜	³⁄₁₆	⅜	4	2	2
6..625	0.500	85.1	½	½	³⁄₁₆	½	6	4	4
	0.432	74.5	⁷⁄₁₆	½	³⁄₁₆	½	6	4	4
	0.375	65.0	⅜	½	³⁄₁₆	½	6	4	4
	0.312	54.7	⁵⁄₁₆	½	³⁄₁₆	½	4	4	2
	0.280	49.3	⁵⁄₁₆	½	³⁄₁₆	½	4	4	2
	0.250	44.2	⁵⁄₁₆	⅜	³⁄₁₆	⅜	4	4	2
	0.188	33.4	¼	⅜	³⁄₁₆	⅜	4	2	2
	0.125	22.4	³⁄₁₆	⅜	³⁄₁₆	⅜	2	2	2
6.125	0.500	78.2	½	½	³⁄₁₆	½	6	4	4
	0.375	59.8	⅜	½	³⁄₁₆	½	4	4	4
	0.312	50.4	¼	½	³⁄₁₆	½	4	4	2
	0.250	40.7	¼	⅜	³⁄₁₆	⅜	4	2	2
	0.188	30.7	³⁄₁₆	⅜	³⁄₁₆	⅜	2	2	2

25% of Member Design Strength								

Table 6-1.
Tension Ties
Welded Tee Ends—Round Sections

Designation		Con-nection Strength (kips)	Weld Size W_w (in.)	Tee Flange t_1 (in.)	Weld Size W_s (in.)	Tee Stem t_2 (in.)	No. of Bolts n		
D in.	t in.						¾-in.	⅞-in.	1-in.
6.000	0.500	76.5	7/16	1/2	3/16	1/2	6	4	4
	0.375	58.6	3/8	1/2	3/16	1/2	4	4	4
	0.312	49.3	5/16	1/2	3/16	1/2	4	4	2
	0.280	44.5	5/16	3/8	3/16	3/8	4	4	2
	0.250	39.9	1/4	3/8	3/16	3/8	4	2	2
	0.188	30.1	3/16	3/8	3/16	3/8	2	2	2
	0.125	20.2	3/16	3/8	3/16	3/8	2	2	2
5.563	0.375	54.1	1/4	1/2	3/16	1/2	4	4	2
	0.258	38.1	1/4	3/8	3/16	3/8	4	2	2
	0.188	27.9	3/16	3/8	3/16	3/8	2	2	2
	0.134	20.2	3/16	3/8	3/16	3/8	2	2	2
5.500	0.500	69.6	5/16	1/2	3/16	1/2	6	4	4
	0.375	53.4	1/4	1/2	3/16	1/2	4	4	2
	0.258	37.6	1/4	3/8	3/16	3/8	4	2	2
5.000	0.500	62.6	5/16	1/2	3/16	1/2	4	4	4
	0.375	48.2	5/16	3/8	3/16	3/8	4	4	2
	0.312	40.6	1/4	3/8	3/16	3/8	4	2	2
	0.258	34.0	1/4	3/8	3/16	3/8	4	2	2
	0.250	33.0	1/4	3/8	3/16	3/8	4	2	2
	0.188	24.9	3/16	3/8	3/16	3/8	2	2	2
	0.125	16.8	3/16	3/8	3/16	3/8	2	2	2
4.500	0.337	39.1	1/4	3/8	3/16	3/8	4	2	2
	0.237	28.1	3/16	3/8	3/16	3/8	2	2	2
	0.188	22.3	3/16	3/8	3/16	3/8	2	2	2
	0.125	15.1	3/16	3/8	3/16	3/8	2	2	2
4.000	0.337	34.5	1/4	3/8	3/16	3/8	4	2	2
	0.313	32.0	1/4	3/8	3/16	3/8	4	2	2
	0.250	26.1	3/16	3/8	3/16	3/8	2	2	2
	0.237	24.9	3/16	3/8	3/16	3/8	2	2	2
	0.226	23.7	3/16	3/8	3/16	3/8	2	2	2
	0.220	23.1	3/16	3/8	3/16	3/8	2	2	2
	0.188	19.8	3/16	3/8	3/16	3/8	2	2	2
	0.125	13.4	3/16	3/8	3/16	3/8	2	2	2

50% of Member Design Strength

Table 6-1.
Tension Ties
Welded Tee Ends—Round Sections

Designation		Con-nection Strength (kips)	Weld Size W_w (in.)	Tee Flange t_1 (in.)	Weld Size W_s (in.)	Tee Stem t_2 (in.)	No. of Bolts n		
D in.	t in.						$\frac{3}{4}$-in.	$\frac{7}{8}$-in.	1-in.
8.625	0.500	225	$\frac{5}{8}$	$1\frac{1}{4}$	$\frac{3}{8}$	$\frac{3}{4}$	16	12	8
	0.375	171	$\frac{7}{16}$	$1\frac{1}{4}$	$\frac{5}{16}$	$\frac{5}{8}$	12	8	8
	0.322	147	$\frac{3}{8}$	$1\frac{1}{4}$	$\frac{5}{16}$	$\frac{1}{2}$	10	8	6
	0.250	116	$\frac{5}{16}$	$1\frac{1}{4}$	$\frac{5}{16}$	$\frac{1}{2}$	8	6	6
	0.188	87.3	$\frac{5}{16}$	$1\frac{1}{4}$	$\frac{5}{16}$	$\frac{1}{2}$	6	6	4
7.500	0.500	195	$\frac{9}{16}$	1	$\frac{3}{8}$	$\frac{3}{4}$	14	10	8
	0.375	148	$\frac{7}{16}$	1	$\frac{5}{16}$	$\frac{5}{8}$	10	8	6
	0.312	125	$\frac{3}{8}$	1	$\frac{5}{16}$	$\frac{1}{2}$	8	6	6
	0.250	101	$\frac{5}{16}$	1	$\frac{5}{16}$	$\frac{1}{2}$	8	6	4
	0.188	75.6	$\frac{5}{16}$	1	$\frac{5}{16}$	$\frac{3}{8}$	6	4	4
7.000	0.500	180	$\frac{9}{16}$	$\frac{7}{8}$	$\frac{3}{8}$	$\frac{3}{4}$	12	10	8
	0.375	138	$\frac{7}{16}$	$\frac{7}{8}$	$\frac{5}{16}$	$\frac{5}{8}$	10	8	6
	0.312	116	$\frac{3}{8}$	$\frac{7}{8}$	$\frac{5}{16}$	$\frac{1}{2}$	8	6	6
	0.250	93.6	$\frac{5}{16}$	$\frac{7}{8}$	$\frac{5}{16}$	$\frac{1}{2}$	6	6	4
	0.188	70.5	$\frac{5}{16}$	$\frac{7}{8}$	$\frac{5}{16}$	$\frac{3}{8}$	6	4	4
	0.125	47.4	$\frac{5}{16}$	$\frac{7}{8}$	$\frac{5}{16}$	$\frac{3}{8}$	4	4	2
6.875	0.500	177	$\frac{9}{16}$	$\frac{7}{8}$	$\frac{3}{8}$	$\frac{3}{4}$	12	10	8
	0.375	135	$\frac{7}{16}$	$\frac{7}{8}$	$\frac{5}{16}$	$\frac{5}{8}$	10	8	6
	0.312	114	$\frac{3}{8}$	$\frac{7}{8}$	$\frac{5}{16}$	$\frac{1}{2}$	8	6	6
	0.250	91.9	$\frac{5}{16}$	$\frac{7}{8}$	$\frac{5}{16}$	$\frac{1}{2}$	6	6	4
	0.188	69.2	$\frac{5}{16}$	$\frac{7}{8}$	$\frac{5}{16}$	$\frac{3}{8}$	6	4	4
6.625	0.500	170	$\frac{1}{2}$	$\frac{7}{8}$	$\frac{3}{8}$	$\frac{3}{4}$	12	8	8
	0.432	149	$\frac{7}{16}$	$\frac{7}{8}$	$\frac{5}{16}$	$\frac{3}{4}$	10	8	6
	0.375	130	$\frac{7}{16}$	$\frac{7}{8}$	$\frac{5}{16}$	$\frac{5}{8}$	10	8	6
	0.312	109	$\frac{3}{8}$	$\frac{7}{8}$	$\frac{5}{16}$	$\frac{1}{2}$	8	6	4
	0.280	98.7	$\frac{5}{16}$	$\frac{7}{8}$	$\frac{5}{16}$	$\frac{1}{2}$	8	6	4
	0.250	88.5	$\frac{5}{16}$	$\frac{7}{8}$	$\frac{5}{16}$	$\frac{1}{2}$	6	6	4
	0.188	66.7	$\frac{5}{16}$	$\frac{7}{8}$	$\frac{5}{16}$	$\frac{3}{8}$	6	4	4
	0.125	44.8	$\frac{5}{16}$	$\frac{7}{8}$	$\frac{5}{16}$	$\frac{3}{8}$	4	4	2
6.125	0.500	156	$\frac{1}{2}$	$\frac{3}{4}$	$\frac{3}{8}$	$\frac{3}{4}$	10	8	6
	0.375	120	$\frac{7}{16}$	$\frac{3}{4}$	$\frac{5}{16}$	$\frac{5}{8}$	8	6	6
	0.312	101	$\frac{3}{8}$	$\frac{3}{4}$	$\frac{1}{4}$	$\frac{1}{2}$	8	6	4
	0.250	81.5	$\frac{5}{16}$	$\frac{3}{4}$	$\frac{1}{4}$	$\frac{1}{2}$	6	4	4
	0.188	61.4	$\frac{1}{4}$	$\frac{3}{4}$	$\frac{1}{4}$	$\frac{3}{8}$	4	4	4

Table 6-1.
Tension Ties
Welded Tee Ends—Round Sections

Designation		Con-nection Strength (kips)	Weld Size W_w (in.)	Tee Flange t_1 (in.)	Weld Size W_s (in.)	Tee Stem t_2 (in.)	No. of Bolts n		
D in.	t in.						¾-in.	⅞-in.	1-in.
6.000	0.500	153	7/16	3/4	3/8	3/4	10	8	6
	0.375	117	7/16	3/4	5/16	5/8	8	6	6
	0.312	98.7	3/8	3/4	1/4	1/2	8	6	4
	0.280	89.0	3/8	3/4	1/4	1/2	6	6	4
	0.250	79.8	5/16	3/4	1/4	1/2	6	4	4
	0.188	60.1	1/4	3/4	1/4	3/8	4	4	4
	0.125	40.4	1/4	3/4	1/4	3/8	4	2	2
5.563	0.375	108	7/16	3/4	1/4	5/8	8	6	4
	0.258	76.2	5/16	3/4	1/4	1/2	6	4	4
	0.188	55.8	1/4	3/4	1/4	1/2	4	4	2
	0.134	40.4	1/4	3/4	1/4	3/8	4	2	2
5.500	0.250	139	7/16	7/8	3/8	7/8	10	8	6
	0.375	107	7/16	5/8	1/4	5/8	8	6	4
	0.258	75.2	5/16	5/8	1/4	1/2	6	4	4
5.000	0.500	125	3/8	7/8	5/16	7/8	8	6	6
	0.375	96.4	3/8	5/8	1/4	5/8	8	6	4
	0.312	81.3	3/8	5/8	1/4	1/2	6	4	4
	0.258	68.0	5/16	5/8	1/4	1/2	6	4	4
	0.250	66.0	5/16	5/8	1/4	1/2	6	4	4
	0.188	49.9	1/4	5/8	1/4	1/2	4	4	2
	0.125	33.6	1/4	5/8	1/4	3/8	4	2	2
4.500	0.337	78.2	3/8	1/2	1/4	1/2	6	4	4
	0.237	56.1	1/4	1/2	3/16	1/2	4	4	2
	0.188	44.6	1/4	1/2	3/16	3/8	4	4	2
	0.125	30.2	3/16	1/2	3/16	3/8	2	2	2
4.000	0.337	69.0	5/16	1/2	3/16	1/2	6	4	4
	0.313	64.1	5/16	1/2	3/16	1/2	6	4	4
	0.250	52.2	5/16	1/2	3/16	1/2	4	4	2
	0.237	49.7	5/16	1/2	3/16	1/2	4	4	2
	0.226	47.4	5/16	1/2	3/16	3/8	4	4	2
	0.220	46.1	1/4	1/2	3/16	3/8	4	4	2
	0.188	39.5	1/4	1/2	3/16	3/8	4	2	2
	0.125	26.8	3/16	1/2	3/16	3/8	2	2	2

25% of Member Design Strength		

Table 6-2.
Compression Struts
Welded Tee Ends—Square Sections

Designation			Con-nection Strength (kips)	Weld Size W_w (in.)	Tee Flange t_1 (in.)	Weld Size W_s (in.)	Tee Stem t_2 (in.)	No. of Bolts n		
H in.	B in.	t in.						$\frac{3}{4}$-in.	$\frac{7}{8}$-in.	1-in.
8	8	$\frac{5}{8}$	126	$\frac{7}{16}$	$\frac{1}{2}$	$\frac{5}{16}$	$1\frac{1}{2}$	8	6	6
		$\frac{1}{2}$	104	$\frac{7}{16}$	$\frac{3}{8}$	$\frac{5}{16}$	$1\frac{1}{4}$	8	6	4
		$\frac{3}{8}$	79.8	$\frac{3}{8}$	$\frac{3}{8}$	$\frac{5}{16}$	$\frac{7}{8}$	6	4	4
		$\frac{5}{16}$	67.2	$\frac{3}{8}$	$\frac{3}{8}$	$\frac{1}{4}$	$\frac{3}{4}$	6	4	4
		$\frac{1}{4}$	54.5	$\frac{5}{16}$	$\frac{3}{8}$	$\frac{1}{4}$	$\frac{5}{8}$	4	4	2
		$\frac{3}{16}$	41.2	$\frac{1}{4}$	$\frac{3}{8}$	$\frac{3}{16}$	$\frac{1}{2}$	4	2	2
7	7	$\frac{5}{8}$	107	$\frac{7}{16}$	$\frac{3}{8}$	$\frac{5}{16}$	$1\frac{1}{4}$	8	6	4
		$\frac{1}{2}$	89.0	$\frac{3}{8}$	$\frac{3}{8}$	$\frac{5}{16}$	$1\frac{1}{4}$	6	6	4
		$\frac{3}{8}$	68.8	$\frac{5}{16}$	$\frac{3}{8}$	$\frac{5}{16}$	$\frac{7}{8}$	6	4	4
		$\frac{5}{16}$	58.2	$\frac{5}{16}$	$\frac{3}{8}$	$\frac{1}{4}$	$\frac{3}{4}$	4	4	4
		$\frac{1}{4}$	47.3	$\frac{1}{4}$	$\frac{3}{8}$	$\frac{1}{4}$	$\frac{5}{8}$	4	4	2
		$\frac{3}{16}$	35.8	$\frac{1}{4}$	$\frac{3}{8}$	$\frac{3}{16}$	$\frac{1}{2}$	4	2	2
6	6	$\frac{5}{8}$	89.8	$\frac{3}{8}$	$\frac{3}{8}$	$\frac{5}{16}$	$1\frac{1}{4}$	6	6	4
		$\frac{1}{2}$	74.7	$\frac{3}{8}$	$\frac{3}{8}$	$\frac{5}{16}$	1	6	4	4
		$\frac{3}{8}$	58.2	$\frac{5}{16}$	$\frac{3}{8}$	$\frac{5}{16}$	$\frac{7}{8}$	4	4	4
		$\frac{5}{16}$	49.3	$\frac{1}{4}$	$\frac{3}{8}$	$\frac{1}{4}$	$\frac{3}{4}$	4	4	2
		$\frac{1}{4}$	40.2	$\frac{1}{4}$	$\frac{3}{8}$	$\frac{1}{4}$	$\frac{5}{8}$	4	2	2
		$\frac{3}{16}$	30.5	$\frac{3}{16}$	$\frac{3}{8}$	$\frac{3}{16}$	$\frac{1}{2}$	2	2	2
		$\frac{1}{8}$	20.7	$\frac{3}{16}$	$\frac{3}{8}$	$\frac{3}{16}$	$\frac{3}{8}$	2	2	2
5	5	$\frac{1}{2}$	60.5	$\frac{5}{16}$	$\frac{3}{8}$	$\frac{5}{16}$	1	4	4	4
		$\frac{3}{8}$	47.4	$\frac{1}{4}$	$\frac{3}{8}$	$\frac{1}{4}$	$\frac{3}{4}$	4	4	2
		$\frac{5}{16}$	40.4	$\frac{1}{4}$	$\frac{3}{8}$	$\frac{1}{4}$	$\frac{5}{8}$	4	2	2
		$\frac{1}{4}$	33.0	$\frac{3}{16}$	$\frac{3}{8}$	$\frac{1}{4}$	$\frac{5}{8}$	4	2	2
		$\frac{3}{16}$	25.2	$\frac{3}{16}$	$\frac{3}{8}$	$\frac{3}{16}$	$\frac{1}{2}$	2	2	2
		$\frac{1}{8}$	17.1	$\frac{3}{16}$	$\frac{3}{8}$	$\frac{3}{16}$	$\frac{3}{8}$	2	2	2
$4\frac{1}{2}$	$4\frac{1}{2}$	$\frac{1}{2}$	53.3	$\frac{1}{4}$	$\frac{3}{8}$	$\frac{5}{16}$	$\frac{7}{8}$	4	4	2
		$\frac{3}{8}$	42.0	$\frac{1}{4}$	$\frac{3}{8}$	$\frac{1}{4}$	$\frac{3}{4}$	4	2	2
		$\frac{5}{16}$	35.9	$\frac{3}{16}$	$\frac{3}{8}$	$\frac{1}{4}$	$\frac{5}{8}$	4	2	2
		$\frac{1}{4}$	29.5	$\frac{3}{16}$	$\frac{3}{8}$	$\frac{3}{16}$	$\frac{1}{2}$	2	2	2
		$\frac{3}{16}$	22.5	$\frac{3}{16}$	$\frac{3}{8}$	$\frac{3}{16}$	$\frac{3}{8}$	2	2	2
		$\frac{1}{8}$	15.3	$\frac{3}{16}$	$\frac{3}{8}$	$\frac{3}{16}$	$\frac{3}{8}$	2	2	2

| 25% of Member Design Strength | | | | | | | | | |

Table 6-2.
Compression Struts
Welded Tee Ends—Square Sections

Designation			Con-nection Strength (kips)	Weld Size W_w (in.)	Tee Flange t_1 (in.)	Weld Size W_s (in.)	Tee Stem t_2 (in.)	No. of Bolts n		
H in.	B in.	t in.						¾-in.	⅞-in.	1-in.
4	4	½	46.2	¼	⅜	¼	¾	4	4	2
		⅜	36.7	³⁄₁₆	⅜	¼	⅝	4	2	2
		⁵⁄₁₆	31.5	³⁄₁₆	⅜	³⁄₁₆	½	2	2	2
		¼	25.9	³⁄₁₆	⅜	³⁄₁₆	½	2	2	2
		³⁄₁₆	19.8	³⁄₁₆	⅜	³⁄₁₆	⅜	2	2	2
		⅛	13.6	³⁄₁₆	⅜	³⁄₁₆	⅜	2	2	2
3½	3½	⅜	31.4	³⁄₁₆	⅜	³⁄₁₆	½	2	2	2
		⁵⁄₁₆	27.0	³⁄₁₆	⅜	³⁄₁₆	½	2	2	2
		¼	22.3	³⁄₁₆	⅜	³⁄₁₆	⅜	2	2	2
		³⁄₁₆	17.2	³⁄₁₆	⅜	³⁄₁₆	⅜	2	2	2
		⅛	11.8	³⁄₁₆	⅜	³⁄₁₆	⅜	2	2	2
3	3	⅜	26.0	³⁄₁₆	⅜	³⁄₁₆	½	2	2	2
		⁵⁄₁₆	22.6	³⁄₁₆	⅜	³⁄₁₆	⅜	2	2	2
		¼	18.7	³⁄₁₆	⅜	³⁄₁₆	⅜	2	2	2
		³⁄₁₆	14.5	³⁄₁₆	⅜	³⁄₁₆	⅜	2	2	2
		⅛	10.0	³⁄₁₆	⅜	³⁄₁₆	⅜	2	2	2
2½	2½	⁵⁄₁₆	18.0	³⁄₁₆	⅜	³⁄₁₆	⅜	2	2	2
		¼	15.1	³⁄₁₆	⅜	³⁄₁₆	⅜	2	2	2
		³⁄₁₆	11.8	³⁄₁₆	⅜	³⁄₁₆	⅜	2	2	2
		⅛	8.2	³⁄₁₆	⅜	³⁄₁₆	⅜	2	2	2
2	2	¼	11.6	³⁄₁₆	⅜	³⁄₁₆	⅜	2	2	2
		³⁄₁₆	9.1	³⁄₁₆	⅜	³⁄₁₆	⅜	2	2	2
		⅛	6.4	³⁄₁₆	⅜	³⁄₁₆	⅜	2	2	2

Table 6-2.
Compression Struts
Welded Tee Ends—Square Sections

Designation			Connection Strength (kips)	Weld Size W_w (in.)	Tee Flange t_1 (in.)	Weld Size W_s (in.)	Tee Stem t_2 (in.)	No. of Bolts n		
H in.	B in.	t in.						¾-in.	⅞-in.	1-in.
8	8	⅝	252	½	⅞	⁷⁄₁₆	2¾	16	12	10
		½	207	½	¾	⅜	2¼	14	10	8
		⅜	160	⁷⁄₁₆	¾	⁵⁄₁₆	1¾	12	8	6
		⁵⁄₁₆	134	⁷⁄₁₆	¾	⁵⁄₁₆	1½	10	8	6
		¼	109	⁵⁄₁₆	¾	⁵⁄₁₆	1¼	8	6	4
		³⁄₁₆	82.4	¼	¾	⁵⁄₁₆	⅞	6	4	4
7	7	⅝	215	⁷⁄₁₆	⅞	⁷⁄₁₆	2½	14	10	8
		½	178	⁷⁄₁₆	¾	⅜	2¼	12	10	8
		⅜	138	⁷⁄₁₆	⅝	⁵⁄₁₆	1¾	10	8	6
		⁵⁄₁₆	116	⁷⁄₁₆	⅝	⁵⁄₁₆	1½	8	6	6
		¼	94.7	⁵⁄₁₆	⅝	⁵⁄₁₆	1¼	6	6	4
		³⁄₁₆	71.7	¼	⅝	⁵⁄₁₆	⅞	6	4	4
6	6	⅝	180	⁷⁄₁₆	¾	⁷⁄₁₆	2½	12	10	8
		½	149	⁷⁄₁₆	⅝	⅜	2	10	8	6
		⅜	116	⅜	⅝	⁵⁄₁₆	1½	8	6	6
		⁵⁄₁₆	98.7	⅜	⅝	⁵⁄₁₆	1½	8	6	4
		¼	80.4	⁵⁄₁₆	⅝	⁵⁄₁₆	1¼	6	4	4
		³⁄₁₆	61.1	¼	⅝	⁵⁄₁₆	⅞	4	4	4
		⅛	41.4	¼	⅝	¼	⅝	4	2	2
5	5	½	121	⅜	⅝	⁵⁄₁₆	2	8	6	6
		⅜	94.8	⁵⁄₁₆	½	⁵⁄₁₆	1½	6	6	4
		⁵⁄₁₆	80.7	⁵⁄₁₆	½	⁵⁄₁₆	1¼	6	4	4
		¼	66.0	⁵⁄₁₆	½	⁵⁄₁₆	1¼	6	4	4
		³⁄₁₆	50.3	¼	½	⁵⁄₁₆	⅞	4	4	2
		⅛	34.2	³⁄₁₆	½	¼	⅝	4	2	2
4½	4½	½	107	⅜	½	⁵⁄₁₆	1¾	8	6	4
		⅜	84.1	⁵⁄₁₆	½	⁵⁄₁₆	1½	6	4	4
		⁵⁄₁₆	71.8	⁵⁄₁₆	½	⁵⁄₁₆	1¼	6	4	4
		¼	58.9	⁵⁄₁₆	½	⁵⁄₁₆	1	4	4	4
		³⁄₁₆	45.0	¼	½	¼	¾	4	4	2
		⅛	30.7	³⁄₁₆	½	³⁄₁₆	½	2	2	2

Table 6-2.
Compression Struts
Welded Tee Ends—Square Sections

Designation			Con-nection Strength	Weld Size	Tee Flange	Weld Size	Tee Stem	No. of Bolts n		
H in.	B in.	t in.	(kips)	W_w (in.)	t_1 (in.)	W_s (in.)	t_2 (in.)	¾-in.	⅞-in.	1-in.
4	4	½	92.4	5/16	½	5/16	1½	6	6	4
		3/8	73.4	5/16	3/8	5/16	1¼	6	4	4
		5/16	62.9	5/16	3/8	5/16	1	4	4	4
		¼	51.7	¼	3/8	5/16	7/8	4	4	2
		3/16	39.6	¼	3/8	¼	5/8	4	2	2
		1/8	27.2	3/16	3/8	3/16	½	2	2	2
3½	3½	3/8	62.8	5/16	3/8	5/16	1	4	4	4
		5/16	54.0	¼	3/8	5/16	7/8	4	4	2
		¼	44.7	¼	3/8	¼	¾	4	4	2
		3/16	34.4	3/16	3/8	¼	5/8	4	2	2
		1/8	23.6	3/16	3/8	3/16	3/8	2	2	2
3	3	3/8	52.0	¼	3/8	5/16	7/8	4	4	2
		5/16	45.1	¼	3/8	¼	¾	4	4	2
		¼	37.4	3/16	3/8	¼	5/8	4	2	2
		3/16	29.0	3/16	3/8	3/16	½	2	2	2
		1/8	19.9	3/16	3/8	3/16	3/8	2	2	2
2½	2½	5/16	36.1	3/16	3/8	¼	5/8	4	2	2
		¼	30.2	3/16	3/8	3/16	½	2	2	2
		3/16	23.6	3/16	3/8	3/16	3/8	2	2	2
		1/8	16.4	3/16	3/8	3/16	3/8	2	2	2
2	2	¼	23.2	3/16	3/8	3/16	3/8	2	2	2
		3/16	18.3	3/16	3/8	3/16	3/8	2	2	2
		1/8	12.9	3/16	3/8	3/16	3/8	2	2	2

<table>
<tr><td colspan="11">25% of Member Design Strength</td></tr>
</table>

Table 6-2.
Compression Struts
Welded Tee Ends—Rectangular Sections

Designation			Con-nection Strength	Weld Size W_w	Tee Flange t_1	Weld Size W_s	Tee Stem t_2	No. of Bolts n		
H in.	B in.	t in.	(kips)	(in.)	(in.)	(in.)	(in.)	$3/4$-in.	$7/8$-in.	1-in.
8	6	$5/8$	107	$7/16$	$3/8$	$5/16$	$1 1/4$	8	6	4
		$1/2$	89.0	$7/16$	$3/8$	$5/16$	1	6	6	4
		$3/8$	68.8	$3/8$	$3/8$	$1/4$	$3/4$	6	4	4
		$5/16$	58.2	$5/16$	$3/8$	$1/4$	$5/8$	4	4	4
		$1/4$	47.3	$1/4$	$3/8$	$1/4$	$5/8$	4	4	2
		$3/16$	35.8	$1/4$	$3/8$	$3/16$	$1/2$	4	2	2
8	4	$5/8$	89.8	$7/16$	$3/8$	$5/16$	1	6	6	4
		$1/2$	74.7	$3/8$	$3/8$	$5/16$	$7/8$	6	4	4
		$3/8$	58.2	$5/16$	$3/8$	$1/4$	$5/8$	4	4	4
		$5/16$	49.3	$1/4$	$3/8$	$1/4$	$5/8$	4	4	2
		$1/4$	40.2	$1/4$	$3/8$	$3/16$	$1/2$	4	2	2
		$3/16$	30.5	$3/16$	$3/8$	$3/16$	$3/8$	2	2	2
		$1/8$	20.7	$3/16$	$3/8$	$3/16$	$3/8$	2	2	2
8	3	$1/2$	67.6	$3/8$	$3/8$	$1/4$	$3/4$	6	4	4
		$3/8$	52.8	$5/16$	$3/8$	$1/4$	$5/8$	4	4	2
		$5/16$	44.9	$1/4$	$3/8$	$3/16$	$1/2$	4	4	2
		$1/4$	36.6	$1/4$	$3/8$	$3/16$	$1/2$	4	2	2
		$3/16$	27.9	$3/16$	$3/8$	$3/16$	$3/8$	2	2	2
		$1/8$	18.9	$3/16$	$3/8$	$3/16$	$3/8$	2	2	2
8	2	$3/8$	47.4	$1/4$	$3/8$	$1/4$	$5/8$	4	4	2
		$5/16$	40.4	$1/4$	$3/8$	$3/16$	$1/2$	4	2	2
		$1/4$	33.0	$3/16$	$3/8$	$3/16$	$1/2$	4	2	2
		$3/16$	25.2	$3/16$	$3/8$	$3/16$	$3/8$	2	2	2
		$1/8$	17.1	$3/16$	$3/8$	$3/16$	$3/8$	2	2	2
6	4	$1/2$	60.5	$5/16$	$3/8$	$5/16$	$7/8$	4	4	4
		$3/8$	47.4	$1/4$	$3/8$	$1/4$	$3/4$	4	4	2
		$5/16$	40.4	$1/4$	$3/8$	$1/4$	$5/8$	4	2	2
		$1/4$	33.0	$3/16$	$3/8$	$3/16$	$1/2$	4	2	2
		$3/16$	25.2	$3/16$	$3/8$	$3/16$	$3/8$	2	2	2
		$1/8$	17.1	$3/16$	$3/8$	$3/16$	$3/8$	2	2	2

25% of Member Design Strength		

Table 6-2.
Compression Struts
Welded Tee Ends—Rectangular Sections

Designation			Connection Strength	Weld Size W_w	Tee Flange t_1	Weld Size W_s	Tee Stem t_2	No. of Bolts n		
H in.	B in.	t in.	(kips)	(in.)	(in.)	(in.)	(in.)	¾-in.	⅞-in.	1-in.
6	3	½	53.3	5/16	3/8	¼	¾	4	4	2
		3/8	42.0	¼	3/8	¼	5/8	4	2	2
		5/16	35.9	¼	3/8	3/16	½	4	2	2
		¼	29.5	3/16	3/8	3/16	½	2	2	2
		3/16	22.5	3/16	3/8	3/16	3/8	2	2	2
		⅛	15.3	3/16	3/8	3/16	3/8	2	2	2
6	2	3/8	36.7	3/16	3/8	¼	5/8	4	2	2
		5/16	31.5	3/16	3/8	3/16	½	2	2	2
		¼	25.9	3/16	3/8	3/16	3/8	2	2	2
		3/16	19.8	3/16	3/8	3/16	3/8	2	2	2
		⅛	13.6	3/16	3/8	3/16	3/8	2	2	2
5	3	½	46.2	¼	3/8	¼	¾	4	4	2
		3/8	36.7	3/16	3/8	¼	5/8	4	2	2
		5/16	31.5	3/16	3/8	3/16	½	2	2	2
		¼	25.9	3/16	3/8	3/16	½	2	2	2
		3/16	19.8	3/16	3/8	3/16	3/8	2	2	2
		⅛	13.6	3/16	3/8	3/16	3/8	2	2	2
5	2	3/8	31.4	3/16	3/8	3/16	½	2	2	2
		5/16	27.0	3/16	3/8	3/16	½	2	2	2
		¼	22.3	3/16	3/8	3/16	3/8	2	2	2
		3/16	17.2	3/16	3/8	3/16	3/8	2	2	2
		⅛	11.8	3/16	3/8	3/16	3/8	2	2	2
4	3	3/8	31.4	3/16	3/8	3/16	½	2	2	2
		5/16	27.0	3/16	3/8	3/16	½	2	2	2
		¼	22.3	3/16	3/8	3/16	3/8	2	2	2
		3/16	17.2	3/16	3/8	3/16	3/8	2	2	2
		⅛	11.8	3/16	3/8	3/16	3/8	2	2	2
4	2	3/8	26.0	3/16	3/8	3/16	½	2	2	2
		5/16	22.6	3/16	3/8	3/16	3/8	2	2	2
		¼	18.7	3/16	3/8	3/16	3/8	2	2	2
		3/16	14.5	3/16	3/8	3/16	3/8	2	2	2
		⅛	10.0	3/16	3/8	3/16	3/8	2	2	2

Table 6-2.
Compression Struts
Welded Tee Ends—Rectangular Sections

Designation			Con-nection Strength (kips)	Weld Size W_w (in.)	Tee Flange t_1 (in.)	Weld Size W_s (in.)	Tee Stem t_2 (in.)	No. of Bolts n		
H in.	B in.	t in.						¾-in.	⅞-in.	1-in.
8	6	⅝	215	½	¾	⅜	2¼	14	10	8
		½	178	⁷⁄₁₆	¾	⁵⁄₁₆	2	12	10	8
		⅜	138	⁷⁄₁₆	⅝	⁵⁄₁₆	1½	10	8	6
		⁵⁄₁₆	116	⁷⁄₁₆	⅝	⁵⁄₁₆	1¼	8	6	6
		¼	94.7	⁵⁄₁₆	⅝	⁵⁄₁₆	1	6	6	4
		³⁄₁₆	71.7	¼	⅝	⁵⁄₁₆	⅞	6	4	4
8	4	⅝	180	½	⅝	⁵⁄₁₆	2	12	10	8
		½	149	⁷⁄₁₆	⅝	⁵⁄₁₆	1¾	10	8	6
		⅜	116	⁷⁄₁₆	⅝	⁵⁄₁₆	1¼	8	6	6
		⁵⁄₁₆	98.7	⅜	⅝	⁵⁄₁₆	1¼	8	6	4
		¼	80.4	⁵⁄₁₆	⅝	⁵⁄₁₆	⅞	6	4	4
		³⁄₁₆	61.1	¼	⅝	¼	¾	4	4	4
		⅛	41.4	¼	⅝	³⁄₁₆	½	4	2	2
8	3	½	135	⁷⁄₁₆	⅝	⁵⁄₁₆	1½	10	8	6
		⅜	106	⁷⁄₁₆	⅝	⁵⁄₁₆	1¼	8	6	4
		⁵⁄₁₆	89.8	⁷⁄₁₆	⅝	⁵⁄₁₆	1	6	6	4
		¼	73.2	⁵⁄₁₆	⅝	⁵⁄₁₆	⅞	6	4	4
		³⁄₁₆	55.7	¼	⅝	¼	⅝	4	4	2
		⅛	37.7	¼	⅝	³⁄₁₆	½	4	2	2
8	2	⅜	94.8	⁷⁄₁₆	½	⁵⁄₁₆	1	6	6	4
		⁵⁄₁₆	80.7	⅜	½	⁵⁄₁₆	⅞	6	4	4
		¼	66.0	⅜	½	¼	¾	6	4	4
		³⁄₁₆	50.3	¼	½	¼	⅝	4	4	2
		⅛	34.2	³⁄₁₆	½	³⁄₁₆	½	4	2	2
6	4	½	121	⅜	½	⁵⁄₁₆	1¾	8	6	6
		⅜	94.8	⅜	½	⁵⁄₁₆	1¼	6	6	4
		⁵⁄₁₆	80.7	⁵⁄₁₆	½	⁵⁄₁₆	1¼	6	4	4
		¼	66.0	⁵⁄₁₆	½	⁵⁄₁₆	⅞	6	4	4
		³⁄₁₆	50.3	¼	½	¼	¾	4	4	2
		⅛	34.2	³⁄₁₆	½	³⁄₁₆	½	4	2	2

Table 6-2.
Compression Struts
Welded Tee Ends—Rectangular Sections

Designation			Con-nection Strength (kips)	Weld Size W_w (in.)	Tee Flange t_1 (in.)	Weld Size W_s (in.)	Tee Stem t_2 (in.)	No. of Bolts n		
H in.	B in.	t in.						¾-in.	⅞-in.	1-in.
6	3	½	107	⅜	½	5/16	1½	8	6	4
		⅜	84.1	5/16	½	5/16	1¼	6	4	4
		5/16	71.8	5/16	½	5/16	1	6	4	4
		¼	58.9	5/16	½	5/16	⅞	4	4	4
		3/16	45.0	¼	½	¼	⅝	4	4	2
		⅛	30.7	3/16	½	3/16	½	2	2	2
6	2	⅜	73.4	⅜	⅜	5/16	1	6	4	4
		5/16	62.9	5/16	⅜	5/16	⅞	4	4	4
		¼	51.7	¼	⅜	¼	¾	4	4	2
		3/16	39.6	¼	⅜	¼	⅝	4	2	2
		⅛	27.2	3/16	⅜	3/16	⅜	2	2	2
5	3	½	92.4	5/16	½	5/16	1½	6	6	4
		⅜	73.4	5/16	⅜	5/16	1¼	6	4	4
		5/16	62.9	5/16	⅜	5/16	1	4	4	4
		¼	51.7	¼	⅜	5/16	⅞	4	4	2
		3/16	39.6	¼	⅜	¼	⅝	4	2	2
		⅛	27.2	3/16	⅜	3/16	½	2	2	2
5	2	⅜	62.8	5/16	⅜	5/16	1	4	4	4
		5/16	54.0	¼	⅜	5/16	⅞	4	4	2
		¼	44.7	¼	⅜	¼	¾	4	4	2
		3/16	34.4	3/16	⅜	¼	⅝	4	2	2
		⅛	23.6	3/16	⅜	3/16	⅜	2	2	2
4	3	⅜	62.8	5/16	⅜	5/16	1	4	4	4
		5/16	54.0	¼	⅜	5/16	⅞	4	4	2
		¼	44.7	¼	⅜	¼	¾	4	4	2
		3/16	34.4	3/16	⅜	¼	⅝	4	2	2
		⅛	23.6	3/16	⅜	3/16	⅜	2	2	2
4	2	⅜	52.0	¼	⅜	5/16	⅞	4	4	2
		5/16	45.1	¼	⅜	¼	¾	4	4	2
		¼	37.4	3/16	⅜	¼	⅝	4	2	2
		3/16	29.0	3/16	⅜	3/16	½	2	2	2
		⅛	19.9	3/16	⅜	3/16	⅜	2	2	2

25% of Member Design Strength

Table 6-2.
Compression Struts
Welded Tee Ends—Round Sections

Designation		Con-nection Strength (kips)	Weld Size W_w (in.)	Tee Flange t_1 (in.)	Weld Size W_s (in.)	Tee Stem t_2 (in.)	No. of Bolts n		
D in.	t in.						¾-in.	⅞-in.	1-in.
8.625	0.500	91.3	7/16	3/8	5/16	1	6	6	4
	0.375	69.6	3/8	3/8	1/4	3/4	6	4	4
	0.322	59.8	5/16	3/8	1/4	5/8	4	4	4
	0.250	47.1	5/16	3/8	3/16	1/2	4	4	2
	0.188	35.4	1/4	3/8	3/16	1/2	4	2	2
7.500	0.500	79.0	3/8	3/8	5/16	7/8	6	4	4
	0.375	60.2	5/16	3/8	1/4	3/4	4	4	4
	0.312	50.6	5/16	3/8	1/4	5/8	4	4	2
	0.250	40.8	1/4	3/8	3/16	1/2	4	2	2
	0.188	30.7	3/16	3/8	3/16	3/8	2	2	2
7.000	0.500	73.3	3/8	3/8	5/16	7/8	6	4	4
	0.375	55.9	5/16	3/8	1/4	3/4	4	4	2
	0.312	47.0	1/4	3/8	1/4	5/8	4	4	2
	0.250	38.0	1/4	3/8	3/16	1/2	4	2	2
	0.188	28.6	3/16	3/8	3/16	3/8	2	2	2
	0.125	19.3	3/16	3/8	3/16	3/8	2	2	2
6.875	0.500	71.8	3/8	3/8	5/16	7/8	6	4	4
	0.375	54.9	5/16	3/8	1/4	3/4	4	4	2
	0.312	46.2	1/4	3/8	1/4	5/8	4	4	2
	0.250	37.3	1/4	3/8	3/16	1/2	4	2	2
	0.188	28.1	3/16	3/8	3/16	3/8	2	2	2
6.625	0.500	69.1	5/16	3/8	5/16	7/8	6	4	4
	0.432	60.5	5/16	3/8	1/4	3/4	4	4	4
	0.375	52.8	5/16	3/8	1/4	3/4	4	4	2
	0.312	44.4	1/4	3/8	1/4	5/8	4	4	2
	0.280	40.1	1/4	3/8	1/4	5/8	4	2	2
	0.250	35.9	1/4	3/8	3/16	1/2	4	2	2
	0.188	27.1	3/16	3/8	3/16	3/8	2	2	2
	0.125	18.2	3/16	3/8	3/16	3/8	2	2	2
6.125	0.500	63.5	5/16	3/8	5/16	7/8	4	4	4
	0.375	48.6	1/4	3/8	1/4	3/4	4	4	2
	0.312	40.9	1/4	3/8	1/4	5/8	4	2	2
	0.250	33.1	3/16	3/8	3/16	1/2	4	2	2
	0.188	24.9	3/16	3/8	3/16	3/8	2	2	2

Table 6-2.
Compression Struts
Welded Tee Ends—Round Sections

Designation		Con-nection Strength (kips)	Weld Size W_w (in.)	Tee Flange t_1 (in.)	Weld Size W_s (in.)	Tee Stem t_2 (in.)	No. of Bolts n		
D in.	t in.						¾-in.	⅞-in.	1-in.
6.000	0.500	62.1	$\frac{5}{16}$	$\frac{3}{8}$	$\frac{5}{16}$	$\frac{7}{8}$	4	4	4
	0.375	47.6	$\frac{1}{4}$	$\frac{3}{8}$	$\frac{1}{4}$	$\frac{3}{4}$	4	4	2
	0.312	40.1	$\frac{1}{4}$	$\frac{3}{8}$	$\frac{1}{4}$	$\frac{5}{8}$	4	2	2
	0.280	36.1	$\frac{1}{4}$	$\frac{3}{8}$	$\frac{3}{16}$	$\frac{1}{2}$	4	2	2
	0.250	32.4	$\frac{3}{16}$	$\frac{3}{8}$	$\frac{3}{16}$	$\frac{1}{2}$	4	2	2
	0.188	24.4	$\frac{3}{16}$	$\frac{3}{8}$	$\frac{3}{16}$	$\frac{3}{8}$	2	2	2
	0.125	16.4	$\frac{3}{16}$	$\frac{3}{8}$	$\frac{3}{16}$	$\frac{3}{8}$	2	2	2
5.563	0.375	43.9	$\frac{1}{4}$	$\frac{3}{8}$	$\frac{1}{4}$	$\frac{5}{8}$	4	4	2
	0.258	30.9	$\frac{3}{16}$	$\frac{3}{8}$	$\frac{3}{16}$	$\frac{1}{2}$	2	2	2
	0.188	22.6	$\frac{3}{16}$	$\frac{3}{8}$	$\frac{3}{16}$	$\frac{3}{8}$	2	2	2
	0.134	16.4	$\frac{3}{16}$	$\frac{3}{8}$	$\frac{3}{16}$	$\frac{3}{8}$	2	2	2
5.500	0.500	56.5	$\frac{5}{16}$	$\frac{3}{8}$	$\frac{5}{16}$	$\frac{7}{8}$	4	4	2
	0.375	43.4	$\frac{1}{4}$	$\frac{3}{8}$	$\frac{1}{4}$	$\frac{5}{8}$	4	4	2
	0.258	30.5	$\frac{3}{16}$	$\frac{3}{8}$	$\frac{3}{16}$	$\frac{1}{2}$	2	2	2
5.000	0.500	50.8	$\frac{1}{4}$	$\frac{3}{8}$	$\frac{5}{16}$	$\frac{7}{8}$	4	4	2
	0.375	39.1	$\frac{1}{4}$	$\frac{3}{8}$	$\frac{1}{4}$	$\frac{5}{8}$	4	2	2
	0.312	33.0	$\frac{3}{16}$	$\frac{3}{8}$	$\frac{1}{4}$	$\frac{5}{8}$	4	2	2
	0.258	27.6	$\frac{3}{16}$	$\frac{3}{8}$	$\frac{3}{16}$	$\frac{1}{2}$	2	2	2
	0.250	26.8	$\frac{3}{16}$	$\frac{3}{8}$	$\frac{3}{16}$	$\frac{1}{2}$	2	2	2
	0.188	20.3	$\frac{3}{16}$	$\frac{3}{8}$	$\frac{3}{16}$	$\frac{3}{8}$	2	2	2
	0.125	13.7	$\frac{3}{16}$	$\frac{3}{8}$	$\frac{3}{16}$	$\frac{3}{8}$	2	2	2
4.500	0.337	31.8	$\frac{3}{16}$	$\frac{3}{8}$	$\frac{3}{16}$	$\frac{1}{2}$	2	2	2
	0.237	22.8	$\frac{3}{16}$	$\frac{3}{8}$	$\frac{3}{16}$	$\frac{3}{8}$	2	2	2
	0.188	18.1	$\frac{3}{16}$	$\frac{3}{8}$	$\frac{3}{16}$	$\frac{3}{8}$	2	2	2
	0.125	12.3	$\frac{3}{16}$	$\frac{3}{8}$	$\frac{3}{16}$	$\frac{3}{8}$	2	2	2
4.000	0.337	28.0	$\frac{3}{16}$	$\frac{3}{8}$	$\frac{3}{16}$	$\frac{1}{2}$	2	2	2
	0.313	26.0	$\frac{3}{16}$	$\frac{3}{8}$	$\frac{3}{16}$	$\frac{1}{2}$	2	2	2
	0.250	21.2	$\frac{3}{16}$	$\frac{3}{0}$	$\frac{3}{16}$	$\frac{3}{8}$	2	2	2
	0.237	20.2	$\frac{3}{16}$	$\frac{3}{8}$	$\frac{3}{16}$	$\frac{3}{8}$	2	2	2
	0.226	19.3	$\frac{3}{16}$	$\frac{3}{8}$	$\frac{3}{16}$	$\frac{3}{8}$	2	2	2
	0.220	18.7	$\frac{3}{16}$	$\frac{3}{8}$	$\frac{3}{16}$	$\frac{3}{8}$	2	2	2
	0.188	16.0	$\frac{3}{16}$	$\frac{3}{8}$	$\frac{3}{16}$	$\frac{3}{8}$	2	2	2
	0.125	10.9	$\frac{3}{16}$	$\frac{3}{8}$	$\frac{3}{16}$	$\frac{3}{8}$	2	2	2

Table 6-2.
Compression Struts
Welded Tee Ends—Round Sections

Designation		Con-nection Strength (kips)	Weld Size W_w (in.)	Tee Flange t_1 (in.)	Weld Size W_s (in.)	Tee Stem t_2 (in.)	No. of Bolts n		
D in.	t in.						¾-in.	⅞-in.	1-in.
8.625	0.500	183	½	¾	5/16	1¾	12	10	8
	0.375	139	7/16	¾	5/16	1½	10	8	6
	0.322	120	⅜	¾	5/16	1¼	8	6	6
	0.250	94.2	5/16	¾	5/16	1	6	6	4
	0.188	70.9	¼	¾	¼	¾	6	4	4
7.500	0.500	158	7/16	⅝	5/16	1¾	10	8	6
	0.375	120	7/16	⅝	5/16	1½	8	6	6
	0.312	101	⅜	⅝	5/16	1¼	8	6	4
	0.250	81.6	5/16	⅝	5/16	1	6	4	4
	0.188	61.4	¼	⅝	¼	¾	4	4	4
7.000	0.500	147	7/16	⅝	5/16	1¾	10	8	6
	0.375	112	⅜	⅝	5/16	1½	8	6	4
	0.312	94.1	⅜	⅝	5/16	1¼	6	6	4
	0.250	76.0	5/16	⅝	5/16	1	6	4	4
	0.188	57.2	¼	⅝	¼	¾	4	4	4
	0.125	38.5	¼	⅝	3/16	½	4	2	2
6.875	0.500	144	7/16	⅝	5/16	1¾	10	8	6
	0.375	110	⅜	⅝	5/16	1½	8	6	4
	0.312	92.4	⅜	⅝	5/16	1¼	6	6	5
	0.250	74.6	5/16	⅝	5/16	1	6	4	4
	0.188	56.2	¼	⅝	¼	¾	4	4	2
6.625	0.500	138	7/16	⅝	5/16	1¾	10	8	6
	0.432	121	7/16	⅝	5/16	1½	8	6	6
	0.375	106	⅜	⅝	5/16	1½	8	6	4
	0.312	88.8	⅜	⅝	5/16	1¼	6	6	4
	0.280	80.1	5/16	⅝	5/16	1	6	4	4
	0.250	71.8	5/16	⅝	5/16	⅞	6	4	4
	0.188	54.2	¼	⅝	¼	¾	4	4	2
	0.125	36.4	¼	⅝	3/16	½	4	2	2
6.125	0.500	127	7/16	½	5/16	1¾	8	6	6
	0.375	97.1	⅜	½	5/16	1¼	8	6	4
	0.312	81.8	⅜	½	5/16	1¼	6	4	4
	0.250	66.1	¼	½	5/16	⅞	6	4	4
	0.188	49.9	¼	½	¼	¾	4	4	2

Table 6-2.
Compression Struts
Welded Tee Ends—Round Sections

Designation		Con-nection Strength (kips)	Weld Size W_w (in.)	Tee Flange t_1 (in.)	Weld Size W_s (in.)	Tee Stem t_2 (in.)	No. of Bolts n		
D in.	t in.						¾-in.	⅞-in.	1-in.
6.000	0.500	124	$7/16$	$1/2$	$5/16$	$1\frac{3}{4}$	8	6	6
	0.375	95.1	$3/8$	$1/2$	$5/16$	$1\frac{1}{4}$	6	6	4
	0.312	80.1	$3/8$	$1/2$	$5/16$	$1\frac{1}{4}$	6	4	4
	0.280	72.3	$5/16$	$1/2$	$5/16$	1	6	4	4
	0.250	64.8	$5/16$	$1/2$	$5/16$	$7/8$	6	4	4
	0.188	48.8	$1/4$	$1/2$	$1/4$	$3/4$	4	4	2
	0.125	32.8	$3/16$	$1/2$	$3/16$	$1/2$	4	2	2
5.563	0.375	87.8	$5/16$	$1/2$	$5/16$	$1\frac{1}{4}$	6	6	4
	0.258	61.8	$5/16$	$1/2$	$5/16$	$7/8$	4	4	4
	0.188	45.3	$1/4$	$1/2$	$1/4$	$3/4$	4	4	2
	0.134	32.8	$3/16$	$1/2$	$3/16$	$1/2$	4	2	2
5.500	0.500	113	$3/8$	$1/2$	$5/16$	$1\frac{3}{4}$	8	6	4
	0.375	86.7	$5/16$	$1/2$	$5/16$	$1\frac{1}{4}$	6	6	4
	0.258	61.1	$5/16$	$3/8$	$5/16$	$7/8$	4	4	4
5.000	0.500	102	$5/16$	$1/2$	$5/16$	$1\frac{3}{4}$	8	6	4
	0.375	78.3	$5/16$	$1/2$	$5/16$	$1\frac{1}{4}$	6	4	4
	0.312	66.0	$5/16$	$1/2$	$5/16$	$1\frac{1}{4}$	6	4	4
	0.258	55.2	$1/4$	$1/2$	$5/16$	$7/8$	4	4	2
	0.250	53.6	$1/4$	$1/2$	$5/16$	$7/8$	4	4	2
	0.188	40.5	$1/4$	$1/2$	$1/4$	$3/4$	4	2	2
	0.125	27.3	$3/16$	$1/2$	$3/16$	$1/2$	2	2	2
4.500	0.337	63.5	$5/16$	$3/8$	$5/16$	1	4	4	4
	0.237	45.6	$1/4$	$3/8$	$1/4$	$3/4$	4	4	2
	0.188	36.2	$3/16$	$3/8$	$1/4$	$5/8$	4	2	2
	0.125	24.6	$3/16$	$3/8$	$3/16$	$1/2$	2	2	2
4.000	0.337	56.0	$1/4$	$3/8$	$5/16$	$7/8$	4	4	2
	0.313	52.0	$1/4$	$3/8$	$5/16$	$7/8$	4	4	2
	0.250	42.4	$1/4$	$3/8$	$1/4$	$3/4$	4	2	2
	0.237	40.4	$1/4$	$3/8$	$1/4$	$5/8$	4	2	2
	0.226	38.5	$1/4$	$3/8$	$1/4$	$5/8$	4	2	2
	0.220	37.4	$3/16$	$3/8$	$1/4$	$5/8$	4	2	2
	0.188	32.1	$3/16$	$3/8$	$1/4$	$5/8$	4	2	2
	0.125	21.8	$3/16$	$3/8$	$3/16$	$3/8$	2	2	2

SLOTTED HSS/GUSSET PLATE TABLES

General Notes

The Slotted HSS/Gusset Plate Tables can be used for selecting the weld size and length, plate thickness, and number of bolts required to develop the Connection Strength listed on the left side of the table. Table 6-3 is used for tension ties and Table 6-4 is used for compression struts. For tension-compression members, select the more conservative design from the two tables. The connection design strength is based on a percentage of the member design strength of the HSS. Tables are provided for 25 percent and 50 percent of the member design strength. For each HSS, the member design strength is computed using the specifications found in the AISC LRFD Manual, Second Edition. For the purposes of calculation, a maximum KL/r value of 60 was used to compute the design compressive strength of the HSS. The tables are presented for 2 rows of $\frac{3}{4}$-, $\frac{7}{8}$-, and 1-in. A325-N bolts.

Material Specifications

The tables are presented for commonly used square, rectangular, and round HSS manufactured under ASTM A500 Grade B. Plates are manufactured to ASTM A36 with $F_y = 36$ ksi and $F_u = 58$ ksi. The weld metal used has a minimum specified strength, $F_{EXX} = 70$ ksi. All bolts are manufactured to ASTM A325 Specifications.

Design Procedure

All connection results are calculated following the limit state procedures outlined in Chapter 6, Slotted HSS/Gusset Plate Connections. For connection strength calculations in Table 6-4, the length of the support gusset is assumed to be the same as the gusset plate. The length of the gusset plate is calculated by

$$L_e + (n/n_r - 1)p + L_e + 1$$

where

L_e = edge distance
n_r = number of rows of bolts
p = bolt pitch
n = number of $\frac{3}{4}$-in. bolts used in the connection

The thickness of the support gusset is assumed to be the same as the gusset plate. For longer lengths or thinner support gussets, the user is directed to AISC LRFD Volume II, Part 11, Connections For Tension and Compression, to evaluate the strength of the support gusset.

The value of n is determined by dividing the percent of member strength by ϕR_n, from AISC LRFD Table 8-11 for single shear, and rounding up to the next multiple of n_r.

The rounded value of n for 1 in. bolts is then used to calculate the minimum required thickness of the gusset plate, t_1, based on bearing, yielding, rupture, and block shear rupture. For the struts, an iterative procedure based on buckling of the gusset plate and support gusset is also performed using the length as explained above. For thicknesses < 1 in. and > 1 in., t_1 is rounded up to the nearest $\frac{1}{8}$-in. and $\frac{1}{4}$-in., respectively.

The minimum required size of the fillet weld connecting the gusset plate to the HSS, W_w, is calculated based on shear in the gusset plate and rounded up to the nearest $\frac{1}{16}$-in. The weld size, W_w, is then checked for the minimum size shown in AISC LRFD Table J2.4 based on the gusset plate thickness, and increased as required. The value of W_w, is then increased by $\frac{1}{16}$-in. to account for the root gap created by the slot in the HSS and the gusset plate.

With the values of t_1 and W_w determined, the minimum required length of the weld, L_w, is calculated based on shear in the HSS wall and strength of the weld. This value is then increased, if required, to a minimum of the HSS depth or outer diameter dimension, and rounded to the nearest inch. To ensure a conservative weld length when calculating L_w, the weld size used does not include the $\frac{1}{16}$-in. increase.

The values of W_w, L_w, t_1, and n are shown in the tables.

Design Parameters

The following values are held constant for calculations in Tables 6-3 and 6-4:

Plates: $F_y = 36$ ksi
            $F_u = 58$ ksi

Welds: $F_{EXX} = 70$ ksi
            $F_w = 42$ ksi

HSS: Square and Rectangular   Round
           $F_y = 46$ ksi $F_y = 42$ ksi
           $F_u = 58$ ksi $F_u = 58$ ksi

Bolts: ASTM A325 $p = 3.0$ in
                           $g = 3.0$ in.
                           $L_e = 1.5$ in.

For different parameters, the user is directed to Slotted HSS/Gusset Plate Connections for evaluating the connection design strength.

Table 6-3.
Tension Ties
Slotted HSS/Gusset Plate—Square Sections

Designation			Con-nection Strength (kips)	Weld Size W_w (in.)	Weld Length L_w (in.)	Gusset Plate t_1 (in.)	No. of Bolts n		
H in.	B in.	t in.					¾-in.	⅞-in.	1-in.
8	8	5/8	170	5/16	9.0	5/8	12	8	6
		1/2	140	1/4	9.0	1/2	10	8	6
		3/8	108	1/4	8.0	1/2	8	6	4
		5/16	90.7	1/4	8.0	1/2	6	6	4
		1/4	73.5	1/4	8.0	1/2	6	4	4
		3/16	55.6	1/4	8.0	1/2	4	4	2
7	7	5/8	145	5/16	8.0	5/8	10	8	6
		1/2	120	1/4	8.0	1/2	8	6	6
		3/8	92.8	1/4	7.0	1/2	6	6	4
		5/16	78.6	1/4	7.0	3/8	6	4	4
		1/4	63.9	1/4	7.0	3/8	6	4	4
		3/16	48.3	1/4	7.0	3/8	4	4	2
6	6	5/8	121	5/16	7.0	5/8	8	6	6
		1/2	101	1/4	7.0	1/2	8	6	4
		3/8	78.5	1/4	6.0	1/2	6	4	4
		5/16	66.6	1/4	6.0	3/8	6	4	4
		1/4	54.2	1/4	6.0	1/2	4	4	2
		3/16	41.2	1/4	6.0	3/8	4	2	2
		1/8	27.9	1/4	6.0	3/8	2	2	2
5	5	1/2	81.6	1/4	5.0	1/2	6	4	4
		3/8	64.0	1/4	5.0	1/2	6	4	4
		5/16	54.4	1/4	5.0	1/2	4	4	2
		1/4	44.5	1/4	5.0	3/8	4	4	2
		3/16	33.9	1/4	5.0	3/8	4	2	2
		1/8	23.1	1/4	5.0	3/8	2	2	2
4½	4½	1/2	71.8	1/4	5.0	1/2	6	4	4
		3/8	56.7	1/4	5.0	3/8	4	4	4
		5/16	48.4	1/4	5.0	3/8	4	4	2
		1/4	39.7	1/4	5.0	3/8	4	2	2
		3/16	30.3	1/4	5.0	3/8	2	2	2
		1/8	20.7	1/4	5.0	3/8	2	2	2

25% of Member Design Strength

Table 6-3.
Tension Ties
Slotted HSS/Gusset Plate—Square Sections

Designation			Con- nection Strength (kips)	Weld Size W_w (in.)	Weld Length L_w (in.)	Gusset Plate t_1 (in.)	No. of Bolts n		
H in.	B in.	t in.					¾-in.	⅞-in.	1-in.
4	4	½	61.7	¼	4.0	½	4	4	4
		⅜	49.1	¼	4.0	½	4	4	2
		5/16	42.2	¼	4.0	⅜	4	2	2
		¼	34.7	¼	4.0	⅜	4	2	2
		3/16	26.6	¼	4.0	⅜	2	2	2
		⅛	18.3	¼	4.0	⅜	2	2	2
3½	3½	⅜	41.6	¼	4.0	⅜	4	2	2
		5/16	35.9	¼	4.0	⅜	4	2	2
		¼	29.7	¼	4.0	⅜	2	2	2
		3/16	22.9	¼	4.0	⅜	2	2	2
		⅛	15.8	¼	4.0	⅜	2	2	2
3	3	⅜	34.0	¼	3.0	⅜	4	2	2
		5/16	29.6	¼	3.0	⅜	2	2	2
		¼	24.6	¼	3.0	⅜	2	2	2
		3/16	19.1	¼	3.0	⅜	2	2	2
		⅛	13.2	¼	3.0	⅜	2	2	2
2½	2½	5/16	23.2	¼	3.0	⅜	2	2	2
		¼	19.5	¼	3.0	⅜	2	2	2
		3/16	15.3	¼	3.0	⅜	2	2	2
		⅛	10.7	¼	3.0	⅜	2	2	2
2	2	¼	14.5	¼	2.0	⅜	2	2	2
		3/16	11.5	¼	2.0	⅜	2	2	2
		⅛	8.2	¼	2.0	⅜	2	2	2

Table 6-3.
Tension Ties
Slotted HSS/Gusset Plate—Square Sections

Designation			Con-nection Strength (kips)	Weld Size W_w (in.)	Weld Length L_w (in.)	Gusset Plate t_1 (in.)	No. of Bolts n		
H in.	B in.	t in.					3/4-in.	7/8-in.	1-in.
8	8	5/8	339	3/8	13.0	1 1/4	22	16	12
		1/2	279	3/8	11.0	1	18	14	10
		3/8	215	5/16	10.0	3/4	14	10	8
		5/16	181	5/16	9.0	5/8	12	10	8
		1/4	147	5/16	8.0	5/8	10	8	6
		3/16	111	1/4	8.0	1/2	8	6	4
7	7	5/8	290	3/8	11.0	1 1/4	20	14	12
		1/2	240	3/8	9.0	1	16	12	10
		3/8	186	5/16	9.0	3/4	12	10	8
		5/16	157	5/16	8.0	3/4	10	8	6
		1/4	128	5/16	7.0	5/8	10	6	6
		3/16	96.7	1/4	7.0	1/2	8	6	4
6	6	5/8	242	3/8	9.0	1 1/4	16	12	10
		1/2	202	3/8	8.0	1	14	10	8
		3/8	157	3/8	6.0	7/8	10	8	6
		5/16	133	5/16	6.0	3/4	10	8	6
		1/4	108	5/16	6.0	5/8	8	6	4
		3/16	82.4	1/4	6.0	1/2	6	4	4
		1/8	55.9	1/4	6.0	1/2	4	4	2
5	5	1/2	163	3/8	6.0	1	12	8	6
		3/8	128	3/8	5.0	7/8	10	6	6
		5/16	109	5/16	5.0	3/4	8	6	4
		1/4	89.0	5/16	5.0	5/8	6	6	4
		3/16	67.9	1/4	5.0	1/2	6	4	4
		1/8	46.2	1/4	5.0	3/8	4	4	2
4 1/2	4 1/2	1/2	144	3/8	6.0	1	10	8	6
		3/8	113	5/16	6.0	3/4	8	6	6
		5/16	96.9	5/16	5.0	5/8	8	6	4
		1/4	79.5	1/4	5.0	1/2	6	4	4
		3/16	60.7	1/4	5.0	3/8	4	4	4
		1/8	41.4	1/4	5.0	3/8	4	2	2

Table 6-3.
Tension Ties
Slotted HSS/Gusset Plate—Square Sections

Designation			Con-nection Strength (kips)	Weld Size W_w (in.)	Weld Length L_w (in.)	Gusset Plate t_1 (in.)	No. of Bolts n		
H in.	B in.	t in.					¾-in.	⅞-in.	1-in.
4	4	½	123	⅜	5.0	⅞	8	6	6
		⅜	98.3	⁵⁄₁₆	5.0	⅝	8	6	4
		⁵⁄₁₆	84.4	⁵⁄₁₆	5.0	⅝	6	4	4
		¼	69.5	¼	5.0	½	6	4	4
		³⁄₁₆	53.3	¼	4.0	½	4	4	2
		⅛	36.6	¼	4.0	⅜	4	2	2
3½	3½	⅜	83.3	⁵⁄₁₆	5.0	⅝	6	4	4
		⁵⁄₁₆	71.8	¼	5.0	½	6	4	4
		¼	59.5	¼	5.0	⅜	4	4	4
		³⁄₁₆	45.9	¼	4.0	⅜	4	4	2
		⅛	31.6	¼	4.0	⅜	2	2	2
3	3	⅜	68.0	¼	5.0	½	6	4	4
		⁵⁄₁₆	59.2	¼	5.0	½	4	4	4
		¼	49.3	¼	3.0	½	4	4	2
		³⁄₁₆	38.3	¼	3.0	⅜	4	2	2
		⅛	26.4	¼	3.0	⅜	2	2	2
2½	2½	⁵⁄₁₆	46.4	¼	4.0	⅜	4	4	2
		¼	39.0	¼	3.0	⅜	4	2	2
		³⁄₁₆	30.7	¼	3.0	⅜	2	2	2
		⅛	21.4	¼	3.0	⅜	2	2	2
2	2	¼	29.0	¼	2.0	⅜	2	2	2
		³⁄₁₆	23.0	¼	2.0	⅜	2	2	2
		⅛	16.4	¼	2.0	⅜	2	2	2

25% of Member Design Strength									

Table 6-3.
Tension Ties
Slotted HSS/Gusset Plate—Rectangular Sections

Designation			Con-nection Strength (kips)	Weld Size W_w (in.)	Weld Length L_w (in.)	Gusset Plate t_1 (in.)	No. of Bolts n		
H in.	B in.	t in.					¾-in.	⅞-in.	1-in.
8	6	$\frac{5}{8}$	145	$\frac{5}{16}$	8.0	$\frac{1}{2}$	10	8	6
		$\frac{1}{2}$	120	$\frac{1}{4}$	8.0	$\frac{1}{2}$	8	6	6
		$\frac{3}{8}$	92.8	$\frac{1}{4}$	8.0	$\frac{1}{2}$	6	6	4
		$\frac{5}{16}$	78.6	$\frac{1}{4}$	8.0	$\frac{3}{8}$	6	4	4
		$\frac{1}{4}$	63.9	$\frac{1}{4}$	8.0	$\frac{3}{8}$	6	4	4
		$\frac{3}{16}$	48.3	$\frac{1}{4}$	8.0	$\frac{3}{8}$	4	4	2
8	4	$\frac{5}{8}$	121	$\frac{5}{16}$	8.0	$\frac{1}{2}$	8	6	6
		$\frac{1}{2}$	101	$\frac{1}{4}$	8.0	$\frac{1}{2}$	8	6	4
		$\frac{3}{8}$	78.5	$\frac{1}{4}$	8.0	$\frac{1}{2}$	6	4	4
		$\frac{5}{16}$	66.6	$\frac{1}{4}$	8.0	$\frac{1}{2}$	6	4	4
		$\frac{1}{4}$	54.2	$\frac{1}{4}$	8.0	$\frac{1}{2}$	4	4	2
		$\frac{3}{16}$	41.2	$\frac{1}{4}$	8.0	$\frac{3}{8}$	4	2	2
		$\frac{1}{8}$	27.9	$\frac{1}{4}$	8.0	$\frac{3}{8}$	2	2	2
8	3	$\frac{1}{2}$	91.2	$\frac{1}{4}$	8.0	$\frac{1}{2}$	6	6	4
		$\frac{3}{8}$	71.2	$\frac{1}{4}$	8.0	$\frac{1}{2}$	6	4	4
		$\frac{5}{16}$	60.5	$\frac{1}{4}$	8.0	$\frac{1}{2}$	4	4	4
		$\frac{1}{4}$	49.4	$\frac{1}{4}$	8.0	$\frac{1}{2}$	4	4	2
		$\frac{3}{16}$	37.6	$\frac{1}{4}$	8.0	$\frac{3}{8}$	4	2	2
		$\frac{1}{8}$	25.5	$\frac{1}{4}$	8.0	$\frac{3}{8}$	2	2	2
8	2	$\frac{3}{8}$	64.0	$\frac{1}{4}$	8.0	$\frac{1}{2}$	6	4	4
		$\frac{5}{16}$	54.4	$\frac{1}{4}$	8.0	$\frac{1}{2}$	4	4	2
		$\frac{1}{4}$	44.5	$\frac{1}{4}$	8.0	$\frac{3}{8}$	4	4	2
		$\frac{3}{16}$	33.9	$\frac{1}{4}$	8.0	$\frac{3}{8}$	4	2	2
		$\frac{1}{8}$	23.1	$\frac{1}{4}$	8.0	$\frac{3}{8}$	2	2	2
6	4	$\frac{1}{2}$	81.6	$\frac{1}{4}$	6.0	$\frac{1}{2}$	6	4	4
		$\frac{3}{8}$	64.0	$\frac{1}{4}$	6.0	$\frac{1}{2}$	6	4	4
		$\frac{5}{16}$	54.4	$\frac{1}{4}$	6.0	$\frac{1}{2}$	4	4	2
		$\frac{1}{4}$	44.5	$\frac{1}{4}$	6.0	$\frac{3}{8}$	4	4	2
		$\frac{3}{16}$	33.9	$\frac{1}{4}$	6.0	$\frac{3}{8}$	4	2	2
		$\frac{1}{8}$	23.1	$\frac{1}{4}$	6.0	$\frac{3}{8}$	2	2	2

Table 6-3.
Tension Ties
Slotted HSS/Gusset Plate—Rectangular Sections

Designation			Con- nection Strength (kips)	Weld Size W_w (in.)	Weld Length L_w (in.)	Gusset Plate t_1 (in.)	No. of Bolts n		
H in.	B in.	t in.					¾-in.	⅞-in.	1-in.
6	3	½	71.8	¼	6.0	⅜	6	4	4
		⅜	56.7	¼	6.0	⅜	4	4	4
		5⁄16	48.4	¼	6.0	⅜	4	4	2
		¼	39.7	¼	6.0	⅜	4	2	2
		3⁄16	30.3	¼	6.0	⅜	2	2	2
		⅛	20.7	¼	6.0	⅜	2	2	2
6	2	⅜	49.1	¼	6.0	½	4	4	2
		5⁄16	42.2	¼	6.0	⅜	4	2	2
		¼	34.7	¼	6.0	⅜	4	2	2
		3⁄16	26.6	¼	6.0	⅜	2	2	2
		⅛	18.3	¼	6.0	⅜	2	2	2
5	3	½	61.7	¼	5.0	½	4	4	4
		⅜	49.1	¼	5.0	½	4	4	2
		5⁄16	42.2	¼	5.0	⅜	4	2	2
		¼	34.7	¼	5.0	⅜	4	2	2
		3⁄16	26.6	¼	5.0	⅜	2	2	2
		⅛	18.3	¼	5.0	⅜	2	2	2
5	2	⅜	41.6	¼	5.0	⅜	4	2	2
		5⁄16	35.9	¼	5.0	⅜	4	2	2
		¼	29.7	¼	5.0	⅜	2	2	2
		3⁄16	22.9	¼	5.0	⅜	2	2	2
		⅛	15.8	¼	5.0	⅜	2	2	2
4	3	⅜	41.6	¼	4.0	⅜	4	2	2
		5⁄16	35.9	¼	4.0	⅜	4	2	2
		¼	29.7	¼	4.0	⅜	2	2	2
		3⁄16	22.9	¼	4.0	⅜	2	2	2
		⅛	15.8	¼	4.0	⅜	2	2	2
4	2	⅜	34.0	¼	4.0	⅜	4	2	2
		5⁄16	29.6	¼	4.0	⅜	2	2	2
		¼	24.6	¼	4.0	⅜	2	2	2
		3⁄16	19.1	¼	4.0	⅜	2	2	2
		⅛	13.2	¼	4.0	⅜	2	2	2

Table 6-3.
Tension Ties
Slotted HSS/Gusset Plate—Rectangular Sections

Designation			Con-nection Strength (kips)	Weld Size W_w (in.)	Weld Length L_w (in.)	Gusset Plate t_1 (in.)	No. of Bolts n		
H in.	B in.	t in.					¾-in.	⅞-in.	1-in.
8	6	⅝	290	⅜	11.0	1	20	14	12
		½	240	⅜	9.0	⅞	16	12	10
		⅜	186	5/16	9.0	¾	12	10	8
		5/16	157	5/16	8.0	⅝	10	8	6
		¼	128	¼	8.0	½	10	6	6
		3/16	96.7	¼	8.0	½	8	6	4
8	4	⅝	242	⅜	10.0	⅞	16	12	10
		½	202	5/16	10.0	¾	14	10	8
		⅜	157	5/16	8.0	⅝	10	8	6
		5/16	133	¼	8.0	½	10	8	6
		¼	108	¼	8.0	½	8	6	4
		3/16	82.4	¼	8.0	½	6	4	4
		⅛	55.9	¼	8.0	½	4	4	2
8	3	½	182	5/16	9.0	¾	12	10	8
		⅜	142	¼	9.0	½	10	8	6
		5/16	121	¼	8.0	½	8	6	6
		¼	98.7	¼	8.0	½	8	6	4
		3/16	75.1	¼	8.0	½	6	4	4
		⅛	50.9	¼	8.0	½	4	4	2
8	2	⅜	128	¼	8.0	½	10	6	6
		5/16	109	¼	8.0	½	8	6	4
		¼	89.0	¼	8.0	½	6	6	4
		3/16	67.9	¼	8.0	⅜	6	4	4
		⅛	46.2	¼	8.0	⅜	4	4	2
6	4	½	163	⅜	6.0	⅞	12	8	6
		⅜	128	5/16	6.0	⅝	10	6	6
		5/16	109	5/16	6.0	⅝	8	6	4
		¼	89.0	¼	6.0	½	6	6	4
		3/16	67.9	¼	6.0	⅜	6	4	4
		⅛	46.2	¼	6.0	⅜	4	4	2

Table 6-3.
Tension Ties
Slotted HSS/Gusset Plate—Rectangular Sections

Designation			Con-nection Strength (kips)	Weld Size W_w (in.)	Weld Length L_w (in.)	Gusset Plate t_1 (in.)	No. of Bolts n		
H in.	B in.	t in.					$\frac{3}{4}$-in.	$\frac{7}{8}$-in.	1-in.
6	3	$\frac{1}{2}$	144	$\frac{5}{16}$	7.0	$\frac{3}{4}$	10	8	6
		$\frac{3}{8}$	113	$\frac{5}{16}$	6.0	$\frac{5}{8}$	8	6	6
		$\frac{5}{16}$	96.9	$\frac{1}{4}$	6.0	$\frac{1}{2}$	8	6	4
		$\frac{1}{4}$	79.5	$\frac{1}{4}$	6.0	$\frac{1}{2}$	6	4	4
		$\frac{3}{16}$	60.7	$\frac{1}{4}$	6.0	$\frac{3}{8}$	4	4	4
		$\frac{1}{8}$	41.4	$\frac{1}{4}$	6.0	$\frac{3}{8}$	4	2	2
6	2	$\frac{3}{8}$	98.3	$\frac{1}{4}$	6.0	$\frac{1}{2}$	8	6	4
		$\frac{5}{16}$	84.4	$\frac{1}{4}$	6.0	$\frac{1}{2}$	6	4	4
		$\frac{1}{4}$	69.5	$\frac{1}{4}$	6.0	$\frac{1}{2}$	6	4	4
		$\frac{3}{16}$	53.3	$\frac{1}{4}$	6.0	$\frac{1}{2}$	4	4	2
		$\frac{1}{8}$	36.6	$\frac{1}{4}$	6.0	$\frac{3}{8}$	4	2	2
5	3	$\frac{1}{2}$	123	$\frac{3}{8}$	5.0	$\frac{7}{8}$	8	6	6
		$\frac{3}{8}$	98.3	$\frac{5}{16}$	5.0	$\frac{5}{8}$	8	6	4
		$\frac{5}{16}$	84.4	$\frac{5}{16}$	5.0	$\frac{5}{8}$	6	4	4
		$\frac{1}{4}$	69.5	$\frac{1}{4}$	5.0	$\frac{1}{2}$	6	4	4
		$\frac{3}{16}$	53.3	$\frac{1}{4}$	5.0	$\frac{1}{2}$	4	4	2
		$\frac{1}{8}$	36.6	$\frac{1}{4}$	5.0	$\frac{3}{8}$	4	2	2
5	2	$\frac{3}{8}$	83.3	$\frac{5}{16}$	5.0	$\frac{5}{8}$	6	4	4
		$\frac{5}{16}$	71.8	$\frac{1}{4}$	5.0	$\frac{1}{2}$	6	4	4
		$\frac{1}{4}$	59.5	$\frac{1}{4}$	5.0	$\frac{3}{8}$	4	4	4
		$\frac{3}{16}$	45.9	$\frac{1}{4}$	5.0	$\frac{3}{8}$	4	4	2
		$\frac{1}{8}$	31.6	$\frac{1}{4}$	5.0	$\frac{3}{8}$	2	2	2
4	3	$\frac{3}{8}$	83.3	$\frac{5}{16}$	5.0	$\frac{5}{8}$	6	4	4
		$\frac{5}{16}$	71.8	$\frac{1}{4}$	5.0	$\frac{1}{2}$	6	4	4
		$\frac{1}{4}$	59.5	$\frac{1}{4}$	5.0	$\frac{3}{8}$	4	4	4
		$\frac{3}{16}$	45.9	$\frac{1}{4}$	4.0	$\frac{3}{8}$	4	4	2
		$\frac{1}{8}$	31.6	$\frac{1}{4}$	4.0	$\frac{3}{8}$	2	2	2
4	2	$\frac{3}{8}$	68.0	$\frac{1}{4}$	5.0	$\frac{1}{2}$	6	4	4
		$\frac{5}{16}$	59.2	$\frac{1}{4}$	5.0	$\frac{1}{2}$	4	4	4
		$\frac{1}{4}$	49.3	$\frac{1}{4}$	4.0	$\frac{1}{2}$	4	4	2
		$\frac{3}{16}$	38.3	$\frac{1}{4}$	4.0	$\frac{3}{8}$	4	2	2
		$\frac{1}{8}$	26.4	$\frac{1}{4}$	4.0	$\frac{3}{8}$	2	2	2

<table>
<tr><td colspan="9">25% of Member Design Strength</td></tr>
<tr><td colspan="9" align="center">Table 6-3.
Tension Ties
Slotted HSS/Gusset Plate—Round Sections</td></tr>
</table>

Designation		Con- nection Strength (kips)	Weld Size W_w (in.)	Weld Length L_w (in.)	Gusset Plate t_1 (in.)	No. of Bolts n		
D in.	t in.					¾-in.	⅞-in.	1-in.
8.625	0.500	112	¼	9.0	½	8	6	4
	0.375	85.7	¼	9.0	½	6	4	4
	0.322	73.7	¼	9.0	⅜	6	4	4
	0.250	58.0	¼	9.0	⅜	4	4	4
	0.188	43.7	¼	9.0	⅜	4	4	2
7.500	0.500	97.3	¼	8.0	½	8	6	4
	0.375	74.1	¼	8.0	½	6	4	4
	0.312	62.3	¼	8.0	½	4	4	4
	0.250	50.3	¼	8.0	½	4	4	2
	0.188	37.8	¼	8.0	⅜	4	2	2
7.000	0.500	90.2	¼	7.0	½	6	6	4
	0.375	68.9	¼	7.0	⅜	6	4	4
	0.312	57.9	¼	7.0	⅜	4	4	4
	0.250	46.8	¼	7.0	⅜	4	4	2
	0.188	35.2	¼	7.0	⅜	4	2	2
	0.125	23.7	¼	7.0	⅜	2	2	2
6.875	0.500	88.5	¼	7.0	½	6	6	4
	0.375	67.7	¼	7.0	⅜	6	4	4
	0.312	56.9	¼	7.0	⅜	4	4	4
	0.250	45.9	¼	7.0	⅜	4	4	2
	0.188	34.6	¼	7.0	⅜	4	2	2
6.625	0.500	85.1	¼	7.0	½	6	4	4
	0.432	74.5	¼	7.0	½	6	4	4
	0.375	65.0	¼	7.0	½	6	4	4
	0.312	54.7	¼	7.0	½	4	4	2
	0.280	49.3	¼	7.0	½	4	4	2
	0.250	44.2	¼	7.0	⅜	4	4	2
	0.188	33.4	¼	7.0	⅜	4	2	2
	0.125	22.4	¼	7.0	⅜	2	2	2
6.125	0.500	78.2	¼	7.0	½	6	4	4
	0.375	59.8	¼	7.0	½	4	4	4
	0.312	50.4	¼	7.0	½	4	4	2
	0.250	40.7	¼	7.0	⅜	4	2	2
	0.188	30.7	¼	7.0	⅜	2	2	2

Table 6-3.
Tension Ties
Slotted HSS/Gusset Plate—Round Sections

Designation		Con-nection Strength (kips)	Weld Size W_w (in.)	Weld Length L_w (in.)	Gusset Plate t_1 (in.)	No. of Bolts n		
D in.	t in.					¾-in.	⅞-in.	1-in.
6.000	0.500	76.5	¼	6.0	½	6	4	4
	0.375	58.6	¼	6.0	½	4	4	4
	0.312	49.3	¼	6.0	½	4	4	2
	0.280	44.5	¼	6.0	⅜	4	4	2
	0.250	39.9	¼	6.0	⅜	4	2	2
	0.188	30.1	¼	6.0	⅜	2	2	2
	0.125	20.2	¼	6.0	⅜	2	2	2
5.563	0.375	54.1	¼	6.0	½	4	4	2
	0.258	38.1	¼	6.0	⅜	4	2	2
	0.188	27.9	¼	6.0	⅜	2	2	2
	0.134	20.2	¼	6.0	⅜	2	2	2
5.500	0.500	69.6	¼	6.0	½	6	4	4
	0.375	53.4	¼	6.0	½	4	4	2
	0.258	37.6	¼	6.0	⅜	4	2	2
5.000	0.500	62.6	¼	5.0	½	4	4	4
	0.375	48.2	¼	5.0	⅜	4	4	2
	0.312	40.6	¼	5.0	⅜	4	2	2
	0.258	34.0	¼	5.0	⅜	4	2	2
	0.250	33.0	¼	5.0	⅜	4	2	2
	0.188	24.9	¼	5.0	⅜	2	2	2
	0.125	16.8	¼	5.0	⅜	2	2	2
4.500	0.337	39.1	¼	5.0	⅜	4	2	2
	0.237	28.1	¼	5.0	⅜	2	2	2
	0.188	22.3	¼	5.0	⅜	2	2	2
	0.125	15.1	¼	5.0	⅜	2	2	2
4.000	0.337	34.5	¼	4.0	⅜	4	2	2
	0.313	32.0	¼	4.0	⅜	4	2	2
	0.250	26.1	¼	4.0	⅜	2	2	2
	0.237	24.9	¼	4.0	⅜	2	2	2
	0.226	23.7	¼	4.0	⅜	2	2	2
	0.220	23.1	¼	4.0	⅜	2	2	2
	0.188	19.8	¼	4.0	⅜	2	2	2
	0.125	13.4	¼	4.0	⅜	2	2	2

Table 6-3.
Tension Ties
Slotted HSS/Gusset Plate—Round Sections

Designation		Con-nection Strength (kips)	Weld Size W_w (in.)	Weld Length L_w (in.)	Gusset Plate t_1 (in.)	No. of Bolts n		
D in.	t in.					$\frac{3}{4}$-in.	$\frac{7}{8}$-in.	1-in.
8.625	0.500	225	$\frac{5}{16}$	11.0	$\frac{3}{4}$	16	12	8
	0.375	171	$\frac{5}{16}$	9.0	$\frac{5}{8}$	12	8	8
	0.322	147	$\frac{1}{4}$	9.0	$\frac{1}{2}$	10	8	6
	0.250	116	$\frac{1}{4}$	9.0	$\frac{1}{2}$	8	6	6
	0.188	87.3	$\frac{1}{4}$	9.0	$\frac{1}{2}$	6	6	4
7.500	0.500	195	$\frac{5}{16}$	9.0	$\frac{3}{4}$	14	10	8
	0.375	148	$\frac{5}{16}$	8.0	$\frac{5}{8}$	10	8	6
	0.312	125	$\frac{1}{4}$	8.0	$\frac{1}{2}$	8	6	6
	0.250	101	$\frac{1}{4}$	8.0	$\frac{1}{2}$	8	6	4
	0.188	75.6	$\frac{1}{4}$	8.0	$\frac{3}{8}$	6	4	4
7.000	0.500	180	$\frac{5}{16}$	9.0	$\frac{3}{4}$	12	10	8
	0.375	138	$\frac{5}{16}$	7.0	$\frac{5}{8}$	10	8	6
	0.312	116	$\frac{1}{4}$	7.0	$\frac{1}{2}$	8	6	6
	0.250	93.6	$\frac{1}{4}$	7.0	$\frac{1}{2}$	6	6	4
	0.188	70.5	$\frac{1}{4}$	7.0	$\frac{3}{8}$	6	4	4
	0.125	47.4	$\frac{1}{4}$	7.0	$\frac{3}{8}$	4	4	2
6.875	0.500	177	$\frac{5}{16}$	8.0	$\frac{3}{4}$	12	10	8
	0.375	135	$\frac{5}{16}$	7.0	$\frac{5}{8}$	10	8	6
	0.312	114	$\frac{1}{4}$	7.0	$\frac{1}{2}$	8	6	6
	0.250	91.9	$\frac{1}{4}$	7.0	$\frac{1}{2}$	6	6	4
	0.188	69.2	$\frac{1}{4}$	7.0	$\frac{3}{8}$	6	4	4
6.625	0.500	170	$\frac{5}{16}$	8.0	$\frac{3}{4}$	12	8	8
	0.432	149	$\frac{5}{16}$	7.0	$\frac{3}{4}$	10	8	6
	0.375	130	$\frac{5}{16}$	7.0	$\frac{5}{8}$	10	8	6
	0.312	109	$\frac{1}{4}$	7.0	$\frac{1}{2}$	8	6	4
	0.280	98.7	$\frac{1}{4}$	7.0	$\frac{1}{2}$	8	6	4
	0.250	88.5	$\frac{1}{4}$	7.0	$\frac{1}{2}$	6	6	4
	0.188	66.7	$\frac{1}{4}$	7.0	$\frac{3}{8}$	6	4	4
	0.125	44.8	$\frac{1}{4}$	7.0	$\frac{3}{8}$	4	4	2
6.125	0.500	156	$\frac{5}{16}$	8.0	$\frac{3}{4}$	10	8	6
	0.375	120	$\frac{5}{16}$	7.0	$\frac{5}{8}$	8	6	6
	0.312	101	$\frac{1}{4}$	7.0	$\frac{1}{2}$	8	6	4
	0.250	81.5	$\frac{1}{4}$	7.0	$\frac{1}{2}$	6	4	4
	0.188	61.4	$\frac{1}{4}$	7.0	$\frac{3}{8}$	4	4	4

Table 6-3.
Tension Ties
Slotted HSS/Gusset Plate—Round Sections

Designation		Con-nection Strength (kips)	Weld Size W_w (in.)	Weld Length L_w (in.)	Gusset Plate t_1 (in.)	No. of Bolts n		
D in.	t in.					¾-in.	⅞-in.	1-in.
6.000	0.500	153	5/16	7.0	¾	10	8	6
	0.375	117	5/16	6.0	5/8	8	6	6
	0.312	98.7	¼	6.0	½	8	6	4
	0.280	89.0	¼	6.0	½	6	6	4
	0.250	79.8	¼	6.0	½	6	4	4
	0.188	60.1	¼	6.0	3/8	4	4	4
	0.125	40.4	¼	6.0	3/8	4	2	2
5.563	0.375	108	5/16	6.0	5/8	8	6	4
	0.258	76.2	¼	6.0	½	6	4	4
	0.188	55.8	¼	6.0	½	4	4	2
	0.134	40.4	¼	6.0	3/8	4	2	2
5.500	0.500	139	3/8	6.0	⅞	10	8	6
	0.375	107	5/16	6.0	5/8	8	6	4
	0.258	75.2	¼	6.0	½	6	4	4
5.000	0.500	125	3/8	5.0	⅞	8	6	6
	0.375	96.4	5/16	5.0	5/8	8	6	4
	0.312	81.3	¼	5.0	½	6	4	4
	0.258	68.0	¼	5.0	½	6	4	4
	0.250	66.0	¼	5.0	½	6	4	4
	0.188	49.9	¼	5.0	½	4	4	2
	0.125	33.6	¼	5.0	3/8	4	2	2
4.500	0.337	78.2	¼	5.0	½	6	4	4
	0.237	56.1	¼	5.0	½	4	4	2
	0.188	44.6	¼	5.0	3/8	4	4	2
	0.125	30.2	¼	5.0	3/8	2	2	2
4.000	0.337	69.0	¼	5.0	½	6	4	4
	0.313	64.1	¼	4.0	½	6	4	4
	0.250	52.2	¼	4.0	½	4	4	2
	0.237	49.7	¼	4.0	½	4	4	2
	0.226	47.4	¼	4.0	3/8	4	4	2
	0.220	46.1	¼	4.0	3/8	4	4	2
	0.188	39.5	¼	4.0	3/8	4	2	2
	0.125	26.8	¼	4.0	3/8	2	2	2

25% of Member Design Strength		

Table 6-4.
Compression Struts
Slotted HSS/Gusset Plate—Square Sections

Designation			Con-nection Strength (kips)	Weld Size W_w (in.)	Weld Length L_w (in.)	Gusset Plate t_1 (in.)	No. of Bolts n		
H in.	B in.	t in.					$\frac{3}{4}$-in.	$\frac{7}{8}$-in.	1-in.
8	8	$\frac{5}{8}$	126	$\frac{3}{8}$	8.0	$1\frac{1}{2}$	8	6	6
		$\frac{1}{2}$	104	$\frac{3}{8}$	8.0	$1\frac{1}{4}$	8	6	4
		$\frac{3}{8}$	79.8	$\frac{3}{8}$	8.0	$\frac{7}{8}$	6	4	4
		$\frac{5}{16}$	67.2	$\frac{5}{16}$	8.0	$\frac{3}{4}$	6	4	4
		$\frac{1}{4}$	54.5	$\frac{5}{16}$	8.0	$\frac{5}{8}$	4	4	2
		$\frac{3}{16}$	41.2	$\frac{1}{4}$	8.0	$\frac{1}{2}$	4	2	2
7	7	$\frac{5}{8}$	107	$\frac{3}{8}$	7.0	$1\frac{1}{4}$	8	6	4
		$\frac{1}{2}$	89.0	$\frac{3}{8}$	7.0	$1\frac{1}{4}$	6	6	4
		$\frac{3}{8}$	68.8	$\frac{3}{8}$	7.0	$\frac{7}{8}$	6	4	4
		$\frac{5}{16}$	58.2	$\frac{5}{16}$	7.0	$\frac{3}{4}$	4	4	4
		$\frac{1}{4}$	47.3	$\frac{5}{16}$	7.0	$\frac{5}{8}$	4	4	2
		$\frac{3}{16}$	35.8	$\frac{1}{4}$	7.0	$\frac{1}{2}$	4	2	2
6	6	$\frac{5}{8}$	89.8	$\frac{3}{8}$	6.0	$1\frac{1}{4}$	6	6	4
		$\frac{1}{2}$	74.7	$\frac{3}{8}$	6.0	1	6	4	4
		$\frac{3}{8}$	58.2	$\frac{3}{8}$	6.0	$\frac{7}{8}$	4	4	4
		$\frac{5}{16}$	49.3	$\frac{5}{16}$	6.0	$\frac{3}{4}$	4	4	2
		$\frac{1}{4}$	40.2	$\frac{5}{16}$	6.0	$\frac{5}{8}$	4	2	2
		$\frac{3}{16}$	30.5	$\frac{1}{4}$	6.0	$\frac{1}{2}$	2	2	2
		$\frac{1}{8}$	20.7	$\frac{1}{4}$	6.0	$\frac{3}{8}$	2	2	2
5	5	$\frac{1}{2}$	60.5	$\frac{3}{8}$	5.0	1	4	4	4
		$\frac{3}{8}$	47.4	$\frac{5}{16}$	5.0	$\frac{3}{4}$	4	4	2
		$\frac{5}{16}$	40.4	$\frac{5}{16}$	5.0	$\frac{5}{8}$	4	2	2
		$\frac{1}{4}$	33.0	$\frac{5}{16}$	5.0	$\frac{5}{8}$	4	2	2
		$\frac{3}{16}$	25.2	$\frac{1}{4}$	5.0	$\frac{1}{2}$	2	2	2
		$\frac{1}{8}$	17.1	$\frac{1}{4}$	5.0	$\frac{3}{8}$	2	2	2
$4\frac{1}{2}$	$4\frac{1}{2}$	$\frac{1}{2}$	53.3	$\frac{3}{8}$	5.0	$\frac{7}{8}$	4	4	2
		$\frac{3}{8}$	42.0	$\frac{5}{16}$	5.0	$\frac{3}{4}$	4	2	2
		$\frac{5}{16}$	35.9	$\frac{5}{16}$	5.0	$\frac{5}{8}$	4	2	2
		$\frac{1}{4}$	29.5	$\frac{1}{4}$	5.0	$\frac{1}{2}$	2	2	2
		$\frac{3}{16}$	22.5	$\frac{1}{4}$	5.0	$\frac{3}{8}$	2	2	2
		$\frac{1}{8}$	15.3	$\frac{1}{4}$	5.0	$\frac{3}{8}$	2	2	2

25% of Member Design Strength

Table 6-4.
Compression Struts
Slotted HSS/Gusset Plate—Square Sections

Designation			Con-nection Strength (kips)	Weld Size W_w (in.)	Weld Length L_w (in.)	Gusset Plate t_1 (in.)	No. of Bolts n		
H in.	B in.	t in.					¾-in.	⅞-in.	1-in.
4	4	½	46.2	5/16	4.0	¾	4	4	2
		3/8	36.7	5/16	4.0	5/8	4	2	2
		5/16	31.5	¼	4.0	½	2	2	2
		¼	25.9	¼	4.0	½	2	2	2
		3/16	19.8	¼	4.0	3/8	2	2	2
		1/8	13.6	¼	4.0	3/8	2	2	2
3½	3½	3/8	31.4	¼	4.0	½	2	2	2
		5/16	27.0	¼	4.0	½	2	2	2
		¼	22.3	¼	4.0	3/8	2	2	2
		3/16	17.2	¼	4.0	3/8	2	2	2
		1/8	11.8	¼	4.0	3/8	2	2	2
3	3	3/8	26.0	¼	3.0	½	2	2	2
		5/16	22.6	¼	3.0	3/8	2	2	2
		¼	18.7	¼	3.0	3/8	2	2	2
		3/16	14.5	¼	3.0	3/8	2	2	2
		1/8	10.0	¼	3.0	3/8	2	2	2
2½	2½	5/16	18.0	¼	3.0	3/8	2	2	2
		¼	15.1	¼	3.0	3/8	2	2	2
		3/16	11.8	¼	3.0	3/8	2	2	2
		1/8	8.2	¼	3.0	3/8	2	2	2
2	2	¼	11.6	¼	2.0	3/8	2	2	2
		3/16	9.1	¼	2.0	3/8	2	2	2
		1/8	6.4	¼	2.0	3/8	2	2	2

<table>
<tr><td colspan="11" style="background:#ccc">50% of Member Design Strength</td></tr>
</table>

Table 6-4.
Compression Struts
Slotted HSS/Gusset Plate—Square Sections

Designation			Connection Strength (kips)	Weld Size W_w (in.)	Weld Length L_w (in.)	Gusset Plate t_1 (in.)	No. of Bolts n		
H in.	B in.	t in.					$\frac{3}{4}$-in.	$\frac{7}{8}$-in.	1-in.
8	8	$\frac{5}{8}$	252	$\frac{3}{8}$	13.0	$2\frac{3}{4}$	16	12	10
		$\frac{1}{2}$	207	$\frac{3}{8}$	11.0	$2\frac{1}{4}$	14	10	8
		$\frac{3}{8}$	160	$\frac{3}{8}$	8.0	$1\frac{3}{4}$	12	8	6
		$\frac{5}{16}$	134	$\frac{3}{8}$	8.0	$1\frac{1}{2}$	10	8	6
		$\frac{1}{4}$	109	$\frac{3}{8}$	8.0	$1\frac{1}{4}$	8	6	4
		$\frac{3}{16}$	82.4	$\frac{3}{8}$	8.0	$\frac{7}{8}$	6	4	4
7	7	$\frac{5}{8}$	215	$\frac{3}{8}$	11.0	$2\frac{1}{2}$	14	10	8
		$\frac{1}{2}$	178	$\frac{3}{8}$	9.0	$2\frac{1}{4}$	12	10	8
		$\frac{3}{8}$	138	$\frac{3}{8}$	7.0	$1\frac{3}{4}$	10	8	6
		$\frac{5}{16}$	116	$\frac{3}{8}$	7.0	$1\frac{1}{2}$	8	6	6
		$\frac{1}{4}$	94.7	$\frac{3}{8}$	7.0	$1\frac{1}{4}$	6	6	4
		$\frac{3}{16}$	71.7	$\frac{3}{8}$	7.0	$\frac{7}{8}$	6	4	4
6	6	$\frac{5}{8}$	180	$\frac{3}{8}$	9.0	$2\frac{1}{2}$	12	10	8
		$\frac{1}{2}$	149	$\frac{3}{8}$	8.0	2	10	8	6
		$\frac{3}{8}$	116	$\frac{3}{8}$	6.0	$1\frac{1}{2}$	8	6	6
		$\frac{5}{16}$	98.7	$\frac{3}{8}$	6.0	$1\frac{1}{2}$	8	6	4
		$\frac{1}{4}$	80.4	$\frac{3}{8}$	6.0	$1\frac{1}{4}$	6	4	4
		$\frac{3}{16}$	61.1	$\frac{3}{8}$	6.0	$\frac{7}{8}$	4	4	4
		$\frac{1}{8}$	41.4	$\frac{5}{16}$	6.0	$\frac{5}{8}$	4	2	2
5	5	$\frac{1}{2}$	121	$\frac{3}{8}$	6.0	2	8	6	6
		$\frac{3}{8}$	94.8	$\frac{3}{8}$	5.0	$1\frac{1}{2}$	6	6	4
		$\frac{5}{16}$	80.7	$\frac{3}{8}$	5.0	$1\frac{1}{4}$	6	4	4
		$\frac{1}{4}$	66.0	$\frac{3}{8}$	5.0	$1\frac{1}{4}$	6	4	4
		$\frac{3}{16}$	50.3	$\frac{3}{8}$	5.0	$\frac{7}{8}$	4	4	2
		$\frac{1}{8}$	34.2	$\frac{5}{16}$	5.0	$\frac{5}{8}$	4	2	2
$4\frac{1}{2}$	$4\frac{1}{2}$	$\frac{1}{2}$	107	$\frac{3}{8}$	6.0	$1\frac{3}{4}$	8	6	4
		$\frac{3}{8}$	84.1	$\frac{3}{8}$	5.0	$1\frac{1}{2}$	6	4	4
		$\frac{5}{16}$	71.8	$\frac{3}{8}$	5.0	$1\frac{1}{4}$	6	4	4
		$\frac{1}{4}$	58.9	$\frac{3}{8}$	5.0	1	4	4	4
		$\frac{3}{16}$	45.0	$\frac{5}{16}$	5.0	$\frac{3}{4}$	4	4	2
		$\frac{1}{8}$	30.7	$\frac{1}{4}$	5.0	$\frac{1}{2}$	2	2	2

Table 6-4.
Compression Struts
Slotted HSS/Gusset Plate—Square Sections

Designation			Con-nection Strength (kips)	Weld Size W_w (in.)	Weld Length L_w (in.)	Gusset Plate t_1 (in.)	No. of Bolts n		
H in.	B in.	t in.					¾-in.	⅞-in.	1-in.
4	4	½	92.4	⅜	5.0	1½	6	6	4
		⅜	73.4	⅜	4.0	1¼	6	4	4
		5/16	62.9	⅜	4.0	1	4	4	4
		¼	51.7	⅜	4.0	⅞	4	4	2
		3/16	39.6	5/16	4.0	⅝	4	2	2
		⅛	27.2	¼	4.0	½	2	2	2
3½	3½	⅜	62.8	⅜	4.0	1	4	4	4
		5/16	54.0	⅜	4.0	⅞	4	4	2
		¼	44.7	5/16	4.0	¾	4	4	2
		3/16	34.4	5/16	4.0	⅝	4	2	2
		⅛	23.6	¼	4.0	⅜	2	2	2
3	3	⅜	52.0	⅜	3.0	⅞	4	4	2
		5/16	45.1	5/16	3.0	¾	4	4	2
		¼	37.4	5/16	3.0	⅝	4	2	2
		3/16	29.0	¼	3.0	½	2	2	2
		⅛	19.9	¼	3.0	⅜	2	2	2
2½	2½	5/16	36.1	5/16	3.0	⅝	4	2	2
		¼	30.2	¼	3.0	½	2	2	2
		3/16	23.6	¼	3.0	⅜	2	2	2
		⅛	16.4	¼	3.0	⅜	2	2	2
2	2	¼	23.2	¼	2.0	⅜	2	2	2
		3/16	18.3	¼	2.0	⅜	2	2	2
		⅛	12.9	¼	2.0	⅜	2	2	2

<table>
<tr><td colspan="11">25% of Member Design Strength</td></tr>
<tr><td colspan="11">Table 6-4.
Compression Struts
Slotted HSS/Gusset Plate—Rectangular Sections</td></tr>
</table>

Designation			Con-nection Strength (kips)	Weld Size W_w (in.)	Weld Length L_w (in.)	Gusset Plate t_1 (in.)	No. of Bolts n		
H in.	B in.	t in.					¾-in.	⅞-in.	1-in.
8	6	⅝	107	⅜	8.0	1¼	8	6	4
		½	89.0	⅜	8.0	1	6	6	4
		⅜	68.8	5⁄16	8.0	¾	6	4	4
		5⁄16	58.2	5⁄16	8.0	⅝	4	4	4
		¼	47.3	5⁄16	8.0	⅝	4	4	2
		3⁄16	35.8	¼	8.0	½	4	2	2
8	4	⅝	89.8	⅜	8.0	1	6	6	4
		½	74.7	⅜	8.0	⅞	6	4	4
		⅜	58.2	5⁄16	8.0	⅝	4	4	4
		5⁄16	49.3	5⁄16	8.0	⅝	4	4	2
		¼	40.2	¼	8.0	½	4	2	2
		3⁄16	30.5	¼	8.0	⅜	2	2	2
		⅛	20.7	¼	8.0	⅜	2	2	2
8	3	½	67.6	5⁄16	8.0	¾	6	4	4
		⅜	52.8	5⁄16	8.0	⅝	4	4	2
		5⁄16	44.9	¼	8.0	½	4	4	2
		¼	36.6	¼	8.0	½	4	2	2
		3⁄16	27.9	¼	8.0	⅜	2	2	2
		⅛	18.9	¼	8.0	⅜	2	2	2
8	2	⅜	47.4	5⁄16	8.0	⅝	4	4	2
		5⁄16	40.4	¼	8.0	½	4	2	2
		¼	33.0	¼	8.0	½	4	2	2
		3⁄16	25.2	¼	8.0	⅜	2	2	2
		⅛	17.1	¼	8.0	⅜	2	2	2
6	4	½	60.5	⅜	6.0	⅞	4	4	4
		⅜	47.4	5⁄16	6.0	¾	4	4	2
		5⁄16	40.4	5⁄16	6.0	⅝	4	2	2
		¼	33.0	¼	6.0	½	4	2	2
		3⁄16	25.2	¼	6.0	⅜	2	2	2
		⅛	17.1	¼	6.0	⅜	2	2	2

Table 6-4.
Compression Struts
Slotted HSS/Gusset Plate—Rectangular Sections

Designation			Con-nection Strength (kips)	Weld Size W_w (in.)	Weld Length L_w (in.)	Gusset Plate t_1 (in.)	No. of Bolts n		
H in.	B in.	t in.					$\frac{3}{4}$-in.	$\frac{7}{8}$-in.	1-in.
6	3	$\frac{1}{2}$	53.3	$\frac{5}{16}$	6.0	$\frac{3}{4}$	4	4	2
		$\frac{3}{8}$	42.0	$\frac{5}{16}$	6.0	$\frac{5}{8}$	4	2	2
		$\frac{5}{16}$	35.9	$\frac{1}{4}$	6.0	$\frac{1}{2}$	4	2	2
		$\frac{1}{4}$	29.5	$\frac{1}{4}$	6.0	$\frac{1}{2}$	2	2	2
		$\frac{3}{16}$	22.5	$\frac{1}{4}$	6.0	$\frac{3}{8}$	2	2	2
		$\frac{1}{8}$	15.3	$\frac{1}{4}$	6.0	$\frac{3}{8}$	2	2	2
6	2	$\frac{3}{8}$	36.7	$\frac{5}{16}$	6.0	$\frac{5}{8}$	4	2	2
		$\frac{5}{16}$	31.5	$\frac{1}{4}$	6.0	$\frac{1}{2}$	2	2	2
		$\frac{1}{4}$	25.9	$\frac{1}{4}$	6.0	$\frac{3}{8}$	2	2	2
		$\frac{3}{16}$	19.8	$\frac{1}{4}$	6.0	$\frac{3}{8}$	2	2	2
		$\frac{1}{8}$	13.6	$\frac{1}{4}$	6.0	$\frac{3}{8}$	2	2	2
5	3	$\frac{1}{2}$	46.2	$\frac{5}{16}$	5.0	$\frac{3}{4}$	4	4	2
		$\frac{3}{8}$	36.7	$\frac{5}{16}$	5.0	$\frac{5}{8}$	4	2	2
		$\frac{5}{16}$	31.5	$\frac{1}{4}$	5.0	$\frac{1}{2}$	2	2	2
		$\frac{1}{4}$	25.9	$\frac{1}{4}$	5.0	$\frac{1}{2}$	2	2	2
		$\frac{3}{16}$	19.8	$\frac{1}{4}$	5.0	$\frac{3}{8}$	2	2	2
		$\frac{1}{8}$	13.6	$\frac{1}{4}$	5.0	$\frac{3}{8}$	2	2	2
5	2	$\frac{3}{8}$	31.4	$\frac{1}{4}$	5.0	$\frac{1}{2}$	2	2	2
		$\frac{5}{16}$	27.0	$\frac{1}{4}$	5.0	$\frac{1}{2}$	2	2	2
		$\frac{1}{4}$	22.3	$\frac{1}{4}$	5.0	$\frac{3}{8}$	2	2	2
		$\frac{3}{16}$	17.2	$\frac{1}{4}$	5.0	$\frac{3}{8}$	2	2	2
		$\frac{1}{8}$	11.8	$\frac{1}{4}$	5.0	$\frac{3}{8}$	2	2	2
4	3	$\frac{3}{8}$	31.4	$\frac{1}{4}$	4.0	$\frac{1}{2}$	2	2	2
		$\frac{5}{16}$	27.0	$\frac{1}{4}$	4.0	$\frac{1}{2}$	2	2	2
		$\frac{1}{4}$	22.3	$\frac{1}{4}$	4.0	$\frac{3}{8}$	2	2	2
		$\frac{3}{16}$	17.2	$\frac{1}{4}$	4.0	$\frac{3}{8}$	2	2	2
		$\frac{1}{8}$	11.8	$\frac{1}{4}$	4.0	$\frac{3}{8}$	2	2	2
4	2	$\frac{3}{8}$	26.0	$\frac{1}{4}$	4.0	$\frac{1}{2}$	2	2	2
		$\frac{5}{16}$	22.6	$\frac{1}{4}$	4.0	$\frac{3}{8}$	2	2	2
		$\frac{1}{4}$	18.7	$\frac{1}{4}$	4.0	$\frac{3}{8}$	2	2	2
		$\frac{3}{16}$	14.5	$\frac{1}{4}$	4.0	$\frac{3}{8}$	2	2	2
		$\frac{1}{8}$	10.0	$\frac{1}{4}$	4.0	$\frac{3}{8}$	2	2	2

Table 6-4.
Compression Struts
Slotted HSS/Gusset Plate—Rectangular Sections

Designation			Con-nection Strength (kips)	Weld Size W_w (in.)	Weld Length L_w (in.)	Gusset Plate t_1 (in.)	No. of Bolts n		
H in.	B in.	t in.					³⁄₄-in.	⁷⁄₈-in.	1-in.
8	6	⁵⁄₈	215	³⁄₈	11.0	2¼	14	10	8
		½	178	³⁄₈	9.0	2	12	10	8
		³⁄₈	138	³⁄₈	8.0	1½	10	8	6
		⁵⁄₁₆	116	³⁄₈	8.0	1¼	8	6	6
		¼	94.7	³⁄₈	8.0	1	6	6	4
		³⁄₁₆	71.7	³⁄₈	8.0	⁷⁄₈	6	4	4
8	4	⁵⁄₈	180	³⁄₈	9.0	2	12	10	8
		½	149	³⁄₈	8.0	1¾	10	8	6
		³⁄₈	116	³⁄₈	8.0	1¼	8	6	6
		⁵⁄₁₆	98.7	³⁄₈	8.0	1¼	8	6	4
		¼	80.4	³⁄₈	8.0	⁷⁄₈	6	4	4
		³⁄₁₆	61.1	⁵⁄₁₆	8.0	³⁄₄	4	4	4
		⅛	41.4	¼	8.0	½	4	2	2
8	3	½	135	³⁄₈	8.0	1½	10	8	6
		³⁄₈	106	³⁄₈	8.0	1¼	8	6	4
		⁵⁄₁₆	89.8	³⁄₈	8.0	1	6	6	4
		¼	73.2	³⁄₈	8.0	⁷⁄₈	6	4	4
		³⁄₁₆	55.7	⁵⁄₁₆	8.0	⁵⁄₈	4	4	2
		⅛	37.7	¼	8.0	½	4	2	2
8	2	³⁄₈	94.8	³⁄₈	8.0	1	6	6	4
		⁵⁄₁₆	80.7	³⁄₈	8.0	⁷⁄₈	6	4	4
		¼	66.0	⁵⁄₁₆	8.0	³⁄₄	6	4	4
		³⁄₁₆	50.3	⁵⁄₁₆	8.0	⁵⁄₈	4	4	2
		⅛	34.2	¼	8.0	½	4	2	2
6	4	½	121	³⁄₈	6.0	1¾	8	6	6
		³⁄₈	94.8	³⁄₈	6.0	1¼	6	6	4
		⁵⁄₁₆	80.7	³⁄₈	6.0	1¼	6	4	4
		¼	66.0	³⁄₈	6.0	⁷⁄₈	6	4	4
		³⁄₁₆	50.3	⁵⁄₁₆	6.0	³⁄₄	4	4	2
		⅛	34.2	¼	6.0	½	4	2	2

Table 6-4.
Compression Struts
Slotted HSS/Gusset Plate—Rectangular Sections

Designation			Con-nection Strength (kips)	Weld Size W_w (in.)	Weld Length L_w (in.)	Gusset Plate t_1 (in.)	No. of Bolts n		
H in.	B in.	t in.					¾-in.	⅞-in.	1-in.
6	3	½	107	⅜	6.0	1½	8	6	4
		⅜	84.1	⅜	6.0	1¼	6	4	4
		5/16	71.8	⅜	6.0	1	6	4	4
		¼	58.9	⅜	6.0	⅞	4	4	4
		3/16	45.0	5/16	6.0	⅝	4	4	2
		⅛	30.7	¼	6.0	½	2	2	2
6	2	⅜	73.4	⅜	6.0	1	6	4	4
		5/16	62.9	⅜	6.0	⅞	4	4	4
		¼	51.7	5/16	6.0	¾	4	4	2
		3/16	39.6	5/16	6.0	⅝	4	2	2
		⅛	27.2	¼	6.0	⅜	2	2	2
5	3	½	92.4	⅜	5.0	1½	6	6	4
		⅜	73.4	⅜	5.0	1¼	6	4	4
		5/16	62.9	⅜	5.0	1	4	4	4
		¼	51.7	⅜	5.0	⅞	4	4	2
		3/16	39.6	5/16	5.0	⅝	4	2	2
		⅛	27.2	¼	5.0	½	2	2	2
5	2	⅜	62.8	⅜	5.0	1	4	4	4
		5/16	54.0	⅜	5.0	⅞	4	4	2
		¼	44.7	5/16	5.0	¾	4	4	2
		3/16	34.4	5/16	5.0	⅝	4	2	2
		⅛	23.6	¼	5.0	⅜	2	2	2
4	3	⅜	62.8	⅜	4.0	1	4	4	4
		5/16	54.0	⅜	4.0	⅞	4	4	2
		¼	44.7	5/16	4.0	¾	4	4	2
		3/16	34.4	5/16	4.0	⅝	4	2	2
		⅛	23.6	¼	4.0	⅜	2	2	2
4	2	⅜	52.0	⅜	4.0	⅞	4	4	2
		5/16	45.1	5/16	4.0	¾	4	4	2
		¼	37.4	5/16	4.0	⅝	4	2	2
		3/16	29.0	¼	4.0	½	2	2	2
		⅛	19.9	¼	4.0	⅜	2	2	2

Table 6-4.
Compression Struts
Slotted HSS/Gusset Plate—Round Sections

Designation		Connection Strength (kips)	Weld Size W_w (in.)	Weld Length L_w (in.)	Gusset Plate t_1 (in.)	No. of Bolts n		
D in.	t in.					¾-in.	⅞-in.	1-in.
8.625	0.500	91.3	⅜	9.0	1	6	6	4
	0.375	69.6	5⁄16	9.0	¾	6	4	4
	0.322	59.8	5⁄16	9.0	⅝	4	4	4
	0.250	47.1	¼	9.0	½	4	4	2
	0.188	35.4	¼	9.0	½	4	2	2
7.500	0.500	79.0	⅜	8.0	⅞	6	4	4
	0.375	60.2	5⁄16	8.0	¾	4	4	4
	0.312	50.6	5⁄16	8.0	⅝	4	4	2
	0.250	40.8	¼	8.0	½	4	2	2
	0.188	30.7	¼	8.0	⅜	2	2	2
7.000	0.500	73.3	⅜	7.0	⅞	6	4	4
	0.375	55.9	5⁄16	7.0	¾	4	4	2
	0.312	47.0	5⁄16	7.0	⅝	4	4	2
	0.250	38.0	¼	7.0	½	4	2	2
	0.188	28.6	¼	7.0	⅜	2	2	2
	0.125	19.3	¼	7.0	⅜	2	2	2
6.875	0.500	71.8	⅜	7.0	⅞	6	4	4
	0.375	54.9	5⁄16	7.0	¾	4	4	2
	0.312	46.2	5⁄16	7.0	⅝	4	4	2
	0.250	37.3	¼	7.0	½	4	2	2
	0.188	28.1	¼	7.0	⅜	2	2	2
6.625	0.500	69.1	⅜	7.0	⅞	6	4	4
	0.432	60.5	5⁄16	7.0	¾	4	4	4
	0.375	52.8	5⁄16	7.0	¾	4	4	2
	0.312	44.4	5⁄16	7.0	⅝	4	4	2
	0.280	40.1	5⁄16	7.0	⅝	4	2	2
	0.250	35.9	¼	7.0	½	4	2	2
	0.188	27.1	¼	7.0	⅜	2	2	2
	0.125	18.2	¼	7.0	⅜	2	2	2
6.125	0.500	63.5	⅜	7.0	⅞	4	4	4
	0.375	48.6	5⁄16	7.0	¾	4	4	2
	0.312	40.9	5⁄16	7.0	⅝	4	2	2
	0.250	33.1	¼	7.0	½	4	2	2
	0.188	24.9	¼	7.0	⅜	2	2	2

Table 6-4.
Compression Struts
Slotted HSS/Gusset Plate—Round Sections

Designation		Con-nection Strength (kips)	Weld Size W_w (in.)	Weld Length L_w (in.)	Gusset Plate t_1 (in.)	No. of Bolts n		
D in.	t in.					$\frac{3}{4}$-in.	$\frac{7}{8}$-in.	1-in.
6.000	0.500	62.1	$\frac{3}{8}$	6.0	$\frac{7}{8}$	4	4	4
	0.375	47.6	$\frac{5}{16}$	6.0	$\frac{3}{4}$	4	4	2
	0.312	40.1	$\frac{5}{16}$	6.0	$\frac{5}{8}$	4	2	2
	0.280	36.1	$\frac{1}{4}$	6.0	$\frac{1}{2}$	4	2	2
	0.250	32.4	$\frac{1}{4}$	6.0	$\frac{1}{2}$	4	2	2
	0.188	24.4	$\frac{1}{4}$	6.0	$\frac{3}{8}$	2	2	2
	0.125	16.4	$\frac{1}{4}$	6.0	$\frac{3}{8}$	2	2	2
5.563	0.375	43.9	$\frac{5}{16}$	6.0	$\frac{5}{8}$	4	4	2
	0.258	30.9	$\frac{1}{4}$	6.0	$\frac{1}{2}$	2	2	2
	0.188	22.6	$\frac{1}{4}$	6.0	$\frac{3}{8}$	2	2	2
	0.134	16.4	$\frac{1}{4}$	6.0	$\frac{3}{8}$	2	2	2
5.500	0.500	56.5	$\frac{3}{8}$	6.0	$\frac{7}{8}$	4	4	2
	0.375	43.4	$\frac{5}{16}$	6.0	$\frac{5}{8}$	4	4	2
	0.258	30.5	$\frac{1}{4}$	6.0	$\frac{1}{2}$	2	2	2
5.000	0.500	50.8	$\frac{3}{8}$	5.0	$\frac{7}{8}$	4	4	2
	0.375	39.1	$\frac{5}{16}$	5.0	$\frac{5}{8}$	4	2	2
	0.312	33.0	$\frac{5}{16}$	5.0	$\frac{5}{8}$	4	2	2
	0.258	27.6	$\frac{1}{4}$	5.0	$\frac{1}{2}$	2	2	2
	0.250	26.8	$\frac{1}{4}$	5.0	$\frac{1}{2}$	2	2	2
	0.188	20.3	$\frac{1}{4}$	5.0	$\frac{3}{8}$	2	2	2
	0.125	13.7	$\frac{1}{4}$	5.0	$\frac{3}{8}$	2	2	2
4.500	0.337	31.8	$\frac{1}{4}$	5.0	$\frac{1}{2}$	2	2	2
	0.237	22.8	$\frac{1}{4}$	5.0	$\frac{3}{8}$	2	2	2
	0.188	18.1	$\frac{1}{4}$	5.0	$\frac{3}{8}$	2	2	2
	0.125	12.3	$\frac{1}{4}$	5.0	$\frac{3}{8}$	2	2	2
4.000	0.337	28.0	$\frac{1}{4}$	4.0	$\frac{1}{2}$	2	2	2
	0.313	26.0	$\frac{1}{4}$	4.0	$\frac{1}{2}$	2	2	2
	0.250	21.2	$\frac{1}{4}$	4.0	$\frac{3}{8}$	2	2	2
	0.237	20.2	$\frac{1}{4}$	4.0	$\frac{3}{8}$	2	2	2
	0.226	19.3	$\frac{1}{4}$	4.0	$\frac{3}{8}$	2	2	2
	0.220	18.7	$\frac{1}{4}$	4.0	$\frac{3}{8}$	2	2	2
	0.188	16.0	$\frac{1}{4}$	4.0	$\frac{3}{8}$	2	2	2
	0.125	10.9	$\frac{1}{4}$	4.0	$\frac{3}{8}$	2	2	2

Table 6-4.
Compression Struts
Slotted HSS/Gusset Plate—Round Sections

Designation		Con-nection Strength (kips)	Weld Size W_w (in.)	Weld Length L_w (in.)	Gusset Plate t_1 (in.)	No. of Bolts n		
D in.	t in.					¾-in.	⅞-in.	1-in.
8.625	0.500	183	⅜	9.0	1¾	12	10	8
	0.375	139	⅜	9.0	1½	10	8	6
	0.322	120	⅜	9.0	1¼	8	6	6
	0.250	94.2	⅜	9.0	1	6	6	4
	0.188	70.9	5/16	9.0	¾	6	4	4
7.500	0.500	158	⅜	8.0	1¾	10	8	6
	0.375	120	⅜	8.0	1½	8	6	6
	0.312	101	⅜	8.0	1¼	8	6	4
	0.250	81.6	⅜	8.0	1	6	4	4
	0.188	61.4	5/16	8.0	¾	4	4	4
7.000	0.500	147	⅜	7.0	1¾	10	8	6
	0.375	112	⅜	7.0	1½	8	6	4
	0.312	94.1	⅜	7.0	1¼	6	6	4
	0.250	76.0	⅜	7.0	1	6	4	4
	0.188	57.2	5/16	7.0	¾	4	4	4
	0.125	38.5	¼	7.0	½	4	2	2
6.875	0.500	144	⅜	7.0	1¾	10	8	6
	0.375	110	⅜	7.0	1½	8	6	4
	0.312	92.4	⅜	7.0	1¼	6	6	4
	0.250	74.6	⅜	7.0	1	6	4	4
	0.188	56.2	5/16	7.0	¾	4	4	2
6.625	0.500	138	⅜	7.0	1¾	10	8	6
	0.432	121	⅜	7.0	1½	8	6	6
	0.375	106	⅜	7.0	1½	8	6	4
	0.312	88.8	⅜	7.0	1¼	6	6	4
	0.280	80.1	⅜	7.0	1	6	4	4
	0.250	71.8	⅜	7.0	⅞	6	4	4
	0.188	54.2	5/16	7.0	¾	4	4	2
	0.125	36.4	¼	7.0	½	4	2	2
6.125	0.500	127	⅜	7.0	1¾	8	6	6
	0.375	97.1	⅜	7.0	1¼	8	6	4
	0.312	81.8	⅜	7.0	1¼	6	4	4
	0.250	66.1	⅜	7.0	⅞	6	4	4
	0.188	49.9	5/16	7.0	¾	4	4	2

Table 6-4.
Compression Struts
Slotted HSS/Gusset Plate—Round Sections

Designation		Con-nection Strength (kips)	Weld Size W_w (in.)	Weld Length L_w (in.)	Gusset Plate t_1 (in.)	No. of Bolts n		
D in.	t in.					$\frac{3}{4}$-in.	$\frac{7}{8}$-in.	1-in.
6.000	0.500	124	$\frac{3}{8}$	6.0	$1\frac{3}{4}$	8	6	6
	0.375	95.1	$\frac{3}{8}$	6.0	$1\frac{1}{4}$	6	6	4
	0.312	80.1	$\frac{3}{8}$	6.0	$1\frac{1}{4}$	6	4	4
	0.280	72.3	$\frac{3}{8}$	6.0	1	6	4	4
	0.250	64.8	$\frac{3}{8}$	6.0	$\frac{7}{8}$	6	4	4
	0.188	48.8	$\frac{5}{16}$	6.0	$\frac{3}{4}$	4	4	2
	0.125	32.8	$\frac{1}{4}$	6.0	$\frac{1}{2}$	4	2	2
5.563	0.375	87.8	$\frac{3}{8}$	6.0	$1\frac{1}{4}$	6	6	4
	0.258	61.8	$\frac{3}{8}$	6.0	$\frac{7}{8}$	4	4	4
	0.188	45.3	$\frac{5}{16}$	6.0	$\frac{3}{4}$	4	4	2
	0.134	32.8	$\frac{1}{4}$	6.0	$\frac{1}{2}$	4	2	2
5.500	0.500	113	$\frac{3}{8}$	6.0	$1\frac{3}{4}$	8	6	4
	0.375	86.7	$\frac{3}{8}$	6.0	$1\frac{1}{4}$	6	6	4
	0.258	61.1	$\frac{3}{8}$	6.0	$\frac{7}{8}$	4	4	4
5.000	0.500	102	$\frac{3}{8}$	5.0	$1\frac{3}{4}$	8	6	4
	0.375	78.3	$\frac{3}{8}$	5.0	$1\frac{1}{4}$	6	4	4
	0.312	66.0	$\frac{3}{8}$	5.0	$1\frac{1}{4}$	6	4	4
	0.258	55.2	$\frac{3}{8}$	5.0	$\frac{7}{8}$	4	4	2
	0.250	53.6	$\frac{3}{8}$	5.0	$\frac{7}{8}$	4	4	2
	0.188	40.5	$\frac{5}{16}$	5.0	$\frac{3}{4}$	4	2	2
	0.125	27.3	$\frac{1}{4}$	5.0	$\frac{1}{2}$	2	2	2
4.500	0.337	63.5	$\frac{3}{8}$	5.0	1	4	4	4
	0.237	45.6	$\frac{5}{16}$	5.0	$\frac{3}{4}$	4	4	2
	0.188	36.2	$\frac{5}{16}$	5.0	$\frac{5}{8}$	4	2	2
	0.125	24.6	$\frac{1}{4}$	5.0	$\frac{1}{2}$	2	2	2
4.000	0.337	56.0	$\frac{3}{8}$	4.0	$\frac{7}{8}$	4	4	2
	0.313	52.0	$\frac{3}{8}$	4.0	$\frac{7}{8}$	4	4	2
	0.250	42.4	$\frac{5}{16}$	4.0	$\frac{3}{4}$	4	2	2
	0.237	40.4	$\frac{5}{16}$	4.0	$\frac{5}{8}$	4	2	2
	0.226	38.5	$\frac{5}{16}$	4.0	$\frac{5}{8}$	4	2	2
	0.220	37.4	$\frac{5}{16}$	4.0	$\frac{5}{8}$	4	2	2
	0.188	32.1	$\frac{5}{16}$	4.0	$\frac{5}{8}$	4	2	2
	0.125	21.8	$\frac{1}{4}$	4.0	$\frac{3}{8}$	2	2	2

END PLATE TABLES

General Notes

The End Plate Tables can be used for selecting the weld size, plate thickness, and number of bolts required to develop the Connection Strength listed on the left side of the table. Tables 6-5, 6-6, and 6-7 are used for tension ties. For compression struts see Chapter 6, End Plates (Axial Compression). The connection design strength is based on a percentage of the member design strength of the HSS. Tables 6-5 and 6-6 are provided for 75 percent and 100 percent of the member design strength. Table 6-7 is provided for 25 percent and 50 percent of the member design strength. For each HSS, the member design strength is computed using the specifications found in the AISC LRFD Manual, Second Edition. The tables are presented for $\frac{3}{4}$-, $\frac{7}{8}$-, and 1-in. A325 bolts.

Material Specifications

The tables are presented for commonly used square, rectangular, and round HSS manufactured under ASTM A500 Grade B. Plates are manufactured to ASTM A36 with $F_y = 36$ ksi and $F_u = 58$ ksi. The weld metal used has a minimum specified strength, $F_{EXX} = 70$ ksi. All bolts are manufactured to ASTM A325 Specifications.

Design Procedure

All connection results are calculated following the limit state procedures outlined in Chapter 6, End Plates (Axial Tension).

Table 6-5: End Plates on Round Sections

The values of t_1, n, and b are determined by an iterative procedure which first calculates the required number of bolts based on $b = 1.5$ in. and the percent of member design strength. Then, the distance b is determined so that a minimum bolt spacing of 3.0 in. is maintained. With this new value of b, the number of bolts required is recalculated. This continues until the values converge on a solution.

The thickness of the end plate, t_1, is then calculated. For thicknesses < 1 in. and > 1 in., t_1 is rounded up to the nearest $\frac{1}{8}$-in. and $\frac{1}{4}$-in., respectively.

The size of the weld connecting the end plate to the HSS, W_w, is then determined and rounded up to the nearest $\frac{1}{16}$-in. The weld size, W_w, is then checked for the minimum size shown in AISC LRFD Table J2.4 based on the end plate thickness, and increased as required.

The values of W_w, t_1, n, and b are shown in Table 6-5.

Table 6-6: End Plate Bolted on Four Sides

To minimize the plate thickness required, the maximum number of bolts that can be distributed around the HSS, while maintaining a spacing of 3-inches, is used. With the number of bolts selected, the thickness of the end plate, t_1, is calculated. For thicknesses < 1 in. and > 1 in., t_1 is rounded up to the nearest $\frac{1}{8}$-in. and $\frac{1}{4}$-in., respectively. The plate thicknesses are shown for the different bolt sizes used. The bolts are then checked for prying action.

The size of the weld connecting the end plate to the HSS, W_w, is then determined and rounded up to the nearest $\frac{1}{16}$-in. The weld size, W_w, is then checked for the minimum size shown in AISC LRFD Table J2.4 based on the end plate thickness, and increased as required.

The values of W_w, t_1, and n are shown in Table 6-6. If "—" is shown, the bolt group cannot resist the given percent of member design strength.

Table 6-7: End Plates Bolted on Two Sides

As with end plates bolted on four sides, the maximum number of bolts which can be placed along the face of the HSS, using a bolt spacing of 3 inches is used. For rectangular sections the bolts are distributed along the long face of the HSS. With the number of bolts selected, the thickness of the end plate, t_1, is calculated. The value of t_1 is then limited to a maximum of 1 in. as explained previously. The plate thicknesses are shown for the different bolt sizes used. The bolts are then checked for prying action.

The size of the weld connecting the end plate to the HSS, W_w, is then determined and rounded up to the nearest $\frac{1}{16}$-in. The weld size, W_w, is then checked for the minimum size shown in AISC LRFD Table J2.4 based on the end plate thickness, and increased as required.

The values of W_w, t_1, and n are shown in Table 6-7. If "—" is shown, the bolt group or plate thickness could not resist the given percent of member design strength.

Design Parameters

The following values are held constant for calculations in Tables 6-5, 6-6, and 6-7:

Plates: $F_y = 36$ ksi
           $F_u = 58$ ksi

Welds: $F_{EXX} = 70$ ksi
           $F_w = 63$ ksi

HSS: Square and Rectangular   Round
        $F_y = 46$ ksi $F_y = 42$ ksi
        $F_u = 58$ ksi $F_u = 58$ ksi

Bolts: ASTM A325 $s = 3.0$ in
                        $b = a = 1.5$ in minimum

For different parameters, the user is directed to the general solution presented earlier for evaluating the connection design strength.

75% of Member Design Strength										

Table 6-5.
Tension Ties
Bolted End Plates—Round Sections

Designation		Con-nection Strength (kips)	Weld Size W_w (in.)	End Plate t_1 (in.)	No. of Bolts n					
					3/4-in.		7/8-in.		1-in.	
D in.	t in.				n	$a = b$	n	$a = b$	n	$a = b$
8.625	0.500	337	3/8	1 1/4	18	2 1/4	13	1 7/8	10	1 1/2
	0.375	257	5/16	1	14	1 1/2	10	1 1/2	8	1 1/2
	0.322	221	5/16	7/8	12	1 1/2	9	1 1/2	7	1 1/2
	0.250	174	1/4	3/4	9	1 1/2	7	1 1/2	5	1 1/2
	0.188	131	1/4	5/8	7	1 1/2	5	1 1/2	4	1 1/2
7.500	0.500	292	3/8	1 1/4	15	1 7/8	11	1 5/8	9	1 1/2
	0.375	222	5/16	7/8	12	1 1/2	9	1 1/2	7	1 1/2
	0.312	187	5/16	7/8	10	1 1/2	7	1 1/2	6	1 1/2
	0.250	151	1/4	3/4	8	1 1/2	6	1 1/2	5	1 1/2
	0.188	113	1/4	5/8	6	1 1/2	5	1 1/2	4	1 1/2
7.000	0.500	271	3/8	1 1/4	14	1 5/8	11	1 3/4	8	1 1/2
	0.375	207	5/16	7/8	11	1 1/2	8	1 1/2	6	1 1/2
	0.312	174	5/16	7/8	9	1 1/2	7	1 1/2	5	1 1/2
	0.250	140	1/4	3/4	8	1 1/2	6	1 1/2	4	1 1/2
	0.188	106	1/4	5/8	6	1 1/2	4	1 1/2	4	1 1/2
	0.125	71.2	3/16	1/2	4	1 1/2	4	1 1/2	4	1 1/2
6.875	0.500	265	3/8	1 1/4	14	1 5/8	11	1 1/2	8	1 1/2
	0.375	203	5/16	7/8	11	1 1/2	8	1 1/2	6	1 1/2
	0.312	171	5/16	7/8	9	1 1/2	7	1 1/2	5	1 1/2
	0.250	138	1/4	3/4	7	1 1/2	6	1 1/2	4	1 1/2
	0.188	104	1/4	5/8	6	1 1/2	4	1 1/2	4	1 1/2
6.625	0.500	255	3/8	1 1/4	13	1 3/4	10	1 1/2	8	1 1/2
	0.432	223	3/8	1	12	1 1/2	9	1 1/2	7	1 1/2
	0.375	195	5/16	7/8	10	1 1/2	8	1 1/2	6	1 1/2
	0.312	164	5/16	7/8	9	1 1/2	7	1 1/2	5	1 1/2
	0.280	148	1/4	3/4	8	1 1/2	6	1 1/2	5	1 1/2
	0.250	133	1/4	3/4	7	1 1/2	5	1 1/2	4	1 1/2
	0.188	100	1/4	5/8	5	1 1/2	4	1 1/2	4	1 1/2
	0.125	67.2	3/16	1/2	4	1 1/2	4	1 1/2	4	1 1/2
6.125	0.500	234	3/8	1 1/4	12	1 5/8	9	1 1/2	7	1 1/2
	0.375	179	5/16	7/8	10	1 1/2	7	1 1/2	6	1 1/2
	0.312	151	5/16	7/8	8	1 1/2	6	1 1/2	5	1 1/2
	0.250	122	1/4	3/4	7	1 1/2	5	1 1/2	4	1 1/2
	0.188	92.1	1/4	5/8	5	1 1/2	4	1 1/2	4	1 1/2

Table 6-5.
Tension Ties
Bolted End Plates—Round Sections

Designation		Con-nection Strength (kips)	Weld Size W_w (in.)	End Plate t_1 (in.)	No. of Bolts n					
					¾-in.		⅞-in.		1-in.	
D in.	t in.				n	$a = b$	n	$a = b$	n	$a = b$
6.000	0.500	229	⅜	1	12	1½	9	1½	7	1½
	0.375	176	⁵⁄₁₆	⅞	9	1½	7	1½	6	1½
	0.312	148	⁵⁄₁₆	⅞	8	1½	6	1½	5	1½
	0.280	134	¼	¾	7	1½	5	1½	4	1½
	0.250	120	¼	¾	6	1½	5	1½	4	1½
	0.188	90.2	¼	⅝	5	1½	4	1½	4	1½
	0.125	60.7	³⁄₁₆	½	4	1½	4	1½	4	1½
5.563	0.375	162	⁵⁄₁₆	⅞	9	1½	7	1½	5	1½
	0.258	114	¼	¾	6	1½	5	1½	4	1½
	0.188	83.6	¼	⅝	5	1½	4	1½	4	1½
	0.134	60.7	³⁄₁₆	½	4	1½	4	1½	4	1½
5.500	0.500	209	⅜	1	11	1½	8	1½	7	1½
	0.375	160	⁵⁄₁₆	⅞	9	1½	6	1½	5	1½
	0.258	113	¼	¾	6	1½	5	1½	4	1½
5.000	0.500	188	⅜	1	10	1½	8	1½	6	1½
	0.375	145	⁵⁄₁₆	⅞	8	1½	6	1½	5	1½
	0.312	122	¼	¾	7	1½	5	1½	4	1½
	0.258	102	¼	¾	6	1½	4	1½	4	1½
	0.250	98.9	¼	¾	5	1½	4	1½	4	1½
	0.188	74.8	¼	⅝	4	1½	4	1½	4	1½
	0.125	50.5	³⁄₁₆	½	4	1½	4	1½	4	1½
4.500	0.337	117	⁵⁄₁₆	⅞	6	1½	5	1½	4	1½
	0.237	84.2	¼	¾	5	1½	4	1½	4	1½
	0.188	66.9	¼	⅝	4	1½	4	1½	4	1½
	0.125	45.4	³⁄₁₆	½	4	1½	4	1½	4	1½
4.000	0.337	103	¼	¾	6	1½	4	1½	4	1½
	0.313	96.1	¼	¾	5	1½	4	1½	4	1½
	0.250	78.2	¼	¾	4	1½	4	1½	4	1½
	0.237	74.6	¼	⅝	4	1½	4	1½	4	1½
	0.226	71.2	¼	⅝	4	1½	4	1½	4	1½
	0.220	69.2	¼	⅝	4	1½	4	1½	4	1½
	0.188	59.3	¼	⅝	4	1½	4	1½	4	1½
	0.125	40.3	³⁄₁₆	½	4	1½	4	1½	4	1½

Table 6-5.
Tension Ties
Bolted End Plates—Round Sections

Designation		Con-nection Strength (kips)	Weld Size W_w (in.)	End Plate t_1 (in.)	No. of Bolts n					
					$\frac{3}{4}$-in.		$\frac{7}{8}$-in.		1-in.	
D in.	t in.				n	$a = b$	n	$a = b$	n	$a = b$
8.625	0.500	450	$\frac{1}{2}$	$1\frac{3}{4}$	22	4	17	$2\frac{7}{8}$	13	$2\frac{1}{2}$
	0.375	343	$\frac{7}{16}$	$1\frac{1}{4}$	18	$2\frac{1}{4}$	13	$1\frac{1}{2}$	10	$1\frac{1}{2}$
	0.322	295	$\frac{3}{8}$	1	15	$1\frac{1}{2}$	11	$1\frac{1}{2}$	9	$1\frac{1}{2}$
	0.250	232	$\frac{5}{16}$	$\frac{7}{8}$	12	$1\frac{1}{2}$	9	$1\frac{1}{2}$	7	$1\frac{1}{2}$
	0.188	175	$\frac{1}{4}$	$\frac{3}{4}$	9	$1\frac{1}{2}$	7	$1\frac{1}{2}$	5	$1\frac{1}{2}$
7.500	0.500	389	$\frac{1}{2}$	$1\frac{3}{4}$	20	$3\frac{1}{2}$	15	$2\frac{5}{8}$	12	2.125
	0.375	296	$\frac{7}{16}$	$1\frac{1}{4}$	15	2	12	$1\frac{1}{2}$	9	$1\frac{1}{2}$
	0.312	249	$\frac{3}{8}$	1	13	$1\frac{1}{2}$	10	$1\frac{1}{2}$	8	$1\frac{1}{2}$
	0.250	201	$\frac{5}{16}$	$\frac{7}{8}$	11	$1\frac{1}{2}$	8	$1\frac{1}{2}$	6	$1\frac{1}{2}$
	0.188	151	$\frac{1}{4}$	$\frac{3}{4}$	8	$1\frac{1}{2}$	6	$1\frac{1}{2}$	5	$1\frac{1}{2}$
7.000	0.500	361	$\frac{1}{2}$	$1\frac{1}{2}$	18	3	14	$2\frac{3}{8}$	11	$1\frac{7}{8}$
	0.375	276	$\frac{7}{16}$	$1\frac{1}{4}$	14	$1\frac{7}{8}$	11	$1\frac{1}{2}$	8	$1\frac{1}{2}$
	0.312	232	$\frac{3}{8}$	1	12	$1\frac{1}{2}$	9	$1\frac{1}{2}$	7	$1\frac{1}{2}$
	0.250	187	$\frac{5}{16}$	$\frac{7}{8}$	10	$1\frac{1}{2}$	7	$1\frac{1}{2}$	6	$1\frac{1}{2}$
	0.188	141	$\frac{1}{4}$	$\frac{3}{4}$	8	$1\frac{1}{2}$	6	$1\frac{1}{2}$	4	$1\frac{1}{2}$
	0.125	94.9	$\frac{1}{4}$	$\frac{5}{8}$	5	$1\frac{1}{2}$	4	$1\frac{1}{2}$	4	$1\frac{1}{2}$
6.875	0.500	354	$\frac{1}{2}$	$1\frac{1}{2}$	18	3.125	13	$2\frac{1}{2}$	11	$1\frac{7}{8}$
	0.375	271	$\frac{7}{16}$	$1\frac{1}{4}$	14	$1\frac{5}{8}$	11	$1\frac{1}{2}$	8	$1\frac{1}{2}$
	0.312	228	$\frac{3}{8}$	1	12	$1\frac{1}{2}$	9	$1\frac{1}{2}$	7	$1\frac{1}{2}$
	0.250	184	$\frac{5}{16}$	$\frac{7}{8}$	10	$1\frac{1}{2}$	7	$1\frac{1}{2}$	6	$1\frac{1}{2}$
	0.188	138	$\frac{1}{4}$	$\frac{3}{4}$	7	$1\frac{1}{2}$	6	$1\frac{1}{2}$	4	$1\frac{1}{2}$
6.625	0.500	340	$\frac{1}{2}$	$1\frac{1}{2}$	17	$2\frac{7}{8}$	13	2.125	10	2
	0.432	298	$\frac{7}{16}$	$1\frac{1}{4}$	15	$2\frac{1}{4}$	12	$1\frac{3}{4}$	9	$1\frac{5}{8}$
	0.375	260	$\frac{3}{8}$	$1\frac{1}{4}$	14	$1\frac{3}{4}$	10	$1\frac{1}{2}$	8	$1\frac{1}{2}$
	0.312	219	$\frac{3}{8}$	1	12	$1\frac{1}{2}$	9	$1\frac{1}{2}$	7	$1\frac{1}{2}$
	0.280	197	$\frac{5}{16}$	$\frac{7}{8}$	10	$1\frac{1}{2}$	8	$1\frac{1}{2}$	6	$1\frac{1}{2}$
	0.250	177	$\frac{5}{16}$	$\frac{7}{8}$	9	$1\frac{1}{2}$	7	$1\frac{1}{2}$	5	$1\frac{1}{2}$
	0.188	133	$\frac{1}{4}$	$\frac{3}{4}$	7	$1\frac{1}{2}$	5	$1\frac{1}{2}$	4	$1\frac{1}{2}$
	0.125	89.6	$\frac{1}{4}$	$\frac{5}{8}$	5	$1\frac{1}{2}$	4	$1\frac{1}{2}$	4	$1\frac{1}{2}$
6.125	0.500	313	$\frac{1}{2}$	$1\frac{1}{2}$	16	$2\frac{3}{4}$	12	2	9	$1\frac{3}{4}$
	0.375	239	$\frac{3}{8}$	$1\frac{1}{4}$	13	$1\frac{5}{8}$	9	$1\frac{1}{2}$	7	$1\frac{1}{2}$
	0.312	201	$\frac{3}{8}$	1	11	$1\frac{1}{2}$	8	$1\frac{1}{2}$	6	$1\frac{1}{2}$
	0.250	163	$\frac{5}{16}$	$\frac{7}{8}$	9	$1\frac{1}{2}$	6	$1\frac{1}{2}$	5	$1\frac{1}{2}$
	0.188	123	$\frac{1}{4}$	$\frac{3}{4}$	7	$1\frac{1}{2}$	5	$1\frac{1}{2}$	4	$1\frac{1}{2}$

Table 6-5.
Tension Ties
Bolted End Plates—Round Sections

Designation		Con-nection Strength (kips)	Weld Size W_w (in.)	End Plate t_1 (in.)	No. of Bolts n					
					¾-in.		⅞-in.		1-in.	
D in.	t in.				n	$a = b$	n	$a = b$	n	$a = b$
6.000	0.500	306	½	1½	16	2¾	12	2.125	9	1¾
	0.375	234	⅜	1	12	1½	9	1½	7	1½
	0.312	197	⅜	1	10	1½	8	1½	6	1½
	0.280	178	5⁄16	⅞	9	1½	7	1½	6	1½
	0.250	160	5⁄16	⅞	8	1½	6	1½	5	1½
	0.188	120	¼	¾	6	1½	5	1½	4	1½
	0.125	80.9	¼	⅝	4	1½	4	1½	4	1½
5.563	0.375	216	⅜	1	11	1½	9	1½	7	1½
	0.258	152	5⁄16	⅞	8	1½	6	1½	5	1½
	0.188	112	¼	¾	6	1½	5	1½	4	1½
	0.134	80.9	¼	⅝	4	1½	4	1½	4	1½
5.500	0.500	278	½	1½	14	2⅜	11	1⅞	8	1⅝
	0.375	214	⅜	1	11	1½	8	1½	7	1½
	0.258	150	5⁄16	⅞	8	1½	6	1½	5	1½
5.000	0.500	250	½	1¼	13	2¼	10	1¾	8	1¾
	0.375	193	⅜	1	10	1½	8	1½	6	1½
	0.312	163	5⁄16	⅞	9	1½	6	1½	5	1½
	0.258	136	5⁄16	⅞	7	1½	5	1½	4	1½
	0.250	132	5⁄16	⅞	7	1½	5	1½	4	1½
	0.188	99.8	¼	¾	5	1½	4	1½	4	1½
	0.125	67.3	¼	⅝	4	1½	4	1½	4	1½
4.500	0.337	156	⅜	1	8	1½	6	1½	5	1½
	0.237	112	¼	¾	6	1½	5	1½	4	1½
	0.188	89.2	¼	¾	5	1½	4	1½	4	1½
	0.125	60.5	¼	⅝	4	1½	4	1½	4	1½
4.000	0.337	138	⅜	⅞	7	1½	6	1½	4	1½
	0.313	128	5⁄16	⅞	7	1½	5	1½	4	1½
	0.250	104	¼	¾	6	1½	4	1½	4	1½
	0.237	99.4	¼	¾	5	1½	4	1½	4	1½
	0.226	94.9	¼	¾	5	1½	4	1½	4	1½
	0.220	92.2	¼	¾	5	1½	4	1½	4	1½
	0.188	79.0	¼	¾	4	1½	4	1½	4	1½
	0.125	53.7	¼	⅝	4	1½	4	1½	4	1½

Table 6-6.
Tension Ties
End Plates Bolted on Four Sides—Square Sections

Designation			Con-nection Strength (kips)	Weld Size W_w (in.)	Cap Plate, t_1 (in.) for bolt size			No. of Bolts/side
H in.	B in.	t in.			$\frac{3}{4}$-in.	$\frac{7}{8}$-in.	1-in.	n
8	8	$\frac{5}{8}$	509	$\frac{1}{2}$	—	—	$1\frac{1}{4}$	3
		$\frac{1}{2}$	419	$\frac{7}{16}$	—	$1\frac{1}{4}$	$1\frac{1}{4}$	3
		$\frac{3}{8}$	323	$\frac{5}{16}$	$1\frac{1}{4}$	1	1	3
		$\frac{5}{16}$	272	$\frac{5}{16}$	$\frac{7}{8}$	$\frac{7}{8}$	$\frac{7}{8}$	3
		$\frac{1}{4}$	220	$\frac{5}{16}$	$\frac{7}{8}$	$\frac{7}{8}$	$\frac{7}{8}$	3
		$\frac{3}{16}$	167	$\frac{1}{4}$	$\frac{3}{4}$	$\frac{3}{4}$	$\frac{3}{4}$	3
7	7	$\frac{5}{8}$	435	$\frac{1}{2}$	—	$1\frac{1}{2}$	$1\frac{1}{4}$	3
		$\frac{1}{2}$	360	$\frac{7}{16}$	$1\frac{1}{2}$	$1\frac{1}{4}$	$1\frac{1}{4}$	3
		$\frac{3}{8}$	279	$\frac{5}{16}$	$1\frac{1}{4}$	$1\frac{1}{4}$	1	3
		$\frac{5}{16}$	236	$\frac{5}{16}$	1	1	1	3
		$\frac{1}{4}$	192	$\frac{5}{16}$	$\frac{7}{8}$	$\frac{7}{8}$	$\frac{7}{8}$	3
		$\frac{3}{16}$	145	$\frac{1}{4}$	$\frac{3}{4}$	$\frac{3}{4}$	$\frac{3}{4}$	3
6	6	$\frac{5}{8}$	363	$\frac{1}{2}$	$1\frac{3}{4}$	$1\frac{1}{2}$	$1\frac{1}{2}$	3
		$\frac{1}{2}$	302	$\frac{7}{16}$	$1\frac{1}{2}$	$1\frac{1}{2}$	$1\frac{1}{2}$	3
		$\frac{3}{8}$	235	$\frac{5}{16}$	$1\frac{1}{4}$	$1\frac{1}{4}$	$1\frac{1}{4}$	3
		$\frac{5}{16}$	200	$\frac{5}{16}$	$1\frac{1}{4}$	$1\frac{1}{4}$	$1\frac{1}{4}$	3
		$\frac{1}{4}$	163	$\frac{5}{16}$	1	1	1	3
		$\frac{3}{16}$	124	$\frac{5}{16}$	$\frac{7}{8}$	$\frac{7}{8}$	$\frac{7}{8}$	3
5	5	$\frac{1}{2}$	245	$\frac{3}{8}$	$1\frac{1}{2}$	$1\frac{1}{4}$	1	2
		$\frac{3}{8}$	192	$\frac{5}{16}$	1	1	$\frac{7}{8}$	2
		$\frac{5}{16}$	163	$\frac{5}{16}$	$\frac{7}{8}$	$\frac{7}{8}$	$\frac{7}{8}$	2
		$\frac{1}{4}$	134	$\frac{1}{4}$	$\frac{3}{4}$	$\frac{3}{4}$	$\frac{3}{4}$	2
		$\frac{3}{16}$	102	$\frac{1}{4}$	$\frac{3}{4}$	$\frac{3}{4}$	$\frac{3}{4}$	2
		$\frac{1}{8}$	69.2	$\frac{1}{4}$	$\frac{5}{8}$	$\frac{5}{8}$	$\frac{5}{8}$	2
$4\frac{1}{2}$	$4\frac{1}{2}$	$\frac{1}{2}$	216	$\frac{3}{8}$	$1\frac{1}{4}$	$1\frac{1}{4}$	1	2
		$\frac{3}{8}$	170	$\frac{5}{16}$	1	1	1	2
		$\frac{5}{16}$	145	$\frac{5}{16}$	$\frac{7}{8}$	$\frac{7}{8}$	$\frac{7}{8}$	2
		$\frac{1}{4}$	119	$\frac{5}{16}$	$\frac{7}{8}$	$\frac{3}{4}$	$\frac{3}{4}$	2
		$\frac{3}{16}$	91.0	$\frac{1}{4}$	$\frac{3}{4}$	$\frac{3}{4}$	$\frac{3}{4}$	2
		$\frac{1}{8}$	62.1	$\frac{1}{4}$	$\frac{5}{8}$	$\frac{5}{8}$	$\frac{5}{8}$	2

Table 6-6.
Tension Ties
End Plates Bolted on Four Sides—Square Sections

Designation			Con-nection Strength (kips)	Weld Size W_w (in.)	Cap Plate, t_1 (in.) for bolt size			No. of Bolts/side
H in.	B in.	t in.			$\frac{3}{4}$-in.	$\frac{7}{8}$-in.	1-in.	n
4	4	$\frac{1}{2}$	187	$\frac{3}{8}$	$1\frac{1}{4}$	$1\frac{1}{4}$	$1\frac{1}{4}$	2
		$\frac{3}{8}$	148	$\frac{5}{16}$	1	1	1	2
		$\frac{5}{16}$	127	$\frac{5}{16}$	$\frac{7}{8}$	$\frac{7}{8}$	$\frac{7}{8}$	2
		$\frac{1}{4}$	105	$\frac{5}{16}$	$\frac{7}{8}$	$\frac{7}{8}$	$\frac{3}{4}$	2
		$\frac{3}{16}$	80.1	$\frac{1}{4}$	$\frac{3}{4}$	$\frac{3}{4}$	$\frac{3}{4}$	2
		$\frac{1}{8}$	55.0	$\frac{1}{4}$	$\frac{5}{8}$	$\frac{5}{8}$	$\frac{5}{8}$	2
$3\frac{1}{2}$	$3\frac{1}{2}$	$\frac{3}{8}$	127	$\frac{5}{16}$	1	1	1	2
		$\frac{5}{16}$	109	$\frac{5}{16}$	$\frac{7}{8}$	$\frac{7}{8}$	$\frac{7}{8}$	2
		$\frac{1}{4}$	90.4	$\frac{5}{16}$	$\frac{7}{8}$	$\frac{7}{8}$	$\frac{7}{8}$	2
		$\frac{3}{16}$	69.6	$\frac{1}{4}$	$\frac{3}{4}$	$\frac{3}{4}$	$\frac{3}{4}$	2
		$\frac{1}{8}$	47.8	$\frac{1}{4}$	$\frac{5}{8}$	$\frac{5}{8}$	$\frac{5}{8}$	2
3	3	$\frac{3}{8}$	105	$\frac{5}{16}$	1	1	1	2
		$\frac{5}{16}$	91.3	$\frac{5}{16}$	$\frac{7}{8}$	$\frac{7}{8}$	$\frac{7}{8}$	2
		$\frac{1}{4}$	75.8	$\frac{5}{16}$	$\frac{7}{8}$	$\frac{7}{8}$	$\frac{7}{8}$	2
		$\frac{3}{16}$	58.7	$\frac{1}{4}$	$\frac{3}{4}$	$\frac{3}{4}$	$\frac{3}{4}$	2
		$\frac{1}{8}$	40.4	$\frac{1}{4}$	$\frac{5}{8}$	$\frac{5}{8}$	$\frac{5}{8}$	2
$2\frac{1}{2}$	$2\frac{1}{2}$	$\frac{5}{16}$	73.0	$\frac{5}{16}$	$\frac{7}{8}$	$\frac{7}{8}$	$\frac{7}{8}$	1
		$\frac{1}{4}$	61.2	$\frac{1}{4}$	$\frac{3}{4}$	$\frac{3}{4}$	$\frac{3}{4}$	1
		$\frac{3}{16}$	47.8	$\frac{1}{4}$	$\frac{3}{4}$	$\frac{3}{4}$	$\frac{5}{8}$	1
		$\frac{1}{8}$	33.2	$\frac{1}{4}$	$\frac{5}{8}$	$\frac{5}{8}$	$\frac{5}{8}$	1
2	2	$\frac{1}{4}$	46.9	$\frac{1}{4}$	$\frac{3}{4}$	$\frac{3}{4}$	$\frac{3}{4}$	1
		$\frac{3}{16}$	36.9	$\frac{1}{4}$	$\frac{3}{4}$	$\frac{3}{4}$	$\frac{3}{4}$	1
		$\frac{1}{8}$	26.1	$\frac{1}{4}$	$\frac{5}{8}$	$\frac{5}{8}$	$\frac{5}{8}$	1

Table 6-6.
Tension Ties
End Plates Bolted on Four Sides—Square Sections

Designation			Con-nection Strength (kips)	Weld Size W_w (in.)	Cap Plate, t_1 (in.) for bolt size			No. of Bolts/side
H in.	B in.	t in.			$\frac{3}{4}$-in.	$\frac{7}{8}$-in.	1-in.	n
8	8	$\frac{5}{8}$	—	—	—	—	—	—
		$\frac{1}{2}$	559	$\frac{9}{16}$	—	—	$1\frac{1}{2}$	3
		$\frac{3}{8}$	431	$\frac{7}{16}$	—	$1\frac{1}{4}$	$1\frac{1}{4}$	3
		$\frac{5}{16}$	363	$\frac{3}{8}$	$1\frac{1}{2}$	1	1	3
		$\frac{1}{4}$	294	$\frac{5}{16}$	1	1	1	3
		$\frac{3}{16}$	222	$\frac{5}{16}$	$\frac{7}{8}$	$\frac{7}{8}$	$\frac{7}{8}$	3
7	7	$\frac{5}{8}$	580	$\frac{5}{8}$	—	—	$1\frac{3}{4}$	3
		$\frac{1}{2}$	480	$\frac{9}{16}$	—	$1\frac{3}{4}$	$1\frac{1}{2}$	3
		$\frac{3}{8}$	371	$\frac{7}{16}$	—	$1\frac{1}{4}$	$1\frac{1}{4}$	3
		$\frac{5}{16}$	314	$\frac{3}{8}$	$1\frac{1}{4}$	$1\frac{1}{4}$	$1\frac{1}{4}$	3
		$\frac{1}{4}$	255	$\frac{5}{16}$	1	1	1	3
		$\frac{3}{16}$	193	$\frac{5}{16}$	$\frac{7}{8}$	$\frac{7}{8}$	$\frac{7}{8}$	3
6	6	$\frac{5}{8}$	484	$\frac{5}{8}$	—	2	$1\frac{3}{4}$	3
		$\frac{1}{2}$	403	$\frac{9}{16}$	—	$1\frac{1}{2}$	$1\frac{1}{2}$	3
		$\frac{3}{8}$	314	$\frac{7}{16}$	$1\frac{1}{2}$	$1\frac{1}{2}$	$1\frac{1}{2}$	3
		$\frac{5}{16}$	266	$\frac{3}{8}$	$1\frac{1}{4}$	$1\frac{1}{4}$	$1\frac{1}{4}$	3
		$\frac{1}{4}$	217	$\frac{5}{16}$	$1\frac{1}{4}$	$1\frac{1}{4}$	$1\frac{1}{4}$	3
		$\frac{3}{16}$	165	$\frac{5}{16}$	1	1	1	3
5	5	$\frac{1}{2}$	326	$\frac{1}{2}$	—	$1\frac{1}{2}$	$1\frac{1}{4}$	2
		$\frac{3}{8}$	256	$\frac{7}{16}$	—	$1\frac{1}{4}$	$1\frac{1}{4}$	2
		$\frac{5}{16}$	218	$\frac{3}{8}$	$1\frac{1}{4}$	1	1	2
		$\frac{1}{4}$	178	$\frac{5}{16}$	$\frac{7}{8}$	$\frac{7}{8}$	$\frac{7}{8}$	2
		$\frac{3}{16}$	136	$\frac{5}{16}$	$\frac{7}{8}$	$\frac{3}{4}$	$\frac{3}{4}$	2
		$\frac{1}{8}$	92.3	$\frac{1}{4}$	$\frac{5}{8}$	$\frac{5}{8}$	$\frac{5}{8}$	2
$4\frac{1}{2}$	$4\frac{1}{2}$	$\frac{1}{2}$	288	$\frac{1}{2}$	—	$1\frac{1}{2}$	$1\frac{1}{4}$	2
		$\frac{3}{8}$	227	$\frac{7}{16}$	$1\frac{1}{2}$	$1\frac{1}{4}$	$1\frac{1}{4}$	2
		$\frac{5}{16}$	194	$\frac{3}{8}$	1	1	1	2
		$\frac{1}{4}$	159	$\frac{5}{16}$	$\frac{7}{8}$	$\frac{7}{8}$	$\frac{7}{8}$	2
		$\frac{3}{16}$	121	$\frac{5}{16}$	$\frac{7}{8}$	$\frac{7}{8}$	$\frac{3}{4}$	2
		$\frac{1}{8}$	82.8	$\frac{1}{4}$	$\frac{3}{4}$	$\frac{5}{8}$	$\frac{5}{8}$	2

Table 6-6.
Tension Ties
End Plates Bolted on Four Sides—Square Sections

Designation			Con- nection Strength (kips)	Weld Size W_w (in.)	Cap Plate, t_1 (in.) for bolt size			No. of Bolts/side
H in.	B in.	t in.			$\frac{3}{4}$-in.	$\frac{7}{8}$-in.	1-in.	n
4	4	$\frac{1}{2}$	249	$\frac{1}{2}$	—	$1\frac{1}{4}$	$1\frac{1}{4}$	2
		$\frac{3}{8}$	198	$\frac{3}{8}$	$1\frac{1}{4}$	$1\frac{1}{4}$	$1\frac{1}{4}$	2
		$\frac{5}{16}$	170	$\frac{3}{8}$	1	1	1	2
		$\frac{1}{4}$	140	$\frac{5}{16}$	1	$\frac{7}{8}$	$\frac{7}{8}$	2
		$\frac{3}{16}$	107	$\frac{5}{16}$	$\frac{7}{8}$	$\frac{7}{8}$	$\frac{7}{8}$	2
		$\frac{1}{8}$	73.3	$\frac{1}{4}$	$\frac{3}{4}$	$\frac{3}{4}$	$\frac{3}{4}$	2
$3\frac{1}{2}$	$3\frac{1}{2}$	$\frac{3}{8}$	169	$\frac{3}{8}$	$1\frac{1}{4}$	$1\frac{1}{4}$	$1\frac{1}{4}$	2
		$\frac{5}{16}$	146	$\frac{5}{16}$	1	1	1	2
		$\frac{1}{4}$	120	$\frac{5}{16}$	1	1	1	2
		$\frac{3}{16}$	92.7	$\frac{5}{16}$	$\frac{7}{8}$	$\frac{7}{8}$	$\frac{7}{8}$	2
		$\frac{1}{8}$	63.8	$\frac{1}{4}$	$\frac{3}{4}$	$\frac{3}{4}$	$\frac{3}{4}$	2
3	3	$\frac{3}{8}$	140	$\frac{3}{8}$	$1\frac{1}{4}$	$1\frac{1}{4}$	$1\frac{1}{4}$	2
		$\frac{5}{16}$	122	$\frac{5}{16}$	1	1	$1\frac{1}{4}$	2
		$\frac{1}{4}$	101	$\frac{5}{16}$	1	1	1	2
		$\frac{3}{16}$	78.2	$\frac{5}{16}$	$\frac{7}{8}$	$\frac{7}{8}$	$\frac{7}{8}$	2
		$\frac{1}{8}$	53.8	$\frac{1}{4}$	$\frac{3}{4}$	$\frac{3}{4}$	$\frac{3}{4}$	2
$2\frac{1}{2}$	$2\frac{1}{2}$	$\frac{5}{16}$	97.3	$\frac{5}{16}$	1	1	1	1
		$\frac{1}{4}$	81.6	$\frac{5}{16}$	$\frac{7}{8}$	$\frac{7}{8}$	$\frac{7}{8}$	1
		$\frac{3}{16}$	63.8	$\frac{1}{4}$	$\frac{3}{4}$	$\frac{3}{4}$	$\frac{3}{4}$	1
		$\frac{1}{8}$	44.3	$\frac{1}{4}$	$\frac{5}{8}$	$\frac{5}{8}$	$\frac{5}{8}$	1
2	2	$\frac{1}{4}$	62.5	$\frac{5}{16}$	$\frac{7}{8}$	$\frac{7}{8}$	$\frac{7}{8}$	1
		$\frac{3}{16}$	49.3	$\frac{1}{4}$	$\frac{3}{4}$	$\frac{3}{4}$	$\frac{3}{4}$	1
		$\frac{1}{8}$	34.8	$\frac{1}{4}$	$\frac{5}{8}$	$\frac{5}{8}$	$\frac{5}{8}$	1

Table 6-6.
Tension Ties
End Plates Bolted on Four Sides—Rectangular Sections

Designation			Con-nection Strength (kips)	Weld Size W_w (in.)	Cap Plate, t_1 (in.) for bolt size			No. of Bolts/side
H in.	B in.	t in.			¾-in.	⅞-in.	1-in.	n
8	6	⅝	435	½	—	1¾	1¾	3 / 3
		½	360	⁷⁄₁₆	1¾	1½	1½	3 / 3
		⅜	279	⁵⁄₁₆	1¼	1¼	1¼	3 / 3
		⁵⁄₁₆	236	⁵⁄₁₆	1¼	1¼	1¼	3 / 3
		¼	192	⁵⁄₁₆	1¼	1¼	1¼	3 / 3
		³⁄₁₆	145	⁵⁄₁₆	1	1	1	3 / 3
8	4	⅝	363	½	—	1½	1¼	3 / 2
		½	302	⁷⁄₁₆	1½	1¼	1¼	3 / 2
		⅜	235	⁵⁄₁₆	1¼	1¼	1¼	3 / 2
		⁵⁄₁₆	200	⁵⁄₁₆	1	1	1	3 / 2
		¼	163	⁵⁄₁₆	⅞	⅞	⅞	3 / 2
		³⁄₁₆	124	¼	¾	¾	¾	3 / 2
		⅛	83.8	¼	⅝	⅝	⅝	3 / 2
8	3	½	274	⅜	1½	1½	1½	3 / 2
		⅜	214	⁵⁄₁₆	1¼	1¼	1¼	3 / 2
		⁵⁄₁₆	182	⁵⁄₁₆	1¼	1¼	1¼	3 / 2
		¼	148	⁵⁄₁₆	1	1	1	3 / 2
		³⁄₁₆	113	⁵⁄₁₆	⅞	⅞	⅞	3 / 2
		⅛	76.4	¼	¾	¾	¾	3 / 2
8	2	⅜	192	⁵⁄₁₆	1¼	1¼	1¼	3 / 1
		⁵⁄₁₆	163	⁵⁄₁₆	1	1	1	3 / 1
		¼	134	⁵⁄₁₆	⅞	⅞	⅞	3 / 1
		³⁄₁₆	102	⁵⁄₁₆	⅞	¾	¾	3 / 1
		⅛	69.2	¼	⅝	⅝	⅝	3 / 1
6	4	½	245	⅜	1½	1½	1½	3 / 2
		⅜	192	⁵⁄₁₆	1¼	1¼	1¼	3 / 2
		⁵⁄₁₆	163	⁵⁄₁₆	1¼	1¼	1¼	3 / 2
		¼	134	⁵⁄₁₆	1	1	1	3 / 2
		³⁄₁₆	102	⁵⁄₁₆	⅞	⅞	⅞	3 / 2
		⅛	69.2	¼	¾	¾	¾	3 / 2

75% of Member Design Strength

Table 6-6.
Tension Ties
End Plates Bolted on Four Sides—Rectangular Sections

Designation			Con- nection Strength (kips)	Weld Size W_w (in.)	Cap Plate, t_1 (in.) for bolt size			No. of Bolts/side
H in.	B in.	t in.			¾-in.	⅞-in.	1-in.	n
6	3	½	216	⅜	1¼	1¼	1¼	3 / 2
		⅜	170	⁵⁄₁₆	1¼	1¼	1¼	3 / 2
		⁵⁄₁₆	145	⁵⁄₁₆	1	1	1	3 / 2
		¼	119	⁵⁄₁₆	1	1	1	3 / 2
		³⁄₁₆	91.0	⁵⁄₁₆	⅞	⅞	⅞	3 / 2
		⅛	62.1	¼	¾	¾	¾	3 / 2
6	2	⅜	148	⁵⁄₁₆	1¼	1¼	1¼	3 / 1
		⁵⁄₁₆	127	⁵⁄₁₆	1¼	1¼	1¼	3 / 1
		¼	105	⁵⁄₁₆	1	1	1	3 / 1
		³⁄₁₆	80.1	⁵⁄₁₆	⅞	⅞	⅞	3 / 1
		⅛	55.0	¼	¾	¾	¾	3 / 1
5	3	½	187	⅜	1¼	1¼	1¼	2 / 2
		⅜	148	⁵⁄₁₆	1¼	1¼	1¼	2 / 2
		⁵⁄₁₆	127	⁵⁄₁₆	1¼	1¼	1¼	2 / 2
		¼	105	⁵⁄₁₆	1	1	1	2 / 2
		³⁄₁₆	80.1	⁵⁄₁₆	⅞	⅞	⅞	2 / 2
		⅛	55.0	¼	¾	¾	¾	2 / 2
5	2	⅜	127	⁵⁄₁₆	1	1	1	2 / 1
		⁵⁄₁₆	109	⁵⁄₁₆	1	1	1	2 / 1
		¼	90.4	⁵⁄₁₆	⅞	⅞	⅞	2 / 1
		³⁄₁₆	69.6	¼	¾	¾	¾	2 / 1
		⅛	47.8	¼	⅝	⅝	⅝	2 / 1
4	3	⅜	127	⁵⁄₁₆	1¼	1¼	1¼	2 / 2
		⁵⁄₁₆	109	⁵⁄₁₆	1	1	1	2 / 2
		¼	90.4	⁵⁄₁₆	⅞	⅞	⅞	2 / 2
		³⁄₁₆	69.6	⁵⁄₁₆	⅞	⅞	⅞	2 / 2
		⅛	47.8	¼	⅝	¾	¾	2 / 2
4	2	⅜	105	⁵⁄₁₆	1	1	⅞	2 / 1
		⁵⁄₁₆	91.3	⁵⁄₁₆	⅞	⅞	⅞	2 / 1
		¼	75.8	¼	¾	¾	¾	2 / 1
		³⁄₁₆	58.7	¼	¾	¾	¾	2 / 1
		⅛	40.4	¼	⅝	⅝	⅝	2 / 1

Table 6-6.
Tension Ties
End Plates Bolted on Four Sides—Rectangular Sections

Designation			Con-nection Strength (kips)	Weld Size W_w (in.)	Cap Plate, t_1 (in.) for bolt size			No. of Bolts/side
H in.	B in.	t in.			$\frac{3}{4}$-in.	$\frac{7}{8}$-in.	1-in.	n
8	6	$\frac{5}{8}$	580	$\frac{5}{8}$	—	—	2	3 / 3
		$\frac{1}{2}$	480	$\frac{9}{16}$	—	2	$1\frac{3}{4}$	3 / 3
		$\frac{3}{8}$	371	$\frac{7}{16}$	—	$1\frac{1}{2}$	$1\frac{1}{2}$	3 / 3
		$\frac{5}{16}$	314	$\frac{3}{8}$	$1\frac{1}{2}$	$1\frac{1}{2}$	$1\frac{1}{2}$	3 / 3
		$\frac{1}{4}$	255	$\frac{5}{16}$	$1\frac{1}{4}$	$1\frac{1}{4}$	$1\frac{1}{4}$	3 / 3
		$\frac{3}{16}$	193	$\frac{5}{16}$	$1\frac{1}{4}$	$1\frac{1}{4}$	$1\frac{1}{4}$	3 / 3
8	4	$\frac{5}{8}$	484	$\frac{5}{8}$	—	—	$1\frac{3}{4}$	3 / 2
		$\frac{1}{2}$	403	$\frac{9}{16}$	—	$1\frac{3}{4}$	$1\frac{1}{2}$	3 / 2
		$\frac{3}{8}$	314	$\frac{7}{16}$	—	$1\frac{1}{4}$	$1\frac{1}{4}$	3 / 2
		$\frac{5}{16}$	266	$\frac{3}{8}$	$1\frac{1}{4}$	$1\frac{1}{4}$	$1\frac{1}{4}$	3 / 2
		$\frac{1}{4}$	217	$\frac{5}{16}$	1	1	1	3 / 2
		$\frac{3}{16}$	165	$\frac{5}{16}$	$\frac{7}{8}$	$\frac{7}{8}$	$\frac{7}{8}$	3 / 2
		$\frac{1}{8}$	112	$\frac{1}{4}$	$\frac{3}{4}$	$\frac{3}{4}$	$\frac{3}{4}$	3 / 2
8	3	$\frac{1}{2}$	365	$\frac{1}{2}$	—	$1\frac{3}{4}$	$1\frac{3}{4}$	3 / 2
		$\frac{3}{8}$	285	$\frac{7}{16}$	$1\frac{3}{4}$	$1\frac{1}{2}$	$1\frac{1}{2}$	3 / 2
		$\frac{5}{16}$	242	$\frac{3}{8}$	$1\frac{1}{2}$	$1\frac{1}{2}$	$1\frac{1}{2}$	3 / 2
		$\frac{1}{4}$	197	$\frac{5}{16}$	$1\frac{1}{4}$	$1\frac{1}{4}$	$1\frac{1}{4}$	3 / 2
		$\frac{3}{16}$	150	$\frac{5}{16}$	1	1	1	3 / 2
		$\frac{1}{8}$	102	$\frac{5}{16}$	$\frac{7}{8}$	$\frac{7}{8}$	$\frac{7}{8}$	3 / 2
8	2	$\frac{3}{8}$	256	$\frac{7}{16}$	—	$1\frac{1}{4}$	$1\frac{1}{4}$	3 / 1
		$\frac{5}{16}$	218	$\frac{3}{8}$	$1\frac{1}{2}$	$1\frac{1}{4}$	$1\frac{1}{4}$	3 / 1
		$\frac{1}{4}$	178	$\frac{5}{16}$	1	1	1	3 / 1
		$\frac{3}{16}$	136	$\frac{5}{16}$	$\frac{7}{8}$	$\frac{7}{8}$	$\frac{7}{8}$	3 / 1
		$\frac{1}{8}$	92.3	$\frac{1}{4}$	$\frac{3}{4}$	$\frac{3}{4}$	$\frac{3}{4}$	3 / 1
6	4	$\frac{1}{2}$	326	$\frac{1}{2}$	—	$1\frac{1}{2}$	$1\frac{1}{2}$	3 / 2
		$\frac{3}{8}$	256	$\frac{7}{16}$	$1\frac{1}{2}$	$1\frac{1}{2}$	$1\frac{1}{2}$	3 / 2
		$\frac{5}{16}$	218	$\frac{3}{8}$	$1\frac{1}{4}$	$1\frac{1}{4}$	$1\frac{1}{4}$	3 / 2
		$\frac{1}{4}$	178	$\frac{5}{16}$	$1\frac{1}{4}$	$1\frac{1}{4}$	$1\frac{1}{4}$	3 / 2
		$\frac{3}{16}$	136	$\frac{5}{16}$	1	1	1	3 / 2
		$\frac{1}{8}$	92.3	$\frac{5}{16}$	$\frac{7}{8}$	$\frac{7}{8}$	$\frac{7}{8}$	3 / 2

Table 6-6.
Tension Ties
End Plates Bolted on Four Sides—Rectangular Sections

Designation			Connection Strength (kips)	Weld Size W_w (in.)	Cap Plate, t_1 (in.) for bolt size			No. of Bolts/side
H in.	B in.	t in.			$\frac{3}{4}$-in.	$\frac{7}{8}$-in.	1-in.	n
6	3	$\frac{1}{2}$	288	$\frac{1}{2}$	$1\frac{3}{4}$	$1\frac{1}{2}$	$1\frac{1}{2}$	3 / 2
		$\frac{3}{8}$	227	$\frac{7}{16}$	$1\frac{1}{4}$	$1\frac{1}{4}$	$1\frac{1}{4}$	3 / 2
		$\frac{5}{16}$	194	$\frac{3}{8}$	$1\frac{1}{4}$	$1\frac{1}{4}$	$1\frac{1}{4}$	3 / 2
		$\frac{1}{4}$	159	$\frac{5}{16}$	$1\frac{1}{4}$	$1\frac{1}{4}$	$1\frac{1}{4}$	3 / 2
		$\frac{3}{16}$	121	$\frac{5}{16}$	1	1	1	3 / 2
		$\frac{1}{8}$	82.8	$\frac{1}{4}$	$\frac{3}{4}$	$\frac{3}{4}$	$\frac{3}{4}$	3 / 2
6	2	$\frac{3}{8}$	198	$\frac{3}{8}$	$1\frac{1}{2}$	$1\frac{1}{2}$	$1\frac{1}{2}$	3 / 1
		$\frac{5}{16}$	170	$\frac{3}{8}$	$1\frac{1}{4}$	$1\frac{1}{4}$	$1\frac{1}{4}$	3 / 1
		$\frac{1}{4}$	140	$\frac{5}{16}$	$1\frac{1}{4}$	$1\frac{1}{4}$	$1\frac{1}{4}$	3 / 1
		$\frac{3}{16}$	107	$\frac{5}{16}$	1	1	1	3 / 1
		$\frac{1}{8}$	73.3	$\frac{5}{16}$	$\frac{7}{8}$	$\frac{7}{8}$	$\frac{7}{8}$	3 / 1
5	3	$\frac{1}{2}$	249	$\frac{1}{2}$	—	$1\frac{1}{2}$	$1\frac{1}{2}$	2 / 2
		$\frac{3}{8}$	198	$\frac{3}{8}$	$1\frac{1}{2}$	$1\frac{1}{2}$	$1\frac{1}{2}$	2 / 2
		$\frac{5}{16}$	170	$\frac{3}{8}$	$1\frac{1}{4}$	$1\frac{1}{4}$	$1\frac{1}{4}$	2 / 2
		$\frac{1}{4}$	140	$\frac{5}{16}$	$1\frac{1}{4}$	$1\frac{1}{4}$	$1\frac{1}{4}$	2 / 2
		$\frac{3}{16}$	107	$\frac{5}{16}$	1	1	1	2 / 2
		$\frac{1}{8}$	73.3	$\frac{5}{16}$	$\frac{7}{8}$	$\frac{7}{8}$	$\frac{7}{8}$	2 / 2
5	2	$\frac{3}{8}$	169	$\frac{3}{8}$	$1\frac{1}{2}$	$1\frac{1}{4}$	$1\frac{1}{4}$	2 / 1
		$\frac{5}{16}$	146	$\frac{5}{16}$	$1\frac{1}{4}$	$1\frac{1}{4}$	$1\frac{1}{4}$	2 / 1
		$\frac{1}{4}$	120	$\frac{5}{16}$	1	1	1	2 / 1
		$\frac{3}{16}$	92.7	$\frac{5}{16}$	$\frac{7}{8}$	$\frac{7}{8}$	$\frac{7}{8}$	2 / 1
		$\frac{1}{8}$	63.8	$\frac{1}{4}$	$\frac{3}{4}$	$\frac{3}{4}$	$\frac{3}{4}$	2 / 1
4	3	$\frac{3}{8}$	169	$\frac{3}{8}$	$1\frac{1}{4}$	$1\frac{1}{4}$	$1\frac{1}{4}$	2 / 2
		$\frac{5}{16}$	146	$\frac{5}{16}$	$1\frac{1}{4}$	$1\frac{1}{4}$	$1\frac{1}{4}$	2 / 2
		$\frac{1}{4}$	120	$\frac{5}{16}$	1	1	1	2 / 2
		$\frac{3}{16}$	92.7	$\frac{5}{16}$	$\frac{7}{8}$	$\frac{7}{8}$	$\frac{7}{8}$	2 / 2
		$\frac{1}{8}$	63.8	$\frac{1}{4}$	$\frac{3}{4}$	$\frac{3}{4}$	$\frac{3}{4}$	2 / 2
4	2	$\frac{3}{8}$	140	$\frac{3}{8}$	$1\frac{1}{4}$	$1\frac{1}{4}$	$1\frac{1}{4}$	2 / 1
		$\frac{5}{16}$	122	$\frac{5}{16}$	1	1	1	2 / 1
		$\frac{1}{4}$	101	$\frac{5}{16}$	$\frac{7}{8}$	$\frac{7}{8}$	$\frac{7}{8}$	2 / 1
		$\frac{3}{16}$	78.2	$\frac{5}{16}$	$\frac{7}{8}$	$\frac{7}{8}$	$\frac{3}{4}$	2 / 1
		$\frac{1}{8}$	53.8	$\frac{1}{4}$	$\frac{3}{4}$	$\frac{3}{4}$	$\frac{5}{8}$	2 / 1

25% of Member Design Strength

Table 6-7.
Tension Ties
End Plates Bolted on Two Sides—Square Sections

Designation			Con-nection Strength (kips)	Weld Size W_w (in.)	Cap Plate, t_1 (in.) for bolt size			No. of Bolts/side
H in.	B in.	t in.			¾-in.	⅞-in.	1-in.	n
8	8	⅝	170	⅜	—	1	1	3
		½	140	⁵⁄₁₆	—	1	1	3
		⅜	108	⁵⁄₁₆	1	1	1	3
		⁵⁄₁₆	90.7	⁵⁄₁₆	1	1	1	3
		¼	73.5	⁵⁄₁₆	1	⅞	⅞	3
		³⁄₁₆	55.6	⁵⁄₁₆	⅞	¾	¾	3
7	7	⅝	145	⁵⁄₁₆	—	1	1	3
		½	120	⁵⁄₁₆	1	1	1	3
		⅜	92.8	⁵⁄₁₆	1	1	1	3
		⁵⁄₁₆	78.6	⁵⁄₁₆	1	1	1	3
		¼	63.9	⁵⁄₁₆	1	1	1	3
		³⁄₁₆	48.3	⁵⁄₁₆	⅞	⅞	¾	3
6	6	⅝	121	⁵⁄₁₆	—	1	1	3
		½	101	⁵⁄₁₆	1	1	1	3
		⅜	78.5	⁵⁄₁₆	1	1	1	3
		⁵⁄₁₆	66.6	⁵⁄₁₆	1	1	1	3
		¼	54.2	⁵⁄₁₆	1	1	1	3
		³⁄₁₆	41.2	⁵⁄₁₆	⅞	⅞	⅞	3
5	5	½	81.6	⁵⁄₁₆	1	1	1	2
		⅜	64.0	⁵⁄₁₆	1	1	1	2
		⁵⁄₁₆	54.4	⁵⁄₁₆	1	1	1	2
		¼	44.5	⁵⁄₁₆	⅞	⅞	⅞	2
		³⁄₁₆	33.9	¼	¾	¾	¾	2
		⅛	23.1	¼	⅝	⅝	⅝	2
4½	4½	½	71.9	⁵⁄₁₆	1	1	1	2
		⅜	56.7	⁵⁄₁₆	1	1	1	2
		⁵⁄₁₆	48.4	⁵⁄₁₆	1	1	1	2
		¼	39.7	⁵⁄₁₆	⅞	⅞	⅞	2
		³⁄₁₆	30.3	¼	¾	¾	¾	2
		⅛	20.7	¼	⅝	⅝	⅝	2

Table 6-7.
Tension Ties
End Plates Bolted on Two Sides—Square Sections

Designation			Connection Strength (kips)	Weld Size W_w (in.)	Cap Plate, t_1 (in.) for bolt size			No. of Bolts/side
H in.	B in.	t in.			$\frac{3}{4}$-in.	$\frac{7}{8}$-in.	1-in.	n
4	4	$\frac{1}{2}$	62.3	$\frac{5}{16}$	1	1	1	2
		$\frac{3}{8}$	49.5	$\frac{5}{16}$	1	1	1	2
		$\frac{5}{16}$	42.4	$\frac{5}{16}$	1	1	1	2
		$\frac{1}{4}$	34.9	$\frac{5}{16}$	$\frac{7}{8}$	$\frac{7}{8}$	$\frac{7}{8}$	2
		$\frac{3}{16}$	26.7	$\frac{1}{4}$	$\frac{3}{4}$	$\frac{3}{4}$	$\frac{3}{4}$	2
		$\frac{1}{8}$	18.3	$\frac{1}{4}$	$\frac{5}{8}$	$\frac{5}{8}$	$\frac{5}{8}$	2
$3\frac{1}{2}$	$3\frac{1}{2}$	$\frac{3}{8}$	42.3	$\frac{5}{16}$	1	1	1	2
		$\frac{5}{16}$	36.4	$\frac{5}{16}$	1	1	1	2
		$\frac{1}{4}$	30.1	$\frac{5}{16}$	$\frac{7}{8}$	$\frac{7}{8}$	$\frac{7}{8}$	2
		$\frac{3}{16}$	23.2	$\frac{1}{4}$	$\frac{3}{4}$	$\frac{3}{4}$	$\frac{3}{4}$	2
		$\frac{1}{8}$	15.9	$\frac{1}{4}$	$\frac{5}{8}$	$\frac{5}{8}$	$\frac{5}{8}$	2
3	3	$\frac{3}{8}$	35.1	$\frac{5}{16}$	1	1	1	2
		$\frac{5}{16}$	30.4	$\frac{5}{16}$	1	1	1	2
		$\frac{1}{4}$	25.3	$\frac{5}{16}$	$\frac{7}{8}$	$\frac{7}{8}$	$\frac{7}{8}$	2
		$\frac{3}{16}$	19.6	$\frac{1}{4}$	$\frac{3}{4}$	$\frac{3}{4}$	$\frac{3}{4}$	2
		$\frac{1}{8}$	13.5	$\frac{1}{4}$	$\frac{5}{8}$	$\frac{5}{8}$	$\frac{5}{8}$	2
$2\frac{1}{2}$	$2\frac{1}{2}$	$\frac{5}{16}$	24.3	$\frac{5}{16}$	1	1	$\frac{7}{8}$	1
		$\frac{1}{4}$	20.4	$\frac{5}{16}$	$\frac{7}{8}$	$\frac{7}{8}$	$\frac{7}{8}$	1
		$\frac{3}{16}$	15.9	$\frac{1}{4}$	$\frac{3}{4}$	$\frac{3}{4}$	$\frac{3}{4}$	1
		$\frac{1}{8}$	11.1	$\frac{1}{4}$	$\frac{5}{8}$	$\frac{5}{8}$	$\frac{5}{8}$	1
2	2	$\frac{1}{4}$	15.6	$\frac{5}{16}$	$\frac{7}{8}$	$\frac{7}{8}$	$\frac{7}{8}$	1
		$\frac{3}{16}$	12.3	$\frac{1}{4}$	$\frac{3}{4}$	$\frac{3}{4}$	$\frac{3}{4}$	1
		$\frac{1}{8}$	8.7	$\frac{1}{4}$	$\frac{5}{8}$	$\frac{5}{8}$	$\frac{5}{8}$	1

Table 6-7.
Tension Ties
End Plates Bolted on Two Sides—Square Sections

Designation			Connection Strength (kips)	Weld Size W_w (in.)	Cap Plate, t_1 (in.) for bolt size			No. of Bolts/side
H in.	B in.	t in.			$\frac{3}{4}$-in.	$\frac{7}{8}$-in.	1-in.	n
8	8	$\frac{5}{8}$	—	—	—	—	—	—
		$\frac{1}{2}$	—	—	—	—	—	—
		$\frac{3}{8}$	215	$\frac{7}{16}$	—	—	1	3
		$\frac{5}{16}$	181	$\frac{3}{8}$	—	1	1	3
		$\frac{1}{4}$	147	$\frac{5}{16}$	1	1	1	3
		$\frac{3}{16}$	111	$\frac{5}{16}$	1	1	1	3
7	7	$\frac{5}{8}$	—	—	—	—	—	—
		$\frac{1}{2}$	—	—	—	—	—	—
		$\frac{3}{8}$	186	$\frac{7}{16}$	—	—	1	3
		$\frac{5}{16}$	157	$\frac{3}{8}$	—	1	1	3
		$\frac{1}{4}$	128	$\frac{5}{16}$	1	1	1	3
		$\frac{3}{16}$	96.7	$\frac{5}{16}$	1	1	1	3
6	6	$\frac{5}{8}$	—	—	—	—	—	—
		$\frac{1}{2}$	202	$\frac{9}{16}$	—	—	1	3
		$\frac{3}{8}$	157	$\frac{7}{16}$	—	1	1	3
		$\frac{5}{16}$	133	$\frac{3}{8}$	—	1	1	3
		$\frac{1}{4}$	108	$\frac{5}{16}$	1	1	1	3
		$\frac{3}{16}$	82.4	$\frac{5}{16}$	1	1	1	3
5	5	$\frac{1}{2}$	—	—	—	—	—	—
		$\frac{3}{8}$	128	$\frac{7}{16}$	—	—	1	2
		$\frac{5}{16}$	109	$\frac{3}{8}$	—	1	1	2
		$\frac{1}{4}$	89.0	$\frac{5}{16}$	1	1	1	2
		$\frac{3}{16}$	67.9	$\frac{5}{16}$	1	1	1	2
		$\frac{1}{8}$	46.2	$\frac{5}{16}$	$\frac{7}{8}$	$\frac{7}{8}$	$\frac{7}{8}$	2
$4\frac{1}{2}$	$4\frac{1}{2}$	$\frac{1}{2}$	144	$\frac{1}{2}$	—	—	1	2
		$\frac{3}{8}$	113	$\frac{7}{16}$	—	1	1	2
		$\frac{5}{16}$	96.9	$\frac{3}{8}$	—	1	1	2
		$\frac{1}{4}$	79.5	$\frac{5}{16}$	1	1	1	2
		$\frac{3}{16}$	60.7	$\frac{5}{16}$	1	1	1	2
		$\frac{1}{8}$	41.4	$\frac{5}{16}$	$\frac{7}{8}$	$\frac{7}{8}$	$\frac{7}{8}$	2

Table 6-7.
Tension Ties
End Plates Bolted on Two Sides—Square Sections

Designation			Connection Strength (kips)	Weld Size W_w (in.)	Cap Plate, t_1 (in.) for bolt size			No. of Bolts/side
H in.	B in.	t in.			$\frac{3}{4}$-in.	$\frac{7}{8}$-in.	1-in.	n
4	4	$\frac{1}{2}$	125	$\frac{1}{2}$	—	—	1	2
		$\frac{3}{8}$	98.9	$\frac{3}{8}$	—	1	1	2
		$\frac{5}{16}$	84.9	$\frac{3}{8}$	1	1	1	2
		$\frac{1}{4}$	69.8	$\frac{5}{16}$	1	1	1	2
		$\frac{3}{16}$	53.4	$\frac{5}{16}$	1	1	1	2
		$\frac{1}{8}$	36.6	$\frac{5}{16}$	$\frac{7}{8}$	$\frac{7}{8}$	$\frac{7}{8}$	2
$3\frac{1}{2}$	$3\frac{1}{2}$	$\frac{3}{8}$	84.7	$\frac{3}{8}$	1	1	1	2
		$\frac{5}{16}$	72.9	$\frac{5}{16}$	1	1	1	2
		$\frac{1}{4}$	60.2	$\frac{5}{16}$	1	1	1	2
		$\frac{3}{16}$	46.4	$\frac{5}{16}$	1	1	1	2
		$\frac{1}{8}$	31.9	$\frac{5}{16}$	$\frac{7}{8}$	$\frac{7}{8}$	$\frac{7}{8}$	2
3	3	$\frac{3}{8}$	70.2	$\frac{3}{8}$	1	1	1	2
		$\frac{5}{16}$	60.9	$\frac{5}{16}$	1	1	1	2
		$\frac{1}{4}$	50.5	$\frac{5}{16}$	1	1	1	2
		$\frac{3}{16}$	39.1	$\frac{5}{16}$	1	1	1	2
		$\frac{1}{8}$	26.9	$\frac{5}{16}$	$\frac{7}{8}$	$\frac{7}{8}$	$\frac{7}{8}$	2
$2\frac{1}{2}$	$2\frac{1}{2}$	$\frac{5}{16}$	48.6	$\frac{5}{16}$	—	1	1	1
		$\frac{1}{4}$	40.8	$\frac{5}{16}$	1	1	1	1
		$\frac{3}{16}$	31.9	$\frac{5}{16}$	1	1	1	1
		$\frac{1}{8}$	22.1	$\frac{5}{16}$	$\frac{7}{8}$	$\frac{7}{8}$	$\frac{7}{8}$	1
2	2	$\frac{1}{4}$	31.3	$\frac{5}{16}$	1	1	1	1
		$\frac{3}{16}$	24.6	$\frac{5}{16}$	1	1	1	1
		$\frac{1}{8}$	17.4	$\frac{5}{16}$	$\frac{7}{8}$	$\frac{7}{8}$	$\frac{7}{8}$	1

25% of Member Design Strength

Table 6-7.
Tension Ties
End Plates Bolted on Two Sides—Rectangular Sections

Designation			Connection Strength (kips)	Weld Size W_w (in.)	Cap Plate, t_1 (in.) for bolt size			No. of Bolts/side
H in.	B in.	t in.			¾-in.	⅞-in.	1-in.	n
8	6	⅝	145	⁵⁄₁₆	—	1	1	3
		½	120	⁵⁄₁₆	1	1	1	3
		⅜	92.8	⁵⁄₁₆	1	1	1	3
		⁵⁄₁₆	78.6	⁵⁄₁₆	1	1	1	3
		¼	63.9	⁵⁄₁₆	⅞	⅞	⅞	3
		³⁄₁₆	48.3	¼	¾	¾	¾	3
8	4	⅝	121	⁵⁄₁₆	1	1	1	3
		½	101	⁵⁄₁₆	1	1	1	3
		⅜	78.5	⁵⁄₁₆	1	1	1	3
		⁵⁄₁₆	66.6	⁵⁄₁₆	⅞	⅞	⅞	3
		¼	54.2	⁵⁄₁₆	⅞	¾	¾	3
		³⁄₁₆	41.2	¼	¾	¾	⅝	3
		⅛	27.9	¼	⅝	⅝	½	3
8	3	½	91.2	⁵⁄₁₆	1	1	1	3
		⅜	71.2	⁵⁄₁₆	1	1	⅞	3
		⁵⁄₁₆	60.5	⁵⁄₁₆	⅞	⅞	⅞	3
		¼	49.4	¼	¾	¾	¾	3
		³⁄₁₆	37.6	¼	⅝	⅝	⅝	3
		⅛	25.5	³⁄₁₆	½	½	½	3
8	2	⅜	64.0	⁵⁄₁₆	⅞	⅞	⅞	3
		⁵⁄₁₆	54.4	⁵⁄₁₆	⅞	⅞	¾	3
		¼	44.5	¼	¾	¾	¾	3
		³⁄₁₆	33.9	¼	⅝	⅝	⅝	3
		⅛	23.1	³⁄₁₆	½	½	½	3
6	4	½	81.6	⁵⁄₁₆	1	1	1	3
		⅜	64.0	⁵⁄₁₆	1	1	1	3
		⁵⁄₁₆	54.4	⁵⁄₁₆	1	1	1	3
		¼	44.5	⁵⁄₁₆	1	⅞	⅞	3
		³⁄₁₆	33.9	⁵⁄₁₆	⅞	¾	¾	3
		⅛	23.1	¼	⅝	⅝	⅝	3

Table 6-7.
Tension Ties
End Plates Bolted on Two Sides—Rectangular Sections

Designation			Con-nection Strength (kips)	Weld Size W_w (in.)	Cap Plate, t_1 (in.) for bolt size			No. of Bolts/side
H in.	B in.	t in.			¾-in.	⅞-in.	1-in.	n
6	3	½	71.9	5/16	1	1	1	3
		3/8	56.7	5/16	1	1	1	3
		5/16	48.4	5/16	1	1	1	3
		¼	39.7	5/16	7/8	7/8	7/8	3
		3/16	30.3	¼	¾	¾	¾	3
		1/8	20.7	¼	5/8	5/8	5/8	3
6	2	3/8	49.5	5/16	1	1	1	3
		5/16	42.4	5/16	1	7/8	7/8	3
		¼	34.9	5/16	7/8	7/8	¾	3
		3/16	26.7	¼	¾	¾	¾	3
		1/8	18.3	¼	5/8	5/8	5/8	3
5	3	½	62.3	5/16	1	1	1	2
		3/8	49.5	5/16	1	1	1	2
		5/16	42.4	5/16	7/8	7/8	7/8	2
		¼	34.9	¼	¾	¾	¾	2
		3/16	26.7	¼	¾	5/8	5/8	2
		1/8	18.3	¼	5/8	5/8	½	2
5	2	3/8	42.3	5/16	7/8	7/8	7/8	2
		5/16	36.4	5/16	7/8	7/8	¾	2
		¼	30.1	¼	¾	¾	¾	2
		3/16	23.2	¼	5/8	5/8	5/8	2
		1/8	15.9	3/16	½	½	½	2
4	3	3/8	42.3	5/16	1	1	1	2
		5/16	36.4	5/16	7/8	7/8	7/8	2
		¼	30.1	5/16	7/8	7/8	¾	2
		3/16	23.2	¼	¾	¾	¾	2
		1/8	15.9	¼	5/8	5/8	5/8	2
4	2	3/8	35.1	5/16	7/8	7/8	7/8	2
		5/16	30.4	5/16	7/8	7/8	7/8	2
		¼	25.3	¼	¾	¾	¾	2
		3/16	19.6	¼	5/8	5/8	5/8	2
		1/8	13.5	3/16	½	½	½	2

Table 6-7.
Tension Ties
End Plates Bolted on Two Sides—Rectangular Sections

Designation			Connection Strength (kips)	Weld Size W_w (in.)	Cap Plate, t_1 (in.) for bolt size			No. of Bolts/side
H in.	B in.	t in.			¾-in.	⅞-in.	1-in.	n
8	6	⅝	—	—	—	—	—	—
		½	—	—	—	—	—	—
		⅜	186	⅜	—	1	1	3
		5/16	157	5/16	—	1	1	3
		¼	128	5/16	1	1	1	3
		3/16	96.7	5/16	1	1	1	3
8	4	⅝	—	—	—	—	—	—
		½	202	7/16	—	—	1	3
		⅜	157	5/16	—	1	1	3
		5/16	133	5/16	1	1	1	3
		¼	108	5/16	1	1	1	3
		3/16	82.4	5/16	1	1	⅞	3
		⅛	55.9	¼	¾	¾	¾	3
8	3	½	182	⅜	—	—	1	3
		⅜	142	5/16	1	1	1	3
		5/16	121	5/16	1	1	1	3
		¼	98.7	5/16	1	1	1	3
		3/16	75.1	5/16	⅞	⅞	⅞	3
		⅛	50.9	¼	¾	¾	¾	3
8	2	⅜	128	5/16	1	1	1	3
		5/16	109	5/16	1	1	1	3
		¼	89.0	5/16	1	1	1	3
		3/16	67.9	5/16	⅞	⅞	⅞	3
		⅛	46.2	¼	¾	¾	¾	3
6	4	½	163	7/16	—	1	1	3
		⅜	128	⅜	1	1	1	3
		5/16	109	5/16	1	1	1	3
		¼	89.0	5/16	1	1	1	3
		3/16	67.9	5/16	1	1	1	3
		⅛	46.2	5/16	⅞	⅞	⅞	3

Table 6-7.
Tension Ties
End Plates Bolted on Two Sides—Rectangular Sections

Designation			Con-nection Strength (kips)	Weld Size W_w (in.)	Cap Plate, t_1 (in.) for bolt size			No. of Bolts/side
H in.	B in.	t in.			¾-in.	⅞-in.	1-in.	n
6	3	½	144	⅜	—	1	1	3
		⅜	113	5/16	1	1	1	3
		5/16	96.9	5/16	1	1	1	3
		¼	79.5	5/16	1	1	1	3
		3/16	60.7	5/16	1	1	1	3
		⅛	41.4	5/16	⅞	⅞	⅞	3
6	2	⅜	98.9	5/16	1	1	1	3
		5/16	84.9	5/16	1	1	1	3
		¼	69.8	5/16	1	1	1	3
		3/16	53.4	5/16	1	1	1	3
		⅛	36.6	5/16	⅞	¾	¾	3
5	3	½	125	⅜	—	—	1	2
		⅜	98.9	5/16	—	1	1	2
		5/16	84.9	5/16	1	1	1	2
		¼	69.8	5/16	1	1	1	2
		3/16	53.4	5/16	1	1	⅞	2
		⅛	36.6	¼	¾	¾	¾	2
5	2	⅜	84.7	5/16	1	1	1	2
		5/16	72.9	5/16	1	1	1	2
		¼	60.2	5/16	1	1	1	2
		3/16	46.4	5/16	⅞	⅞	⅞	2
		⅛	31.9	¼	¾	¾	¾	2
4	3	⅜	84.7	⅜	1	1	1	2
		5/16	72.9	5/16	1	1	1	2
		¼	60.2	5/16	1	1	1	2
		3/16	46.4	5/16	1	1	1	2
		⅛	31.9	5/16	⅞	¾	¾	2
4	2	⅜	70.2	5/16	1	1	1	2
		5/16	60.9	5/16	1	1	1	2
		¼	50.5	5/16	1	1	1	2
		3/16	39.1	5/16	⅞	⅞	⅞	2
		⅛	26.9	¼	¾	¾	¾	2

REFERENCES

Birkemoe, P. C., and Packer, J. A., 1986, "Ultimate Strength Design of Bolted Tubular Tension Connections," *Proceedings, Steel Structures—Recent Research Advances and their Applications to Design*, Budva, Yugoslavia, pp. 153–168.

Igarashi, S., Wakiyama, K., Inque, K., Matsumoto, T., and Murase, Y., 1985, "Limit Design of High Strength Bolted Tube Flange Joint, Parts 1 and 2," *Journal of Structural and Construction Engineering Transactions of AIJ, Department of Architecture Reports*, Osaka University, Japan.

Packer, J. A., Bruno, L., and Birkemoe, P. C., 1989, "Limit Analysis of Bolted RHS Flange Plate Joints," *Journal of Structural Engineering*, American Society of Civil Engineers, 115(9):2226–2242.

Packer, J. A., and Henderson, J. E., 1992, *Design Guide for Hollow Structural Section Connections*, Canadian Institute of Steel Construction.

Struik, J. H. A., and de Back, J., 1969, "Tests on Bolted T-stubs With Respect to a Bolted Beam-to-Column Connection," *Stevin Laboratory Report 6-69-13*, Delft University of Technology, The Netherlands.

CHAPTER 7

CAP PLATES, BASE PLATES, AND SPLICES

OVERVIEW

Chapter 7 contains general information, design considerations, and examples for the design of cap plates and base plates for HSS. The information is based on references, the 1993 AISC LRFD Specification and basic engineering principles.

Following is a list of the topics addressed.

CAP PLATES

The simplest form of attachment to an HSS is to connect the framing member to the top of the HSS. The cap plate serves as a bearing device to transfer the reactions from the framing member into the HSS. As illustrated in the Chapter on Moment Connections the cap plate may also be used to transfer moment to the HSS column. The moment transfer is through a force couple which consists of compressive and tensile reactions delivered to the cap plate.

Cap Plates Subjected to Compressive Reactions

There are three limit states when a concentrated force is applied to a cap plate. These are:

1. Flexural strength of the cap plate.
2. Local compression yielding of the HSS wall.
3. Local compression crippling of the HSS wall.

1. Flexural strength of the cap plate:

The strength of the cap plate, in terms of reaction resistance, can be determined from the equation:

$$\phi R_n = \phi \, \frac{B t_1^2}{4\left(\dfrac{N_r}{2} + a - \dfrac{H}{2}\right)} \, F_{yc} \tag{7-1}$$

where

$\phi\ = 0.90$
$B\ =$ the HSS width, in.
$t_1\ =$ the cap plate thickness, in.
$N_r =$ the required bearing length for the attached member, in.
$a\ =$ the distance from the HSS centroid to the end of the attached member, in.
$H\ =$ the HSS depth, in.
$F_{yc} =$ the specified yield strength of the cap plate, ksi

See Figure 7-1 for further definition of terms.

This equation applies only if the cap plate is subjected to cantilever bending. This occurs when the beam or joist reaction point is outside of the HSS face.

If a stiffener is used in the beam, and is positioned over the HSS wall, then Equation 7-1 does not apply, since the cap plate is not subjected to bending. Also if the denominator of Equation 7-1 results in a negative number, bending of the cap plate can be disregarded.

2. Compression yielding of the HSS wall:

The compression yield strength of the HSS wall can be determined from the AISC HSS Specification expression:

$$\phi R_n = \phi(5t_1 + N)F_y t \le \phi B F_y t \tag{7-2}$$

where

$\phi = 1.0$
$N =$ the bearing length of the load on top of the cap plate across the width of the HSS
$\quad =$ two times the "k" distance for W shapes without stiffeners
$\quad =$ the width of the joist or joist girder seat, N_b, as shown in Figure 7-1
$F_y =$ the specified yield strength of the HSS, ksi
$t\ =$ the HSS wall thickness, in.
$B\ =$ the HSS width
$t_1 =$ the thickness of cap plate, in.

3. *Compression crippling of the HSS wall:*

Compression crippling of the HSS wall subjected to a concentrated load on a column cap was treated in Chapter 5. The procedure is repeated here for completeness.

Using the AISC HSS Specification expression for local crippling

$$\phi R_n = \phi 0.80 t^2 \left[1 + 3 \left(\frac{N}{B/2} \right) \left(\frac{t}{t_1} \right)^{1.5} \right] \sqrt{EF_y (t_1 / t)} \tag{7-3}$$

where

$\phi = 0.75$
t = the HSS wall thickness
N = the bearing length of the load as used in the wall yielding expression
B = the HSS width
t_1 = the thickness of the cap plate
F_y = the specified yield strength of the HSS
E = modulus of elasticity of the HSS

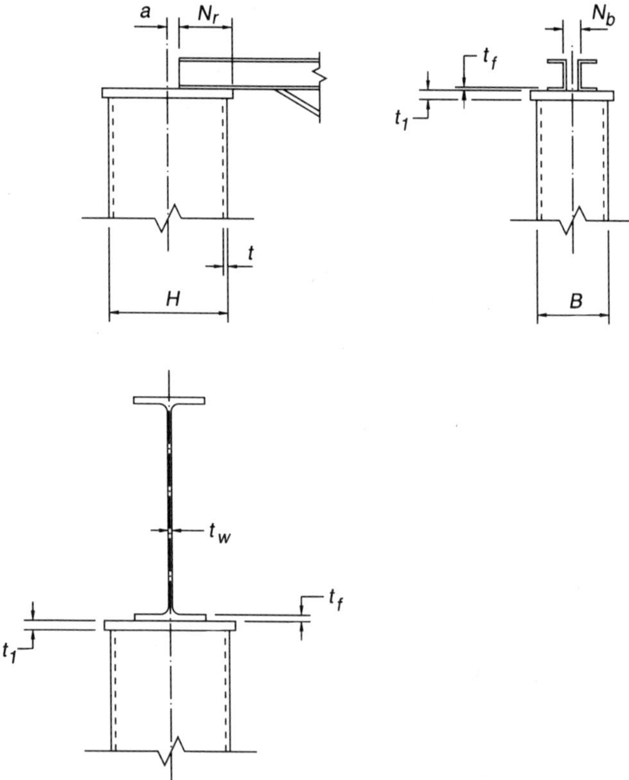

Fig. 7-1. Column cap terms.

Example 7-1

Determine if the connection shown in Figure Ex. 7-1 can adequately support a factored beam reaction of 50 kips. The reaction is delivered by a W18×40, $F_y = 36$ ksi.

Solution:

1. Flexural strength of the cap plate.

First determine if the beam reaction causes cantilever bending of the cap plate.

Bending will not occur unless $N > H - 2a$.

$N > 10 - (2)(0.5) = 9$ in.

The N for the cap plate shown is 7.5 in. so cantilever bending will not occur provided the 7.5 in. cap plate dimension is not increased to more than 9 in.

Check the W18×40 to determine if the bearing length of 7.5 in. is adequate.

From AISC LRFD Specification Eq. K1-3:

$$\phi R_n = \phi(2.5k + N)F_{yw}t_w$$

$$= (1.0)[(2.5)(1.19) + 7.5](36)(0.315)$$

$$= 118.8 \text{ kips} > 50 \text{ kips}$$

From AISC LRFD Specification Eq. K1-5b

$$R_n = \phi 68 t_w^2 \left[1 + \left(\frac{4N}{d} - 0.2 \right) \left(\frac{t_w}{t_f} \right)^{1.5} \right] \sqrt{\frac{F_{yw}t_f}{t_w}}$$

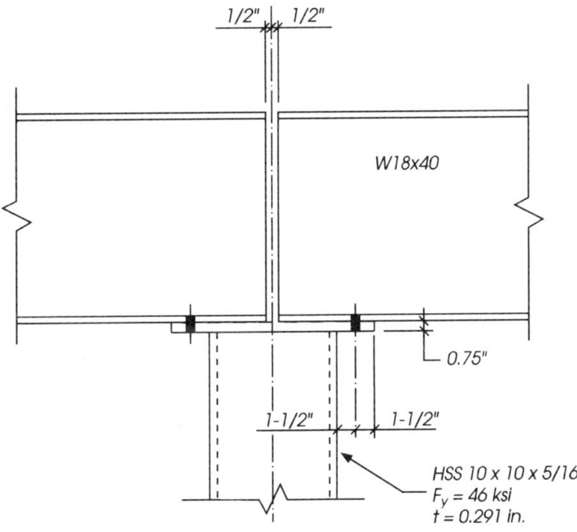

Figure Ex. 7-1.

$$= (0.75)(68)(0.315)^2 \left[1 + \left(\frac{(4)(7.5)}{17.9} - 0.2 \right) \left(\frac{0.315}{0.525} \right)^{1.5} \right] \sqrt{\frac{(36)(0.525)}{(0.315)}}$$

$$= 66.1 \text{ kips} > 50 \text{ kips}$$

The 7.5 in. bearing length is adequate.

2. Compression yielding of the HSS wall.

$$\phi R_n = \phi [5t_1 + N] F_y t \leq \phi B F_y t$$

$$= 1.0[5(0.75) + 2(1.19)](46)(0.291)$$

$$= 82.0 \text{ kips} > 50 \text{ kips} \quad \textbf{o.k.}$$

3. Compression crippling of the HSS wall.

$$\phi R_n = \phi 0.80 t^2 \left[1 + 3 \left(\frac{N}{B/2} \right) \left(\frac{t}{t_1} \right)^{1.5} \right] \sqrt{E F_y (t_1 / t)}$$

$$= 0.75(0.80)(0.291)^2 \left[1 + 3 \left(\frac{2.38}{10/2} \right) \left(\frac{0.291}{0.75} \right)^{1.5} \right] \sqrt{29,000(46)(0.75 / 0.291)}$$

$$= 127 \text{ kips} > 50 \text{ kips} \quad \textbf{o.k.}$$

Cap Plates Subjected to Tensile Concentrated Loads

Concentrated tension loads are typically delivered to cap plate through the bolts which connect the supported member to the HSS. The limit states are:

1. Bending of the cap plate at the HSS wall.
2. Tension yield of the HSS wall.
3. Tension rupture of the cap-plate weld.

1. Bending of the cap plate at the HSS wall.

The required thickness of the cap plate can be determined based on cantilever bending theory. Using a 45 degree load distribution width from the connecting bolt centerline, and using two bolts to connect the framing member to the cap plate, the required cap plate thickness equals:

$$t_1 = \sqrt{\frac{4 P_u e}{l_e \phi F_{yc}}}$$

where

$\phi = 0.90$
P_u = the factored tensile load, kips
e = the distance of the bolts from the HSS wall, in.
l_e = effective width of the cap plate, in.
   = the lesser of $(2e + g)$ or $4e$
F_{yc} = the specified yield strength of the cap plate, ksi
g = the bolt gage, in.

2. Tensile yield of the HSS wall.

$$\phi R_n = \phi l_e F_y t \le \phi B F_y t$$

where

  $\phi = 1.0$
  l_e = the effective width of the HSS wall, in.
    = the lesser of $(2e + g)$ or $4e$
  F_y = the specified yield strength of the HSS, ksi
  t = the HSS wall thickness, in.
  B = the HSS width, in.

3. Tensile rupture of the cap-plate weld.

$$\phi R_n = \phi l_e (0.6 F_{EXX}) 0.707 w$$

where

  ϕ = 0.75
  l_e = the effective width of the HSS wall, in.
    = the lesser of $(2e + g)$ or $4e$
  F_{EXX} = electrode strength, ksi
  w = weld size, in.

Example 7-2

Determine the required cap plate thickness for the load and connection shown below. See Figure Ex. 7-2.

Solution:

1. Bending of the cap plate at the HSS wall.

$$t_1 = \sqrt{\frac{4 P_u e}{l_e \phi F_{yc}}}$$

where

  P_u = 15 kips
  e = 1.5 in.
  g = 4.0 in.
  l_e = $2(1.5) + 4.0 \le 4(1.5)$
    = 6.0 in.
  F_{yc} = 36 ksi

$$t_c = \sqrt{\frac{4(15)(1.5)}{6.0(0.9)(36)}}$$

    = 0.680 in.

Use ¾-in. A36 Plate

2. Check tension yielding of the HSS wall.

$$\phi R_n = \phi l_e F_y t \leq \phi B F_y t$$

where

F_y = 46 ksi
t = 0.174 in.
B = 8.0 in.
w = 6.0 in. (< B)

$$\phi R_n = 1.0(6.0)(46)(0.174)$$

$$= 48.0 \text{ kips} > 15 \text{ kips} \quad \textbf{o.k.}$$

3. Check tension rupture of the cap-plate weld.

$$\phi R_n = \phi l_e(0.6 F_{EXX})0.707w$$

where

F_{EXX} = 70 ksi
w_{min} = 15/[0.75(6.0)(0.6)(70)(0.707)]
= 0.112 in.

AISC minimum $\frac{5}{16}$-in. weld size controls, use w = $\frac{5}{16}$-in.

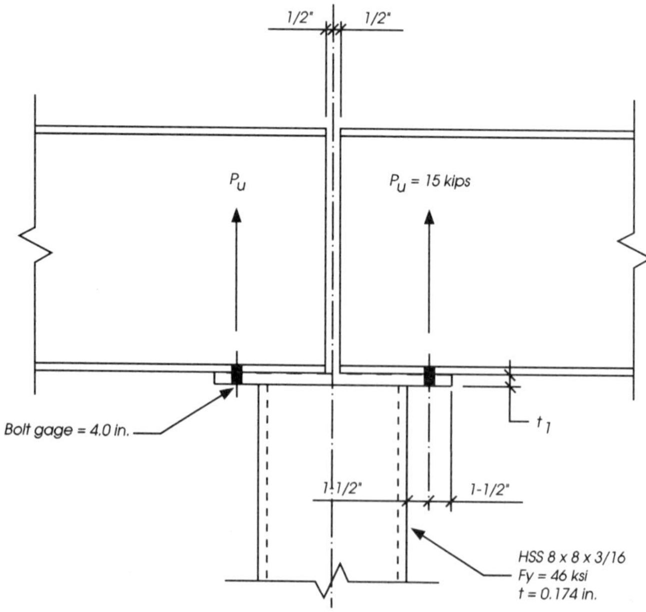

Figure Ex. 7-2.

BASE PLATES

Axial Compression Loading:

HSS column base plates can be designed based on the cantilever projection of the base plate from the walls of the HSS.

A design procedure identical to that contained in the AISC LRFD Manual can be used provided adjustments are made for the "m" and "n" distances for round HSS. DeWolf and Ricker (1990) point out that for round HSS the cantilever projection can be determined using a cantilever projection based on 0.8 times the outside diameter, and for rectangular or square HSS 0.95 times the outside dimension of the HSS. These design dimensions are shown in Fig. 7-2.

Axial Tensile Loading

For base plates subjected to tensile loads base plate thicknesses can be determined using cantilever beam theory in a manner identical to that used for the design of cap plates subjected to tensile concentrated loads. However, in cases where the anchor rods are positioned as shown in Figure 7-3 an alternate procedure must be used.

Unfortunately published solutions to determine base plate thickness or weld design strength for the base plate-anchor rod condition shown in Figure 7-3 do not exist. By examining Figure 7-3 it is obvious that the weld at the flange tip is subjected to a concentration of load because of the location of the anchor rod. The following design procedure is suggested for this condition.

1. The effective width of the base plate, b_e, should be taken as $2L$.
2. The maximum effective width to be used is five inches.
3. A maximum weld length of four inches can be used to transmit load between the base plate and the column section.
4. The base plate must be thick enough so as not to over strain the welds. Equation 7-8 can be used to determine minimum plate thickness.

In equation format the design strength for a single anchor rod can be expressed as follows:
Based on the plate effective width:

$$\phi R_n = \phi_p b_e t_p^2 F_{yp} / (4L) = \phi_p t_p^2 F_{yp} / 2 \le \phi_p 1.25 t_p^2 F_{yp} / L$$

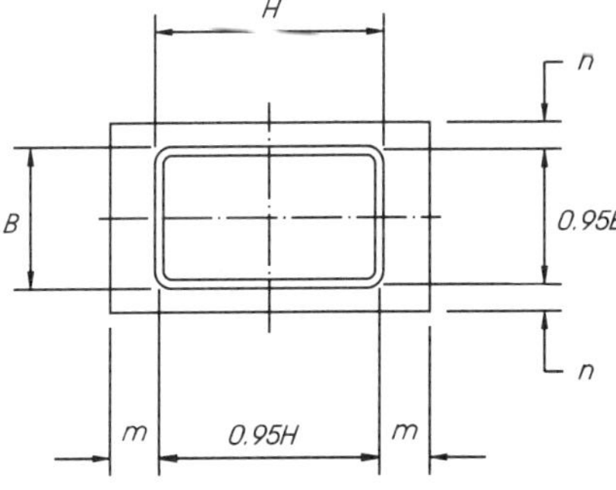
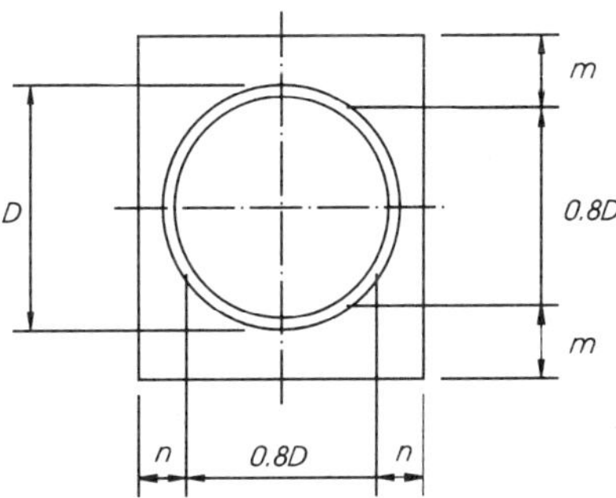

Fig. 7-2. m and n distances.

Based on weld strength:

$$\phi R_n = \phi_w F_w W_t(4)$$

Based on weld strain:

$$\phi R_n = \phi_p(50)W_t\, t_p^{1.5}$$

where

ϕ_p = 0.90
ϕ_w = 0.75
b_e = the effective plate width, in.

$2L \le 5$ in.

L = the distance from the anchor rod to the HSS corner, in.
W_t = the throat width of the weld, in.
  = weld size $\times$ 0.707
t_p = the base plate thickness, in.
F_{yp} = the specified yield strength for the base plate, ksi
F_w = the nominal weld stress, ksi
  = $1.5(0.6F_{EXX})$, ksi (1.5 factor is for 90° loading)
F_{EXX} = electrode classification number ksi

Additional information concerning finishing requirements, holes for anchor rods, grouting, and leveling methods can be found in the AISC LRFD Manual Part 11.

COLUMN SPLICES

Field splices for HSS can be either welded or bolted. For welded splices the connection between the HSS can be either full penetration or partial penetration. Welding details for these splices are provided in Chapter 2. The section can be held in position by temporarily bolting the sections together. A typical temporary bolting pattern is shown in Figure 7-4.

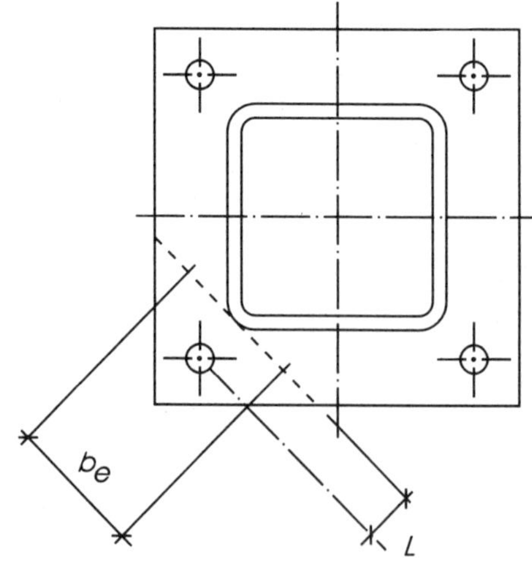

Fig. 7-3. HSS with corner anchor rods.

Examples of several splice connections are shown in Figure 7-5.

REFERENCES
DeWolf, John T., and Ricker, David T., 1990, *Column Base Plates, AISC Design Guide I*, American Institute of Steel Construction, Chicago, IL.

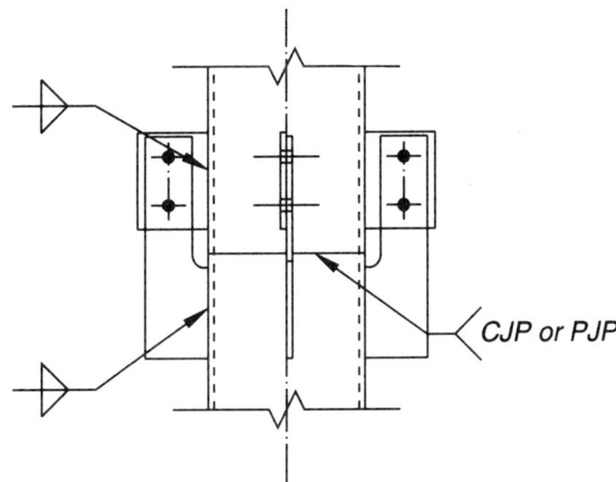

NOTE: If the erection plates are to be removed
the contract documents must so indicate.

Figure 7-4.

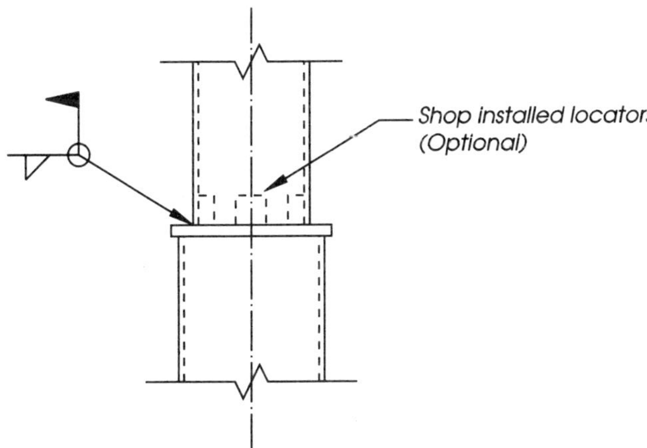

Fig. 7-5. HSS splice connections.

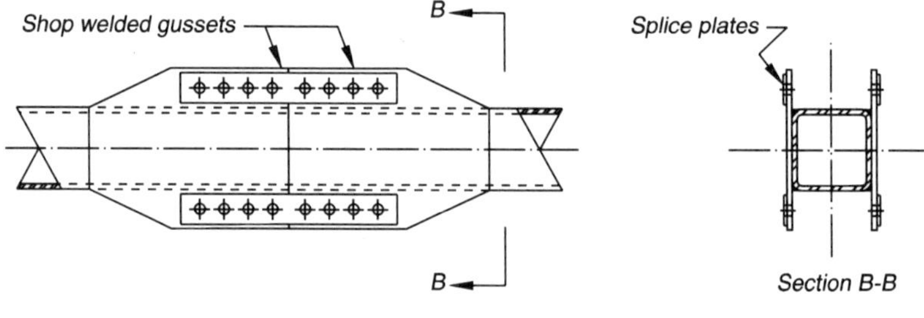

(a) Splice plates on HSS

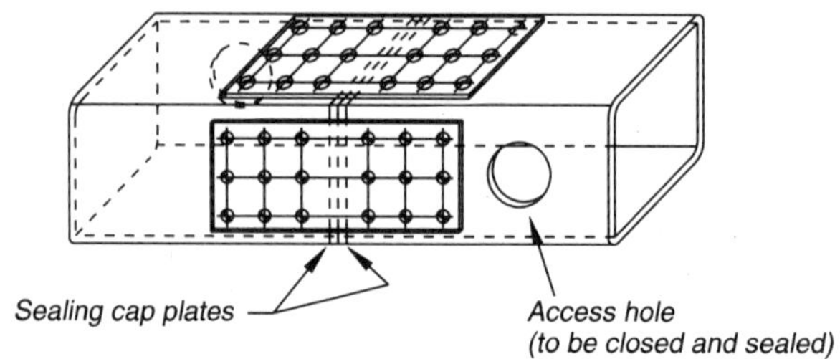

(b) Splice plates on HSS

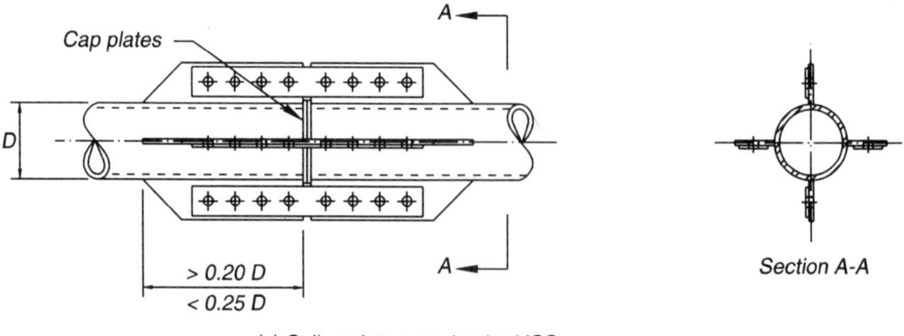

(c) Splice plates on circular HSS

Figure 7-6.

CHAPTER 8

WELDED TRUSS CONNECTIONS

OVERVIEW

Chapter 8 contains design information for HSS-to-HSS truss connections in which branch (web) members are directly welded to a continuous chord that passes through a connection. All connections are assumed to be planar, predominantly statically loaded, to have non-overlapping branches and to be primarily axially loaded. Connection types covered include unreinforced T-, Y-, gapped K-, and Cross-connections, as well as some connections reinforced by plate stiffeners. Design information is based primarily on AWS D1.1-96, supplemented by selected design information from Packer and Henderson (1992) and the AISC HSS Specification. Design information for overlapped K-connections can be found in AWS D1.1-96. Information relative to stiffening of joints using concrete filling can be found in Packer and Henderson (1992).

Following is a list of the Chapter contents:

TERMINOLOGY AND ECCENTRICITY

This Chapter uses terminology and connection symbols adopted in the AISC *Specification for the Design of Steel Hollow Structural Sections.* Figure 8-1 illustrates many of these symbols applied to a typical gapped K-connection in a Warren truss.

Eccentricity (e in Figure 8-1), is positive when measured towards the outside of a chord, and negative when measured towards the inside. (Alternatively, eccentricity is positive when measured away from the branches in a planar truss). The gap dimension (g) or overlap ($-g$) may be calculated from the following equations, which are useful for calculating eccentricities, gaps and overlaps.

$$g = (e + 0.5H) / C - (A + B)$$

or, alternatively, this equation can be rewritten as:

$$e = C(A + B + g) - 0.5H$$

where

$$A = H_{b1} / 2\sin\theta_1$$
$$B = H_{b2} / 2\sin\theta_2$$
$$C = \sin\theta_1\sin\theta_2 / \sin(\theta_1 + \theta_2)$$

These equations can also be used for panel points that have a stiffening plate on the surface of the chord when the terms $0.5H$ in the above equations are replaced with $(0.5H + t_p)$, where t_p = stiffening plate thickness.

Some other common and relevant connection parameters are:

β = diameter or width ratio between branch (or web) member(s) and chord
     = D_b / D for round HSS, and
     = B_b / B for rectangular HSS

β_{eff} = $2(B_{b1} + B_{b2} + H_{b1} + H_{b2}) / 8B$ for gapped K-connections between rectangular HSS

γ = half width to thickness ratio of the chord

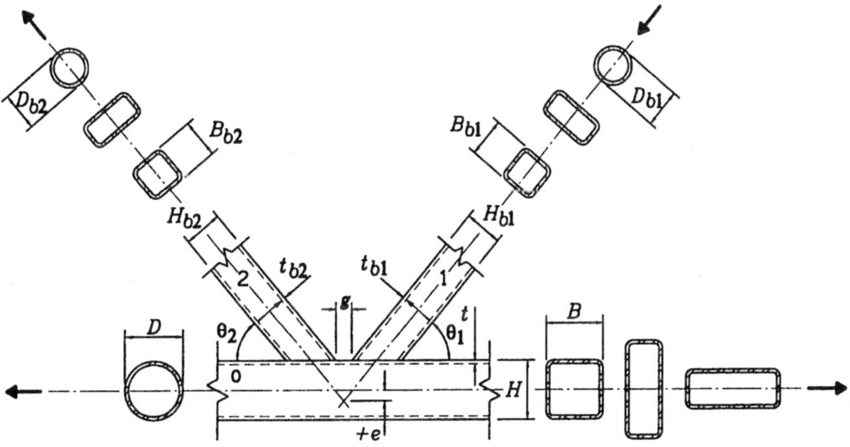

Fig. 8-1. Standard notation for gapped
K-connections between HSS.

$\quad$ = $D / 2t$ for round HSS and $B / 2t$ for rectangular HSS

$\eta\quad$ = length of contact of the branch with the chord, divided by the chord width (for rectangular HSS)

$\quad$ = $(H_b / \sin\theta) / B$

T-, Y-, GAPPED K-, AND CROSS-CONNECTIONS

Generally, the word "rectangular" in reference to HSS is meant to include both square and rectangular sections (as distinct from "round"). It is essential that the designer have an appreciation of factors that make it possible for HSS to be connected at truss panel points without extensive (and expensive) reinforcement. Apparent economies by selecting members according to minimum weights will quickly vanish at the connections if a designer does not have a knowledge of the critical considerations that influence the efficiency of a connection. Some important principles to bear in mind are as follows:

1. Chord members should generally have thick walls rather than thin walls. The stiffer walls resist loads from the branch members more effectively, and the connection design strength thereby increases as the width (or diameter) to thickness ratio of the chord decreases. For the compression chord however, a large thin section is more efficient in providing buckling resistance, so for this member the final HSS wall slenderness will be a compromise between connection strength and buckling strength, and relatively stocky sections will usually be chosen.

2. Branch (or web) members should have thin walls rather than thick walls, as connection design strength increases as the ratio of chord wall thickness to branch wall thickness increases. In addition, thin branch member walls will require smaller fillet welds to develop the full yield strength of the connected member wall.

3. All compression members should have D / t, B / t or H / t values that are less than the limiting λ_r values in the AISC HSS Specification.

4. Ideally, HSS branch members should be narrower than the chord member to avoid difficult welding details at the sides. For rectangular HSS chord members, equal width branches present an awkward flare bevel weld situation (possibly with backing bars) for the joint at the corner of the chord section. A preferred arrangement is branch members just sufficiently narrower than the chord to permit the branch member and some of the fillet weld to sit on the flat of the rectangular HSS chord member. In the absence of more specific information, the flat width of a cold-formed rectangular HSS member can be taken as $B - 3t$ (or $H - 3t$), which assumes that the outside corner radius of a rectangular HSS is typically $1.5t$. One exception to the foregoing, where branch members should have the same width as the chord member, is in Vierendeel trusses.

5. Gapped K-connections are preferred to overlapped K-connections because the members are easier to prepare, fit and weld. Consequently, design criteria for overlapped K-connections are not covered within the scope of this Chapter. If overlapped K-connections are to be used, they will prove to be more expensive yet will likely have a greater connection strength and stiffness. Design information for such connections can be found elsewhere (AWS D1.1-96 and Packer and Henderson, 1992).

6. An angle of less than 30° between a branch member and a chord creates significant welding and inspection difficulties, and is not covered by the scope of these recommendations.

Gapped K-Connections

The majority of HSS truss connections have one compression web member and one tension web member welded to the chord as shown in Figure 8-1. The Warren arrangement is referred to as a K connection and the Pratt as an N connection. The latter is a special case of the former; and both can be either gapped or overlapped connections.

Limit States for Gapped K-Connections

Experimental research (Wardenier and Stark 1978, Kurobane *et al.* 1980, and Kurobane 1981) on HSS welded truss connections has shown that different limit states can exist depending on the type of connection, loading conditions, and various geometric parameters. Limit states have been described (Wardenier 1982) for square and rectangular HSS as illustrated in Figure 8-2:

Limit state A

Plastification of the chord wall (one branch member pushing the face in and the other pulling it out)

Limit state B

Punching shear rupture of the chord face around the perimeter of a branch (or web) member in either compression or tension.

Limit state C

Rupture of the tension branch member or its weld.

Limit state D

Local buckling of the compression branch member.

Limit state E

Shear yielding of the chord member in the gap region.

Limit state F

Chord wall bearing or local buckling under the compression branch member.

Limit state G

Local buckling of the chord face behind the heel of the tension branch member.

Failure in test specimens has also been observed to be a combination of more than one limit state.

It should be noted that Limit states C and D are generally combined under the term "uneven load distribution" and are treated identically, since the connection resistance in both cases is determined by the effective cross section of the critical branch member, with some branch member walls possibly being only partially effective.

Plastification of the chord wall (limit state A) is the most common controlling limit state for gapped connections with small to medium ratios of branch member widths to chord width. For medium width ratios ($\beta \approx 0.6$ to 0.8), this limit state generally occurs together with tearing in the chord (limit state B) or the tension branch member (limit state C), although the latter only occurs in connections with relatively thin-walled branch members. Limit state D, involving local buckling of the compression branch member, is the most common limit state for overlapped connections. Shear yielding of the entire chord section (limit state E) is observed in gapped connections where the width (or diameter) of the branch members is close to that of the chord ($\beta \approx 1.0$). Local buckling (limits states F and G) occurs occasionally in square and rectangular HSS connections with high chord width (or depth) to thickness ratios (B/t or H/t).

Design Formulas for Gapped K-Connections

Various formulas exist for the aforementioned connection limit states. Some have been derived theoretically, while others are primarily empirical. The general criterion for design strength is an ultimate strength limitation, but the recommendations presented

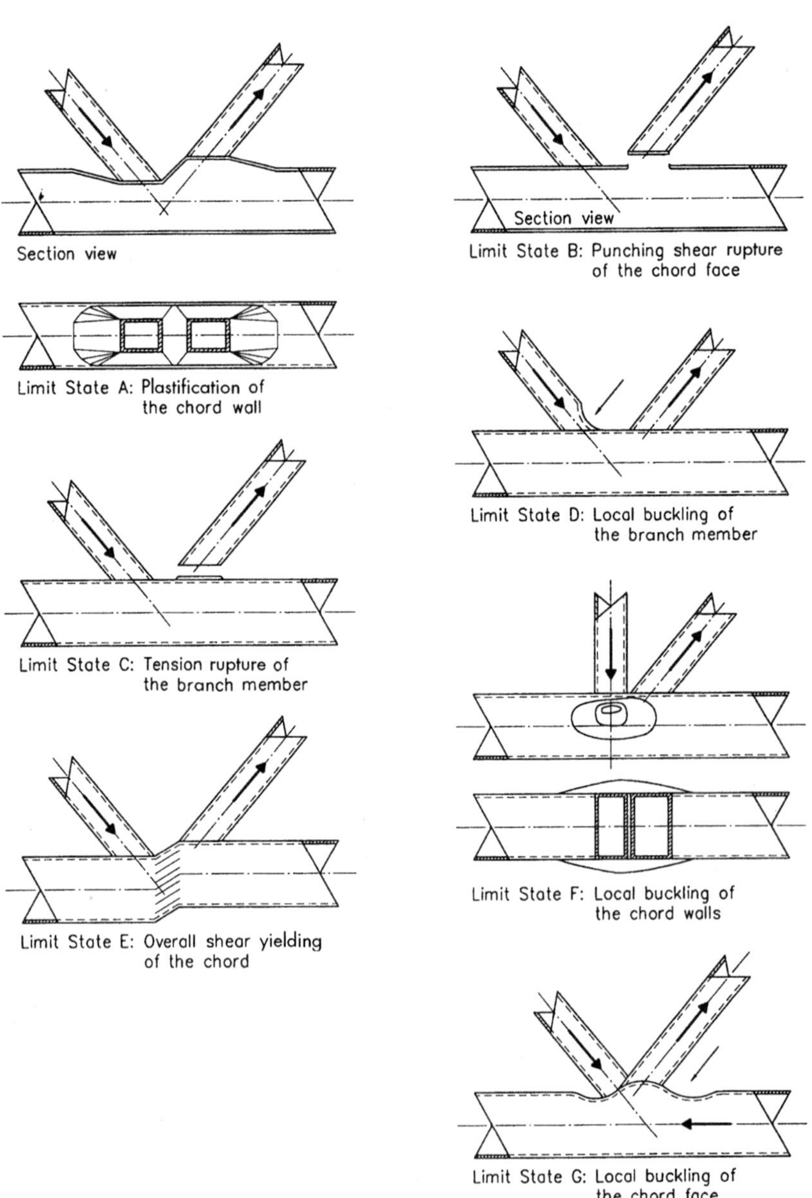

Fig. 8-2. Limit states for K-connections in rectangular HSS trusses.

herein (and their limits of applicability) have been set so that a deformation limit state should not be exceeded at service loads.

LRFD recommendations for axially-loaded connections are summarized in Table 8-1 for round chords and Table 8-2 for rectangular (or square) chords. Connection design strengths are expressed as a component of the branch member axial force normal to the chord member. Branch members in trusses other than Vierendeel trusses can be treated as pin-connected to the chord member. The chord can be considered either as pin-connected at each panel point or as a continuous member. Classification of a connection as a K-connection depends not only on the arrangement of the members but also on the mechanism of load transfer. To qualify as a K-connection, the force component normal to the chord from one branch member must be balanced substantially by the normal component of another branch member force, on the same side of the chord. Some comments on Tables 8-1 and 8-2 follow:

1. A common design criterion for all gapped K-connections is limit state A, plastificaiton of the chord wall. The plastic moment resistance of the chord face per unit length is represented by the term $(F_y\, t^2\, /\, 4)$.
2. Table 8-2 shows that the design strength of a gapped K-connection with rectangular HSS is largely independent of the gap size (i.e. there is no gap size parameter, g). For similar connections between round HSS, Table 8-1 includes a modest influence of the gap size by means of the α term in the function Q_q. As the gap size g increases for round HSS connections the connection design strength decreases.
3. Table 8-2, for rectangular HSS, is simplified for most connections if square members are used. The reason is that gapped K-connections with square HSS need only be examined for the limit state of chord wall plastification (Figure 8-2 limit state A), whereas those with rectangular HSS must be considered for limit states B, C or D, and E as well. This approach has allowed the creation of graphical design charts, which are presented later for connections between square HSS.
4. Tables 8-1 and 8-2 both include a punching shear rupture check (Figure 8-2 limit state B). This limit state may be critical for thin chord members and is possible beneath both the tension and compression branches. The physical condition required for punching shear rupture to occur requires the width of the branch member (plus fillet welds if used), transverse to the chord direction, to be less than the width of the chord member, so that a piece of the chord face can punch out around the "footprint" of the branch member.
5. Table 8-2 includes a check for uneven load distribution (sometimes referred to as an effective width check), which covers limit states C and D in Figure 8-2. In this limit state the branch fails prematurely because the walls transverse to the chord direction are highly overstressed near their corners and understressed near the mid-wall. The overstress occurs because of the varying stiffness across the chord connecting face. The vertical walls of the chord member act like stiffeners and thus much of the branch force is transferred into the sides of the chord member. Experience with laboratory tests has shown that this limit state will be preceded by chord wall plastification, for all β values if square HSS are used.

T-, Y-, and Cross-Connections

The T- connection is a special case of the Y-connection. The basic difference between T- or Y-connections and K-connections is that the component of load normal to the chord in T- and Y-connections is resisted by shear and bending in the chord, whereas with K-connections the normal component from one branch member is balanced primarily by the similar component in the other. Cross-connections are characterized by load (tension or compression) being transferred through the chord member, regardless of the geometrical appearance of the connection. Thus, for example, in a typical simply-supported, symmetrical, Warren roof truss, with gravity loads acting at each top chord panel point, the K-connection on the top chord at the truss centerline is a Cross-connection since both branch members are in compression and this compressive force is transferred through the chord member to balance the panel point load.

Design Formulas for T-, Y-, and Cross-Connections

LRFD recommendations are summarized in Tables 8-1 (for round HSS) and Table 8-2 (for rectangular HSS). The recommendations presented herein (and their limits of applicability) have been set so that a deformation limit state should not be exceeded at service loads. Only branch axial load cases are given and, as with K- and N-connections, design strengths are expressed as a component of the branch member axial force normal to the chord member. Some comments on Tables 8-1 and 8-2 follow:

1. The most common governing limit state for T-, Y-. and Cross-connections is chord wall plastification. This is the only applicable design strength equation in Table 8-2 for $\beta \leq 0.85$, and is based on a yield line mechanism in the flexible rectangular HSS chord face. By limiting connection required strength under factored loads to the connection yield strength, one ensures that deformations will be acceptable at service loads.

2. For T-, Y-, and Cross-connections with moderate to high branch-to-chord width ratios (β), connection flexibility is no longer a problem and design strength is based on either punching shear (see point 4 in Design Formulas for Gapped K-Connections) or the strength of the chord section. When the branch is loaded in compression the chord member may be subject to "general collapse" (as termed for round HSS) or "side wall yielding" / "side wall buckling" (as termed for rectangular HSS). The criteria for chord sidewall buckling in Table 8-2 are based on AISC LRFD Specification Sections K1.3, K1.4 and K1.6. These sections have been modified to account for the two side walls of an HSS chord and are expressed in a non-dimensional form. When the branch is loaded in tension the chord member may be subject to sidewall yielding beneath the "footprint" of the brace member.

3. All design strength expressions in Tables 8-1 and 8-2 pertaining to compressive collapse of the chord section have a relatively low resistance factor ϕ of 0.80 or less, recognizing the greater scatter in resistance prediction associated with a compression limit state. For T-, Y-, and Cross-connections between rectangular HSS (Table 8-2) with β values between 0.85 (the limit of application for the chord wall plastification limit state) and 1.0 (the β value for application of the chord sidewall limit states), a combination of these limit states is likely to occur. One should interpolate linearly between these design strengths. For example, for a rectangular HSS Cross-connection with a β value of 0.925 and compression-loaded branches, one first determines the governing strength from limit states 3 and 4b in Table 8-2, and then takes the average of this value and that from limit state 1. The lower of this calculated resistance and that from limit state 2 (punching shear in Table 8-2) and that from limit state 5 (uneven load distribution in Table 8-2) is the connection design strength.

4. All rectangular HSS T-, Y-, and Cross-connections with high branch width to chord width ratios ($\beta > 0.85$) are also checked for the uneven load distribution limit state and for punching shear rupture of the chord face. For this range of width ratios, the branch member loads are carried largely by their sidewalls parallel to the chords while the walls transverse to the chords carry relatively little load. The upper limit of $\beta = 1 - 1/\gamma$ for checking punching shear rupture is set by the physical possibility

of such a failure, when one considers that the shear has to be between the outer limits of the web width and the inner face of the chord wall.

LIMITS OF APPLICABILITY

The expressions for design strength in Tables 8-1 and 8-2 are valid only within specific ranges of various dimensional parameters for the connected members, since some of the limit state expressions (or their calibrations) are partly empirical. Although many hundreds of laboratory connection tests and truss tests have been undertaken worldwide, and these are the basis for the design recommendations, the formulas may not be reliable outside the parameter range for which they have been validated.

These parameter ranges of validity are expressed in mathematical terms in Tables 8-1 and 8-2. For rectangular HSS, the limits of applicability in Table 8-2 are those given by IIW (1989), which have also been adopted (as have the associated connection design formulas) by AWS D1.1 and Packer and Henderson (1992). AWS D1.1, whose connection design formulas for round HSS are adopted in Table 8-1, does not contain a set of limits of applicability for round HSS, except that the yield stress is limited to 60 ksi, which is satisfied by the material requirements in the AISC HSS Specification. Nevertheless, for connections between round HSS it would be prudent to also use a set of parameter limits that reflects the bounds of most test results. Hence, for round HSS the limits of application from IIW (1989) and Packer and Henderson (1992) are also included in Table 8-1.

GRAPHICAL DESIGN CHARTS

Figure 8-3 (Packer, 1986) allows a quick evaluation of whether a gapped K-connection can be configured from proposed square branch sections joined to a rectangular or square chord section. The chart indicates the maximum value of β (average width of branch members relative to the chord width) which can be accommodated while remaining

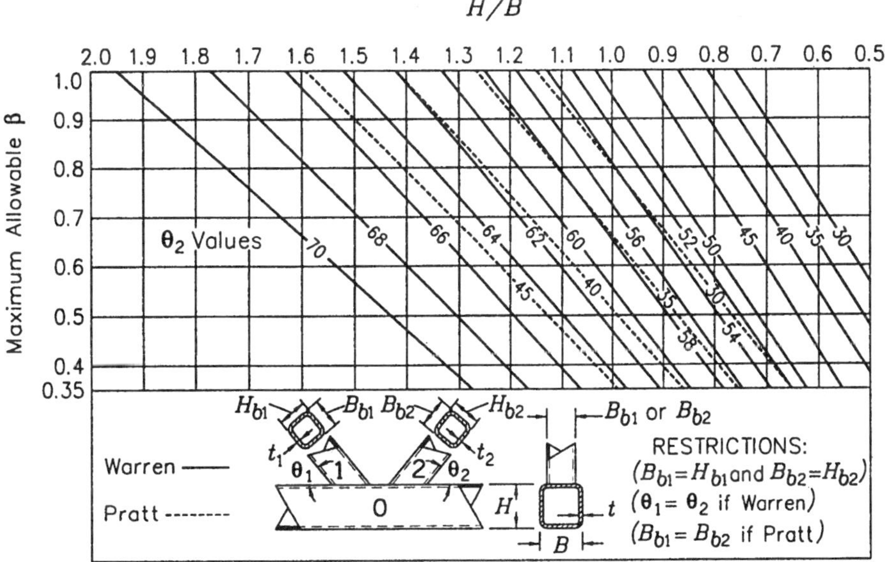

Fig. 8-3. Maximum β for rectangular HSS gapped K-connections, based on allowable eccentricity limits, chord aspect ratio and inclination of web members.

within the allowable bounds of nodal eccentricity. Input parameters are chord to branch member angles and chord aspect ratio H/B. One enters the chart on the horizontal axis with a particular H/B ratio, reads down to the appropriate diagonal line for the branch member angles, then reads across to the vertical axis for the maximum permissible β value. If the intersection point appears to lie above the top of the chart then the maximum β value is just 1.0.

Three further design charts, which relate to the formulas in Table 8-2, are available for the preliminary design of certain connections between square HSS (Reusink and Wardenier, 1989, Packer and Henderson, 1992). These are given in Figures 8-4 to 8-6.

Figures 8-4 and 8-5 will be conservative if compared to the corresponding formulas in Table 8-2 (Packer and Henderson, 1992). These charts are entered along the horizontal axis, one then reads up to the appropriate curve, then across to the vertical axis to obtain the uncorrected connection efficiency. The connection efficiency (defined as the connection design strength divided by the full-section yield strength of the particular branch or web member being considered) is employed in Figures 8-4 and 8-5. That is, connection efficiency ($C_{T,Y,X}$ or C_K) is equal to $\phi P_n / (A_b F_{yb})$. These efficiencies then need to be multiplied by three factors to obtain the final connection efficiency in each case:

a. The first factor, which corrects for different strengths between the chord and the branch member is ($F_y t / F_{yb} t_b$). It is normal, however, to use the same grade of HSS throughout a truss so this term usually reduces to t / t_b.

b. The second factor, which adjusts for the angle between the branch member and the chord is $1 / \sin\theta$, where θ in a K-connection refers to the particular branch being checked. In general, the connection design strength must be calculated for each

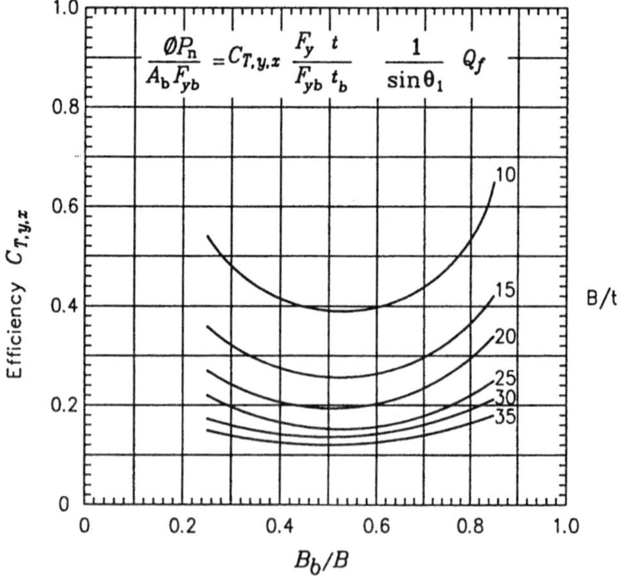

Fig. 8-4. Branch member efficiency for square HSS T, Y, and
Cross connections, with $0.25 <\!<= \beta <\!<= 0.85.$

branch member in a K-connection and each design strength compared with the factored axial load in that member.

c. The third factor, which corrects for the influence of compressive longitudinal stresses on the connection strength, is Q_f. This is defined in Table 8-2 and shown graphically in Figure 8-6. One should note that there is no strength reduction effect for tension chords and Q_f should be taken as unity in this case. The definition of Q_f in Table 8-2 shows that it is dependent upon the level of both the axial and bending stresses in the chord adjacent to a connection. If the axial compression load or bending moment in the chord is different on each side of a connection, the quantity U (in Table 8-2) should be calculated for the most critical combination of compression stresses.

DESIGN OF REINFORCED CONNECTIONS

Instances may occur when a truss connection has inadequate strength, and a designer must resort to some form of connection reinforcement. Labor costs associated with connection reinforcement are significant, and the resulting structure may lose its aesthetic appeal. However, when only one or just a few connections of a truss are inadequate, the reinforcement of these connections may be an acceptable solution.

K-Connections Stiffened with Plates

The most common method of strengthening HSS connections is to weld a stiffening plate or plates to the HSS chord member. It is particularly applicable to gapped K-connections with rectangular chord members, although an unstiffened overlapped K-connection is often preferable from the viewpoints of economy and fatigue. However, a gapped K-connection with a stiffening plate eliminates the necessity for double cuts on the web

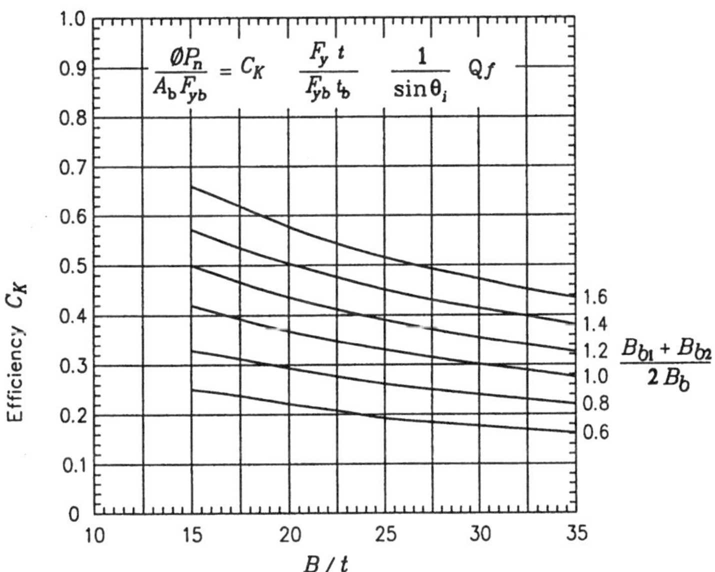

Fig. 8-5. Web member efficiency for square HSS K gap connections.

members, and may prove more acceptable to the fabricator. The addition of a flat plate welded to the connecting face of the chord member greatly reduces local deformations of the connection. Consequently the overall truss deformation is reduced. It also permits a more uniform stress distribution in the truss branch members thereby removing the "uneven load distribution" limit state in Table 8-2.

The type of reinforcement required depends upon the governing limit state that causes the inadequate connection strength. Two types of plate reinforcement are shown in Figure 8-7(a) welded to the chord connecting face, and (b) welded to the chord side walls. Both are applicable to connections with rectangular HSS chord members. An alternative to stiffening a connection with plates is to insert a length of chord with the same outer dimension(s) but thicker walls at the connection. This is equivalent to the use of a "joint can" as in offshore steel structures. The joint can would be butt welded to the rest of the chord and AWS D1.1 contains provisions for the length of the joint can.

The strength of rectangular HSS gapped K-connections is typically controlled by the criteria for either chord wall plastification or chord shear in the gap region (see Table 8-2). When plastification of the chord wall controls, the connection strength can be increased by using flange plate reinforcement as shown in Figure 8-7(a). This will usually occur when $\beta < 1.0$ and the members are square. When chord shear controls, the connection strength can be increased by reinforcing with a pair of side plates as shown in Figure 8-7(b). This limit state will usually govern when $\beta = 1.0$ and $H < B$. In order to calculate the necessary stiffening plate thickness t_p the connection design strength expressions in Table 8-2 (for limit states 1, 2 and 3) can be used by considering t_p as the chord face thickness and neglecting t. The plate yield stress should also be used. The length of the stiffening plate L_p should be in accordance with the AWS D1.1 provisions for joint can length as it is a similar reinforcement principle.

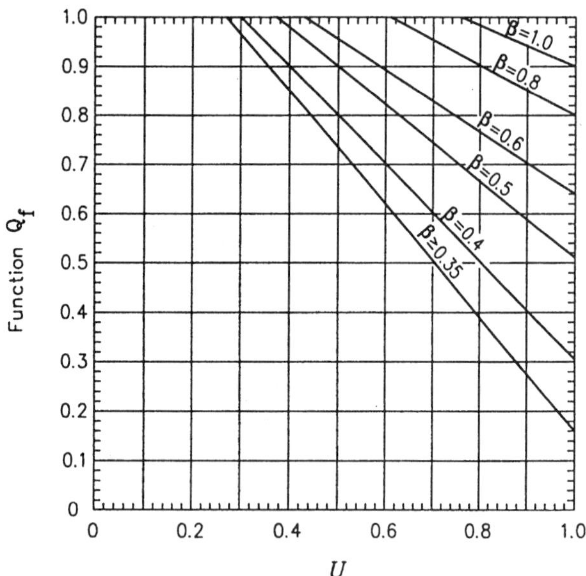

Fig. 8-6. Function Q_f which describes the influence of chord compressive stress, for square HSS connections.

A minimum gap between the web members, just sufficient to permit welding of the web members independently to the stiffening plate, is suggested. (See Figure 8-7.) All-around welding is generally required to connect the stiffening plate to the chord member, and to prevent corrosion on the two inner surfaces in exposed applications. It may also be advisable to drill a small hole in the stiffening plate under a web member to allow entrapped air to escape prior to closing the weld. This will prevent the expanding heated air from causing voids in the closing weld (Stelco, 1981).

In order to avoid partial overlapping of one branch member onto another in a K-connection, fabricators may elect to weld each branch member to a vertical stiffener as shown in Figure 8-8(a). Another variation of this concept is to use the reinforcement shown in Figure 8-8(b). For both of these connections, $t_p \geq 2t_{b1} + 2t_{b2}$ is recommended (Dutta and Würker, 1988). The K-connection shown in Figure 8-8(c) is not acceptable, as it does not develop the strength of an overlapped K-connection where one branch member is welded only to the chord face. Also, it is difficult to create and ensure an effective saddle weld between the two branch members.

If the strength of a gapped K-connection is inadequate and chord shear is the governing limit state, then as mentioned before, one should stiffen with side plate reinforcement, as

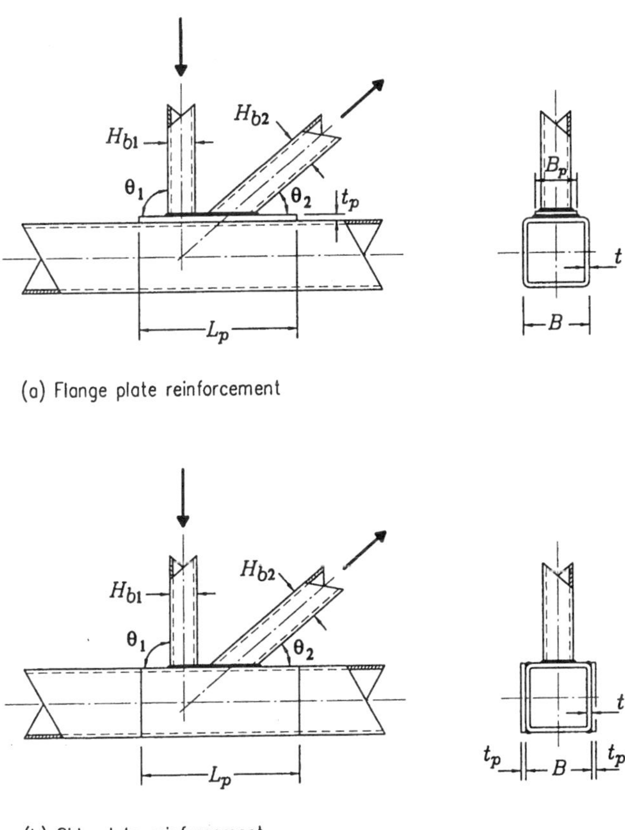

(a) Flange plate reinforcement

(b) Side plate reinforcement

Fig. 8-7. Truss connections with plate stiffening.

shown in Figure 8-7(b). The stiffening plates should again have a minimum length as dictated by AWS D1.1 joint-can requirements and have the same depth as the chord member.

For round-to-round HSS K-connections, stiffening plates shaped to saddle the chord member have been used, particularly where Cross-connection loading exists (Figure 8-9). It is suggested that the K-connection design strength expressions in Table 8-1 be used, by again considering t_p as the chord wall thickness and neglecting t.

T-, Y-, and Cross-Connections Stiffened with Plates

Under tension or compression brace loading, the strength of a rectangular HSS T-, Y-, and Cross-connection is typically controlled by either chord wall plastification or chord sidewall strength (see Table 8-2). When plastification of the chord face controls, the connection strength can be increased by using flange plate reinforcement similar to the connection shown in Figure 8-7(a). This will usually occur when $\beta \leq 0.85$. When chord sidewall strength controls, the connection strength can be increased by reinforcing with a pair of side plates similar to the connection shown in Figure 8-7(b). This limit state will usually govern when $\beta \approx 1.0$. As with reinforced K-connection, to calculate the necessary stiffening plate thickness the connection design strength expressions in Table 8-2, suitably modified, can be used. The length of the stiffening plate should also be in accordance with the AWS D1.1 provisions for joint can length.

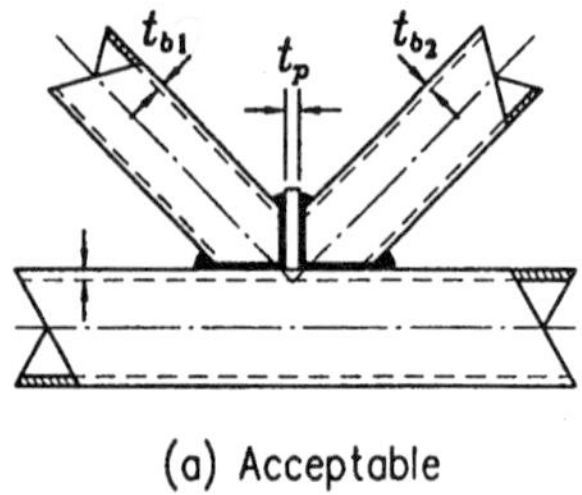

(a) Acceptable

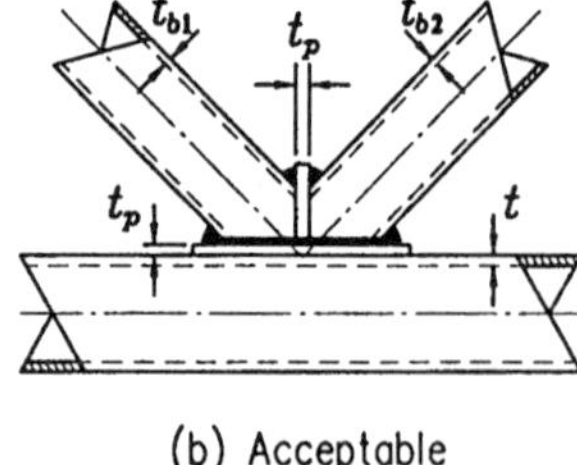

(b) Acceptable

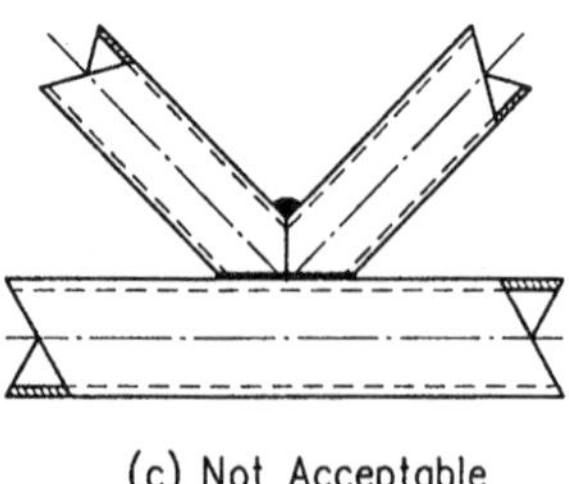

(c) Not Acceptable

Fig. 8-8. Some acceptable and unacceptable
non-standard K-connections.

<table>
<tr><td colspan="3" align="center">

Table 8-1.
Design Strength (ϕP_n) for Welded Connections Between
Round HSS with Axial Loads in the Branches

</td></tr>
<tr>
<th>Type of
Connection</th>
<th>Connection Design
Strength</th>
<th>Limits of
Applicability</th>
</tr>
<tr>
<td align="center">gapped K</td>
<td>

Lesser of the following limit state strengths:

1. Chord Plastification
In Q_q term use $\alpha = 1.0 + 0.7g / D_b$
$\quad 1.0 \le \alpha < 1.7$

2. Punching Shear (Controls only if
$\quad D_b < D - 2t$)

</td>
<td align="center">

$-0.55D \le e \le 0.25D$

$\theta \ge 30°$

$D/t, \; D_b/t_b, \le 50$

$0.2 < D_b/D \le 1.0$

$g \ge t_{b1} + t_{b2}$

</td>
</tr>
<tr>
<td align="center">T and Y</td>
<td>

Lesser of the following limit state strengths:

1. Chord Plastification
In Q_q term use $\alpha = 1.7$

2. Punching Shear (Only if $D_b < D - 2t$)

</td>
<td align="center">

$\theta \ge 30°$

$D/t, \; D_b/t_b, \le 50$

$0.2 < D_b/D \le 1.0$

</td>
</tr>
<tr>
<td align="center">Cross</td>
<td>

Lesser of the following limit state strengths:

1. Chord Plastification
In Q_q term use $\alpha = 2.4$

2. Punching Shear (Only if $D_b < D - 2t$)

3. General Collapse of Chord
(Only if branches loaded in compression)

</td>
<td align="center">

$\theta \ge 30°$

$D_b/t_b \le 50$

$D/t \le 40$

$0.2 < D_b/D \le 1.0$

</td>
</tr>
<tr><td colspan="3" align="center">

Limit States

</td></tr>
<tr>
<td>1. Chord Plastification:</td>
<td align="center">$\phi P_n \sin\theta = \phi F_y t^2 (6\pi\beta Q_q) Q_f$</td>
<td align="center">, $\phi = 0.8$</td>
</tr>
<tr>
<td>2. Punching Shear:</td>
<td align="center">$\phi P_n \sin\theta = \phi\pi D_b t (0.6 F_y)$</td>
<td align="center">, $\phi = 0.95$</td>
</tr>
<tr>
<td>3. General Collapse of Chord</td>
<td align="center">$\phi P_n \sin\theta = \phi F_y t^2 (1.9 + 7.2\beta) Q_\beta Q_f$</td>
<td align="center">, $\phi = 0.8$</td>
</tr>
<tr><td colspan="3" align="center">

Parameters and Functions

</td></tr>
<tr><td colspan="3">

$Q_q = (1.7 / \alpha + 0.18 / \beta) Q_\beta^{0.7(\alpha-1)}$ $\beta \;\;= D_b / D$

                                                       $\gamma \;\;= D / 2t$

$Q_\beta = 1.0$, for $\beta \le 0.6$ $P_u \;\;= $ axial force in chord

                                                      $A_g \;\;= $ chord area

$\quad = 0.3 / [\beta(1 - 0.833\beta)]$, for $\beta > 0.6$ $F_y \;\;= $ chord yield strength

                                                      $t \;\;\;= $ chord thickness

$Q_f = 1.0 - 0.03\gamma U^2 \le 1.0$ for compression chord $M_u \;\;= $ moment in chord

                                                      $S \;\;\;= $ chord elastic section modulus

$\quad = 1.0$ for tension chord $\alpha \;\;\;= $ chord ovalization parameter

$U^2 = (P_n / A_g F_y)^2 + (M_n / S F_y)^2$

</td></tr>
</table>

Table 8-2a.
Design Strength (ϕP_n) for Welded Connections Between Rectangular HSS with Axial Loads in the Branches

Type of Connection	Connection Design Strength	Limits of Applicability
gapped K	Lesser of the following limit state strengths: 1. Chord Plastification: $\phi P_n \sin\theta = \phi F_y t^2 (9.8\beta_{eff}\gamma^{0.5})Q_f$ $\phi = 0.9$ 2. Punching Shear: (Only if branches are not square or of unequal width) $\phi P_n \sin\theta = \phi 0.6 F_y tB(2\eta + \beta_{gap} + \beta_{eop})$ $\phi = 0.95$ 3. Chord Shear in the Gap Region: (Necessary only if $H < B$) Follow AISC HSS Specification Section 5.2 4. Uneven Load Distribution: (Only if members are not square or of unequal width) $\phi P_n \sin\theta = \phi F_{yb} t_b (2H_b + b_{gap} + b_{eoi} - 4t_b)$ $\phi = 0.95$	$-0.55H \le e \le 0.25H$ $\theta \ge 30°$ $B/t, H/t, B_b/t_b,$ $H_b/t_b \le 35$ $(B_b/t_b, H_b/t_b)$ for compression branch $\le 1.25(E/F_{yb})^{0.5}$ $F_y, F_{yb} \le 52$ ksi $0.5 \le H/B \le 2.0$ $F_y/F_u, F_{yb}/F_{ub} \le 0.8$ $\zeta \ge 0.5(1 - \beta)$ $\beta_{eff} \ge 0.35$ $B_b/B \ge 0.1 + \gamma/50$

Parameters and Functions

$\beta = B_b/B$

$\beta_{eff} = 2(B_{b1} + B_{b2} + H_{b1} + H_{b2})/8B$

$\beta_{eop} = 5\beta/\gamma \le \beta$

$Q_f = 1.3 - 0.4U/\beta \le 1$ for compression chord

$\quad = 1.0$ for tension chord

$U = |P_u/A_g F_y| + |M_u/SF_y|$

$b_{eoi} = (5B_b/\gamma)(F_y/F_{yb})(t/t_b) \le B_b$

P_n = axial force in chord

A_g = chord area

M_u = moment in chord

S = chord elastic section modulus

$\gamma = B/2t$

$\eta = H_b/B\sin\theta$

$\zeta = g/B$

$\beta_{gap} = \beta$ when $\zeta \le 1.5(1 - \beta)$, otherwise $\beta_{gap} = \beta_{eop}$

$b_{gap} = B_b$ when $\zeta \le 1.5(1 - \beta)$, otherwise $b_{gap} = b_{eoi}$

Table 8-2b.
Design Strength (ϕP_n) for Welded Connections Between Rectangular HSS with Axial Loads in the Branches

Type of Connection	Connection Design Strength	Limits of Applicability
T, Y and Cross	Lesser of the following limit state strengths:	$\theta \geq 30°$

T, Y and Cross — Lesser of the following limit state strengths:

1. Chord Plastification:
(Only if $\beta \leq 0.85$)
$$\phi P_n \sin\theta = \phi F_y t^2 [2\eta / (1 - \beta) + 4 / (1 - \beta)^{0.5}] Q_f$$
$$\phi = 1.0$$

2. Punching Shear:
(Only for $0.85 \leq \beta \leq 1 - 1/\gamma$)
$$\phi P_n \sin\theta = \phi 0.6 F_y t B (2\eta + 2\beta_{eop})$$
$$\phi = 0.95$$

3. Chord Side Wall Yielding:
(Only for $\beta \cong 1.0$)
$$\phi P_n \sin\theta = \phi 2 F_y t (5k + N)$$
$$\phi = 1.0 \text{ for tension-loaded branches}$$
$$\phi = 0.8 \text{ for compression-loaded branches}$$

4. Chord Side Wall Buckling:
(Only for $\beta \cong 1.0$ and compression-loaded branches)
4(a) For T and Y connections
$$\phi P_n \sin\theta = \phi 1.6 t^2 (1 + 3N/H)(EF_y)^{0.5} Q_f$$
$$\phi = 0.75$$

4(b) For Cross connections
$$\phi P_n \sin\theta = \phi [48 t^3 / (H - 4t)](EF_y)^{0.5} Q_f$$
$$\phi = 0.8$$

5. Uneven Load Distribution:
(Only for $\beta > 0.85$)
$$\phi P_n \sin\theta = \phi F_{yb} t_b (2H_b + 2b_{eoi} - 4t_b)$$
$$\phi = 0.95$$

Limits of Applicability:

$\theta \geq 30°$

$B/t, H/t, B_b/t_b,$
$H_b/t_b \leq 35$

$(B_b/t_b, H_b/t_b)$
for compression branch
$\leq 1.25(E/F_{yb})^{0.5}$

$F_y, F_{yb} \leq 52$ ksi

$0.5 \leq H/B \leq 2.0$

$F_y/F_u, F_{yb}/F_{ub} \leq 0.8$

$B_b/B \geq 0.25$

Parameters and Functions

$\beta = B_b/B$

$\beta_{eop} = 5\beta/\gamma \leq \beta$

$Q_f = 1.3 - 0.4U/\beta \leq 1$ for compression chord

$\quad = 1.0$ for tension chord

$U = |P_u / A_g F_y| + |M_u / S F_y|$

$b_{eoi} = (5B_b/\gamma)(F_y/F_{yb})(t/t_b) \leq B_b$
P_n — axial force in chord
A_g = chord area
S = chord elastic section modulus
M_u = moment in chord
k = outside corner radius of the HSS (may be taken as $1.5t$)
$\eta = H_b / \sin\theta$

REFERENCES

AISC, 1993, *Load and Resistance Factor Design Specification for Structural Steel Buildings*, AISC, Chicago, IL.

AISC, 1997, *Specification for the Design of Steel Hollow Structural Sections*, AISC, Chicago, IL.

AWS, 1996, *Structural Welding Code—Steel, ANSI/AWS D1.1-96*, AWS, Miami, FL.

Dutta, D., and Würker, K., 1988, *Handbuch Hohlprofile in Stahlkonstruktionen*, Verlag TÜV Rheinland GmbH, Köln, Germany.

IIW, 1989, *Design Recommendations for Hollow Section Joints—Predominantly Statically Loaded*, 2nd. Ed., IIW, XV-E, IIW Doc. XV-701-89, IIW Annual Assembly, Helsinki, Finland.

Kurobane, Y., 1981, *New Developments and Practices in Tubular Joint Design*, IIW Doc. XV-488-81 + Addendum, IIW Annual Assembly, Oporto, Portugal.

Kurobane, Y., Makino, Y., and Mitsui, Y., 1980, *Re-analysis of Ultimate Strength Data for Truss Connections in Circular Hollow Sections*, IIW Doc. XV-461-80, Kumamoto University, Japan.

Packer, J. A., 1995, "Concrete-Filled HSS Connections," *Journal of Structural Engineering*, ASCE, Vol. 121, No. 3, pp. 458–467.

Packer, J. A. and Henderson, J. E., 1992, *Design Guide for Hollow Structural Section Connections*, CISC, Willowdale, Ontario, Canada.

Packer, J. A., 1986, "Design Examples for HSS Trusses," *Canadian Journal of Civil Engineering*, Vol. 13, No. 4, pp. 460–473.

Reusink, J. H., and Wardenier, J., 1989, "Simplified Design Charts for Axially Loaded Joints of Square Hollow Sections," *3rd International Symposium on Tubular Structures, Lappeenranta, Finland, Proceedings*, pp. 54–61.

STELCO, 1981, *Hollow Structural Sections—Design Manual for Connections*, 2nd. Ed., Stelco Inc., Hamilton, Canada.

Wardenier, J., 1982, *Hollow Section Joints*, Delft University Press, Delft, The Netherlands.

Wardenier, J., and Stark, J. W. B., 1978, "The Static Strength of Welded Lattice Girder Joints in Structural Hollow Sections: Parts 1 to 10," *CIDECT Final Report 5Q/78/4*, Delft University of Technology, The Netherlands.

Wardenier, J., Kurobane, Y., Packer, J. A., Dutta, D., and Yeomans, N., 1991, *Design guide for circular hollow section (CHS) joints under predominantly static loading*, CIDECT (Ed.) and Verlag TÜV Rheinland GmbH, Köln, Federal Republic of Germany.

CHAPTER 9

TRUSS DESIGN EXAMPLES

OVERVIEW

Chapter 9 contains general design and economic guidelines, preliminary member selection considerations, and three truss design examples. The examples have been selected to illustrate the use of the AISC HSS Specification relative to HSS-to-HSS connections. The examples address the design of a Warren truss spanning 90 feet and loaded along the top chord at five-foot intervals as illustrated in Figure 9-1. In addition to designing the appropriate members for the truss, the design of a support detail is illustrated in the first two examples. In one case, the support connection is to the side of an HSS column; in the other case, the support connection is seated atop an HSS column. The first example illustrates the equations relative to the connection of rectangular HSS to rectangular HSS. The second example presents the use of equations relative to the connection of round HSS to round HSS. The third example illustrates the alternative of not using some or all of the verticals shown in Figure 9-1.

GENERAL DESIGN AND ECONOMICAL CONSIDERATIONS

No absolute statements can be made about which truss configuration will provide the most economical solution for a particular situation; however, the following statements can be made regarding truss design:

1. Span-to-depth ratios of 12 to 20 generally prove to be economical; however, shipping depth limitations should be considered so that shop fabrication can be maximized. The maximum depth for shipping is conservatively about 14 feet. Greater depths will require the branch members to be field bolted, which will increase erection costs.

2. The length between splice points is also limited by shipping lengths. The maximum shippable length varies according to the destination of the trusses; but lengths of 80 feet are generally shippable and lengths of 100 feet are often shippable. Since the maximum available mill length is approximately 60 feet, the distance between splice points is normally set at 60 feet.

3. If trusses are analyzed using frame analysis software and rigid joints are assumed, secondary bending moments will result in the analysis. Secondary stresses should not be neglected if the beneficial effects of continuity are utilized in the design process, e.g. for determination of effective length. The designer must be consistent. That is, if the joints are considered as pins for the determination of forces, then they should also be considered as pins in the design process. The assumption of rigid joints may in some cases result in an unconservative estimate of the deflection of the truss.

4. Repetition is beneficial and economical. Use as few different truss depths as possible. It is often more economical to vary the chord size instead of the truss depth.

5. For symmetrical trusses, use a center splice to simplify fabrication even though forces may be larger than for an offset splice. This results in two identical fabrication pieces for every truss.

6. End plates are efficient for compression splices.

7. It is often less expensive to locate the work point of the end diagonal at the face of the supporting member rather than designing the connection for the eccentricity between the column centerline and the face of the column.

8. Use gapped connections whenever possible.

9. Use fillet welds for HSS-to-HSS connections whenever possible.

10. Select branch (web) members slightly narrower $(B - 3t)$ than the chords for ease of welding.

11. Generally, try to limit wall thickness to ½-in., as larger thicknesses have limited availability.

PRELIMINARY MEMBER SELECTION CONSIDERATIONS

Time will be saved when selecting member sizes by keeping in mind the basic constraints or "limits of applicability" of various dimensional parameters which must be met for connections. These were shown in Chapter 8, Tables 8-1, 8-2a and 8-2b. Also, it can be expedient to select over-strength members (by as much as a third in some cases) since high efficiency connections are not always achievable when joining minimum-size HSS.

These constraints place extensive limits upon the selection of members. However, once the constraints are met, determination of the connection strength is greatly simplified. For most gapped connections, only the chord wall plastification limit state (limit state A) needs to be considered.

Example 1: Warren Truss Using Rectangular and Square HSS

Design the truss shown in Figure 9-1 using rectangular or square HSS. The truss loads (and the final design) are given in Figure 9-2. Truss joints are identified for reference. The top chord of the truss is laterally braced 60 in. on center. The members used shall conform to ASTM A500 Gr B which has F_y = 46 ksi for rectangular and square HSS.

Preliminary Sizing of the Chord Members

The chord members should be designed using shapes with relatively thick walls, and the minimum size consistent with slenderness considerations.

Compute the required tension chord area using the forces shown in Figure 9-2:

$$P_u = 476 + 10.5\cos55.4 = 482 \text{ kips}$$

$$A_g = \frac{P_u}{\phi F_y} = \frac{482}{0.9(46)} = 11.6 \text{ in.}^2$$

Use an HSS7×7×½ or an HSS8×6×½ with A_g = 11.6 in.² from the property tables in Chapter 1.

One could consider use of an HSS7×7×⅜ with A_g = 8.97 in.² near the end beyond the splice location shown in Figure 9-2.

$$\phi P_n = \phi F_y A_g = 0.9(46)(8.97) = 371 \text{ kips} > 333 \text{ kips}$$

The HSS7×7×⅜ could be used in the last three panels.

For the compression chord, the chord is laterally braced at each purlin/joist load point. Using K = 0.9, the effective length Kl = (0.9)(60) = 54 in.

One could use Table 6-4 for struts where either one-half or one-quarter of the compressive strength at Kl/r = 60 is present as a means for obtaining a preliminary size. However, it may be as easy to select by trial after selection of the tension chord.

Consider an HSS8×8×½ with A_g = 13.5 in.² and r = 3.04 in.

$$Kl/r = 54/3.04 = 17.8$$

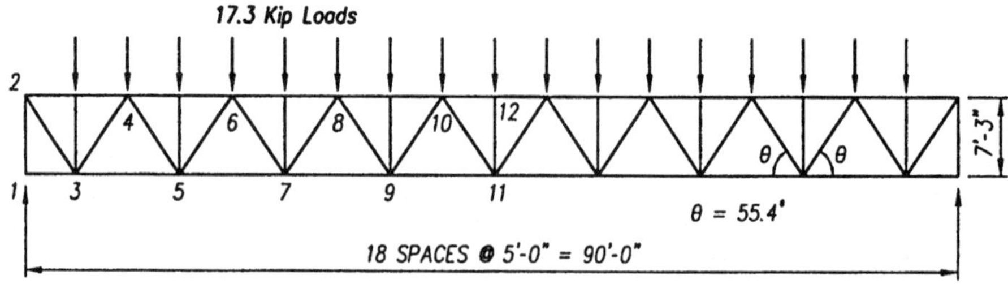

Fig. 9-1. Warren truss design example.

Obtain $\phi_c F_{cr}$ from Table 9-1 to determine

$$\phi P_n = \phi_c F_{cr} A_g = 38.3(13.5) = 517 \text{ kips} > 483 \text{ kips}$$

Use the HSS8×8×½ for the top chord.

Beyond the splice it may be possible to use the HSS8×8×⅜ with $A_g = 10.4$ in.^{2} in conjunction with the HSS8×8×½ in the central portion of the truss.

$$Kl/r \approx 17.7; \quad \phi_c P_n = \phi_c F_{cr} A_g = 38.3(10.4) = 398 \text{ kips}$$

The HSS8×8×⅜ could be used for the first three top chord panels.

Note that there may be some advantage in fabrication economy if the top and bottom chords are selected with the same outside dimensions.

Preliminary Selection of the Branch Members.

The area required for the first tension diagonal with $P_u = 178$ kips is

$$A_g = \frac{P_u}{\phi F_y} = \frac{178}{0.9(46)} = 4.30 \text{ in.}^2$$

The HSS4×4×⅜ with $A_g = 4.78$ in.^{2} is one option. Other alternatives are:

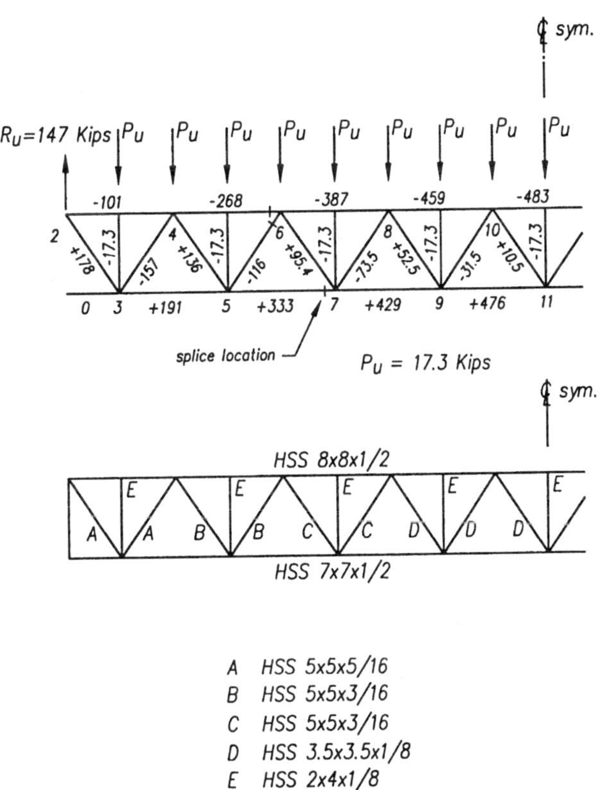

Fig. 9-2. Truss forces and square HSS design sizes.

HSS5×4×$\frac{5}{16}$ and HSS4½×4½×$\frac{5}{16}$; A_g = 4.68 in.^{2}
HSS6×4×¼ and HSS5×5×¼; A_g = 4.30 in.^{2}
HSS6×5×¼ and HSS5½×5½×$\frac{5}{16}$; A_g = 4.77 in.^{2}

Since it is less likely that the 4- and 4½-in.-wide sections will work at the connection, examine the HSS6×4×¼ for the first compressive web member which has a P_u = 157 kips. The effective length for all web diagonals will be taken using $K = 0.75$ per AISC HSS Specification Section 4.1.1. The centerline length equals 94.9 in.

$$KL = (0.75)(94.9) = 71.2 \text{ in.}$$

For the HSS6×4×¼ r_y = 1.61 in. and $Kl/r = 71.2/1.61 = 44.2$

From Table 9-1, $\phi_c F_{cr}$ = 34.3 and

$$\phi P_n = \phi_c F_{cr} A_g = 34.3(4.30) = 147 \text{ kips} < 157 \text{ kips}$$

Since HSS6×4×¼ doesn't work examine the HSS6×5×¼

where

$$Kl/r = 71.2/1.98 = 36.0 \text{ and}$$

$$\phi P_n = \phi_c F_{cr} A_g = 35.8(4.77) = 171 \text{ kips} > 157 \text{ kips}$$

Consider either HSS6×5×¼ or HSS5½×5½×¼ for the first two web members.

Branch Connection Criteria

Check first the various limits of applicability for use of the HSS Specification criteria in Section 9.4.3a. Use of ASTM A500 Gr B rectangular HSS meets the strength criteria with $F_y \leq 52$ ksi and the ductility criteria with $F_y/F_u \leq 0.8$. The square chords satisfy the chord aspect ratio range of 0.5 to 2.0.

The truss being examined has a branch angle $\theta = 55.4$ in. which is greater than 30°. The HSS width-thickness ratio for the chords and tension branches must be less or equal to 35 (See Table 8-2) which is satisfied as shown below:

$$\frac{B}{t} = \frac{8}{0.349} = 22.9 \text{ for the top chord}$$

$$\frac{B}{t} = \frac{6}{0.174} = 34.5 \text{ for a branch with } \tfrac{3}{16}\text{-in. nominal wall thickness and 6-in. width}$$

For the compressively loaded branch members, the width-thickness ratio must be less than

$$1.25\sqrt{E/F_{yb}} = 1.25\sqrt{29,000/46} = 31.4$$

which is satisfied using a 5-in. wide branch with a $\frac{3}{16}$-in. wall or a 3½-in. wide branch with ⅛-in. wall.

The branch width should be no more than the flat width of the chord (as noted in Chapter 8) to permit fillet welding on the sides, although some encroachment on the corner radii is permitted. If an HSS7×7×⅜ is used as a bottom chord an HSS6×5×¼ branch can be used since the flat width is $7 - 3(0.375) = 5.87$ inches. However, if the chord is an HSS7×7×½ it would be better to limit the width of the branch to 5.5 inches.

The last concern regarding applicability relates to the joint eccentricity e which must satisfy

$$-0.55H \le e \le 0.25H$$

where

H = chord depth

e is positive measured away from the branches

In Figure 9-3 the situation is illustrated for the Warren Truss with the applicable relationships between the gap, the eccentricity e and the work point eccentricity e_{wp}. The expressions in Figure 9-3 are a simplification for the Warren Truss with $\theta_1 = \theta_2 = \theta$ of the general relationships given in Chapter 8.

The maximum gap can be determined at the 7-in.-deep bottom chord first panel point with two 5-in.-wide branch members as:

$$\frac{2e + H}{\tan\theta} - \frac{H_1 + H_2}{2\sin\theta} = \frac{2(0.25)(7) + 7}{\tan 55.4} - \frac{5 + 5}{2\sin 55.4} = 1.17 \text{ in.}$$

To maintain a gapped connection at the bottom chord with the presence of the truss vertical, the use of the detail with a plate shown in Figure 9-4 will be recommended. The 1.17-in. gap computed is considered to be about the practical minimum for the fabrication of this detail, so 5 inches will be the maximum branch depth used.

The initial check should typically be of the branch strength based on chord wall plastification. This is usually followed by examining the strength considering uneven load distribution. Punching shear limits on branch strength should then be checked for rectangular branches or where the branches are of two different widths. Lastly, shear in the chord at the gap location should be examined.

There are four nondimensional parameters used in the evaluation of the branch strength of rectangular sections.

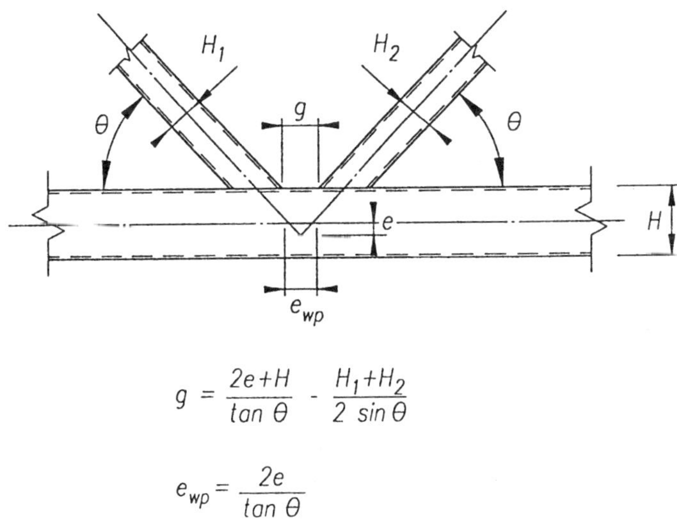

$$g = \frac{2e + H}{\tan\theta} - \frac{H_1 + H_2}{2\sin\theta}$$

$$e_{wp} = \frac{2e}{\tan\theta}$$

Fig. 9-3. Gap and working point geometry.

1. β, branch to chord width ratio $= B_b / B$
2. γ, chord slenderness or half width/thickness $= B / 2t$
3. η, load length of branch on chord/chord width $= H_b / B\sin\theta$
4. ζ, the gap ratio $= g / B$

Bottom Chord Branch Connection

Examine the first interior bottom chord joint, which is identified as joint 3 in Figure 9-2. With HSS7×7×⅜ bottom chord and HSS6×5×¼ branch members

$$\beta = \frac{6}{7} = 0.857$$

$$\gamma = \frac{7}{2(0.349)} = 10.0$$

Chord wall plastification occurs when:

1. $\dfrac{B_b}{B} \geq 0.1 + \gamma / 50$

$0.857 > 0.1 + 10.0/50 = 0.30$

Also,

β_{eff} = (sum of perimeters of the two branches)/(8 times the chord width)

Actual $\beta_{eff} = \dfrac{2(6 + 5)(2)}{8(7)} = 0.786 > 0.35$ (min. allowable β_{eff})

Note: $\beta_{eff} = \beta$ when there are two square branches of equal size.

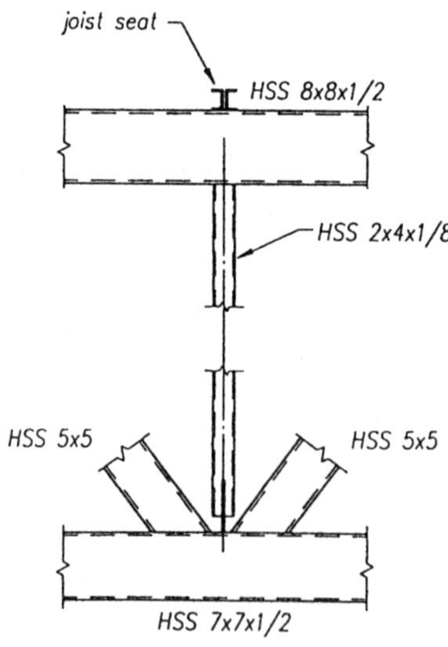

Fig. 9-4. Truss vertical details—Square HSS.

2. $\zeta = g / B \geq 0.5 (1 - \beta)$

$1.17/7 = 0.167 \geq 0.5(1 - 0.786) = 0.107$

Chord plastification strength:

$$\phi P_n \sin\theta = \phi F_y t^2 [9.8\beta_{eff}\sqrt{\gamma}]Q_f$$

where

$\phi P_n \sin\theta$ = maximum shear in the chord at the panel point (Joint 3) in question. This occurs just to the left of the vertical since the force in the vertical plus the vertical component of the force in the right diagonal are resisted by the vertical component of the force in the left diagonal.

ϕ = 0.9
F_y = 46 ksi for the chord
t = 0.349 inches for the nominal ⅜-in. wall
Q_f = 1 for the tension chord

$$\phi P_n \sin\theta = 0.9(46)(0.349)^2[9.8(0.786)\sqrt{10.0}](1)$$
$$= 123 \text{ kips}$$

The critical applied shear load at the first bottom chord panel is $178\sin55.4° = 147$ kips which is greater than the 123 kip strength. The best alternative appears to be an HSS7×7×½ bottom chord with HSS5×5 web members, i.e. HSS5×5×¼ or HSS5×5×⁵⁄₁₆ for the first tensional diagonal and HSS5×5×⁵⁄₁₆ for the first compression diagonal.

With HSS5×5 branch members and HSS7×7×½ bottom chord, $\beta = 5/7 = 0.714$ and $\gamma = 7 / (2 \times 0.465) = 7.53$

$$\zeta = g / B = 1.17/7 = 0.167 \geq 0.5(1 - \beta) = 0.5(1 - 0.714) = 0.143$$

With two square branches of equal size, $\beta_{eff} = \beta = 0.714 \geq 0.1 + \gamma / 50 = 0.251$

Chord plastification strength:

$$\phi P_n \sin\theta = \phi F_y t^2 [9.8\beta_{eff}\sqrt{\gamma}]Q_f$$
$$= 0.9(46)(0.465)^2[9.8(0.714)\sqrt{7.53}](1) = 172 \text{ kips} > 147 \text{ kips}$$

At this point check the remaining criteria from Table 8-2a. With equal-width square branches, punching shear and uneven load distribution do not need to be checked.

Chord Shear Strength (check not necessary since $H < B$):

The HSS5×5 branch members work at the bottom chord and the HSS7×7×½ chord is satisfactory.

Top Chord Branch Connection

The top chord connection for the HSS will be more critical as the HSS8×8×½ is wider and subject to a Q_f reduction due to the top chord compression.

At the top chord

$\beta = 5 / 8 = 0.625$
$\gamma = 8/(2 \times 0.465) = 8.60$

At maximum eccentricity e, the gap

$$g = \frac{2(0.25)8 + 8}{\tan 55.4} - \frac{5 + 5}{2\sin 55.4} = 2.20 \text{ in.}$$

$\zeta = g / B \geq 0.5(1 - \beta)$ gives $g \geq 0.5(1 - 0.625)8 = 1.50$ inches

Using the 1.50-in. gap and $\beta_{eff} = \beta$, compute

$$\phi P_n \sin\theta = \phi F_y t^2 [9.8\beta_{eff}\sqrt{\gamma}]Q_f$$
$$= 0.9(46)(0.465)^2[9.8(0.625)\sqrt{8.60}]Q_f = 161Q_f$$

At panel point 2 (See Figure 9-2)

$$U = \frac{P_u}{A_g F_y} = \frac{101}{13.5(46)} = 0.163$$

$Q_f = 1.3 - 0.4U / \beta = 1.3 - 0.4(0.163) / 0.625 = 1.23 > 1 \ (Q_{f\max} = 1.0)$

Use $Q_f = 1$ and $\phi P_n \sin\theta = 161$ kips > 147 kips required.

For the branch connection in Figure 9-4 at the support detail where there is only one branch member, use twice the perimeter of the branch member in determining β_{eff} according to the definition given earlier.

At panel point 4

$$U = \frac{P_u}{A_g F_y} = \frac{268}{13.5(46)} = 0.432$$

$Q_f = 1.3 - 0.4(0.432) / 0.625 = 1.02 > 1$; Use $Q_f = 1$

At panel point 10 with 5×5 branch members

$$U = 0.777 \text{ and } Q_f = 0.802$$

The branch design strength equals

$$\phi P_n = 161(0.802)/\sin 55.4° = 157 \text{ kips}$$

Therefore, HSS5×5 branch members will work throughout the entire truss.

Branch Member Selection—Interior

Determine thickness required for the third and fourth diagonals.

For **HSS5×5×³⁄₁₆**

In tension, $\phi P_n = \phi F_y A_g = 0.9(46)(3.28) = 136$ kips

$KL / r = (0.75)(94.9)/1.96 = 36.3$

From Table 9-1 $\phi_c F_{cr} = 35.8$ and compressive $\phi P_n = 35.8(3.28) = 117$ kips
   HSS5×5×³⁄₁₆ are okay by inspection for the remaining interior diagonals.
   If it were necessary to reduce steel weight further, the use of a smaller size such as an HSS3½×3½ could be evaluated. However, labor savings due to repetition and uniformity may offset the cost of additional steel weight.

For tension diagonals:

   HSS3½×3½×³⁄₁₆; $\phi P_n = 0.9(46)(2.24) = 92.7$ kips
   HSS3½×3½×¹⁄₈; $\phi P_n = 0.9(46)(1.54) = 63.7$ kips

For compression diagonals:

$Kl/r = 71.2/1.35 = 52.7$ and $\phi_c F_{cr} = 32.7$ from Table 9-1
HSS$3\frac{1}{2}\times3\frac{1}{2}\times\frac{3}{16}$; $\phi P_n = 32.7(2.24) = 73.2$ kips
HSS$3\frac{1}{2}\times3\frac{1}{2}\times\frac{1}{8}$; $\phi P_n = 32.7(1.54) = 50.4$ kips

Based on the members above, use of HSS$3\frac{1}{2}\times3\frac{1}{2}$ can begin at panel point 8. HSS$3\frac{1}{2}\times3\frac{1}{2}\times\frac{1}{8}$ can be used for the six central web members of the truss provided that the connection strength check is satisfactory. Punching shear and uneven load distribution will have to be checked per specification requirements at panel point 8 since there are two branch member sizes.

Chord Plastification:

$$\beta_{eff} = 4(3.5 + 5.0)/(8 \times 8.0) = 0.531$$
$$\gamma = 7.53 \text{ as computed earlier for the top chord}$$
$$U = P_u / A_g F_y = 459/(13.5 \times 46) = 0.739$$
$$\beta = B_b / B = 3.5/8 = 0.438$$
$$Q_f = 1.3 - 0.4U / \beta = 1.3 - 0.4(0.739)/0.438 = 0.625$$
$$\phi P_n \sin\theta = \phi F_y t^2[9.8\beta_{eff}\sqrt{\gamma}]Q_f = 0.9(46)(0.465)^2[9.8(0.531)\sqrt{7.53}]0.625$$
$$= 79.9 \text{ kips} > 73.5\sin55.4 = 60.5 \text{ kips} \quad \textbf{o.k.}$$

If gap is kept at 1.5 in.

$$\zeta = g / B = 1.5/3.5 \geq 0.5(1 - \beta) = 0.5(1 - 0.438)$$

$$= 0.429 > 0.281$$

and

$$\beta_{eff} = 0.531 \geq 0.1 + \gamma / 50 = 0.251$$

Punching Shear:

$$\phi P_n \sin\theta = \phi(0.6F_y)tB[2\eta + \beta_{gap} + \beta_{eop}]$$

where

$$\eta = H_b / B\sin\theta = 3.5/(8 \times \sin55.4°) = 0.532$$

Since $\zeta \leq 1.5(1 - \beta)$, $\beta_{gap} = \beta$

$$\beta_{eop} = 5\beta / \gamma = 5(0.438)/7.53 = 0.291 < \beta = 0.438$$

$$\phi = 0.95$$

$$\phi P_n \sin\theta = (0.95)(0.6 \times 46)(0.465)(8)[2(0.532) + 0.438 + 0.291] = 175 \text{ kips}$$

Uneven load distribution for the HSS$3\frac{1}{2}\times3\frac{1}{2}\times\frac{3}{16}$:

$$\phi P_n = \phi F_{yb} t_b[2H_b + b_{gap} + b_{eoi} - 4t_b]$$

$$t_b = 0.174 \text{ in.}$$

$$F_{yb} = 46 \text{ ksi}$$

$$B_b = H_b = 3.5 \text{ in.}$$

Since $\zeta \leq 1.5(1 - \beta)$, $b_{gap} = B_b$

$$b_{eoi} = \frac{10}{B\,/\,t}\,\frac{F_y t}{F_{yb} t_b}\,B_b \leq B_b$$

$$= \frac{10}{8\,/\,0.465}\,\frac{46(0.465)}{46(0.174)}\,(3.5) = 5.43 > B_b;\ \text{Use } b_{eoi} = 3.5 \text{ in.}$$

Therefore

$$\phi P_n = 0.95(46)(0.174)[2(3.5) + 3.5 + 3.5 - 4(0.174)] = 101 \text{ kips} > 52.5 \text{ kips}$$

Thus, the change from HSS5×5 to HSS3½×3½ at panel point 8 is satisfactory. The connection at panel points 9 and 10 should present no problems. However, check chord plastification at panel point 10,

$$\phi P_n \sin\theta = \phi F_y t^2 [9.8\beta_{eff}\sqrt{\gamma}]Q_f$$

$$\beta_{eff} = \beta = 0.438$$

$$U = 483/(13.5 \times 46) = 0.778$$

$$Q_f = 1.3 - 0.4(0.778)/0.438 = 0.589$$

and

$$\phi P_n \sin\theta = 0.9(46)(0.465)^2[9.8(0.438)\sqrt{7.53}]0.589 = 62.1 \text{ kips} \quad \textbf{o.k.}$$

Design of Vertical Member

The vertical members of the truss must carry a compressive load $P_u = 17.3$ kips between the top and bottom chord. Consider the unbraced length to be the clear distance between chords or $87 - (7 + 8)/2 = 79.5$ in. Consider the use of an HSS 2×2×3⁄16.

$$KL\,/\,r = (1.0)(79.5)/0.732 = 109$$

From Table 9-1

$$\phi_c F_{cr} = 17.7 \text{ and } \phi P_n = \phi_c F_{cr} A_g = 17.7(1.19) = 21.0 \text{ kips} > 17.3 \text{ kips}$$

The HSS2×2×3⁄16 is satisfactory provided the chord connection is satisfactory. At the top chord a joist or purlin sits atop the chord and its reaction is transmitted through the chord to the HSS2×2×3⁄16 attached to the bottom face. Thus this represents a cross connection and the provisions of HSS Specification Section 9.4.3.(b), which are summarized in Figure 8-2b, apply.

First check chord plastification where

$$\phi P_n \sin\theta = \phi F_y t^2 [2\eta\,/\,(1 - \beta) + 4\,/\,\sqrt{1 - \beta}]Q_f$$

with

$$\begin{aligned}
\sin\theta &= 1 \\
F_y &= 46 \text{ ksi} \\
t &= 0.465 \text{ in.} \\
\eta &= H_b\,/\,B\sin\theta = 2/(8 \times 1) = 0.25 \\
\beta &= B_b\,/\,B = 2/8 = 0.25 = \text{lower limit imposed} \\
\phi &= 1.0
\end{aligned}$$

which gives

$$\phi P_n = (1.0)(46)(0.465)^2 \left[\frac{2(0.25)}{(1-0.25)} + \frac{4}{\sqrt{1-0.25}} \right] Q_f = 52.6 Q_f$$

The strength is adequate if $Q_f \geq 17.3/52.6 = 0.329$

With $Q_f = 1.3 - 0.4U/\beta$ and $\beta = 0.25$ one finds that $U \leq (1.3 - 0.329)(0.25/0.4) = 0.607$ would be okay. Thus $P_u \leq 0.607(13.5)(46) = 377$ kips, which means the HSS2×2×³⁄₁₆ will work as illustrated in Fig. 9-4 only for the two outermost verticals.

Select an HSS4×2×⅛ as an alternative vertical

$Kl/r_y = 79.5/0.830 = 95.8;$

From Table 9-1 $\phi_c F_{cr} = 21.1$ and $\phi P_n = \phi_c F_{cr} A_g = 21.1(1.30) = 27.4$ kips.

Check chord plastification using this section. While η remains unchanged, $\beta = 4/8 = 0.5$. With $U = 483/(13.5 \times 46) = 0.778$ for the center vertical,

$$Q_f = 1.3 - 0.4(0.778)/0.5 = 0.678$$

and

$$\phi P_n = \phi F_y t^2 [2\eta/(1-\beta) + 4/\sqrt{1-\beta}\,] Q_f$$

$$= 1.0(46)(0.465)^2 \left[\frac{2(0.25)}{(1-0.5)} + \frac{4}{\sqrt{1-0.5}} \right] 0.678$$

$$= 44.9 \text{ kips} > 17.3 \text{ kips}$$

Punching shear, chord side wall yielding, chord side wall buckling and uneven load distribution need not be checked for the value of β obtained using the 4-in. wide branch.

To avoid the use of an overlapped joint detail at the bottom chord where the 5×5 HSS branch members occur, a transverse plate will be employed as illustrated in Figure 9-4. The strength of this connection will be evaluated using the effective width of transverse plate expression from the HSS Specification:

$$\phi R_n = \phi \frac{10 F_y t}{B/t} B_b$$

where

$\phi = 1.0$
$F_y = 46$ ksi
$t = 0.465$
$B = 7$ in.
$B_b =$ plate width; use 5 inches

$$\phi R_n = 1.0 \frac{10(46)(0.465)}{7/0.465}(5) = 71.0 \text{ kips} > 17.3 \text{ kips}$$

The 71.0 kip value is undoubtedly conservative due to the proximity of the branch members to the 5-in. wide plate.

Design of Truss Support

The detail to be used for the support of this truss is illustrated in Figure 9-5.

Design of the Plate to Chord Connection

The vertical reaction of 147 kips must be transferred from the HSS chord to the end plate.

Check the shear strength of the HSS chord

$$\phi V_n = \phi(0.6F_y)2Ht$$

$$= 0.9(0.6)(46)2(8)(0.465)$$

$$= 185 \text{ kips} > 147 \text{ kips} \quad \textbf{o.k.}$$

Weld Size

$$\phi R_n = \phi F_w A_w$$

$$\phi = 0.75$$

$$F_w = 0.6F_{EXX} = (0.60)(70) = 42 \text{ ksi}$$

with

D = weld size in sixteenths

$$A_w = 0.707(D/16)(2H) = 0.088DH$$

Equating ϕR_n to 147 kips with $H = 8$ inches

$$147 = (0.75)(42)(0.088D)(8)$$

$$D = 6.63$$

Use a $7/16$-in. fillet weld.

Determine number of bolts required

Shear = 147 kips

Try 6 - $7/8$-in. A325X Bolts.

From AISC LRFD Manual Table 8-11

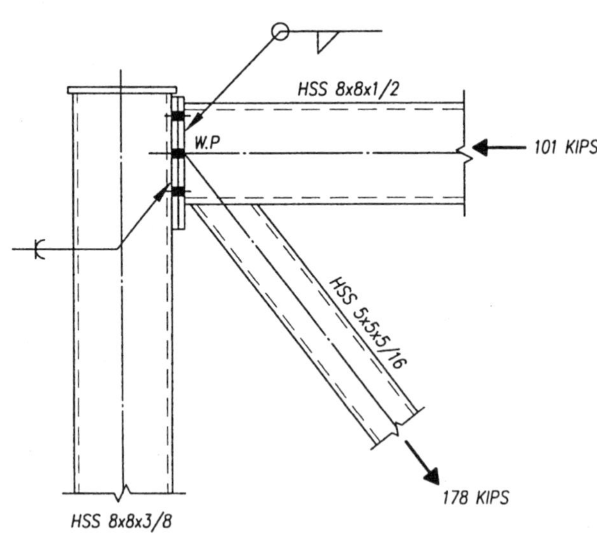

Fig. 9-5. Truss face support detail—Square HSS.

$$\phi R_n = (6)(27.1) = 163 \text{ kips} > 147 \text{ kips} \quad \textbf{o.k.}$$

Use 6 - $\frac{7}{8}$-in. A325X Bolts.

Determine end plate thickness

Try $\frac{1}{2}$-in. End Plate.

From AISC LRFD Manual Table 8-13, the design bearing strength per inch of thickness = 91.4 kips.

Therefore,

$$\phi R_n = (6)(91.4)(0.5) = 274 \text{ kips} > 147 \text{ kips} \quad \textbf{o.k.}$$

Use flare bevel welds to connect a plate to the column face as illustrated in Figure 9-5.

From AISC LRFD Manual Table 8-36 (page 8-153) the effective throat area of a flare bevel weld equals $\frac{5}{8}$ of the HSS wall thickness. Therefore:

$$\phi = 0.8$$
$$\phi F_{EXX} = 70 \text{ ksi}$$
$$E = \tfrac{5}{8}t = 0.218 \text{ in.}$$
$$t = \text{HSS wall thickness, in.} = 0.349 \text{ in.}$$
$$L = \text{Length of weld, in.}$$

Try $L = 10$ in.

$$\phi R_n = (0.8)(0.6)(70)(0.218)(2 \times 10) = 147 \text{ kips} \quad \textbf{o.k.}$$

Use: End Plate $\frac{1}{2}$-in. $\times$ 10 in. $\times$ 1 ft.-2 in.

Note: With the above connection design procedure, the column must be designed to resist the eccentric moment caused by the work point at the face of the column. Alternatively, the work point can be selected at the column centerline if the connection is designed to resist the eccentric moment.

Example 2: Warren Truss Using Round HSS

Redesign the truss shown in Figure 9-1 using round HSS rather than square and rectangular HSS. The truss loads and the final design for this truss are given in Figure 9-6. The round HSS conforming to ASTM A500 Gr B have the same ultimate strength of 58 ksi, but the yield strength value is 42 ksi instead of 46 ksi. These round HSS satisfy the criteria for use in HSS trusses.

Preliminary Sizing of the Chord Members

The maximum tensile chord force is 483 kips. Therefore the area of the tensile chord must be

$$A_g \geq \frac{P_u}{\phi F_y} = \frac{483}{0.9(42)} = 12.8 \text{ in.}^2$$

A review of the section properties in Chapter 1 will reveal that the 9.625-in. diameter HSS with a nominal wall thickness of 0.500 inches at 13.4 in.^{2} of area is the smallest section that is acceptable. An HSS12.500×0.375 would be an alternative at a lesser thickness. However, use of a smaller diameter chord of greater thickness will allow the use of smaller branch or web members in the truss.

The compression chord of the truss must carry a force of 483 kips over an effective unbraced length of 54 inches. Consider the HSS10.000×0.500 as a possible section.

$$KL/r = 54/3.38 = 16.0$$

From the Design Stress for Compression Members with F_y = 42 ksi, Table 9-2, $\phi_c F_{cr}$ = 35.1 ksi. Therefore,

$$\phi_c P_n = \phi_c F_{cr} A_g = 35.1(13.9) = 488 \text{ kips} > 483 \text{ kips}$$

Based on this, it is decided to use an HSS10.000×0.500 for both the top and bottom chord. Use of the same diameter chords will simplify the branch fabrication.

Selection of Branch Diameter

Compute first the range of diameters to consider based on the desire to have a gapped K-connection.

$$g = \frac{2(e + 0.5D)}{\tan 55.4^\circ} - \frac{D_b}{\sin 55.4^\circ}$$

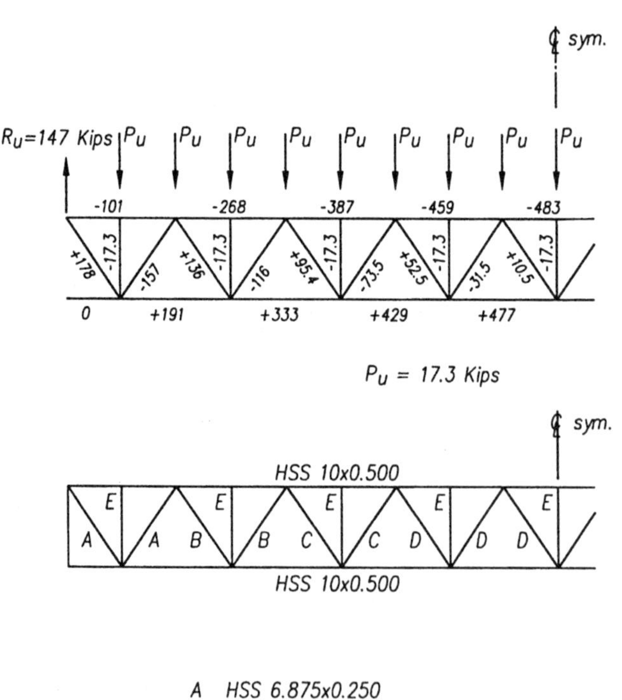

Fig. 9-6. Truss force round HSS design sizes.

The maximum g occurs at the maximum $e = 0.25D$ allowed in order to use the equations in the HSS Specification.

$$g_{max} = \frac{2(0.75)(10)}{1.45} - \frac{D_b}{0.823} = 10.3 - \frac{D_b}{0.823}$$

The minimum g as prescribed in the HSS Specification is the sum of the branch wall thicknesses, which is estimated as $0.349 + 0.349 = 0.698$ in. $= g_{min}$.

Equating g_{max} to g_{min} reveals that the limiting branch diameter for a gapped K-connection is

$$D_b \leq 0.823(10.3 - 0.698) = 7.90 \text{ in.}$$

i.e. a maximum 7.500 in. diameter size.

The first tension diagonal requires an area

$$A_g \geq \frac{178}{0.9(42)} = 4.71 \text{ in.}^2$$

An HSS6.875×0.250 is the lightest round HSS with ¼-in. nominal wall thickness that is satisfactory.

Next select the required size for the first compression diagonal which must carry a compressive load of 157 kips. With a center-to-center diagonal length of 94.9 in. and using $K = 0.75$,

$$KL = 0.75(94.9) = 71.1 \text{ in.}$$

Consider use of an HSS 6.875×0.250 for this diagonal

$$KL / r = 71.1/2.35 = 30.3$$

Thus, $\phi P_n = \phi_c F_{cr} A_g = 33.7(4.86) = 164$ kips using Table 9-2. The HSS6.875×0.250 is satisfactory for consideration pending its connection design strength.

Connection Design for First Bottom Chord Panel Point

For gapped K-connections, chord plastification and punching shear must be checked as indicated in Table 8-1. Check first the punching shear strength

$$\phi P_n \sin\theta = \phi\pi D_b t(0.6F_y)$$

where

$\phi \quad = 0.95$
$t \quad = 0.465$ inches for the chord
$F_y \quad = 42$ ksi
$\theta \quad = 55.4°$

which gives

$$\phi P_n \sin\theta = 0.95\pi D_b(0.465)(0.6 \times 42) = 35.0 D_b$$

The minimum branch diameter

$$D_b = \frac{P_u \sin\theta}{35.0} = \frac{178\sin55.4}{35.0} = 4.18 \text{ inches}$$

which means that the 6.875-in. diameter branch section can be checked for its strength as limited by chord plastification.

The gap at maximum eccentricity of $0.25D$ is

$$g = 10.3 - \frac{6.875}{\sin 55.4°} = 1.95 \text{ in.}$$

$$e_{wp} = \frac{2(0.25 \times 6.875)}{\tan 55.4°} = 2.37 \text{ in.}$$

Use e_{wp} = 2 in. which makes the gap $1.99 - 2.37 + 2 = 1.62$ in. = g

$$\therefore \ \alpha = 1.0 + 0.7g / D_b = 1.0 + 0.7(1.62)/6.875 = 1.16$$

$$\beta = D_b / D = 6.875/10 = 0.688$$

Since $\beta > 0.6$ then

$$Q_\beta = 0.3 / [\beta(1 - 0.83\beta)] = 0.3 / [0.688(1 - 0.833(0.688))] = 1.02$$

$$Q_q = \left(\frac{1.7}{\alpha} + \frac{0.18}{\beta} \right) Q_\beta^{0.7(\alpha-1)} = \left(\frac{1.7}{1.165} + \frac{0.18}{0.688} \right) 1.02^{0.7(0.165)} = 1.73$$

Since the $Q_\beta^{0.7(\alpha-1)}$ term is usually very nearly 1.0, one could use the value of 1 to obtain a conservative Q_q especially when making the initial check of the strength.

The branch strength based on the chord plastification limit state at panel point 3 is:

$$\phi P_n \sin\theta = \phi F_y t^2 (6\pi\beta Q_q)Q_f$$

where

$$\phi \ = 0.80$$
$$F_y = 42 \text{ ksi}$$
$$t \ \ = 0.465 \text{ in.}$$

and

$$Q_f = 1.0 \text{ at the tension chord}$$

$$\phi P_n \sin\theta = 0.80(42)(0.465)^2(6\pi)(0.688)(1.73)(1.0) = 162 \text{ kips}$$

which is greater than

$$P_u \sin\theta = 178\sin 55.4° = 147 \text{ kips.}$$

Had $\phi P_n \sin\theta$ been less than 147 kips, the expression would have to be reevaluated for a 7.000- or 7.500-in. branch diameter. Another option would have been to increase the chord wall thickness to 0.625-in. or reinforce the wall with a saddle as illustrated in Figure 8-9. However, the use of a saddle reinforcement or an increase in chord thickness is usually not economical.

If this truss had a seat-type connection such that the first diagonal was also welded to the face of the compression chord (See Figure 9-8), the strength based on chord plastification would need to be checked for the same 178-kip force at panel point 2, but with Q_f evaluated as

$$Q_f = 1.0 - 0.03\gamma U^2$$
$$U^2 = (\phi P_u / A_g F_y)^2 + (M_u / SF_y)^2$$
$$\gamma \ = D / 2t = 10/(2 \times 0.465) = 10.8$$
$$P_u = \text{axial force in chord} = 101 \text{ kips}$$
$$A_g = \text{chord area} = 13.9 \text{ in.}^2$$
$$F_y = \text{chord yield strength} = 42 \text{ ksi}$$
$$M_u = \text{moment in chord} = 0$$

S = chord elastic section modulus

which gives

$$U^2 = (101.3/13.9/42)^2 = 0.0301$$

and

$$Q_f = 1.0 - 0.03(10.8)(0.0301) = 0.990$$

Therefore at the top chord adjacent to the support

$$\phi P_n \sin\theta = 162 Q_f = 162(0.990) = 160 \text{ kips} > 147 \text{ kips}$$

Thus the branch-to-chord connection as illustrated in Fig. 9-8 is satisfactory using the HSS6.875×0.250.

Examine the next tensile and compressive diagonals

For a tensile force of 136 kips the required

$$A_g = \frac{136}{0.9(42)} = 3.60 \text{ in.}^2$$

This tension diagonal could be an **HSS6.875×0.188** which has an area of 3.66 in.^{2} Check this section for the 116 kip force in compression.

$$KL/r = 71.1/2.37 = 30.0$$

Using Table 9-2 $\phi P_n = \phi_c F_{cr} A_g = 33.8(3.66) = 124$ kips which is greater than the 116 kips required.

One could use **HSS6.875×0.188** members for all other inner diagonals of the truss. Even though Q_f decreases as one approaches the maximum compressive chord force, its minimum value with $U^2 = 1$ is 0.678. The diagonals near the center have a factored load considerably less than $\phi P_n = 197(0.678) = 134$ kips.

Examination of Smaller Web Members

The use of a smaller web member is possible for the fifth through ninth diagonals of the truss. Consider using 5.000-in. diameter HSS for the central diagonals.

Check the connection design strength first. Based on the results of the initial punching shear check, this check does not have to be repeated.

Determine the gap to use for the 5.000-in. diameter (D_b) HSS. Consider using zero eccentricity ($e = 0$) in which case the gap

$$g = \frac{2(0.5D)}{1.45} - \frac{D_b}{0.823} = \frac{2(0.5)(10)}{1.45} - \frac{5}{0.823} = 0.821 \text{ in.}$$

To permit the easy installation of the plate for the truss vertical make $e_{wp} = 0.50$ inches which would increase g to

$$g = 0.821 + 0.50 = 1.32 \text{ in.}$$

Now

$$\alpha = 1 + 0.7g/D_b = 1 + 0.7(1.32)/5 = 1.19$$

and

$$\beta = D_b/D = 5/10 = 0.50$$

Since $\beta < 0.6$, Q_β is defined as 1.0 and $Q_\beta^{0.7(\alpha-1)} = 1.0$

Thus

$$Q_q = \left(\frac{1.7}{\alpha} + \frac{0.18}{\beta}\right)(1.0) = \frac{1.7}{1.19} + \frac{0.18}{0.5} = 1.80$$

The chord plastification expression for ϕP_n is

$$\phi P_n \sin\theta = \phi F_y t^2 (6\pi\beta Q_q)Q_f = 0.8(42)(0.465)^2(6\pi)(0.5)(1.80)Q_f = 123Q_f$$

At the tension chord $\phi P_n \sin\theta = 123$ kips and at the compression chord $\phi P_n \sin\theta > 123(0.678) = 83.3$ kips using the minimum Q_f for the chord, which was computed earlier.

The 83.3 kips is greater than $94.5\sin55.4° = 77.8$ kip factored shear force at panel point 6. Thus the 5.000-in. diameter can be considered without computing an exact Q_f for the location. The area required in the fifth diagonal is

$$A_g = \frac{P_u}{\phi F_{yb}} = \frac{94.5}{0.9(42)} = 2.50 \text{ in.}^2$$

Since the area of the HSS5.000×0.188 is 2.64 in.^{2} this section can be used for the fifth diagonal member.

Next check the compression diagonal with $P_u = 73.5$ kips. Considering an HSS5.000×0.188,

$$KL/r = 71.1/1.71 = 41.6$$

Obtaining $\phi_c F_{cr}$ from Table 9-2

$$\phi P_n = \phi_c F_{cr} A_g = 32.1(2.64) = 84.7 \text{ kips} > 73.5 \text{ kips}$$

HSS5.000×0.188 can be used for this and all other diagonals with lesser axial load.

It would be possible to use HSS5×0.125 web members if desired since $D_b/t_b = 5/0.116 = 43.1 < 50$.

Compressive $\phi P_n = \phi_c F_{cr} A_g = 32.1(1.78) = 57.1$ kips

Tensile $\phi P_n = \phi F_y A_g = 0.9(42)(1.78) = 67.3$ kips

Based on these values, an HSS5.000×0.125 will work for the seventh through ninth diagonals.

The summary of the truss design, showing the minimum sizes that were selected above, is given in Figure 9-6.

Design of Verticals

Select a round HSS to carry the applied purlin load $P_u = 17.3$ kips from the top to bottom chord (Figure 9-7). Consider use of an HSS1.900×0.145 which has a radius of gyration $r = 0.626$. The effective length can be taken as the distance between the faces of the chords i.e. $L = 87 - 10 = 77$ in. This produces

$$KL/r = 77/0.626 = 123 \text{ and } \phi_c F_{cr} = 14.1$$

$$\phi P_n = \phi_c F_{cr} A_g = 14.1(0.75) = 10.6 \text{ kips} < 17.3 \text{ kips}$$

Since this is not satisfactory, consider an HSS2.375×0.154.

$$KL/r = 77/0.791 = 97.3 \text{ and } \phi_c F_{cr} = 20.0(1.00) = 20.0 \text{ kips} > 17.3 \text{ kips}$$

This section is okay provided that the connection design strength is adequate at the compression chord. Since the load is from a purlin or joist sitting atop the chord and transmitted through the chord to the HSS2.375×0.154 as illustrated in Figure 9-7(a), it is necessary to consider this a Cross-connection rather than a T-connection (wherein the load would be transferred along the chord by shear).

From Table 8-1,

$\alpha = 2.4$

$\beta = 2.375/10 = 0.238$

$Q_\beta = 1.0$

$$Q_q = \left(\frac{1.7}{2.4} + \frac{0.18}{0.238}\right)(1.0) = 1.47$$

With $U = 1$; $Q_f = 1 - 0.03\gamma U^2 = 1 - 0.03(10.8)(1) = 0.676$

Based on chord plastification with $\theta = 90°$ and $\phi = 0.8$:

$$\phi P_n \sin\theta = \phi P_n = \phi F_y t^2 (6\pi\beta Q_q)Q_f$$

$$= 0.8(42)(0.465)^2(6\pi)(0.238)(1.47)(0.676) = 32.3 \text{ kips}$$

Based on general collapse of the chord with $\theta = 90°$ and $\phi = 0.8$:

$$\phi P_n \sin\theta = \phi P_n = \phi F_y t^2 [1.9 + 7.2\beta]Q_\beta Q_f$$

$$= 0.8(42)(0.465)^2[1.9 + 7.2(0.238)](1.0)(0.676) = 17.7 \text{ kips}$$

Based on punching shear with $\theta = 90°$ and $\phi = 0.95$:

$$\phi P_n \sin\theta = \phi P_n = 0.95\pi D_b t(0.6F_y) = 0.95\pi(2.38)(0.465)(0.6)(42) = 83.2 \text{ kips}$$

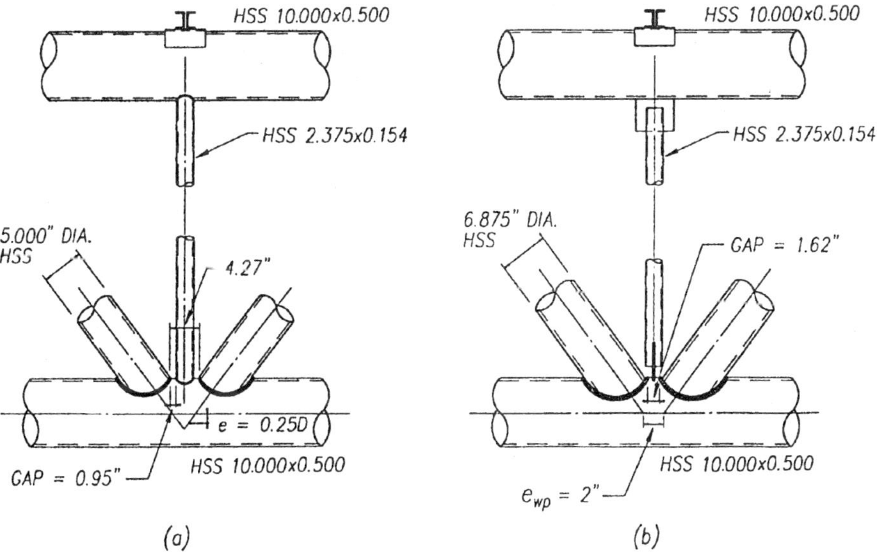

Fig. 9-7. Truss vertical details—Round HSS.

Since all three ϕP_n exceed 17.3 kips, the connection of the HSS2.375×0.154 at the top chord is satisfactory throughout the truss.

The connection at the bottom chord with the vertical member present represents a combined K- and T-connection. With 5-in. diameter web members, the maximum gap between web members (when $e = 0.25D$) is

$$g_{max} = \frac{2(0.75D)}{1.45} - \frac{D_b}{0.823} = \frac{2(0.75)(10)}{1.45} - \frac{5}{0.823} = 4.27 \text{ in.}$$

This permits the installation of the HSS2.375×0.154 vertical leaving a 0.95-in. gap as shown in Figure 9-7(b).

Consider the use of a ³⁄₈-in. thick by 4-in. wide plate to connect the vertical member at the bottom chord as illustrated in Figure 9-7(b). This permits the use of the gapped connection where the 6.875-in. diameter web members are employed.

Use of the chord strength expression for transverse plates from the AISC HSS Specification results in a relatively conservative value for the connection strength to the chord.

$$\phi P_n = \frac{\phi 5 F_y t^2 Q_f}{1 - 0.81 b_1 / D} = \frac{0.8(5)(42)(0.465)^2(1)}{1 - 0.81(4) / 10} = 53.7 \text{ kips} > 17.3 \text{ kips}$$

An alternative connection for the top chord could be a plate positioned longitudinally as illustrated in Fig. 9-7(b). Consider a plate ³⁄₈-in. thick and 6 inches long. Using AISC HSS Specification Section 8.2 for this case

$$\phi P_n = \phi 5 F_y t^2 (1 + 0.25 h_1 / D) Q_f$$

$$= 0.8(5)(42)(0.465)^2 (1 + 0.25(6)/10) Q_f = 41.8 Q_f$$

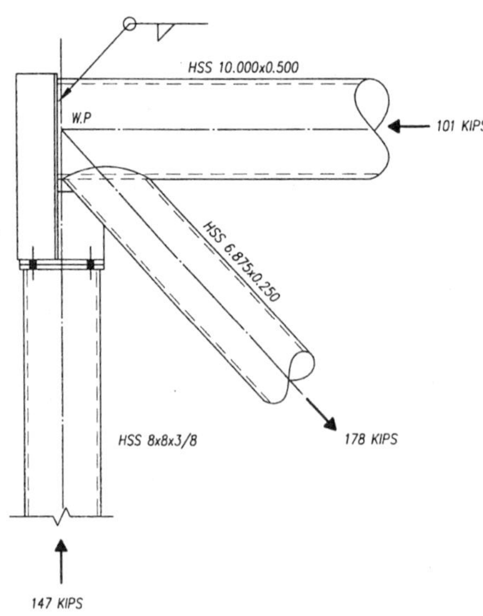

Fig. 9-8. Truss seat support detail—Round HSS.

$$U = \frac{P_u}{A_g F_y} = \frac{483}{13.9(42)} = 0.827$$

$$Q_f = 1 - 0.3(0.827)(1 + 0.827) = 0.547$$

$$\phi P_n = 41.8(0.547) = 22.9 \text{ kips} > 17.3 \text{ kips}$$

Although this longitudinal plate is satisfactory it is obvious that the transverse plate used at the bottom has significantly better basic strength (before discounting for the effect of Q_f).

Design of Truss Support

The detail to be used at the support is illustrated in Figure 9-8.

Design of the Plate-to-Chord Connection:

The vertical reaction of 147 kips must be transferred from the HSS chord to the vertical plate. First check the shear strength of the HSS chord. For round HSS the area for shear resistance is taken as one-half the cross sectional area. Thus,

$$\phi V_n = \phi(0.6 F_y A_g / 2)$$

$$= (0.9)((0.6)(42)(13.9)/2)$$

$$= 158 \text{ kips} > 147 \text{ kips} \quad \textbf{o.k.}$$

Determine Weld Size

$$\phi R_n = F_w A_w / 2$$

$$\phi = 0.75$$

$$F_w = 0.6 F_{EXX} = (0.60)(70) = 42 \text{ ksi}$$

With D = weld size in sixteenths

$$A_w = (0.707)(D/16)(\pi)(10) = 1.39D \text{ in.}$$

Equating ϕR_n to 147 kips

$$(0.75)(42)(1.39D)/2 = 147$$

$$D = 6.71 \text{ sixteenths}$$

Use $\frac{7}{16}$-in. fillet weld

Check Stiffened Vertical Plate

Try a $\frac{1}{2} \times 14$-in. plate with $\frac{3}{8}$-in. stiffeners.

Neglecting the stiffeners:

Area of plate column = $(0.5)(14) = 7.0 \text{ in.}^2$

Assume zero-height column:

$$\phi P_n = 0.85 F_y A$$

$$= 0.85(36)(7) = 214 \text{ kips} > 147 \text{ kips}$$

By inspection **o.k.** as a short column.

Check Plate Size to Column Cap

From Table 6-2 (50 percent of member strength)

$$\phi R_n = 160 \text{ kips} > 147 \text{ kips}$$

Weld cap plate to column, $W_w = \frac{7}{16}$-in.

Cap plate, $t_1 = \frac{3}{4}$-in.

Weld stem to cap plate, $W_s = \frac{5}{16}$-in.

Refine weld size to HSS:

Try $\frac{5}{16}$-in. fillet.

Strength of the weld connecting the cap plate to the HSS.

$$\phi R_n = \phi F_w A_w$$

$$\phi = 0.75$$

$$F_w = (0.60)(70)(1.5) = 63 \text{ ksi}$$

Note that the 1.5 factor is from LRFD Specification Appendix J2.4.

From Chapter 6

$$A_w = 2(5t_1 + t_2 + 2W_s)(0.707)W_w$$

t_1 can be taken as 1.5 in. since two plates exist.

$$= 2[(5)(1.50) + 0.5 + (2)(0.3125)](0.707)(5/16)$$

$$= 3.81 \text{ in.}^2$$

$$\phi R_n = (0.75)(63)(3.81) = 180 \text{ kips} > 147 \text{ kips} \quad \textbf{o.k.}$$

Use $\frac{5}{16}$-in. fillet to column.

Note: Since the work point of the truss aligns with the column centroid, no column moment is developed; however the stiffened plate connection must be able to resist the eccentric axial force.

Example 3: Truss Designs with Verticals Removed

It would have been possible to design the top chord without the verticals in Examples 2 and 3.

For the truss composed of rectangular and square HSS, the use of an HSS10×8×½ can be shown to be acceptable as a top chord carrying the loads indicated in Figure 9-1 without the presence of the vertical members. Since the chord width and thickness are unchanged, there is little change in the branch connection calculation except that the utilization ratio U would have to include the M_u / SF_y component. H/t is less than 35 and the joint eccentricity remains within the range of acceptability.

For the truss composed of round HSS, a similar approach could be taken. An HSS10.000×0.625 can be shown to be acceptable as a top chord carrying the loads shown in Figure 9-1 with no vertical members in the truss. Even though the utilization factor U increases due to the presence of the moment ratio, the increase in thickness of the HSS top chord more than compensates for any reduction in Q_f such that there would be no problem with the connection design for the diagonals.

The trusses designed in the previous two examples could have several of the verticals removed without affecting the top chord size. Since the top chord closer to the support needed to retain a ½-in. thickness simply to accept the branch forces, the top chord has excess strength in that region. Removal of the two outermost verticals, or in other words, the retention of the five (of the nine) verticals in the center portion of the truss will be examined.

The top chord was analyzed as a continuous beam as illustrated in Figure 9-9. The truss analyzed with pinned diagonals and continuous chord could have been used to obtain the chord bending moments. The top chords used in the design examples were found to be satisfactory for the combined flexural and axial loads. Removal of the third vertical for the truss with the HSS10.000×0.500 top chord was found to be just adequate, but the HSS8×8×½ top chord was found to be overloaded.

In reexamining the connection of the branches to the top chord for the example using round HSS, with the two outermost verticals removed, the critical condition occurs at panel point 4 where one has the joint moment of 29.3 ft-kips to consider in conjunction with the 268 kip axial compression.

The utilization ratio U for the HSS8×8×½ is

$$U = \left| \frac{P_u}{A_g F_y} \right| + \left| \frac{M_u}{S F_y} \right| = \frac{268}{13.5(46)} + \frac{29.3(12)}{31.2(46)} = 0.677$$

$$Q_f = 1.3 - 0.4U / \beta = 1.3 - 0.4(0.677)/0.625 = 0.867$$

and $\phi P_n \sin\theta = 161 Q_f$ from earlier calculations.

Thus

$$\phi P_n \sin\theta = 161(0.867) = 139 \text{ kips}$$

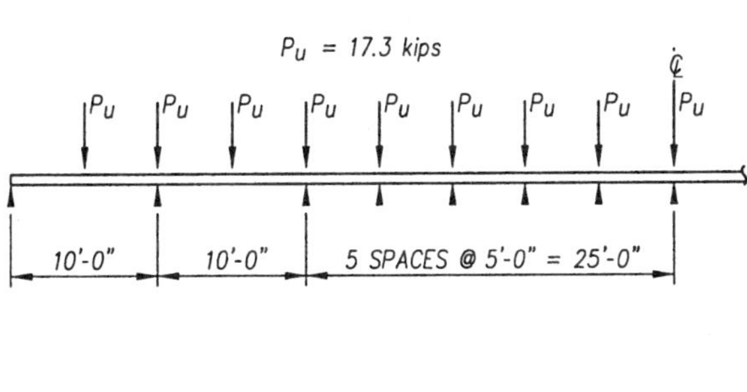

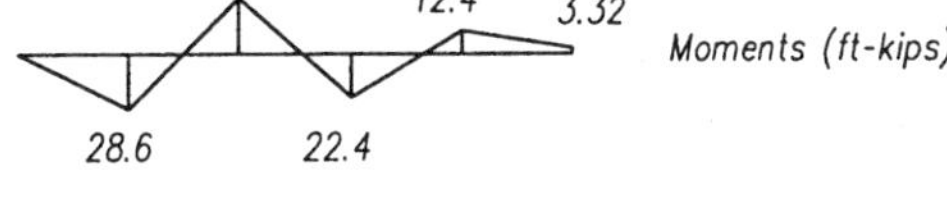

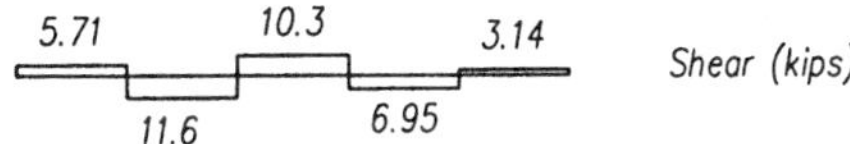

Fig. 9-9. Top chord bending and shear with two outer verticals removed.

The shear on the chord is $157\sin55.4° + 11.6 = 141$ kips with the 11.6 kips being the additional shear from the continuous beam analysis. The 141 kips is about 1 percent over the 139 kip strength of the chord.

A similar evaluation for the HSS10.000×0.500 top chord finds

$$U^2 = \left(\frac{268}{13.9(42)}\right)^2 + \left(\frac{29.3(12)}{31.7(42)}\right)^2 = 0.280$$

$$Q_f = 1 - 0.03\gamma U^2 = 1 - 0.03(10.8)(0.281)^2 = 0.974$$

and $\phi P_n \sin\theta = 163 Q_f$ from earlier calculations.

Thus

$$\phi P_n \sin\theta = 163(0.974) = 159 \text{ kips} > 141 \text{ kips}$$

The HSS10×0.500 top chord is satisfactory with only the five verticals in the center portion of the truss.

$\dfrac{Kl}{r}$	$\phi_c F_{cr}$ ksi	$\dfrac{Kl}{r}$	$\phi_c F_{cr}$ ksi	$\dfrac{Kl}{r}$	$\phi_c F_{cr}$ ksi	$\dfrac{Kl}{r}$	$\phi_c F_{cr}$ ksi	$\dfrac{Kl}{r}$	$\phi_c F_{cr}$ ksi
1	39.10	41	34.92	81	25.15	121	14.57	161	8.23
2	39.09	42	34.73	82	24.87	122	14.33	162	8.13
3	39.08	43	34.53	83	24.60	123	14.10	163	8.03
4	39.06	44	34.33	84	24.32	124	13.88	164	7.93
5	39.03	45	34.12	85	24.05	125	13.66	165	7.84
6	39.01	46	33.91	86	23.77	126	13.44	166	7.74
7	38.97	47	33.70	87	23.50	127	13.23	167	7.65
8	38.93	48	33.49	88	23.22	128	13.02	168	7.56
9	38.89	49	33.27	89	22.95	129	12.82	169	7.47
10	38.84	50	33.05	90	22.67	130	12.62	170	7.38
11	38.78	51	32.82	91	22.40	131	12.43	171	7.30
12	38.72	52	32.60	92	22.13	132	12.25	172	7.21
13	38.66	53	32.37	93	21.85	133	12.06	173	7.13
14	38.59	54	32.14	94	21.58	134	11.88	174	7.05
15	38.51	55	31.90	95	21.31	135	11.71	175	6.97
16	38.43	56	31.66	96	21.03	136	11.54	176	6.89
17	38.35	57	31.42	97	20.76	137	11.37	177	6.81
18	38.26	58	31.18	98	20.49	138	11.20	178	6.73
19	38.16	59	30.94	99	20.22	139	11.04	179	6.66
20	38.06	60	30.69	100	19.95	140	10.89	180	6.59
21	37.96	61	30.44	101	19.69	141	10.73	181	6.51
22	37.85	62	30.19	102	19.42	142	10.58	182	6.44
23	37.73	63	29.94	103	19.15	143	10.43	183	6.37
24	37.61	64	29.68	104	18.89	144	10.29	184	6.30
25	37.49	65	29.43	105	18.62	145	10.15	185	6.23
26	37.36	66	29.17	106	18.36	146	10.01	186	6.17
27	37.23	67	28.91	107	18.10	147	9.87	187	6.10
28	37.09	68	28.65	108	17.84	148	9.74	188	6.04
29	36.95	69	28.38	109	17.58	149	9.61	189	5.97
30	36.80	70	28.12	110	17.33	150	9.48	190	5.91
31	36.65	71	27.86	111	17.07	151	9.36	191	5.85
32	36.50	72	27.59	112	16.82	152	9.23	192	5.79
33	36.34	73	27.32	113	16.56	153	9.11	193	5.73
34	36.17	74	27.05	114	16.31	154	9.00	194	5.67
35	36.01	75	26.78	115	16.06	155	8.88	195	5.61
36	35.84	76	26.51	116	15.82	156	8.77	196	5.55
37	35.66	77	26.24	117	15.57	157	8.66	197	5.50
38	35.48	78	25.97	118	15.32	158	8.55	198	5.44
39	35.30	79	25.70	119	15.07	159	8.44	199	5.39
40	35.11	80	25.42	120	14.82	160	8.33	200	5.33

Table 9-2.
Design Stress for Compression Members of 42 ksi Specified Yield Stress Steel, $\phi_c = 0.85$

$\dfrac{Kl}{r}$	$\phi_c F_{cr}$ ksi	$\dfrac{Kl}{r}$	$\phi_c F_{cr}$ ksi	$\dfrac{Kl}{r}$	$\phi_c F_{cr}$ ksi	$\dfrac{Kl}{r}$	$\phi_c F_{cr}$ ksi	$\dfrac{Kl}{r}$	$\phi_c F_{cr}$ ksi
1	35.70	41	32.20	81	23.86	121	14.53	161	8.23
2	35.69	42	32.03	82	23.62	122	14.31	162	8.13
3	35.68	43	31.87	83	23.38	123	14.10	163	8.03
4	35.66	44	31.70	84	23.15	124	13.88	164	7.93
5	35.65	45	31.52	85	22.91	125	13.66	165	7.84
6	35.62	46	31.35	86	22.67	126	13.44	166	7.74
7	35.59	47	31.17	87	22.43	127	13.23	167	7.65
8	35.56	48	30.99	88	22.19	128	13.02	168	7.56
9	35.52	49	30.81	89	21.95	129	12.82	169	7.47
10	35.48	50	30.62	90	21.71	130	12.62	170	7.38
11	35.44	51	30.43	91	21.47	131	12.43	171	7.30
12	35.39	52	30.24	92	21.23	132	12.25	172	7.21
13	35.33	53	30.04	93	20.99	133	12.06	173	7.13
14	35.27	54	29.85	94	20.75	134	11.88	174	7.05
15	35.21	55	29.65	95	20.51	135	11.71	175	6.97
16	35.14	56	29.45	96	20.27	136	11.54	176	6.89
17	35.07	57	29.24	97	20.03	137	11.37	177	6.81
18	35.00	58	29.04	98	19.79	138	11.20	178	6.73
19	34.92	59	28.83	99	19.55	139	11.04	179	6.66
20	34.83	60	28.62	100	19.32	140	10.89	180	6.59
21	34.75	61	28.41	101	19.08	141	10.73	181	6.51
22	34.65	62	28.19	102	18.84	142	10.58	182	6.44
23	34.56	63	27.98	103	18.61	143	10.43	183	6.37
24	34.46	64	27.76	104	18.37	144	10.29	184	6.30
25	34.36	65	27.54	105	18.14	145	10.15	185	6.23
26	34.25	66	27.32	106	17.90	146	10.01	186	6.17
27	34.14	67	27.10	107	17.67	147	9.87	187	6.10
28	34.02	68	26.87	108	17.44	148	9.74	188	6.04
29	33.90	69	26.65	109	17.21	149	9.61	189	5.97
30	33.78	70	26.42	110	16.98	150	9.48	190	5.91
31	33.65	71	26.19	111	16.75	151	9.36	191	5.85
32	33.52	72	25.97	112	16.52	152	9.23	192	5.79
33	33.39	73	25.73	113	16.30	153	9.11	193	5.73
34	33.25	74	25.50	114	16.07	154	9.00	194	5.67
35	33.11	75	25.27	115	15.85	155	8.88	195	5.61
36	32.97	76	25.04	116	15.62	156	8.77	196	5.55
37	32.82	77	24.80	117	15.40	157	8.66	197	5.50
38	32.67	78	24.57	118	15.18	158	8.55	198	5.44
39	32.52	79	24.33	119	14.96	159	8.44	199	5.39
40	32.36	80	24.10	120	14.74	160	8.33	200	5.33

Specification for the Design of Steel Hollow Structural Sections

April 15, 1997

AMERICAN INSTITUTE OF STEEL CONSTRUCTION, INC.
One East Wacker Drive, Suite 3100, Chicago, IL 60601-2001

Table of Contents

Commentary

Symbols

A	area used to calculate A_e
A_g	gross area of cross-section
A_e	effective net area for tension member
A_n	net area
A_w	web area
B	overall width of rectangular HSS
B_b	overall width of rectangular HSS branch member in a truss connection
C	HSS torsional constant
D	outside diameter of round HSS
D_b	outside diameter of round HSS branch member in a truss connection
E	modulus of elasticity
F_{cr}	critical stress for column buckling
	critical stress for local buckling
	critical stress for torsion
F_n	nominal stress for rectangular HSS shear resistance
F_u	minimum specified tensile strength of the HSS
F_{vcr}	critical shear stress
F_y	minimum specified yield stress of the HSS
F_{y1}	minimum specified yield strength of plate or connecting element that is welded to an HSS
F_{yb}	minimum specified yield strength of HSS branch member in a truss connections
H	overall height of rectangular HSS
H_b	overall height of rectangular HSS branch member in a truss connection
K	compression member effective length factor
L	connection length
	length of weld
L_b	unbraced length
L_e	effective weld length
L_{pd}	maximum unbraced length for plastic moment M_p in plastic analysis
L_r	maximum unbraced length for yield moment M_r
M_n	nominal flexural strength
M_p	plastic moment of section
M_r	yield moment of section
M_u	required flexural strength
M_{ur}	resultant required flexural strength for round HSS
M_1	smaller moment at end of an unbraced length in plastic analysis
M_2	larger moment at end of an unbraced length in plastic analysis
N	bearing length of concentrated load along length of HSS
P_n	nominal axial strength
P_u	required axial strength
P_{ua}	additional axial load for the moment in a rectangular HSS branch
P_y	axial yield load
Q	effective area factor
Q_f	connection resistance reduction factor for compression in HSS (Section 8.1), parameter used for truss connections as defined in Section 9.4

Q_q parameter used for truss connections as defined in Section 9.4

Q_β parameter used for truss connections as defined in Section 9.4

R_f reduction factor for wind forces on exposed HSS

R_n nominal strength of HSS and connections to HSS

S elastic section modulus

S_{eff} effective elastic section modulus for thin-walled rectangular HSS

T_n nominal torsional strength

T_u required torsional strength

U shear lag factor, parameter used for truss connections as defined in Section 9.4

V_n nominal shear strength

V_u required shear strength

Z plastic section modulus

a length of essentially constant shear in a beam

b flat width of rectangular HSS flange or side, which is permitted to be taken as $B - 3t$

b_1 width of plate or connecting element that is welded to an HSS

b_{eoi} parameter used for truss connections as defined in Section 9.4

b_{gap} parameter used for truss connections as defined in Section 9.4

c constant for bending in rectangular HSS branches of truss connections

d bolt diameter

e eccentricity in K-connections

f stress

g gap between branch members in a gapped K-connection

h flat width of rectangular HSS web or side, which is permitted to be taken as $H - 3t$

k distance from point of application of concentrated force to critical section of HSS

l member length

r radius of gyration

r_y radius of gyration about the y-axis

t design HSS wall thickness as given in Section 1.2

t_1 thickness of plate or connecting element that is welded to an HSS

t_b thickness of branch member in an HSS truss connection

t_p plate thickness (Section 8.3.3)

$\bar{x}$ eccentricity for shear lag

θ branch angle with respect to chord as illustrated in Figure C9.4-1

α parameter used for truss connections as defined in Section 9.4

β parameter used for truss connections as defined in Section 9.4

β_{eff} parameter used for truss connections as defined in Section 9.4

β_{eop} parameter used for truss connections as defined in Section 9.4

β_{gap} parameter used for truss connections as defined in Section 9.4

γ parameter used for truss connections as defined in Section 9.4

ϕ resistance factor

η parameter used for truss connections as defined in Section 9.4

λ wall slenderness

       constant for type of load in round HSS truss connections

λ_c column slenderness

λ_p maximum wall slenderness for compact section

λ_r maximum wall slenderness for non-compact section

θ angle between branch and chord

ζ parameter used for truss connections as defined in Section 9.4

PREFACE

The AISC *Load and Resistance Factor Design (LRFD) Specification for Structural Steel Buildings* is intended to cover the common design criteria in routine office practice. Accordingly, it is not feasible to also cover the many special and unique problems encountered within the full range of structural design practice. This AISC *Specification for the Design of Steel Hollow Structural Sections* is a separate document that addresses one such topic: the design and construction of building systems that utilize steel hollow structural sections (HSS). A list of Symbols and a non-mandatory Commentary with background information are provided.

The AISC Committee on Specifications, Task Committee 118—Structural Tubing is responsible for its ongoing development. Additionally, the AISC Committee on Specification has enhanced these provisions through careful scrutiny, discussion, suggestion for improvements, and endorsement.

By the AISC Committee on Specifications, Task Committee 118—Structural Tubing,

D. R. Sherman, Chairman	F. J. Palmer
R. C. Kaehler	J. M. Ricles
L. A. Kloiber	J. A. Yura
J. A. Packer	C. J. Carter, Secretary

APPENDIX

AISC SPECIFICATION
FOR THE DESIGN OF STEEL
HOLLOW STRUCTURAL SECTIONS

Specification for the Design of Steel Hollow Structural Sections

April 15, 1997

1.1. SCOPE

This Specification is intended for the design of round and rectangular hollow structural sections (HSS) that are used as structural members in buildings and the design of connections to HSS. HSS include: (1) prismatic structural shapes and (2) products of a pipe or tubing mill that meet the geometric tolerances, tensile requirements and chemical requirements of a standard specification. Rectangular HSS include square and rectangular cross-sections that have rounded corners within the tolerances of an appropriate product specification. Only unstiffened non-composite HSS in non-fatigue applications are considered in this Specification.

This Specification is a supplement to the AISC *Load and Resistance Factor Design Specification for Structural Steel Buildings* (AISC, 1993), hereinafter referred to as the LRFD Specification. In some cases, criteria taken from the LRFD Specification have been modified to appear in non-dimensional form and to apply directly to rectangular HSS, which have two webs. For situations that are not covered in this Specification, the criteria in the LRFD Specification shall apply. In seismic applications, HSS shall also be designed to meet the requirements of the AISC *Seismic Provisions for Structural Steel Buildings* (AISC, 1997).

1.2. MATERIAL

1. Structural Steel

HSS material that meets the requirements in one of the following ASTM specifications is approved for use under this Specification:

ASTM A53-96 Grade B, "Standard Specification for Pipe, Steel, Black and Hot-Dipped, Zinc-Coated, Welded and Seamless"

ASTM A500-93, "Standard Specification for Cold-Formed Welded and Seamless Carbon Steel Structural Tubing in Rounds and Shapes"

ASTM A501-93, "Standard Specification for Hot-Formed Welded and Seamless Carbon Steel Structural Tubing"

ASTM A618-95, "Standard Specification for Hot-Formed Welded and Seamless High-Strength Low-Alloy Structural Tubing"

ASTM A847-93, "Standard Specification for Cold-Formed Welded and Seamless High Strength, Low Alloy Structural Tubing with Improved Atmospheric Corrosion Resistance"

Certified mill test reports or certified reports of tests made by the fabricator or a qualified testing laboratory that meet the requirements in ASTM A370 and the governing specification shall constitute sufficient evidence of conformity with one of the above ASTM standards. If requested, the fabricator shall provide an affidavit stating that the structural steel furnished meets the requirements of the grade specified.

2. Design Wall Thickness

The design wall thickness t shall be used in calculations involving the HSS wall thickness. When the design wall thickness is not known, it is permitted to be taken as 0.93 times the nominal wall thickness.

1.3. LOADS AND LOAD COMBINATIONS

In the absence of other applicable building code provisions, the load combinations of LRFD Specification Section A4 shall apply.

If permitted by the applicable building code, wind forces on the projected areas of exposed HSS are permitted to be reduced by the factor R_f from the forces on frameworks with similar configurations but using sections or shapes with flat elements. R_f shall be taken as follows:

(a) For round HSS, $R_f = 2/3$.

(b) For rectangular HSS with outside corner radii that are greater than or equal to 0.05 times the width B and wind force acting on the short side (B), $R_f = 0.4 + 0.6B/H \le 2/3$, where H is the depth of the HSS.

(c) For rectangular HSS under other conditions, $R_f = 1.0$.

2.1. EFFECTIVE NET AREA FOR TENSION MEMBERS

The effective net area A_e for tension members shall be determined as follows:

$$A_e = AU \tag{2.1-1}$$

(a) For a welded connection that is continuous around the perimeter, $A = A_g$, where A_g is the gross area and $U = 1$.

(b) For connections with concentric gusset plates and slotted HSS, $A = A_n$, where the net area A_n at the end of the gusset plate is the gross area minus the product of the thickness and total width of material that is removed to form the slots, and

$$U = 1 - (\overline{x}/L) \le 0.9 \tag{2.1-2}$$

Table 2.2-1
Limiting Wall Slenderness for
Compression Elements

Element	λ	Limiting Wall Slenderness	
		λ_p	λ_r

round HSS
  for axial compression
  for flexure
  for plastic analysis
	D/t
	[a]
		n.a.
		$0.0714E/F_y$
		$0.0448E/F_y$
			$0.114E/F_y$
			$0.309E/F_y$
			n.a.

rectangular HSS wall
  for uniform compression
 	
  for plastic analysis
	b/t
	or
	h/t
		$1.12\sqrt{E/F_y}$
		
		$0.939\sqrt{E/F_y}$
			$1.40\sqrt{E/F_y}$

rectangular HSS wall as a web in flexural compression	h/t	$3.76\sqrt{E/F_y}$	$5.70\sqrt{E/F_y}$
rectangular HSS wall as a web in combined flexure and axial compression	h/t	[b]	$5.70\sqrt{E/F_y}\left(1-\dfrac{0.75P_u}{\phi_b P_y}\right)$

[a] D/t must be less than or equal to $0.448E/F_y$.
[b] For $P_u/\phi_b P_y \le 0.125$

$$3.76\sqrt{E/F_y}\left(1-\frac{2.75P_u}{\phi_b P_y}\right)$$

for $P_u/\phi_b P_y > 0.125$

$$1.12\sqrt{E/F_y}\left(2.33-\frac{P_u}{\phi_b P_y}\right) \ge 1.49\sqrt{E/F_y}$$

In the above equation, $\bar{x}$ is the perpendicular distance from the weld to the centroid of the cross-sectional area that is tributary to the weld.

For round HSS with a single concentric gusset plate

$$\bar{x} = \frac{D}{\pi} \qquad (2.1\text{-}3)$$

For rectangular HSS with a single concentric gusset plate

$$\bar{x} = \frac{B^2 + 2BH}{4(B+H)} \qquad (2.1\text{-}4)$$

(c) For connections with rectangular HSS and a pair of side gusset plates, $A = A_g$, where A_g is the gross area and U shall be calculated using Equation 2.1-2 with

$$\bar{x} = \frac{B^2}{4(B + H)} \tag{2.1-5}$$

where

L = length of the connection in the direction of loading
D = outside diameter of round HSS
B = overall width of rectangular HSS
H = overall height of rectangular HSS

Larger values of U are permitted to be used when justified by tests or rational criteria. For other end-connection configurations, U shall be determined by tests or rational criteria.

2.2   LOCAL BUCKLING

1.   Classification of Steel Sections

HSS are classified for local buckling of the wall in compression as compact, non-compact, or slender-element cross-sections according to the limiting wall slenderness ratios λ_p and λ_r in Table 2.2-1. For an HSS to qualify as compact, the wall slenderness ratio λ must be less than or equal to λ_p. If λ exceeds λ_p but is less than or equal to λ_r, the HSS is non-compact. If λ exceeds λ_r, the HSS is a slender-element cross-section. The wall slenderness ratio shall be calculated as follows:

(a) For round HSS, $\lambda = D/t$, where D is the outside diameter and t is the wall thickness. This Specification is applicable only to round HSS with λ less than or equal to $0.448E/F_y$, where E is the modulus of elasticity and F_y is the minimum specified yield stress.
(b) For flanges of rectangular HSS, $\lambda = b/t$, where b is the clear distance between webs less the inside corner radius at each web and t is the wall thickness. If the corner radius is not known, b is permitted to be taken as the overall flange width B minus three times the wall thickness t.
(c) For webs of rectangular HSS, $\lambda = h/t$, where h is the clear distance between flanges less the inside corner radius at each flange and t is the wall thickness. If the corner radius is not known, h is permitted to be taken as the overall web depth H minus three times the wall thickness t.

2.   Design by Plastic Analysis

Design by plastic analysis is permitted when λ is less than or equal to λ_p for plastic analysis as defined in Table 2.2-1.

3.   Design in Seismic Applications

In seismic applications, λ shall also meet the requirements in the AISC *Seismic Provisions for Structural Steel Buildings* (AISC, 1997).

2.3.   LIMITING SLENDERNESS RATIOS

For compression members, the slenderness ratio Kl/r preferably should not exceed 200.

For tension members, the slenderness ratio l/r preferably should not exceed 300.

Members that are primarily tension members but that are subject to some compression under other load conditions need not satisfy the compression slenderness limit.

For bracing members in seismic applications, l/r shall meet the requirements in AISC *Seismic Provisions for Structural Steel Buildings.*

3.1. DESIGN FOR TENSION

The design strength of tension members $\phi_t P_n$ shall be the lower value obtained according to the limit states of yielding in the gross section and fracture in the net section.

(a) For yielding on the gross area:

$$\phi_t = 0.9$$

$$P_n = F_y A_g \tag{3.1-1}$$

(b) For rupture on the net effective area:

$$\phi_t = 0.75$$

$$P_n = F_u A_e \tag{3.1-2}$$

where

$\quad A_e$ = effective net area
$\quad A_g$ = gross area of HSS
$\quad F_y$ = specified minimum yield strength
$\quad F_u$ = specified minimum tensile strength
$\quad P_n$ = nominal axial strength

4.1. EFFECTIVE LENGTH AND SLENDERNESS LIMITATIONS

1. Effective Length

The effective length factor K for compression members shall be taken as follows or as determined by rational analysis:

(a) In trusses that are made with HSS branch (web) members that are welded around their full perimeter to continuous HSS chord members, the effective length factor K used to modify the length between panel points for in-plane buckling, or between locations of lateral bracing for out-of-plane buckling, shall be not less than:

$\quad K = 0.75$ for branch members
$\quad K = 0.9$ for chord members

(b) In trusses that are made with HSS branch members that do not meet the requirements in Section 4.1.1(a) or with non-HSS branch members connected to continuous HSS chord members, the effective length factor K used to modify the length between panel points for in-plane buckling shall be not less than:

$\quad K = 1.0$ for branch members
$\quad K = 0.9$ for chord members

(c) In frames for which lateral stability is provided by diagonal bracing, shear walls or equivalent means, K shall be taken as unity, unless a lesser value can be justified by rational analysis.

(d) In frames for which lateral stability is dependant upon the flexural stiffness of rigidly connected beams and columns, K shall be determined by rational analysis.

2. Design by Plastic Analysis

Design by plastic analysis is permitted if the column slenderness parameter λ_c is less than or equal to $1.5K$ and the axial force in columns of unbraced frames due to factored gravity loads plus factored lateral loads does not exceed ϕ_c times $0.75F_y A_g$.

4.2. DESIGN FOR COMPRESSION

The design strength for flexural buckling of compression members $\phi_c P_n$ shall be determined as follows:

$$\phi_c = 0.85$$

$$P_n = F_{cr} A_g \tag{4.2-1}$$

F_{cr} shall be determined as follows:

(a) For $\lambda_c \sqrt{Q} \leq 1.5$

$$F_{cr} = Q(0.658^{Q\lambda_c^2})F_y \tag{4.2-2}$$

(b) For $\lambda_c \sqrt{Q} > 1.5$

$$F_{cr} = \left[\frac{0.877}{\lambda_c^2}\right] F_y \tag{4.2-3}$$

where

$$\lambda_c = \frac{KL}{r\pi}\sqrt{\frac{F_y}{E}} \tag{4.2-4}$$

Q shall be determined as follows:

(a) For $\lambda \leq \lambda_r$ in Section 2.2, $Q = 1$

(b) For $\lambda > \lambda_r$ in Section 2.2, for round HSS with $\lambda < 0.448E/F_y$

$$Q = \frac{0.0379E}{F_y(D/t)} + \frac{2}{3} \tag{4.2-5}$$

For rectangular HSS,

$$Q = \frac{\text{effective area}}{A_g} \tag{4.2-6}$$

where the effective area is equal to the summation of the effective areas of the sides using

$$b_e = 1.91t \sqrt{\frac{E}{f}} \left[1 - \frac{0.381}{(b/t)} \sqrt{\frac{E}{f}} \right] \le b \qquad (4.2\text{-}7)$$

with $f = P_u / A_g$.

5.1. DESIGN FOR FLEXURE

The design flexural strength $\phi_b M_n$ shall be determined as follows:

$$\phi_b = 0.9$$

(a) For round HSS, for $\lambda \le \lambda_p$ in Section 2.2,

$$M_n = M_p = F_y Z \qquad (5.1\text{-}1)$$

for $\lambda_p < \lambda \le \lambda_r$

$$M_n = \left(\frac{0.0207}{D/t} \frac{E}{F_y} + 1 \right) F_y S \qquad (5.1\text{-}2)$$

for $\lambda_r < \lambda \le 0.448E/F_y$

$$M_n = \frac{0.330E}{D/t} S \qquad (5.1\text{-}3)$$

(b) For rectangular HSS, for $\lambda \le \lambda_p$ in Section 2.2,

$$M_n = M_p = F_y Z \qquad (5.1\text{-}4)$$

for $\lambda_p < \lambda \le \lambda_r$

$$M_n = \left[M_p - (M_p - M_r) \left(\frac{\lambda - \lambda_p}{\lambda_r - \lambda_p} \right) \right] \qquad (5.1\text{-}5)$$

where

$$M_r = F_y S$$

for $\lambda > \lambda_r$

$$M_n = F_y S_{eff} \qquad (5.1\text{-}6)$$

where S_{eff} is the effective section modulus with the effective width of the compression flange taken as

$$b_e = 1.91t \sqrt{\frac{E}{F_y}} \left[1 - \frac{0.381}{(b/t)} \sqrt{\frac{E}{F_y}} \right] \le b \qquad (5.1\text{-}7)$$

L_b is not limited for HSS structures designed by elastic analysis. Design by plastic analysis is permitted for compact round HSS with λ less than or equal to $0.0448E/F_y$ and for rectangular HSS with λ less than or equal to λ_p in Section 2.2. For rectangular HSS bent about the major axis, the laterally unbraced length L_b of the compression flange adjacent to plastic hinge locations that are associated with the failure mechanism shall not exceed L_{pd}, where

$$L_{pd} = \frac{5,000 + 3,000(M_1/M_2)}{F_y} r_y \geq \frac{3,000 r_y}{F_y} \qquad (5.1\text{-}8)$$

and

F_y = specified minimum yield stress

M_1 = smaller moment at the end of the unbraced length

M_2 = larger moment at the end of the unbraced length

r_y = radius of gyration about the minor axis

M_1/M_2 is positive when moments cause reverse curvature and negative for single curvature

For seismic applications, refer to AISC *Seismic Provisions for Structural Steel Buildings* (AISC, 1997).

5.2. DESIGN FOR SHEAR

The design shear strength of unstiffened HSS $\phi_v V_n$ shall be determined as follows:

$$\phi_v = 0.9$$

(a) For round HSS with

$$\frac{a}{D} \leq \frac{3.2(E/F_y)^2}{(D/t)^{2.5}} \qquad (5.2\text{-}1)$$

and $\lambda \leq \lambda_r$ in Table 2.2-1,

$$V_n = 0.3 F_y A_g \qquad (5.2\text{-}2)$$

(b) For rectangular HSS,

$$V_n = F_n A_w \qquad (5.2\text{-}3)$$

where

$$A_w = 2Ht \qquad (5.2\text{-}4)$$

F_n shall be determined as follows:

(i) For $h/t \leq 2.45\sqrt{E/F_y}$

$$F_n = 0.6 F_y \qquad (5.2\text{-}5)$$

(ii) For $2.45\sqrt{E/F_y} < h/t \leq 3.07\sqrt{E/F_y}$

$$F_n = 0.6 F_y (2.45\sqrt{E/F_y})/(h/t) \qquad (5.2\text{-}6)$$

(iii) For $3.07\sqrt{E/F_y} < h/t \leq 260$

$$F_n = 0.458\pi^2 E/(h/t)^2 \qquad (5.2\text{-}7)$$

6. DESIGN FOR TORSION

The design torsional strength $\phi_T T_n$ shall be determined as follows:

$$\phi_T = 0.9$$

$$T_n = F_{cr} C \tag{6-1}$$

where C is the HSS torsional constant.

F_{cr} shall be determined as follows:

(a) For round HSS, F_{cr} shall be the larger of

$$\frac{1.23E}{\sqrt{L/D}\ (D/t)^{5/4}} \quad \text{and} \quad \frac{0.6E}{(D/t)^{3/2}} \tag{6-2}$$

but shall not exceed $0.6F_y$.

(b) For rectangular HSS, for $h/t \le 2.45\sqrt{E/F_y}$

$$F_{cr} = 0.6F_y \tag{6-3}$$

for $2.45\sqrt{E/F_y} < h/t \le 3.07\sqrt{E/F_y}$

$$F_{cr} = 0.6F_y (2.45\sqrt{E/F_y}\,)/(h/t) \tag{6-4}$$

for $3.07\sqrt{E/F_y} < h/t \le 260$

$$F_{cr} = 0.458\pi^2 E/(h/t)^2 \tag{6-5}$$

7.1. DESIGN FOR COMBINED FLEXURE AND AXIAL FORCE

The interaction of flexure and axial force shall be limited by Equations 7.1-1 and 7.1-2.

(a) For $P_u/\phi P_n \ge 0.2$

$$\frac{P_u}{\phi P_n} + \frac{8}{9}\left(\frac{M_{ux}}{\phi_b M_{nx}} + \frac{M_{uy}}{\phi_b M_{ny}}\right) \le 1.0 \tag{7.1-1}$$

(b) For $P_u/\phi P_n < 0.2$

$$\frac{P_u}{2\phi P_n} + \left(\frac{M_{ux}}{\phi_b M_{nx}} + \frac{M_{uy}}{\phi_b M_{ny}}\right) \le 1.0 \tag{7.1-2}$$

where

P_u = required axial tensile or compressive strength

P_n = nominal tensile or compressive strength determined in accordance with Sections 3.1 or 4.2

M_u = required flexural strength determined in accordance with LRFD Specification Section C1

M_n = nominal flexural strength determined in accordance with Section 5.1

x = subscript relating symbol to strong axis bending

y = subscript relating symbol to weak axis bending

ϕ = ϕ_t from Section 3.1 for tension

   = 0.85 for compression

ϕ_b = 0.90

For biaxial flexure of round HSS that are laterally unbraced along their length and with end conditions such that the effective length factor K is the same for any direction of bending, the design is permitted to be based upon a single resultant moment M_{ur}, where:

$$M_{ur} = \sqrt{M_{ux}^2 + M_{uy}^2} \qquad (7.1\text{-}3)$$

Alternatively, use of the provisions in LRFD Specification Appendix H.3b is permitted.

7.2. DESIGN FOR COMBINED TORSION, SHEAR, FLEXURE, AND/OR AXIAL FORCE

When the required torsional strength is significant, the interaction of torsion, shear, flexure, and/or axial force shall be limited by Equation 7.2-1.

$$\left(\frac{P_u}{\phi P_n} + \frac{M_u}{\phi_b M_n}\right) + \left(\frac{V_u}{\phi_v V_n} + \frac{T_u}{\phi_T T_n}\right)^2 \le 1.0 \qquad (7.2\text{-}1)$$

In the above equation,

P_u = required axial tensile or compressive strength

P_n = nominal tensile or compressive strength determined in accordance with Sections 3.1 or 4.2

M_u = required flexural strength determined in accordance with LRFD Specification Section C1

M_n = lesser of $F_y S$ and M_n determined in accordance with Section 5.1

S = elastic section modulus

V_u = required shear strength at the section corresponding to M_u

V_n = nominal shear strength determined in accordance with Section 5.2

T_u = required torsional strength

T_n = nominal torsional strength determined in accordance with Section 6

ϕ = ϕ_t from Section 3.1 for tension
= 0.85 for compression

ϕ_b = $\phi_v = \phi_T = 0.90$

8. CONCENTRATED FORCES ON HSS

The design strength ϕR_n at locations of concentrated forces on unstiffened HSS shall be determined from the applicable criteria in Sections 8.1 through 8.3.

8.1. Concentrated Force Distributed Transversely

When a concentrated force is distributed transversely to the axis of the HSS, the design strength ϕR_n shall be determined as follows:

(a) For round HSS,

$$\phi = 1.0$$

$$R_n = \frac{5 F_y t^2}{1 - 0.81 b_1 / D} Q_f \qquad (8.1\text{-}1)$$

where

b_1 = the width of the load

Q_f = 1 for tension in the HSS

= $1 - 0.3f/F_y - 0.3(f/F_y)^2 \le 1$ for compression in the HSS

f = the magnitude of the maximum compression stress in the HSS due to axial force and bending at the location of the concentrated force

(b) For rectangular HSS,

$$\phi = 1.0$$

$$R_n = \frac{10F_y t}{(B/t)} b_1 \le F_{y1} t_1 b_1 \qquad (8.1\text{-}2)$$

where

b_1 = the width of the loaded plate

t_1 = the thickness of the loaded plate

F_{y1} = minimum specified yield strength of the loaded plate

When the force is distributed across the full width of the rectangular HSS, the limit state of local web yielding shall be checked for both tensile and compressive forces and the limit state of web crippling shall be checked for compressive forces. For local web yielding,

$$\phi = 1.0$$

$$R_n = 2F_y t(5k + N) \qquad (8.1\text{-}3)$$

For web crippling,

$$\phi = 0.75$$

$$R_n = 1.6t^2[1 + 3N/h]\sqrt{EF_y} \qquad (8.1\text{-}4)$$

where

k = outside corner radius of the HSS, which if not known is permitted to be taken as $1.5t$

N = bearing length of the load along the length of the HSS

h = flat width of side wall of the HSS as defined in Section 2.2.1

When the force is distributed across a width of the rectangular HSS that is greater than $0.85B$ but less than $B - 2t$, the design strength shall not exceed ϕR_n, where

$$\phi = 1.0$$

$$R_n = 0.6F_y t(2t_1 + 2b_{ep}) \qquad (8.1\text{-}5)$$

where $b_{ep} = 10b_1/(B/t) \le b_1$

When compressive forces coincide on opposite faces of the rectangular HSS, the limit state of compression buckling of the webs shall be checked and the design strength shall not exceed ϕR_n, where

$$\phi = 0.90$$

$$R_n = \frac{48t^3\sqrt{EF_y}}{h} \tag{8.1-6}$$

8.2. Concentrated Force Distributed Longitudinally at the Center of the HSS Face

When a concentrated force is distributed longitudinally along the axis of the HSS at the center of the HSS face, the design strength ϕR_n shall be determined as follows:

(a) For round HSS,

$$\phi = 1.0$$

$$R_n = 5F_y t^2 (1 + 0.25N/D)Q_f \tag{8.2-1}$$

where

N = bearing length of the load along the length of the HSS
Q_f = 1 for tension in the HSS
    = $1 - 0.3f/F_y - 0.3(f/F_y)^2 \leq 1$ for compression in the HSS
f = the magnitude of the maximum compression stress in the HSS due to axial force and bending at the location of the concentrated force

(b) For rectangular HSS,

$$\phi = 1.0$$

$$R_n = \frac{F_y t^2}{1 - t_1/B}\left[\frac{2N}{B} + 4\sqrt{1 - t_1/B}\right]Q_f \tag{8.2-2}$$

where t_1 is the thickness of the loaded plate.

8.3. Concentrated Axial Force on the End of a Rectangular HSS with a Cap Plate

When a concentrated force acts on the end of an HSS with a cap plate and along the axis of the HSS, the design strength ϕR_n shall be determined for each loaded wall as follows. The limit state of local wall yielding shall be checked for both tensile and compressive forces and the limit state of wall crippling shall be checked for compressive forces. For local wall yielding,

$$\phi = 1.0$$

$$R_n = (5t_1 + N)F_y t \leq BF_y t \tag{8.3-1}$$

For wall crippling,

$$\phi = 0.75$$

$$R_n = 0.80t^2\left[1 + 3\left(\frac{N}{B/2}\right)\left(\frac{t}{t_1}\right)^{1.5}\right]\sqrt{EF_y(t_1/t)} \tag{8.3-2}$$

where

t_1 = thickness of cap plate

N = bearing length of the load across the width of the HSS

9.1. GENERAL PROVISIONS FOR CONNECTIONS AND FASTENERS

The provisions of LRFD Specification Section J1.1 through J1.11 and the provisions for bolts and threaded parts in LRFD Specification Section J3.1 through J3.11 shall apply with the following additions and modifications.

1. Through Bolts

When connections are made using bolts that pass completely through an unstiffened HSS, the bolts shall be installed to only the snug-tight condition and the connection shall be considered to be a shear bearing connection. The bearing strength per loaded wall is ϕR_n, where

$$\phi = 0.75$$

$$R_n = 1.8 F_y \, dt \tag{9.1-1}$$

where

F_y = minimum specified yield strength of the HSS

d = bolt diameter

t = HSS wall thickness

2. Special Connectors

The design strength of special connectors other than the bolts considered in LRFD Specification Table J3.2 shall be verified by tests.

3. Tension Connectors

When bolts or other connectors in tension are attached to an HSS wall, the strength of the HSS wall shall be determined by rational analysis.

9.2 WELDS

The non-uniformity of load transfer along the line of weld due to differences in relative flexibility of HSS walls in HSS-to-HSS and similar connections shall be considered in proportioning such connections. In such cases, the strength of fillet welds shall be determined from LRFD Specificaton Section J2.4, excluding the alternative in Appendix J2.4, and the effective weld length L_e of groove and fillet welds shall be limited as follows:

(a) In T-, Y-, and Cross-connections with rectangular HSS as defined in Section 9.4,

$$L_e = 2H_b + B_b \text{ for } \theta \le 50 \text{ degrees} \tag{9.2-1}$$

$$L_e = 2H_b \text{ for } \theta \ge 60 \text{ degrees} \tag{9.2-2}$$

Linear interpolation shall be used to determine L_e for values of θ between 50 and 60 degrees.

(b) In gapped K-connections with rectangular HSS as defined in Section 9.4,

$$L_e = 2H_b + 2B_b \text{ for } \theta \le 50 \text{ degrees} \tag{9.2-3}$$

$$L_e = 2H_b + B_b \text{ for } \theta \geq 60 \text{ degrees} \tag{9.2-4}$$

Linear interpolation shall be used to determine L_e for values of θ between 50 and 60 degrees.

(c) When a transverse plate is welded to the face of an HSS member,

$$L_e = 2\,\frac{10}{B/t}\,\frac{F_y t}{F_{y1} t_1}\,b_1 \leq 2b_1 \tag{9.2-5}$$

where

H_b = width of branch member wall that is parallel to the axis of the chord member, in.

B_b = width of branch member wall that is transverse to the axis of the chord member, in.

θ = least angle between branch member and chord member

B = width of chord member wall to which plate is attached, in.

b_1 = width of attached plate, in.

t = thickness of chord member wall, in.

t_1 = thickness of attached plate, in.

F_y = yield strength of HSS, ksi

F_{y1} = yield strength of plate, ksi

In lieu of the above, other rational criteria are permitted.

9.3. OTHER CONNECTION REQUIREMENTS

The provisions of LRFD Specification Sections J4 through J10 shall apply with the following additions and modifications.

1. Shear Rupture Strength

The design shear rupture strength along a path adjacent to a fillet weld on the HSS wall shall be taken as ϕR_n, where

$$\phi = 0.75$$

$$R_n = 0.6 F_u t L \tag{9.3-1}$$

where

t = HSS wall thickness

L = length of weld

2. Tension Rupture Strength

The design tension rupture strength along a path adjacent to a fillet weld on the HSS wall shall be taken as ϕR_n, where

$$\phi = 0.75$$

$$R_n = F_u t L \tag{9.3-2}$$

where

t = HSS wall thickness

L = length of weld

3. Punching Shear Rupture Strength

When a plate that is parallel to the longitudinal axis of an HSS and projects from the wall is subjected to a load that is parallel but eccentric or has a component perpendicular to the HSS wall,

$$\phi_t f t_p \leq 1.2 \phi_v F_u t \tag{9.3-3}$$

where

$\phi_v = 0.75$

$\phi_t = 0.90$

f = maximum stress in the plate perpendicular to the HSS wall

t_p = plate thickness

F_u = HSS tensile strength

t = HSS wall thickness

4. Eccentric Connections

For trusses that are made with HSS that are connected by welding branch members to chord members, eccentricities within the limits of applicability in Section 9.4 are permitted without consideration of the resulting moments for the design of the connection, except in fatigue applications. In fatigue applications, refer to AWS D1.1.

9.4. HSS-TO-HSS TRUSS CONNECTIONS

HSS-to-HSS truss connections are defined as connections that consist of one or more branch members that are directly welded to a continuous chord that passes through the connection and shall be classified as follows:

(1) When the punching load in a branch member is equilibrated by beam shear in the chord member, the connection shall be classified as a T-connection when the branch is perpendicular to the chord and a Y-connection otherwise.

(2) When the punching load in a branch member is essentially equilibrated by loads in other branch member(s) on the same side of the joint, the connection shall be classified as a K-connection.

(3) When the punching load is transmitted through the chord member and is equilibrated by branch member(s) on the opposite side, the connection shall be classified as a Cross-connection.

When branch members transmit part of their load as K-connections and part of their load as T-, Y-, or Cross-connections, the design strength shall be determined by interpolation on the proportion of each in total.

For the purposes of this Specification, the centerlines of the branch member(s) and the chord members shall lie in a single plane and K-connections shall be used in the gapped configuration. For other configurations such as a multi-planar connection, a connection with a branch member that is offset so that its centerline does not intersect with the centerline of the chord, or when an overlapped K-connection is used, the provisions of AWS D1.1, other verified design procedures, tests, or rational analysis shall be used.

1.   **Definitions of Parameters**

β = the width ratio; the ratio of branch diameter to chord diameter = D_b/D for round HSS; the ratio of overall branch width to chord width = B_b/B for rectangular HSS.

γ = the chord slenderness ratio; the ratio of one-half the diameter to the wall thickness = $D/2t$ for round HSS; the ratio of one-half the width to wall thickness = $B/2t$ for rectangular HSS.

η = the load length parameter, applicable only to rectangular HSS; the ratio of the length of contact of the branch with the chord in the plane of the connection to the chord width = N/B, where $N = H_b/\sin\theta$ and θ is the angle between the branch and chord.

ζ = the gap ratio; the ratio of the gap between the branches of a gapped K-connection to the width of the chord; = g/D for round HSS; = g/B for rectangular HSS.

2.   **Criteria for Round HSS**

The design strength of the branch ϕP_n and/or ϕM_n for axial loads in the branch and for flexure in the branch, respectively, shall be determined from the limit states of chord wall plastification, punching shear rupture and general collapse as applicable below.

The interaction of stress due to chord member forces and local branch connection forces shall be considered. The chord-stress interaction parameter Q_f shall be determined as

$$Q_f = 1.0 - \lambda\gamma U^2 \tag{9.4-1}$$

where U is the utilization ratio given by

$$U^2 = \left(\frac{P_u}{A_g F_y}\right)^2 + \left(\frac{M_u}{S F_y}\right)^2 \tag{9.4-2}$$

and

λ = 0.030 for axial load in the branch

      = 0.044 for in-plane bending in the branch

      = 0.018 for out-of-plane bending in the branch

P_u = required axial strength in chord

A_g = chord gross area

F_y = chord yield stress

M_u = larger required flexural strength in chord and connection

S = chord elastic section modulus

2a.   **Limits of Applicability**

The criteria herein are applicable only when the connection configuration is within the following limits of applicability:

1) joint eccentricity: $-0.55D \leq e \leq 0.25D$, where D is the chord diameter and e is positive away from the branches

2) branch angle: $\theta \geq 30°$

3) wall stiffness: ratio of diameter to wall thickness less than or equal to 50 for

chords and branches in T-, Y- and K-connections and less than or equal to 40 for chords of Cross-connections

4) width ratio: $0.2 < D_b / D \leq 1.0$

5) gap: g greater than or equal to the sum of the branch wall thicknesses

2b. Branches with Axial Loads

For T-, Y-, and gapped K-connections, the design strength of the branch ϕP_n shall be determined from the limit states of chord wall plastification and punching shear rupture. For Cross-connections, the design strength of the branch ϕP_n shall be determined from the limit states of chord wall plastification, punching shear rupture and general collapse.

1. For the limit state of chord wall plastification,

$$\phi = 0.8$$

$$P_n \sin\theta = t^2 F_y \, [6\pi\beta Q_q] Q_f \tag{9.4-3}$$

where

$$Q_q = \left(\frac{1.7}{\alpha} + \frac{0.18}{\beta} \right) Q_b^{0.7(\alpha - 1)}$$

For $\beta \leq 0.6$, $Q_b = 1.0$

For $\beta > 0.6$, $Q_\beta = \dfrac{0.3}{\beta(1 - 0.833\beta)}$

α = chord ovalization parameter

= 1.7 for T- and Y-connections

= $1.0 + 0.7g / D_b$, $1 \leq \alpha < 1.7$ for gapped K-connections

= 2.4 for Cross-connections

2. For the limit state of punching shear rupture,

$$\phi = 0.95$$

$$P_n \sin\theta = \pi D_b t (0.6 F_y) \tag{9.4-4}$$

3. For the limit state of general collapse,

$$\phi = 0.8$$

$$P_n \sin\theta = t^2 F_y (1.9 + 7.2\beta) Q_b Q_f \tag{9.4-5}$$

2c. Branches with Flexure

For T-, Y-, gapped K-, and Cross-connections, the design strength of the branch ϕM_n shall be determined from the limit states of chord wall plastification and punching shear rupture.

1. For the limit state of chord wall plastification,

$$\phi = 0.8$$

$$M_n \sin\theta = t^2 F_y \, [D_b / 4] \, [6\pi\beta Q_q] Q_f \tag{9.4-6}$$

where

$$Q_q = \left(\frac{2.1}{\alpha} + \frac{0.6}{\beta} \right) Q_\beta^{1.2(\alpha - 0.67)}$$

For $\beta \leq 0.6$, $Q_\beta = 1.0$

For $\beta > 0.6$, $Q_\beta = \dfrac{0.3}{\beta(1 - 0.833\beta)}$

α = chord ovalization parameter

= 0.67 for in-plane bending

= 1.5 for out-of-plane bending

For combinations of in-plane and out-of-plane bending, α shall be determined by interpolation and Q_f shall be determined with interpolated values of λ.

2. For the limit state of punching shear rupture,

$$\phi = 0.95$$

$$M_n \sin\theta = D_b^2 t (0.6 F_y) \tag{9.4-7}$$

2d.   Branches with Combined Axial Loads and Flexure

The interaction of combined axial loads and flexure in HSS-to-HSS truss connections shall meet the following requirement:

$$\left(\frac{P_u}{\phi P_n} \right)^{1.75} + \frac{M_u}{\phi M_n} \leq 1.0 \tag{9.4-8}$$

where

P_u = required axial strength of the branch

M_u = required flexural strength of the branch

3.   Criteria For Rectangular HSS

The design axial strength ϕP_n and the design flexural strength ϕM_n of the branch shall be determined from the limit states of chord wall plastification, punching shear rupture, sidewall strength, and uneven load distribution as applicable below.

The interaction of stress due to chord member forces and local branch connection forces shall be considered with the chord-stress interaction parameter Q_f, where

Q_f = 1 when the chord is in tension

= $1.3 - 0.4 U / \beta \leq 1$ when the chord is in compression $\qquad$ (9.4-9)

$$U = \left| \frac{P_u}{A_g F_y} \right| + \left| \frac{M_u}{S F_y} \right| \tag{9.4-10}$$

where

P_u = required axial strength of the chord

M_u = required flexural strength of the chord

3a. Limits of Applicability

The criteria herein are applicable only when the connection configuration is within the following limits:

1) joint eccentricity: $-0.55H \le e \le 0.25H$, where H is the chord depth and e is positive away from the branches
2) branch angle: $\theta \ge 30°$
3) wall stiffness: ratio of wall width to wall thickness less than or equal to 35 for chords and branches; also less than or equal to $1.25\sqrt{E/F_{yb}}$ for branches in compression
4) strength: F_y less than or equal to 52 ksi for chord and branches
5) chord aspect ratio: $0.5 \le$ ratio of depth to width ≤ 2.0
6) ductility: $F_y / F_u \le 0.8$
7) other limits apply for specific criteria

3b. Branches with Axial Loads in T-, Y- and Cross-connections

For T-, Y-, and Cross-connections, the design strength of the branch ϕP_n shall be determined from the limit states of chord wall plastification, punching shear rupture, sidewall strength, and uneven load distribution.

1. For the limit state of chord wall plastification,

$$\phi = 1.0$$

$$P_n \sin\theta = F_y t^2 \left[\frac{2\eta}{1-\beta} + \frac{4}{\sqrt{(1-\beta)}} \right] Q_f \qquad (9.4\text{-}11)$$

This limit state need not be checked when $\beta > 0.85$ nor when $\beta < 0.25$.

2. For the limit state of punching shear rupture,

$$\phi = 0.95$$

$$P_n \sin\theta = 0.6 F_y tBD[2\eta + 2\beta_{eop}] \qquad (9.4\text{-}12)$$

In the above equation, the effective outside punching parameter $\beta_{eop} = 5\beta/\gamma$ shall not exceed β. This limit state need not be checked when $\beta > 1 - 1/\gamma$ nor when $\beta < 0.85$.

3. For the limit state of sidewall strength, the design strength for branches in tension shall be taken as the design strength for local sidewall yielding. For the limit state of sidewall strength, the design strength for branches in compression shall be taken as the lesser of the design strengths for local sidewall yielding and sidewall crippling. This limit state need not be checked unless the chord member and branch member have the same width ($\beta = 1.0$).

For the limit state of local yielding,

$\phi = 1.0$ for a branch in tension
$\quad = 0.8$ for a branch in compression

$$P_n \sin\theta = 2t F_y (5k + N) \qquad (9.4\text{-}13)$$

where

k = outside corner radius of the HSS, which is permitted to be taken as $1.5t$ if unknown

N = bearing length of the load along the length of the HSS, $H_b / \sin\theta$

H_b = height of the branch

For the limit state of sidewall crippling, in T- and Y-connections,

$$\phi = 0.75$$

$$P_n \sin\theta = 1.6t^2[1 + 3N / H] \sqrt{EF_y}\, Q_f \qquad (9.4\text{-}14)$$

For the limit state of sidewall crippling in Cross-connections,

$$\phi = 0.8$$

$$P_n \sin\theta = \left[\frac{48t^3}{H - 4t}\right] \sqrt{EF_{yc}}\, Q_f \qquad (9.4\text{-}15)$$

4. For the limit state of uneven load distribution,

$$\phi = 0.95$$

$$P_n = F_{yb} t_b [2H_b + 2b_{eoi} - 4t_b] \qquad (9.4\text{-}16)$$

where

$$b_{eoi} = \frac{10}{B / t} \frac{F_y t}{F_{yb} t_b} B_b \leq B_b \qquad (9.4\text{-}17)$$

F_{yb} = branch yield strength

t_b = branch thickness

This limit state need not be checked when $\beta < 0.85$.

3c.   Branches with Axial Loads in Gapped K-connections

For gapped K-connections, the design strength of the branch ϕP_n shall be determined from the limit states of chord wall plastification, punching shear rupture, shear yielding, and uneven load distribution.

1. For the limit state of chord wall plastification,

$$\phi = 0.9$$

$$P_n \sin\theta = F_y t^2 [9.8\beta_{eff} \sqrt{\gamma}\,]\, Q_f \qquad (9.4\text{-}18)$$

where β_{eff} is the sum of the perimeters of the two branches divided by 8 times the chord width. This limit state need not be checked when $B_b / B < 0.1 + \gamma / 50$ and $\zeta < 0.5(1 - \beta)$.

2. For the limit state of punching shear rupture,

$$\phi = 0.95$$

$$P_n \sin\theta = (0.6F_y)tBD[2\eta + \beta_{gap} + \beta_{eop}] \qquad (9.4\text{-}19)$$

This limit state need not be checked when $\beta < 0.1 + \gamma / 50$ nor for square branches of the same width as the chord.

3. For the limit state of shear yielding of the chord in the gap, the design strength shall be checked in accordance with Section 5.2.

4. For the limit state of uneven load distribution,

$$\phi = 0.95$$

$$P_n = F_{yb}\, t_b [2H_b + b_{gap} + b_{eoi} - 4t_b] \qquad (9.4\text{-}20)$$

where

$$\beta_{gap} = \beta, \text{ when } \zeta \le 1.5(1 - \beta)$$
$$\quad\quad = \beta_{eop}, \text{ for other connections}$$
$$\beta_{eop} = 5\beta / \gamma \le \beta$$
$$b_{gap} = B_b, \text{when } \zeta \le 1.5(1 - \beta)$$
$$\quad\quad = b_{eoi}, \text{ for other connections}$$

$$b_{eoi} = \frac{10}{B/t}\,\frac{F_y\, t}{F_{yb}\, t_b}\, B_b \le B_b \qquad (9.4\text{-}21)$$

3d. Branches with Bending

Primary bending moments M_u due to applied loads, cantilevered beams, side-sway of unbraced frames, and other sources, shall be considered in the design as an additional axial tension or compression load

$$P_{ua} = \frac{M_u}{c\sin\theta} \qquad (9.4\text{-}22)$$

where

$c = N/4$ for in-plane bending

$c = B_b/4$ for out-of-plane bending

10. GENERAL REQUIREMENTS FOR HSS FABRICATION

The following requirements shall be met in addition to the requirements of LRFD Specification Chapter M.

1. When water can collect inside an HSS, either during construction or during service, HSS shall be sealed, provided with a drain hole at the base, or protected by other suitable means.
2. HSS shall be cleaned with a suitable solvent if paint is specified per LRFD Specification Section M3.1.
3. HSS shall be cleaned with a suitable solvent at locations of welding.

Commentary

April 15, 1997

1.1. SCOPE

For the purposes of this Specification, HSS are defined as hollow structural sections with constant wall thickness and a round, square or rectangular cross-section that is constant along the length of the member. HSS are manufactured by forming skelp (strip or plate) to the desired shape and joining the edges with a continuously welded seam. Although the term pipe is commonly associated with round members that are used for fluid transmission, only steel pipe products that are used for structural purposes are included in the HSS definition. Published information is available describing the details of the various methods used to manufacture HSS (STI, 1996; Graham, 1965).

Because the design requirements for pressure containment systems are more stringent than those for structural members, this Specification does not apply to members for which pressure containment is essential. Several other potential applications are also excluded from the scope of this Specification: (1) buried cylindrical shapes for which soil interaction is an important factor in the required strength; (2) stiffened HSS; (3) composite HSS; and, (4) HSS in fatigue applications. However, it is not intended that HSS with connection elements that also stiffen the cross-section be excluded from the scope.

Non-HSS products such as fabricated pipes and stiffened shells are excluded from the scope of this Specification. These are defined as members that are formed by shaping plates and joining them with one or more longitudinal seam welds, but neither in a tubing mill nor in accordance with a product specification. Although it is certainly possible to fabricate large pipes and shells to the same degree of perfection as is commonly obtained with manufactured HSS, such quality is not universally assured by standard product specifications. Because the buckling strength of cylindrical sections is greatly influenced by geometric imperfections, there is good justification for excluding such products from the scope of this Specification. Accordingly, it is left to the Structural Engineer of Record to determine the suitability of such products for use with this Specification.

HSS are efficient structural members for the resistance of compressive and torsional forces. Consequently, they are increasingly selected in structural applications such as columns and members in plane trusses or space frames. HSS generally have a lower ratio of exposed surface area to volume when compared with other shapes, which results in reduced painting, fireproofing, and maintenance expense. Additionally, their low resistance to external fluid flow provides a distinct advantage for frameworks that are exposed to wind or water currents. The use of HSS has been limited in the past by difficulty in joining members, but modern fabricating technology has overcome this disadvantage.

This Specification combines design guidelines from several sources. The primary basis for the recommendations is the design philosophy and criteria contained in the LRFD Specification. Because much of the LRFD Specification reflects the behavior of wide flange members, modifications have been made where HSS have been shown to behave differently or when interpretation of LRFD criteria to HSS applications can be clarified or simplified. Such modifications are explained in this Commentary. The criteria have also been modified to appear in a non-dimensional form. Following is a general cross-reference from Sections in this Specification to the relevant Sections and related Appendices in the LRFD Specification.

1.1	Scope	A1
1.2	Material	—
	1. Structural Steel	A3.1
	2. Design Wall Thickness	—
1.3	Loads and Load Combinations	A4
2.1	Effective Net Area for Tension Members	B3
2.2	Local Buckling	—
	1. Classification of Steel Sections	B5.1
	2. Design by Plastic Analysis	B5.2
	3. Design in Seismic Applications	—
2.3	Limiting Slenderness Ratios	B7
3.1	Design for Tension	D1
4.1	Effective Length and Slenderness Limitations	E1
4.2	Design for Compression	E2
5.1	Design for Flexure	F1
5.2	Design for Shear	F2
6.	Design for Torsion	—
7.1	Design for Combined Flexure and Axial Force	H1
7.2	Design for Combined Torsion, Shear, Flexure, and/or Axial Force	H2
8.	Concentrated Forces on HSS	K1
9.1	General Provisions for Connections and Fasteners	J1, J3
9.2	Welds	J2
9.3	Other Connection Requirements	—
	1. Shear Rupture Strength	J4.1
	2. Tension Rupture Strength	J4.2
	3. Punching Shear Rupture Strength	—
	4. Eccentric Connections	J5.1
9.4	HSS-to-HSS Truss Connections	—
10.	General Requirements for HSS Fabrication	M

In areas where the LRFD Specification contains little direct guidance for the

Table C1.2-1
Minimum Tensile Properties of HSS Steels

Specification	Grade	F_y, ksi (MPa)	F_u, ksi (MPa)
ASTM A53	B	35 (240)	60 (415)
ASTM A500 (round)	A B C	33 (228) 42 (290) 46 (317)	45 (311) 58 (400) 62 (428)
ASTM A500 (rectangular)	A B C	39 (269) 46 (317) 50 (345)	45 (311) 58 (400) 62 (428)
ASTM A501	—	36 (248)	58 (400)
ASTM A618 (round)	I and II III	50 (345) 50 (345)	70 (483) 65 (448)
ASTM A847	—	50 (345)	70 (483)
CAN3/CSA G40.21	350W	51 (350)	65 (450)

design of buildings with HSS, such as connections, basic research and criteria from other sources have been used. Much of the basic research is taken from the Comite International pour le Developpement et l'Etude de la Construction Tubulaire (CIDECT) programs, which has sponsored many projects in Europe and Canada concerning HSS construction. These have been incorporated into this Specification with modifications to provide design strengths that are comparable with those in the LRFD Specification. Consequently, this Specification is intended for the design of structural members in building structures that are normally encountered in structural engineering practice and experience. When the general uncertainty in loading or quality of control is substantially different, this Specification may not apply.

When HSS are used in the seismic force resisting system of buildings in high-seismic regions, the requirements in the AISC *Seismic Provisions for Structural Steel Buildings* (AISC, 1997) are applicable.

1.2. MATERIAL

ASTM A53 Grade B is included as an approved HSS material specification because it is the most readily available round HSS product in the United States. Other North American HSS products that have properties and characteristics that are similar to the approved ASTM products are produced in Canada under CAN3/CSA-G40.21-M, "Structural Quality Steels." In addition, steel pipe is produced to other specifications that meet the strength, ductility and weldability requirements of the materials in Section 1.2, but may have additional requirements for notch toughness or pressure testing.

Minimum specified yield and tensile strengths are summarized in Table C1.2-1 for various HSS material specifications and grades. Round HSS can be readily obtained in most of the material specifications and grades in Table C1.2-1, although atmospheric-corrosion-resistant material (ASTM A618 and A847) may require a special order. For rectangular HSS, ASTM A500 Grade B is the most commonly available material and a special order would be required for any other material. Depending upon size, either welded or seamless round HSS can be obtained. In North America, however, all ASTM A500 rectangular HSS for structural purposes are welded. Rectangular HSS differ from box sections in that they have uniform thickness except for some thickening in the rounded corners.

Even though ASTM A501 includes rectangular HSS, hot-formed rectangular HSS are not currently produced in the United States. CAN3/CSA G40.21 includes hot-formed Class H and cold-formed Class C. However, Class H rectangular HSS are produced by hot-finishing HSS that were manufactured by cold-forming. Hot-formed HSS have relatively low levels of residual stress, which enhances their performance in compression and may provide better ductility in the corners of rectangular HSS.

ASTM A500 tolerances allow for a wall thickness that is not greater than plus/minus 10 percent of the nominal value. Because the plate and strip from which electric-resistance-welded (ERW) HSS are made are produced to a much smaller thickness tolerance, manufacturers in the United States consistently

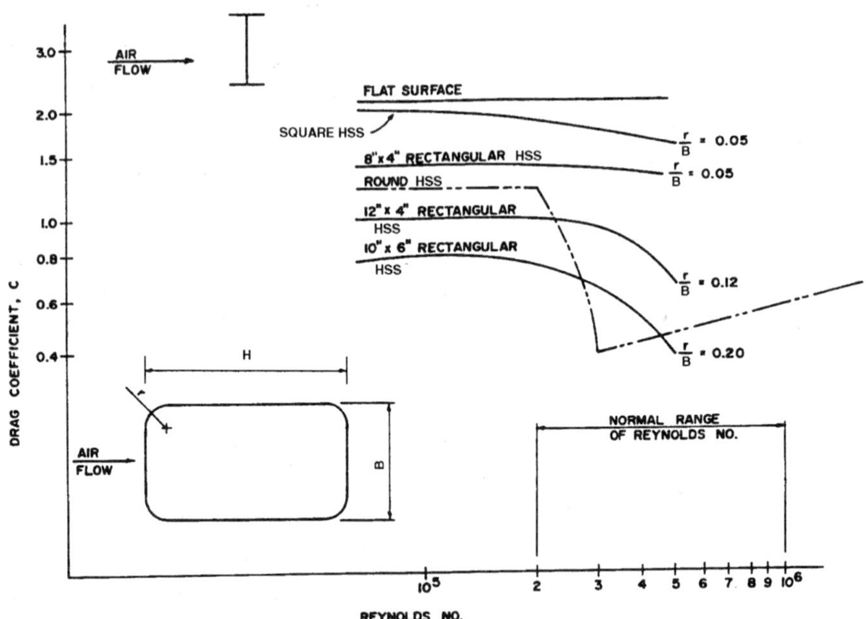

Fig. C1.3-1. Variation in drag coefficient.

Table C1.3-1
Drag Coefficients

Section	Corner Radius	c	c/2.03
Square HSS	0.05B	2.03	1.00
Round HSS	—	1.25	0.62
8-in. × 4-in. HSS	0.05B	1.4	0.69
12-in. × 4-in. HSS	0.12B	1.0	0.49
10-in. × 6-in. HSS	0.20B	0.8	0.39

produce ERW HSS with a wall thickness that is near the lower-bound wall thickness limit. Consequently, AISC and the Steel Tube Institute of North America (STI) recommend that 0.93 times the nominal wall thickness should be used for calculations involving engineering design properties of ERW HSS. This results in a mass variation that is similar to that found in other structural shapes. Submerged-arc-welded (SAW) HSS are produced with a wall thickness that is near the nominal thickness and require no such reduction. The design wall thickness and section properties based upon this thickness have been tabulated in AISC and STI publications since 1997.

1.3. LOADS AND LOAD COMBINATIONS

In many instances, the members in a framing system have no influence on the type or magnitude of the loads that must be considered in design. This is certainly true for dead loads, live loads and impact loads. Horizontal crane forces, when they are present, and wind forces on enclosed structures are also not influenced by the type of members used in the framing system. Consequently, reference is made in this Specification to the applicable building code, ASCE 7 or the

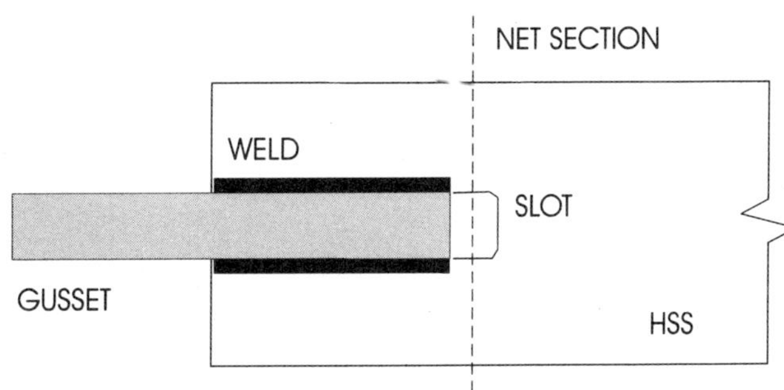

Fig. C2.1-1. Net area through slot for single gusset plate.

provisions of LRFD Specification Section A4. There are, however, two situations in which the use of HSS may allow a reduction in the design forces that must be considered: wind forces on exposed frameworks and pressures created by the enclosed nature of the HSS.

Wind forces on exposed frameworks can occur either in the final structural configuration or during construction. The shape of a round HSS has a lower resistance to fluid flow than shapes with flat elements (e.g., W-shapes) and, therefore, reduces the wind forces. The general determination of wind pressures is given in the applicable building code or ASCE 7 when building codes do not apply. The determination of wind forces on exposed frameworks is a complex problem involving the solidarity ratio, shielding and wind angle. In the absence of other wind-force reduction provisions that consider member shape, the provisions in this Specification can be used.

Wind forces on an exposed profile are proportional to a drag coefficient C, which

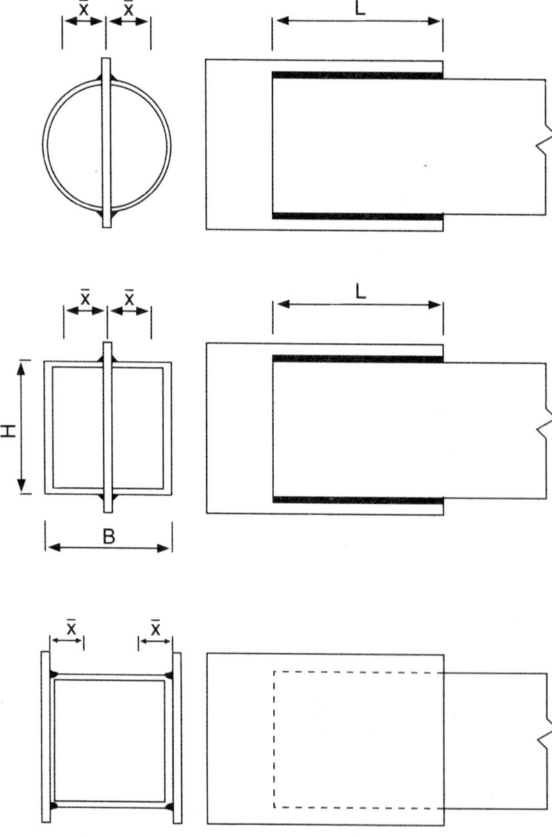

Fig. C2.1-2. Determination of x̄.

varies with the type of profile. The value of C for a square shape with sharp corners is 2.03. Research (Hayus, 1968) indicates that the rounded corners of a square HSS reduce this drag coefficient. As shown in Figure C1.3-1, further reductions occur when rectangular HSS are oriented with the short side perpendicular to the wind. However, C varies considerably with the orientation of the section relative to the wind and with the Reynolds number. The maximum values of C as indicated by the flat portions of the curves in Figure C1.3-1 provide conservative values for use in design. Table C1.3-1 lists these values along with their relative magnitude normalized by 2.03, which is the value of C for the square HSS. The Figure and Table also include the corresponding C for a round HSS. In this case, a one-third reduction (two-thirds factor) from the force on a flat surface is justified.

A similar reduction concept can be applied to wind acting on the short side of an HSS. Using the two data points for HSS with the sharpest corners (square and 8-in. × 4-in. rectangular HSS), a linear reduction factor on the wind force is approximated by $0.4 + 0.6(B / H)$, in which the aspect ratio of the HSS is used as the variable. Conservatively, the reduction has been cut off at the one-third reduction for round HSS, even though the data indicates that with a larger corner radius, C may be less than that for a round HSS. There is no reduction for the wind force on the long side of a rectangular HSS.

2.1. EFFECTIVE NET AREA FOR TENSION MEMBERS

End connections for HSS in tension are commonly made by welding around the perimeter of the HSS. Alternatively, an end connection with gusset plates can be used. Single gusset plates are welded in longitudinal slots that are located at a centerline of the cross-section. Because welding around the end of the gusset plate is not recommended, the net area at the end of the slot in the HSS will be less than the gross area, as illustrated in Figure C2.1-1. A pair of gusset plates can be welded to opposite sides of a rectangular HSS with flare bevel groove welds.

For end connections of these three types, the general provisions of LRFD Specification Section B3 are simplified and the connection eccentricity $\bar{x}$ can be explicitly defined. These types of gusset-plate connections and the definitions of $\bar{x}$ and L are illustrated in Figure C2.1-2.

2.2. LOCAL BUCKLING

The wall slenderness parameters and the slenderness limits λ_p and λ_r in Table 2.2-1 are taken from LRFD Specification Section B5, but have been presented in a non-dimensional form. The design wall thickness as defined in Section 1.2.2 is used to determine slenderness.

The limits for rectangular HSS walls in uniform compression have been used in AISC Specifications since 1969. They are based upon the work of Winter (1947) where adjacent stiffened compression elements in box sections of uniform thickness were observed to provide negligible torsional restraint for one another along their corner edges. The λ_p limit for plastic analysis is adopted from CSA (1994). The web slenderness limits are the same as those used for webs in wide-flange shapes.

The λ_r limit for round HSS in compression was first used by AISC in the 1978 AISC ASD Specification. It was recommended by Schilling (1965) based upon research at Cornell University that produced provisions in the 1968 AISI Cold-Formed Specification (Winter, 1968). The same limit was also used to define a compact shape in bending in the 1978 AISC ASD Specification. However, the limits for λ_p and λ_r were changed in the 1986 AISC LRFD Specification based upon experimental research on round HSS in bending (Sherman, 1985 and Galambos, 1988). Excluding the use of round HSS with $D/t > 0.448E/F_y$ was also recommended by Schilling (1965).

Lower values of λ_r are specified for high-seismic design in the AISC *Seismic Provisions for Structural Steel Buildings* (AISC, 1997) based upon tests (Lui & Goel, 1987) that have shown that rectangular HSS braces subjected to reversed axial load fracture catastrophically under relatively few cycles if a local buckle forms. This was confirmed more recently in tests (Sherman, 1995) when local buckling did not occur with low values of λ_r. Rectangular braces sustained over 500 cycles, even though general column buckling did occur. In test when local buckling did occur (with high values of λ_r), specimens failed in less than 40 cycles. The seismic λ_r is based upon on tests (Lui & Goel, 1987) of HSS that had a small enough b/t so that braces performed satisfactorily for members with reasonable column slenderness. Filling the rectangular HSS with lean concrete (concrete mixed with a low proportion of cement) has been shown to effectively stiffen the HSS walls and improve cyclic performance. Because fracture at a low number of cycles is also possible with cold-formed round HSS braces that form local buckles, the limiting λ_r in compression in the AISC *Seismic Provisions* was chosen to be the same as for plastic analysis.

3.1. DESIGN FOR TENSION

Except for HSS that are subjected to cyclic load reversals, there is no information that the factors governing the strength of HSS in tension differ from those for other structural shapes. Therefore, the criteria in Section 3.1 are identical to those in LRFD Specification Section D1. However, because the number of different end connection types that are practical for HSS is limited, the determination of the net effective area A_e can be simplified as it has been in Section 2.1.

4.1. EFFECTIVE LENGTH AND SLENDERNESS LIMITATIONS

The high torsional strength and stiffness of an HSS provides increased restraint for members that frame into it when compared to that provided by other structural shapes. For example, the connection between an HSS branch member and a continuous HSS chord member is commonly made with a continuous weld around the perimeter of the branch member. In such a connection, the chord then provides considerable end restraint both in-plane and out-of-plane of the truss; the HSS branch member also provides a degree of lateral restraint against rotation of the chord. In both cases, advantage can be taken of the end restraint by using the effective lengths in Section 4.1.1(a). The use of K equal to 0.75 for the branch members and 0.9 for the chord between bracing points is based upon the recommendations of CIDECT research (Rondal, 1992).

It is important to note that even though end restraint is present, it is still reasonable to assume that the truss joints are pinned. Secondary moments due to end fixity may be neglected unless the joint eccentricity exceeds the limits of applicability in Section 9.4 or fatigue is a design consideration. For fatigue applications, refer to AWS D1.1.

The provisions for unbraced frames and braced frames are taken from the LRFD Specification. Values of K for compression members in frames can be determined from the alignment charts and equations in the corresponding LRFD Specification Commentary.

4.2. DESIGN FOR COMPRESSION

The axial compressive strength of an HSS is influenced by its method of production, shape, and dimensions and is further complicated by large differences between theoretical predictions and experimental results for local buckling, especially for round HSS. Rather than repeat the excellent discussions that can be found elsewhere concerning the behavior of various hollow cross-sections (Schilling, 1965; McGuire, 1968; Galambos, 1988; and Sherman, 1992), the results and basis for the design equations is explained herein. Some of the major considerations that must be included in a comprehensive criteria for HSS are as follows:

1. As with any thin-walled member of constant cross-section, either overall flexural buckling or local buckling can be the controlling limit state. Flexural buckling strength under axial load is governed by the slenderness ratio KL/r, the yield strength F_y, residual stresses and initial out-of-straightness. Local buckling of rectangular HSS is based upon the principles of plate buckling theory and is governed by the square of the width-thickness ratio. In very short round sections, local buckling is similar to that for an infinitely

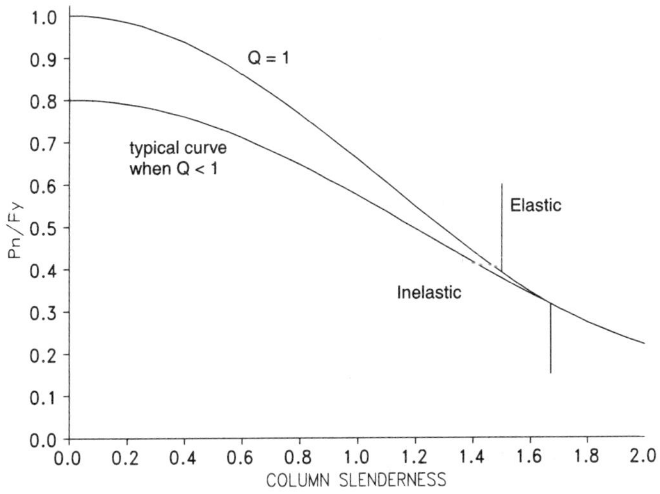

Fig. C4.2-1. Influence of Q on column strength.

Table C4.2-1
Summary of Test Programs
on HSS Columns

Round HSS			
Type	**Symbol in Figure C4.2-2**	**Reference**	**No. of Tests**
Hot-Formed	□	CIDECT#	10
Cold-Formed	■	CIDECT#	65
		Sherman (1980)	4
Fabricated Pipe	✳	Chen (1977)	10
		Yang (1987)	6
Rectangular HSS			
Type	**Symbol in Figure C4.2-3**	**Reference**	**No. of Tests**
Hot-Formed	□	CIDECT#	88
		Estuar (1965)	10
Cold-Formed	■	CIDECT#	132
		Bjorhovde (1979)	1
		Key (1985)	11
Cold-Formed Stress Relieved	✳	Bjorhovde (1979)	19
		Sherman (1969)	2
# No reference cited for CIDECT data			

wide plate and the ratio of the length to the thickness is of prime importance. For longer round sections, the local buckling configuration consists of approximately square waves along the length and around the circumference and the strength is a function of the ratio of the diameter to the thickness. The local buckling strength of thin round sections is extremely sensitive to initial distortion from the perfect cylindrical surface. Because manufactured HSS generally have less initial distortion than a comparable fabricated cylinder, local buckling generally occurs at a higher load.

2. The strength of short compression members is governed by local buckling whereas the strength of long compression members is governed by flexural buckling. In compression members of intermediate length, there is an interaction between local and flexural buckling.

3. Rectangular HSS that are cold-formed from round HSS have a rounded stress-strain curve due to through-thickness residual stresses that vary from

about $0.4F_y$ to $0.8F_y$. This reduces the flexural buckling strength of intermediate-length members to below that of comparable hot-formed HSS with similar yield strength. Residual stresses are not as large in cold-formed round HSS and are negligible in hot-formed HSS.

4. Manufactured HSS tend to have small initial out-of-straightness, which increases the strength of intermediate-length HSS relative to other shapes.

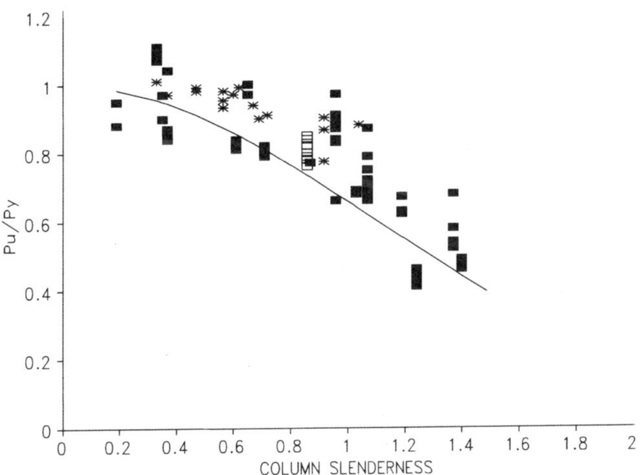

Fig. C4.2-2. Test data for round HSS.

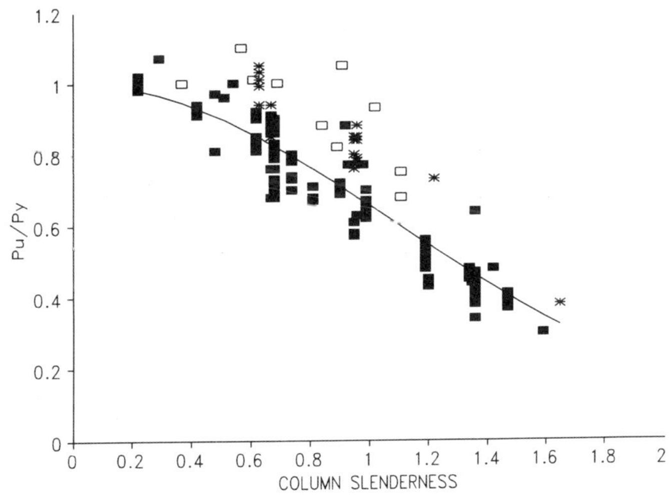

Fig. C4.2-3. Test data for rectangular HSS.

Because many HSS have a wall slenderness ratio that exceeds λ_r, the equations for flexural buckling include the interaction with local buckling as given in LRFD Specification Appendix B. In effect, the local buckling reduction factor Q reduces the yield stress in both the inelastic flexural buckling equation and in the slenderness λ_c. Its influence is illustrated in Figure C4.2-1. Of course, if λ is less than or equal to λ_r, the local buckling reduction factor Q is equal to unity and the equations reduce to the flexural buckling equations of LRFD Specification Chapter E.

For round HSS, Q has its origins in the critical stress from local buckling test data with conservative adjustments in an early AISI Specification (Winter, 1968). The constants have been further adjusted for a non-dimensional form and a design strength format rather than an allowable stress format.

The effective width equation that is used to obtain Q for rectangular HSS is also from an early AISI Specification (Winter, 1968). The constants in this equation were established for closed cross-sections with uniform thickness.

The flexural buckling equations and ϕ_c equal to 0.85 are the same as those used in the LRFD Specification and can be used conservatively for all HSS. However, for hot-formed HSS or ASTM A500 Grade D HSS, which have lower residual stresses but are generally unavailable in the United States, a higher resistance factor could be justified and would give factored resistances that are comparable to those in other specifications that use multiple column curves. This is based upon the extensive CIDECT column test program in Europe and several less extensive studies in North America, as summarized in Table C4.2-1. A large amount of CIDECT data for seamless HSS is not included. The test data for HSS in axial compression are illustrated in Figure C4.2-2 for round HSS and Figure C4.2-3 for rectangular HSS. The curve for the LRFD nominal column strength

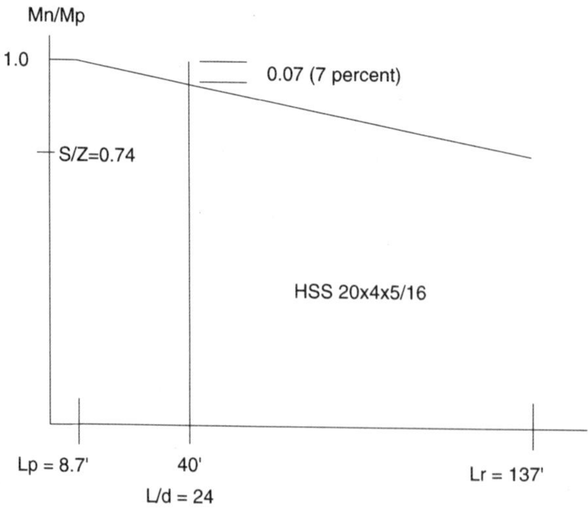

Fig. C5.1-1. Lateral-torsional buckling of rectangular HSS.

(no resistance factor) is superimposed on the Figures. The yield load used to non-dimensionalize the ordinates is the cross-sectional area times the tensile-test yield strength, which was determined by the 0.2 percent offset method for the cold-formed HSS that exhibit rounded stress-strain curves. In the figures, solid squares represent data from tests on cold-formed HSS while open squares represent those on hot-formed HSS. Asterisks represent special cases of fabricated pipe in Figure C4.2-2 and stress-relieved rectangular HSS in Figure C4.2-3. Note that cold-formed ASTM A500 rectangular HSS that are stress relieved to 840 degrees F (450 degrees C) have strengths that are similar to those for hot-formed HSS.

5.1. DESIGN FOR FLEXURE

The provisions for the nominal flexural strength of HSS include the limit states of yielding, inelastic local buckling and elastic local buckling. Neither round HSS nor square HSS nor rectangular HSS bent about the minor axis are subject to lateral-torsional buckling. Furthermore, for rectangular HSS bent about the major axis, the limit state of lateral torsional buckling is not included in this Specification in spite of LRFD Specification provisions that reduce the flexural strength when the unbraced length exceeds limiting unbraced length values.

Because of the high torsional resistance of the closed cross-section, the critical unbraced lengths L_p and L_r that correspond to the development of the plastic moment and the yield moment, respectively, are very large. For example, as shown in Figure C5.1-1, an HSS20×4×⁵⁄₁₆, which has one of the largest depth-width ratios among standard HSS, has L_p of 8.7 ft and L_r of 137 ft as determined in accordance with the LRFD Specification. An extreme deflection limit might correspond to a length-to-depth ratio of 24 or a length of 40 ft for this member. Using the specified linear reduction between the plastic moment and the yield moment for lateral-torsional buckling, the plastic moment is reduced by only 7 percent for the 40-ft length. In most practical designs where the moment gradient C_b is also a factor, the reduction will be nonexistent or insignificant.

The failure modes and post-buckling behavior of round HSS can be grouped into three categories (Galambos, 1988; Sherman, 1992).

1. For low D/t, a long plastic plateau occurs in the moment-rotation curve. The cross-section gradually ovalizes, local wave buckles eventually form, and the moment resistance subsequently decays slowly. Flexural strength may exceed the theoretical plastic moment due to strain hardening.
2. For intermediate D/t, the plastic moment is nearly achieved but a single local buckle develops and the moment resistance decays slowly with little or no plastic plateau region.
3. For thin-walled round HSS, multiple buckles form suddenly with very little ovalization and the flexural strength drops quickly.

The flexural strength criteria for round HSS reflect these three regions of behavior and are based upon five experimental programs involving hot-formed seamless pipe, electric-resistance-welded pipe and fabricated tubing (Galambos, 1988). The criteria are the same as those in LRFD Specification Appendix F.

The criteria for local buckling of rectangular HSS are also the same as those in

LRFD Specification Appendix F. The equation for the effective width of the compression flange when b/t exceeds λ_r is the same as that used for rectangular HSS in axial compression except that the stress is taken as the yield stress. This implies that the stress in the corners of the compression flange is at yield when the ultimate post-buckling strength of the flange is reached. When using the effective width, the nominal flexural strength is determined from the effective section modulus to the compression flange using the distance from the shifted neutral axis. A slightly conservative estimate of the nominal flexural strength can be obtained by using the effective width for both the compression and tension flange, thereby maintaining the symmetry of the cross-section and simplifying the calculation.

The shape factor (Z/S) for HSS is generally between 1.15 and 1.4. Hence, the maximum limit on M_p of $1.5M_y$ in LRFD Specification Chapter F is satisfied for HSS and need not be explicitly checked.

In order to use plastic analysis for design, the HSS must have sufficient rotational capacity at the plastic moment to develop the hinge mechanism. This requires a compact cross-section to prevent premature local buckling and lateral bracing for rectangular HSS that are bent about the major axis to prevent lateral-torsional buckling in the vicinity of hinges. The requirement for L_{pd} is taken from the LRFD Specification. The provisions for the wall slenderness defining the compact section are more restrictive than λ_p, which defines when M_p can be achieved without consideration of rotational capacity. This is due to the higher shape factors for HSS relative to wide-flange shapes where λ_p also applies to plastic analysis. The more restrictive D/t for round HSS is also used in the LRFD Specification and the b/t limit for rectangular HSS is based upon test results for cold-formed sections (Korol, 1972), which are most frequently used in North America.

5.2. DESIGN FOR SHEAR

Little information is available on round HSS subjected to transverse shear and recommendations are based upon criteria for local buckling of cylinders subjected to torsion (Galambos, 1988). Because torsion is generally constant along the member length and transverse shear usually has a gradient, the shear span a in the limit for the nominal shear strength V_n is a length of nearly constant shear. Local buckling due to torsion is dependant upon the length of the cylinder, except for long members. The shear yield strength can be achieved without local buckling when

$$\frac{\tau_y}{E}\left(\frac{D}{t}\right)^{1.25}\left(\frac{L}{D}\right)^{0.5} \le 1.076 \qquad (C5.2\text{-}1)$$

Using τ_y equal to $0.6F_y$ and taking the length L as the shear span, the limit for the shear span to depth ratio is obtained.

In the equation for the nominal shear strength V_n of round HSS, it is assumed that the shear stress at the neutral axis, which is calculated as VQ/Ib, is at the shear yield stress of $0.6F_y$. For a thin round section with radius R and thickness t, I is $\pi R^3 t$, the first moment of area Q is $2R^2 t$, and $b = 2t$. This gives the stress at the centroid as $V/(\pi Rt)$, in which the denominator is recognized as half the area of the round HSS.

For HSS that can achieve the plastic moment, the shear span limit will be much larger than the shear span in any practical design and the equation for V_n will be applicable. Only thin HSS may require a reduction in the shear strength based upon first shear yield, where the designer must perform a rational evaluation based upon the elastic buckling strength of cylinders in torsion. Even in this case, shear will only govern the design of round HSS beams for the case of thin sections with short spans. For more information regarding shear strength when the shear span limit is exceeded, the designer may refer to Brockenbrough and Johnston (1981).

The provisions for the nominal shear strength of rectangular HSS are the same as those for unstiffened webs in the LRFD Specification. For rectangular HSS, the nominal shear strength considers the two webs in the section.

6. DESIGN FOR TORSION

HSS are frequently used in space-frame construction and in other situations wherein significant torsional moments must be resisted by the members. Because of its closed cross-section, an HSS is far more efficient in resisting torsion than an open cross-section such as a W-shape or channel. While normal and shear stresses due to warping of the cross-section are usually significant in shapes of open cross-section, they are insignificant for closed cross-sections and the total torsional moment can be assumed to be resisted by pure torsional shear stresses, which are sometimes called St. Venant torsional stresses.

In HSS, the pure torsional shear stress is assumed to be uniformly distributed and is equal to the torsional moment T_u divided by a torsional shear constant for

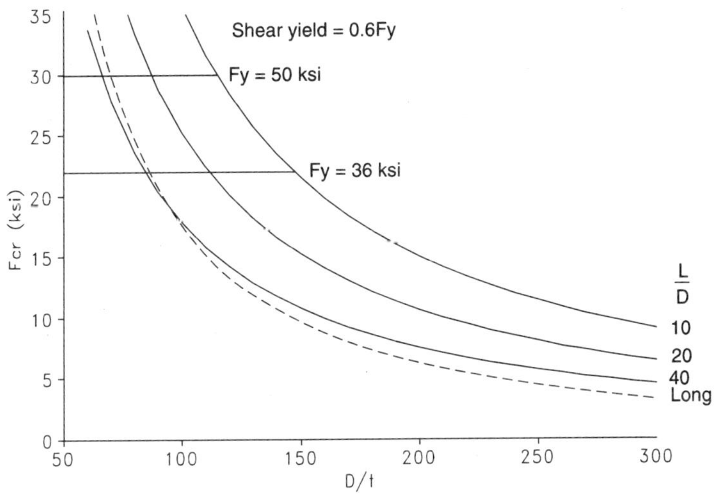

Fig. C6-1. Torsion F_{cr} for round HSS.

the cross-section C. In a limit state format, the nominal torsional resisting moment is the shear constant times the critical shear stress F_{cr}.

For a round HSS, the torsional shear constant is equal to the polar moment of inertia divided by the radius:

$$C = \frac{\pi(D^4 - D_i^4)}{32D/2} \approx \frac{\pi(D - t)^2 t}{2} \tag{C6-1}$$

where D_i is the inside diameter.

For a rectangular HSS, the torsional shear constant is obtained as $2tA_o$ using the membrane analogy (Timoshenko, 1956), where A_o is the area bounded by the midline of the section. Conservatively assuming an outside corner radius that is equal to $2t$, the midline corner radius is $1.5t$ and

$$A_o = (B - t)\,(H - t) - \frac{9(4 - \pi)}{4}\,t^2 \tag{C6-2}$$

which yields

$$C = 2(B - t)\,(H - t)t - 4.5(4 - \pi)t^3 \tag{C6-3}$$

The resistance factor and nominal strengths used for torsion are the same as those used for flexural shear.

When considering local buckling in round HSS subjected to torsion, most structural members will either be long or of moderate length and the criteria for short cylinders will not apply. The elastic local buckling strength of long cylinders is unaffected by end conditions and the critical stress is as given by Galambos (1988) as:

$$F_{cr} = \frac{K_t E}{(D/t)^{3/2}} \tag{C6-4}$$

The theoretical value of K_t is 0.73 but a value of 0.6 is recommended to account for initial imperfections. An equation for the critical elastic local stress for round HSS of moderate length ($L > 5.1D^2/t$) where the edges at the ends are not fixed against rotation is given by Schilling (1965) and Galambos (1988) as:

$$F_{cr} = \frac{1.23E}{\sqrt{L/D}\,(D/t)^{5/4}} \tag{C6-5}$$

This equation includes a 15 percent reduction to account for initial imperfections.

The local buckling equations are plotted in Figure C6-1 with modulus of elasticity E equal to 29,000 ksi. Although there is some inconsistency concerning the division between long and moderately long HSS, it appears from Figure C6-1 that the expression for sections of moderate length is valid for most practical lengths. It is also evident that it would be uneconomical to neglect the increase in torsional strength for members of moderate length. Therefore, in this Specification, the length effect is included for the simple end conditions and the approximately 10 percent increase in buckling strength is neglected for edges fixed at the ends. A limitation is provided so that the shear yield strength is not exceeded.

The critical stress criteria for rectangular HSS are identical to the flexural shear criteria of Section 5.2. The shear distribution due to torsion is uniform in the longest sides of a rectangular HSS, which is the same distribution that is assumed to exist in a beam web. Therefore, it is reasonable that the criteria for buckling is the same in both cases.

7.1. DESIGN FOR COMBINED FLEXURE AND AXIAL FORCE

The provisions for interaction between flexure and axial force are taken directly from LRFD Specification Section H1. As stated in LRFD Specification Section C1, for structures designed on the basis of elastic analysis, the required flexural strength M_u shall be determined from a second-order analysis. Second-order effects must be considered not only for the beam-column member, but also for the other framing members that connect to it and the associated connections. In lieu of a second-order analysis, an approximate second-order procedure is permitted to determine M_u as detailed in LRFD Specification Section C1.

In the case of biaxial flexure on round HSS, which have the same section modulus about any axis, the LRFD Specification interaction equations lead to inconsistent results when compared with design for a resultant moment (Pillai and Ellis, 1971). The interaction equation adds the effects of M_{ux} and M_{uy} which can be conservative by as much as 40 percent when compared to combining them vectorially. Therefore, provision has been included in this Specification for the direct design based upon the resultant moment M_{ur}.

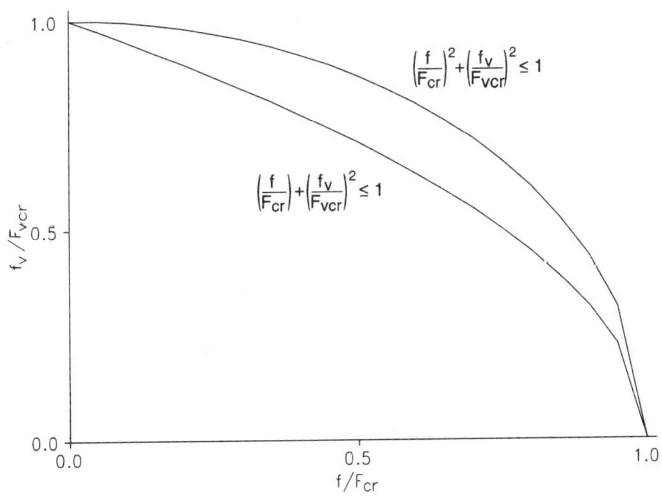

Fig. C7.2-1. Stress interaction of normal and shear effects.

7.2. DESIGN FOR COMBINED TORSION, SHEAR, FLEXURE, AND/OR AXIAL FORCE

Several interaction equation forms have been proposed for load combinations that produce both normal and shear stresses. In one common form, the normal and shear stresses are combined elliptically with the sum of the squares (Felton and Dobbs, 1967).

$$\left(\frac{f}{F_{cr}}\right)^2 + \left(\frac{f_v}{F_{vcr}}\right)^2 \leq 1 \qquad\qquad (C7.2\text{-}1)$$

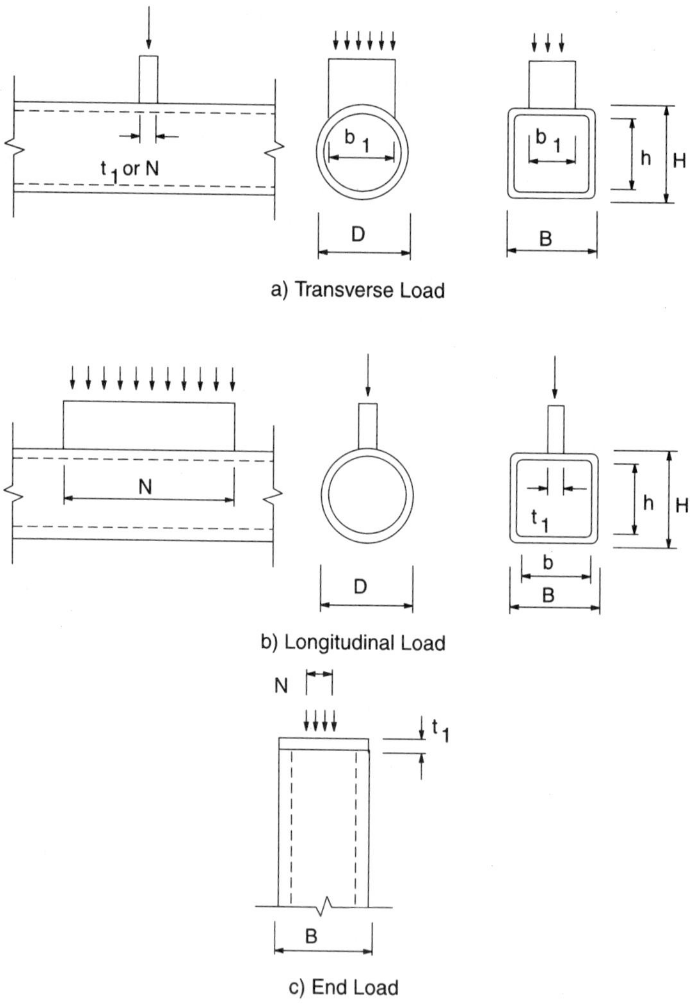

Fig. C8.1-1. Concentrated force configurations on HSS.

In a second form, the first power of the normal stresses is used.

$$\left(\frac{f}{F_{cr}}\right)+\left(\frac{f_v}{F_{vcr}}\right)^2 \le 1 \tag{C7.2-2}$$

These equations are plotted in Figure C7.2-1. The latter form is more conservative but not overly conservative (Schilling, 1965) and is the basis for the interaction equation used in this Specification in a limit-states format. Flexure and axial force effects are combined linearly and then combined with the square of a linear combination of flexural shear and torsion effects. When an axial load is present, the required flexural strength M_u is obtained by second-order analysis. Because the interaction is based upon an elastic combination of stresses, the nominal flexural strength M_n is limited to the yield moment.

8. CONCENTRATED FORCES ON HSS

Concentrated forces result from line loads that are applied to the HSS through a plate, a connecting element that has a flange, or a similar element. The line load can be distributed either transversely or longitudinally as shown in Figure C8.1-1.

Round HSS with either transverse or longitudinal line loads fail by local plastic distortion of the section. The resistances for these loadings are given by Packer & Henderson (1992). Because the resistance equations given by Packer and Henderson account for variations between experimental strengths and theory, the appropriate resistance factor ϕ is 1.0. When there are compressive normal stresses in the HSS in the vicinity of the line load, the resistance to local plastic distortion is reduced and a reduction factor Q_f is applied. In Packer and Henderson, Q_f is determined as a function of f, which is the maximum compressive stress in the chord (a negative number). In this Specification, the Packer and Henderson equation for Q_f has been modified so that the magnitude of f is used (an absolute value).

The limit state for a rectangular HSS depends upon the width of the load relative to the width of the loaded face of the HSS. For transverse loads, the variable stiffness across the face of the HSS causes a non-uniform distribution of the force in the connecting element delivering the load. Failure may occur due to an excessive concentration of force on an effective width of the connecting element at the resistance given by Packer & Henderson (1992). The limit of $F_{y1}\,t_1 b_1$ does not represent the limit state for the axial strength of the loaded plate, but limits the effective width given by Packer & Henderson (1992) from exceeding the width of the HSS. When the force is across nearly the full width of the HSS, the concentration of force at the end of the connecting element can cause a punching shear rupture through the wall of the HSS. Again, the resistance is as given by Packer & Henderson (1992).

When the force is across the full width, failure of the sidewalls of the HSS is possible. The resistance for local sidewall yielding is taken from LRFD Specification Section K1.3, adjusted for the presence of two sidewalls acting as webs. The resistance for sidewall crippling is consistent with LRFD Specification Section K1.4 in a non-dimensional form and accounting for two sidewalls. The same is true for the compression buckling criteria from LRFD Specification

Section K1.6 when compressive forces are applied on opposite faces of the HSS. For the limit states taken from the LRFD Specification, the identical resistance factors are used.

In the case of a longitudinal line load where the width is very narrow relative to the width of the HSS, the limit state is local plastic distortion of the face of the HSS and the resistance is as given by Packer & Henderson (1992). As in the case of round HSS where the limit state is also local plastic distortion, a reduction factor Q_f is applied when the HSS is in compression. Because the resistance equations given by Packer and Henderson include variability, the resistance factor ϕ is 1.0.

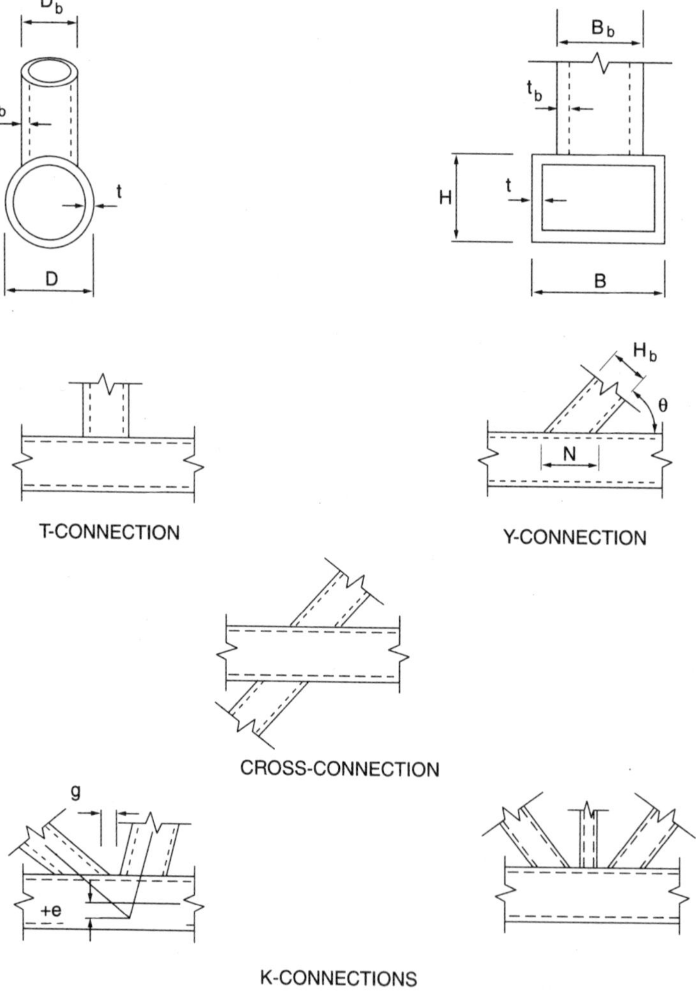

Fig. C9.4-1. HSS-to-HSS truss connection configurations.

Another type of concentrated force on a rectangular HSS is on the end of a capped column supporting a beam or joist as in Figure C8.1-1c. Because the load is on only one side of the HSS, the resistance for local yielding is taken directly from LRFD K1.3. The criteria for crippling of the HSS wall is taken from LRFD Specification Equation K1-4 with the depth taken as one-half the full width of the HSS, $B/2$. This is based upon a model in which the force is assumed to spread out at 45 degrees in both directions from the center point of the load width. The force will be across the full width of the side at a depth no greater than $B/2$, below which crippling will not occur.

9.1. GENERAL PROVISIONS FOR CONNECTIONS AND FASTENERS

Because HSS are frequently combined with other shapes in structures using standard simple shear or moment connections, the general provisions of LRFD Specification Sections J1.1 through J1.11 apply even though several of these provisions are not directly applicable to the HSS.

Although welding is the most frequently used method for making attachments to HSS, there are a variety of methods for using bolts or mechanical fasteners. When ASTM A325, A490 or A307 bolts are used to make an attachment to one wall of an HSS using an open end or access hole to install the nut so that all plies of the connection are in contact within the grip of the bolt, the provision of LRFD Specification Section J3 apply. These provisions also apply to other portions of a structure that includes HSS. However, if a bolt passes completely through the HSS, an attempt to fully tension the bolt would distort the cross section of the HSS wall and the specified minimum installed tension could not be achieved without damaging the member. Therefore, through-bolts should only be installed in the snug-tight condition. The bearing strength of such connections as given in Section 9.1.1 is consistent with that in LRFD Specification Section J8. The designer should note that some connection details do not distribute the load to both walls equally. Additionally, some details may induce bending in the through-bolt.

There are a variety of blind bolts, unique installation methods, welded studs and special connectors that could be used with HSS. Many of these alternatives are still in a developmental stage and general criteria for their use are not available. In such cases, test data must be used to justify their suitability for a particular application.

With any connection configuration where the fasteners transmit a tensile force to the HSS wall, a rational analysis must be used to determine the appropriate limit states. These may include a yield-line mechanism in the HSS wall and/or pull-out through the HSS wall in addition to applicable limit states for the fasteners subject to tension.

9.2. WELDS

These provisions are based upon similar provisions in AWS D1.1. The design strength of a connection, for a member welded to a wall of an HSS, is a function of the geometric parameters of the connected members. It is often less than the strength of the member and in many cases cannot be increased by increasing the weld strength. The weld, however, must be sized to provide for the uneven distribution of load along the weld line at the required strength. The effective length provisions included in this Section are intended to provide for ductile joint behavior through prevention of progressive failure or "unzipping" of the weld. The variables are as illustrated for the Y-connection in Figure C9.4-1.

These effective length provisions were derived from experimental results and the lower-bound weld strength in LRFD Specification Section J2.4, excluding Appendix J2.4 provisions. For this reason, the provisions herein are conservatively limited to the lower-bound weld strength provisions.

Other rational approaches are available in AWS D1.1, such as the use of fillet welds with an effective throat thickness of at least 1.1 times the thickness of the branch member or the use of prequalified fillet and partial-joint-penetration groove welded details that are sized to ensure ductile behavior.

9.3. OTHER CONNECTION REQUIREMENTS

The requirements of LRFD Specification Section J4 through J10 are applicable to connections to HSS as well as to other portions of a structure. The notation

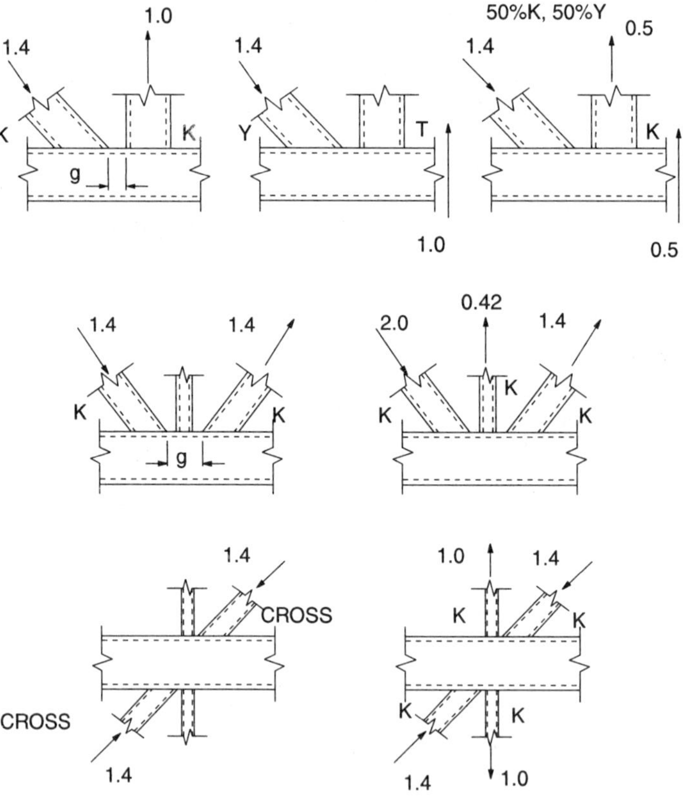

Fig. C9.4-2. Examples of connection criteria classifications
(all inclined members at 45 degrees).

for shear rupture strength and tension rupture strength in LRFD Specification Section J4 have been modified when applied specifically to the HSS. An additional limit state for punching shear has been added based upon failures that were observed in tests of single plate connections to HSS (Sherman, 1996).

The tension force in a unit length of the plate is ft_p. Limiting this force to the through-thickness shear strength of the HSS for the two welded planes on the sides of the plate,

$$ft_p \le 2(0.6F_u)t \tag{C9.3-1}$$

The punching shear rupture criterion includes the standard resistance factors of 0.9 for yielding and 0.75 for rupture. For a single plate connection, the stress f is a bending stress that results from the factored shear acting at an eccentricity (Sherman and Ales, 1991). A simple and conservative criterion is obtained by setting f equal to the yield strength of the plate F_{yp}, which is the maximum possible value. This reduces the above equation (with resistance factors) to

$$t_p < \frac{F_u}{F_{yp}} t \tag{C9.3-2}$$

An exception to LRFD Specification Section J5.1 on eccentric connections is that bending moments caused by eccentricities in HSS-to-HSS truss connections need not be considered for connection design as long as the eccentricities are within the limits of applicability in Section 9.4. These eccentricities have been included in the database upon which the connection strength criteria in Section 9.4 have been based.

9.4. HSS-TO-HSS TRUSS CONNECTIONS

A wide variety of connection configurations is possible where branch members are directly welded to a chord that is continuous through the connection. The criteria in this specification are limited to a few of the configurations that are typically used for planar HSS trusses in building applications: T-, Y-, gapped K- and Cross-connections as illustrated in Figure C9.4-1.

Overlapped K-connections are not covered in this Specification because they are more expensive to fabricate than connections with a gap, especially for round HSS. For this and other connection configurations that are not included in Figure C9.4-1, the design must be based upon the requirements in AWS D1.1, experimental test results or other verified design procedures. Design procedures such as those contained in the recommendations of International Institute of Welding (IIW, 1989) would generally be acceptable.

The criterion that is used for the design strength of a branch of a gapped K-connection is based upon the mechanism of load transfer rather than the configuration of the connection. The gapped K-connection criteria apply only when the branch axial force is equilibrated by that is other branch members on the same side of the chord. If the force is equilibrated by shear in the chord, the criteria for a T- or Y-connection apply. If the load is transferred through the chord, the criteria for a Cross-connection must be used. Examples of these principles are shown in Figure C9.4-2. When the load is equilibrated by a combination of two of the mechanisms, the design strength is determined by interpolation between the two criteria based upon the percentage of the load transfer by each mechanism.

The symbols for key dimensions that are used in HSS-to-HSS truss connection design criteria are also defined in Figure C9.4-1. Parameters derived from ratios of these dimensions that are used in AWS D1.1 and other specifications are defined in Section 9.4.1.

The criteria in Section 9.4.2 for round HSS and Section 9.4.3 for rectangular HSS have been taken from AWS D1.1. The parameter Q_f, which is a strength reduction factor that accounts for the level of axial and bending stress in the chord at the connection, is defined in both sections. The criteria are partly empirical and are based upon numerous worldwide test programs involving

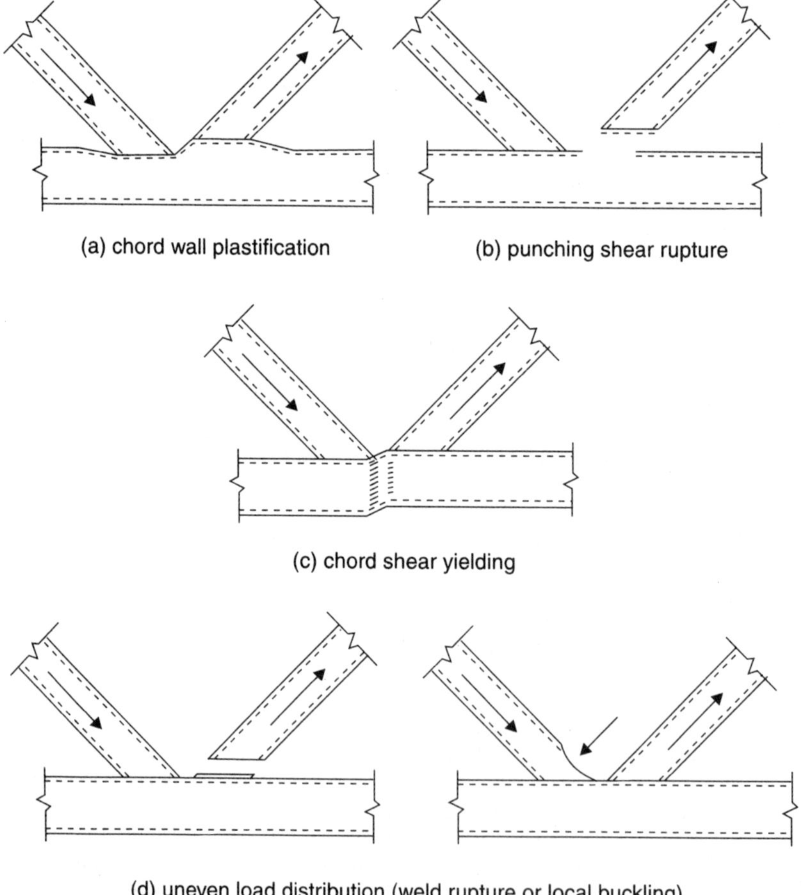

Fig. C9.4-3. Limit states in gapped K-connections.

hundreds of tests. Since the criteria are in some cases the result of empirical curve fitting, they may not be accurate outside the range of tested parameters. AWS D1.1 does not contain a set of limits of applicability for round HSS except for an upper limit on yield strength of 60 ksi. The HSS material specifications listed in Section 3.1 meet this limitation. However, since the database used by various international specifications and design recommendations is essentially the same, the limits of applicability given by Packer and Henderson (1992) for round HSS have been used in Section 9.4.2a.

The criteria in Section 9.4.3 for rectangular HSS are similar to those in other established HSS design specifications and recommendations throughout the world. They include the limits of applicability for rectangular HSS connections in Section 9.4.3a.

The criteria place limits on the magnitude of the component of the branch force that is perpendicular to the chord for various limit states. These limit states include:

1. Chord wall plastification: High local stresses and distortions occur in the vicinity of the joint. These are the result of bending in the wall of the chord.
2. Punching shear rupture: For chords with thin walls, shear stresses through the thickness of the chord around the perimeter of the branch result in a tearing out of material, which can be the controlling limit state.
3. General collapse: This is most common in Cross-connections where the branches are in compression or in other situations when the load is transferred through the chord. It results in a squashing of a round HSS chord or a buckling of the side wall of a rectangular HSS chord. The sidewall strength criteria in Section 9.4.3b. are based upon similar provisions in LRFD Specification Sections K1.3, K1.4 and K1.6. They have been modified to account for the presence of two sidewalls in the HSS and the buckling criteria have been expressed in a non-dimensional form.
4. Uneven load distribution: In connections where the branch is equal or nearly equal in width to the chord, the increase in the stiffness of the chord wall from its center toward its sides concentrates the transfer of the branch force toward the sides of the chord and might cause a premature failure of the branch.

These limit states are illustrated in Figure C9.4-3. In some situations, it can be determined beforehand that a particular limit state will not apply. For example, chord wall plastification will not occur in rectangular HSS T-connections when the branch has the same or nearly the same width as the chord. Hence, ranges of parameters are indicated in Section 9.4.3b and Section 9.4.3c when a limit state need not be checked. In Section 9.4.3c, one such limit is given as $\beta_b / B \geq 0.1 + \gamma / 50$, which is based upon Packer and Henderson (1992). This is a departure from AWS D1.1, which specifies a limit of $\beta_{eff} \geq 0.1 + \gamma / 50$.

Since the criteria place a limit on the force that the branch can carry, it is not always possible to develop the full strength of the branch without reinforcing the chord with stiffeners. It should be noted that such reinforcement involves expensive fabrication. To minimize cost, it may be more desirable to maximize the efficiency of the connection with respect to the force that can be developed in the branch. Consider the following suggestions:

1. Chord members should be relatively thick and branch members should be relatively thin. This is efficient for the strength of the connection and thinner branches reduce the required weld size, if the branch member is joined to develop its wall strength. There may be some compromise required in the design of com pression chord members because larger sizes with thinner walls make efficient compression members. This, however, may be detrimental to the strength of the connection.

2. All members should have D/t or b/t values below the limit that would classify them as thin-walled sections. Tension-only branches are a possible exception.

3. The angle between the branch member and chord member should be greater than 30 degrees. This is efficient for the truss layout and avoids difficult welding and inspection requirements and uncertainties in design criteria as well.

4. Branch member width should be kept narrower than the chord member width to avoid difficult welding details and inspection at the edges.

5. Gapped connections are preferred to overlapped connections. Although the latter are more efficient for connection strength, they increase the fabrication cost due to the increased difficulty in joint preparation, particularly with round HSS.

If the connection is inadequate to carry the branch required strength, the chord and branch members can be resized to improve the connection efficiency, the connection can be locally reinforced, or a joint can may be used. External or internal local reinforcement is expensive to fabricate and the former may be unsatisfactory from a functional or aesthetic viewpoint. However, filling the chord with concrete in the vicinity of the connection is a satisfactory method of reinforcing the joint. Joint cans consist of a segment in the vicinity of the connection that has the same outer dimension as the rest of the chord but a larger wall thickness. This requires butt-welded splices to the rest of the chord, which can significantly increase fabrication cost. AWS D1.1 contains provisions for the design of joint cans including required length.

Section 9.4.3c is taken from AWS D1.1. The additional axial tension or compression load P_{ua} due to flexure in truss members is generally subject to a lower resistance factor than that used in beam bending. The definition of c, although given in a simplified form, is consistent with that given in AWS D1.1.

10.   GENERAL REQUIREMENTS FOR HSS FABRICATION

The general provisions for fabrication, erection, and quality control as specified in LRFD Specification Chapter M are also applicable to HSS. In addition, the following HSS-specific concerns are addressed:

1. Because the interior of an HSS is difficult to inspect, some concern has been expressed regarding internal corrosion. However, good design practice can eliminate the concern and the need for expensive protection.
Corrosion occurs in the presence of oxygen and water. In an enclosed building, it is improbable that there would be sufficient reintroduction of moisture to cause severe corrosion. Therefore, internal corrosion protection is a consideration only in HSS that are exposed to weather.

In a sealed HSS, internal corrosion cannot progress beyond the point where the oxygen or moisture necessary for chemical oxidation is consumed (AISI, 1970). The oxidation depth is insignificant when the corrosion process must stop, even when a corrosive atmosphere exists at the time of sealing. If fine openings exist at connections, moisture and air can enter the HSS through capillary action or by aspiration due to the partial vacuum that is created if the HSS is cooled rapidly (Blodgett, 1967). This can be prevented by providing pressure-equalizing holes in locations that make it impossible for water to flow into the HSS by gravity.

Situations where conservative practice would recommend an internal protective coating include: (1) open HSS where changes in the air volume by ventilation or direct flow of water is possible; and, (2) open HSS subject to a temperature gradient that would cause condensation. In such instances it may also be prudent to use a minimum $\frac{5}{16}$-in. wall thickness.

HSS that are filled or partially filled with concrete should not be sealed. In the event of fire, water in the concrete will vaporize and may create pressure sufficient to burst a sealed HSS. Care should be taken to ensure that water does not remain in the HSS during or after construction, since the expansion caused by freezing can create pressure that is sufficient to burst an HSS.

Galvanized HSS assemblies should not be completely sealed because rapid pressure changes during the galvanizing process tend to burst sealed assemblies.

2. As a result of manufacturing, a light oil coating is generally present on the outer surface of the HSS. If paint is specified, HSS must be cleaned of this oil coating with a suitable solvent; see the *Steel Structures Painting Manual*. (SSPC, 1991)

3. To avoid weld contamination, the light oil coating that is generally present after manufacturing an HSS should be removed with a suitable solvent in locations where welding will be performed. In cases where an external coating has been applied at the mill, the coating should be removed at the location of welding or the manufacturer should be consulted regarding the suitability of welding in the presence of the coating.

References

AISC, *Seismic Provisions for Structural Steel Buildings*, American Institute of Steel Construction, Chicago, IL, 1997.

AISC, *Load and Resistance Design Specification for Structural Steel Buildings*, American Institute of Steel Construction, Chicago, IL, 1993.

ASCE, *Minimum Design Loads for Buildings and Other Structures,* American Society of Civil Engineers, Reston, VA, 1995.

AISI, *Interior Corrosion of Structural Steel Closed Sections*, American Iron and Steel Institute, Bulletin 18, February, Washington, DC., 1970.

Bjorhovde R. and P. C. Birkemoe, "Limit State Design of HSS Columns", Canadian Journal of Civil Engineering, Vol. 6, No. 2, June (pp. 276-291), 1979.

Blodgett, O. W., "The Question of Corrosion in Hollow Steel Sections," *Welding Design Studies in Steel Structures*, Lincoln Electric, D610.163, August, Cleveland, OH, 1967.

Brockenbrough, R. B. and B. G. Johnston, *USS Steel Design Manual*, United States Steel Corporation, Pittsburgh, PA, 1981.

CSA *Standard S16.1 Limit States Design of Steel Structures*, Canadian Standards Association, Rexdale, Ontario, Canada, 1994.

Chen, W. F. and D. A. Ross, "Tests of Fabricated Tubular Columns," *Journal of the Structural Division*, Vol. 103, ST3, paper 12809, ASCE, Reston, VA, 1977.

Estuar, F. R. and L. Tall, "The Column Strength of Hot-Rolled Tubular Shapes-An Experimental Evaluation," *Fritz Engineering Laboratory Report No. 296.1*, Lehigh University, Bethlehem, PA, 1965.

Felton, L. P. and M. W. Dobbs, "Optimum Design of Tubes for Bending and Torsion," *Journal of the Structural Division*, Vol. 93, ST4, paper 5397, ASCE, Reston, VA, 1967.

Galambos, T. V, *Guide to Design Criteria for Metal Compression Members*, Structural Stability Research Council, John Wiley & Sons, New York, NY, 1988.

Graham, R. R, "Manufacture and Use of Structural Tubing," *Journal of Metals*, September, 1965.

Hayus, F, *Drag Measurements on One Square Section and Two Rectangular Sections with Different Corner Radii,* English Translation of CIDECT Report 1-NK1-68-41, September, 1968.

IIW, *Design recommendations for Hollow Structural Section Joints—Predominantly*

Statically Loaded, 2nd ed., International Institute of Welding Subcommission XV-E, Doc. XV- 582-85, IIW Annual Assembly, Helsinki, Finland, 1989.

Key, P. W. and G. J. Hancock, *An Experimental Investigation of the Column Behaviour of Cold-Formed Square Hollow Sections*, Research Report R493, University of Sydney, Sydney, Australia, 1985.

Korol, R. M., *The Plastic Behaviour of Hollow Structural Sections with Implication for Design*, Canadian Structural Engineering Conference, 1972.

Lui, Z. and Goel, S. C., *Investigation of Concrete Filled Steel Tubes under Cyclic Bending and Buckling*, UMCE Report 87-3, University of Michigan, Ann Arbor, MI, 1987.

Packer, J. A. and Henderson, J. E., *Design Guide for Hollow Structural Section Connections*, Canadian Institute of Steel Construction, Willowdale, Ontario, Canada, 1992.

Rondal, J., *Structural Stability of Hollow Sections*, Comite International pour le Developpement et l'Etude de la Construction Tubulaire, Verlag TÜV Rheinland GmbH, Köln, Germany, 1992.

Schilling, C. G., "Buckling Strength of Circular Tubes," *Journal of the Structural Division*, Vol. 91, ST5, paper 4520, ASCE, Reston, VA, 1965.

Sherman, D. R. and Lukas, D. E., "Torsionally Stiff Columns Under Eccentric Loads," *Journal of the Structural Division*, Vol. 96, ST2, paper 7090, ASCE, Reston, VA, 1969.

Sherman, D. R., *Local Buckling Behavior of Tubular Strut Type Beam-Columns*, Civil Engineering Department Report, University of Wisconsin-Milwaukee, Milwaukee, WI, 1980.

Sherman, D. R., "Bending Equations for Circular Tubes," *Proceedings Annual Technical Session*, Structural Stability Research Council, Lehigh University, Bethlehem, PA, 1985.

Sherman, D. R. and Ales, J. M., "The Design of Shear Tabs with Tubular Columns," *Proceedings of the 1991 AISC National Steel Construction Conference*, AISC, Chicago, IL, 1991.

Sherman, D. R., "Tubular Members," *Constructional Steel Design-An International Guide,* edited by P.J. Dowling, J.H. Harding, and R. Bjorhovde, Chapter 2.4, pp. 91–104, Elsevier Applied Science, London, England, 1992.

Sherman, D. R., *Stability Related Deterioration of Structures*, 1995 Theme Conference, Structural Stability Research Council, Lehigh University, Bethlehem, PA, 1995.

SSPC, *Steel Structures Painting Manual*, Steel Structures Painting Council, Pittsburgh, PA, 1991.

STI, *Principle Producers and Capabilities*, Steel Tube Institute, Mentor, OH, 1996.

Timoshenko, S., *Strength of Materials, Part II*, 3rd ed., D. Van Nostrand Company, Inc, 1956.

Winter, G., *Commentary on the Specification for the Design of Cold-Formed Steel Members*, American Iron and Steel Institute, Washington, D.C., 1968.

Yang, X. M., "Study of Carrying Capacity of Fabricated Tubular Columns Under Axial Compression," *Proceedings Annual Technical Session*, Structural Stability Research Council, Lehigh University, Bethlehem, PA, 1987.